NMR IN BIOLOGICAL SYSTEMS

FOCUS ON STRUCTURAL BIOLOGY

Volume 6

Series Editor
ROB KAPTEIN
Bijvoet Center for Biomolecular Research,
Utrecht University, The Netherlands

NMR in Biological Systems

From Molecules to Humans

by

K.V.R. CHARY
Tata Institute of Fundamental Research,
Mumbai, India

and

GIRJESH GOVIL
Tata Institute of Fundamental Research,
Mumbai, India

Springer

ISBN 978-90-481-7698-4
ISBN 978-1-4020-6680-1 (e-book)

Published by Springer,
P.O. Box 17, 3300 AA Dordrecht, The Netherlands.

www.springer.com

Printed on acid-free paper

CONTENTS

CHAPTER 5: PROTEIN NMR: GENERAL PRINCIPLES AND RESONANCE ASSIGNMENTS 163

PREFACE

In the past twenty years, there has been tremendous progress in life sciences. During the same period, NMR has emerged as a major tool, which has catalysed this growth. There are several good books, which deal with the basics of modern NMR and its diverse applications. This book is aimed at providing knowledge of NMR as applied to all aspects of life sciences in an integrated fashion.

When we received a grant to set up a National Facility for High Field NMR in India, one of the conditions of the granting agency was that we undertake training and education of scientists in India in the area of biological NMR. For this purpose, our Facility has been conducting a number of workshops and schools in India. We had the honour of having Professor Richard Ernst and Professor Kurt Wüthrich on the Faculty of some of these schools. The scientists who have attended these schools were either young research scholars aspiring to enter the field of biological NMR or professional biochemists, pharmaceutical chemists and medical doctors who wanted to know how NMR can be helpful in getting more information in their respective fields. In addition, our Institute runs regular Ph. D. programs where too some of our students head for a career in NMR as applied to biological systems. Though, most of these scientists were familiar at handling NMR in applications to Chemical Sciences, they required additional guiding experience when the issues to be addressed involved Biological NMR. This has been the motivation for this monograph.

During the organization of such programs and our constant interaction with researchers, we consistently felt the need for a text-book which can cover modern trends in NMR of biological systems. The available books either cover only the techniques in NMR or cover a particular aspect of biological NMR at a fairly high research level. This book mainly caters to the needs of: (i) graduate level students as a graduate course who mostly learn such techniques from senior post-docs in the laboratory; (ii) those who wish to understand if a particular problem they have in medical and pharmaceutical sciences would be answered by NMR, but have no intention to become professional NMR spectroscopists and (iii) those who are experts in biochemistry and yet do not know how NMR can help to provide them information on structural or functional aspect of proteins, nucleic acids, cells and tissues, human and plant organs and other biological materials. We hope that this book builds a means of knowledge transfer between the beginners and the experts in NMR as applied to all aspects of life sciences.

Mumbai, May 2, 2007

K.V.R. Chary
Girjesh Govil

FOREWORD

Scientific research, which is the basis of prosperity, requires powerful tools for elucidating the secrets of nature. Nuclear Magnetic Resonance (NMR) has proved to be one of the most informative tools of science with applications to nearly all fields from solid state physics, to material science, synthetic and natural products chemistry, molecular biology, clinical medicine, and even to psychology, trying to understand the functioning of the human brain.

Today, research on life processes, in order to prevent and heal diseases, stands at the very forefront of science. Life processes are understood in terms of the function and the interaction of macromolecules in biological systems. NMR is a marvellous instrument for exploring the functioning of biological macromolecules in their native environment. The advanced knowledge necessary to apply NMR successfully to biological systems is presented in a comprehensive manner in this highly valuable book.

The limits between disciplines are becoming more and more translucent in these days. Physics, chemistry, biology, and medicine share common goals and methods of approach. This is put into evidence in this book that covers the basic physical principles of NMR, which are needed to design and apply sophisticated NMR experiments. It describes the spectral parameters that characterize the chemical environment and the interaction of nuclei. It presents in great detail the procedures available for determining the solution structure of biological macromolecules, particularly proteins, nucleic acids, and carbohydrates. It devotes ample space to the discussion of bimolecular function and interaction. Today also solid-state NMR is used more and more in biological research, as is well described in this monograph. Thc last four chapters link NMR to applications in living systems. At first, medical imaging methods are described that have led to spectacular clinical usages of magnetic resonance, especially in the context of functional magnetic resonance imaging (fMRI), where new avenues have been opened for understanding the functioning of our brain. NMR is also a powerful tool for studying the metabolism of living cells, tissue, and organs. Finally, the same techniques can equally well be applied to the study of plant metabolism.

Indeed, the book covers NMR applications to virtually all fields of the life sciences. It is thus forward looking and its study will provide an exceptionally sound basis for progress and success in those important and active fields. It is written in a style that is, at the same time, accurate and easily comprehensive. It will turn out to be useful for novices to the field as well as to the experts. In this sense, it might be a reliable companion on the way into and through the biological sciences, using NMR as a powerful tool of exploration.

The authors, Professor Girjesh Govil and Prof. Kandala Venkata Ramana Chary have together more than 75 years of experience in NMR of biological systems, both having contributed in terms of methodological developments and beautiful applications to biologically relevant molecular processes. The superb knowledge and the enthusiasm of the two authors are felt throughout this book.

Indeed, there is an unbroken chain of knowledge transfer from one of the discoverers of NMR, Felix Bloch at Stanford University, with whose research group S.S. Dharmatti has collaborated. S.S. Dharmatti introduced NMR at the Tata Institute of Fundamental Research in Mumbai, and he also introduced Girjesh Govil into the secrets of NMR. Kandala Venkata Ramana Chary, finally is one of the many brilliant students of Girjesh Govil. Thus, it is not astonishing that from the foundations to the most sophisticated applications, all aspects of NMR are masterfully covered in this marvelous book.

Zurich, 19 February 2007 Richard R. Ernst

ACKNOWLEDGEMENTS

At the outset, we thank our host organization, the Tata Institute of Fundamental Research, Department of Atomic Energy, Government of India, for all its support in making this an achievable task. Dr. Raja Ramanna Fellowship awarded by Department of Atomic Energy, Government of India, to one of us (GG) is gratefully acknowledged.

We sincerely thank the reviews of some of the Chapters by Prof. R.R. Ernst, Dr. Edwin D. Becker and Prof. M. Levitt. A number of our colleagues and friends have helped us by going through the manuscript and providing helpful suggestions towards its improvement. We thank Shankarnarayana, Rukmini Raju, P.K. Madhu, K.D. Narayanan, R.V. Hosur, H.M. Sonawat, N.R. Jagannathan, Anant Patel, Hari Priya, Samavedam Srinivas, K.V. Ramanathan, Hemavathi Ramaswamy, Anil Saran, Sree Priya, Sudha Srivastava, Hanudatta S. Atreya, and Anurag Govil for their critical reading of some of the chapters and their insightful and constructive comments and suggestions. Help provided by post-doctoral fellows (Ritu Bansal, Manjula Devi, and Venkatesh Ramakrishnan) and research scholars (Sourajit, Sulakshana, Ravi, Himanshu, Atul, Kousik, Rajendra and Ashok), summer students (Suryanarayana, Anmol Aggarwal, Abhinav Kandala, Purnima, Shikha Srivastava) and Mamata Joshi and Devidas, is gratefully acknowledged.

We would like to acknowledge the patience and support of our families (Anuradha and Vijay) for their understanding, while we went through repeated rounds of drafts of the book. Grateful acknowledgements to all our teachers, who include our parents and grand parents.

Our thanks to the editorial and production staff at Springer, The Netherlands, who have been patient, prompt and very encouraging.

Finally, we should submit that, the Book was written on a PC using Microsoft Word. Several figures were drawn by the authors using Microsoft Power-point and Sigma-Plot. Some MRI/MRS figures provided by N.R. Jagannathan and few SSNMR figures provided by P.K. Madhu are gratefully acknowledged. Our grateful thanks to all the authors, whose original figures have been incorporated in the book.

K.V.R. Chary
Girjesh Govil

ABBREVIATIONS

2D	Two Dimensional
2QC	Two Quantum Coherences
2QF-COSY	Double-Quantum Filtered COrrelated SpectroscopY
3D	Three Dimensional
3QF-COSY	Three-Quantum Filtered COrrelated SpectroscopY
5-FU	5-Fluoro Uracil
A	Adenine
ACP	Amorphous Calcium Phosphate
ADC	Analogue-to-Digital Converter
ADC	Apparent Diffusion Coefficient
ADEM	Acute Disseminated EncephaloMyelitis
ADP	Adenosine DiPhosphate
AFM	Atomic Force Microscopy
AIDS	Acquired ImmunoDeficiency Syndrome
AMBER	Assisted Model Building with Energy Refinement
aps	antiparallel-stranded
ATP	Adenosine TriPhosphate
BChl	bacterioChlorophyll
bHLH	basic Helix-Loop-Helix
BIV	Bovine Immunodeficiency Virus
BMRB	Biological Magnetic Resonance data Bank
BOLD	Blood Oxygenation Level Dependent
BPS	Branch Point Sequence
BPTI	Basic Pancreatic Trypsin Inhibitor
bR	bacterioRhodopsin
BRU	BRUshite
C	Cytosine
CA	CarbanotohydroxyApatite
CaBP	Calcium Binding Proteins

CD	Circular Dichroism
CE	Contrast Enhancement
CEMRI	Contrast Enhancement MRI
CFE	Cell Free Extracts
CHARMM	Chemistry at HARvard Macromolecular Mechanics
CHD	Coronary Heart Diseases
CHESS	CHEmical Shift Selective water saturation
Chl-a	Chlorophyll a
CNS	Central Nervous System
CoA	Coenzyme A
COSY	COrrelated SpectroscopY
CP	Cross Polarization
CPMG	Carr-Purcell-Meiboom-Gill
CRAMPS	Combined Rotation And Multiple Pulse Spectroscopy
CSA	Chemical-Shift Anisotropy
CSF	Cerebral Spinal Fluid
CSPs	Chemical Shift Perturbations
CT	Computerised Tomography
CW	Continuous-Wave
CYANA	Combined Assignment And DYnamics Algorithm For NMR Applications
DARR	Dipolar Assisted Rotational Resonance
DCIS	Ductal Carcinoma *In Situ*
DD	Dipolar Decoupling
DIR	Double Inversion Recovery
DEPT	Distortionless Enhanced by Polarization Transfer
DG	Distance Geometry
DHFR	Dihydrofolate Reductase
DhU	Dihydro Uracil base
DLS	Dynamic Light Scattering
DMPC	1,2DiMyrositoyl-sn-glycero-3-PhosphoCholine

DMSO	DiMethyl SulphOxide
DNA	DeoxyriboNucleic Acid
DPPC	1,2DiPalmitoyl-sn-glycero-3-PhosphoCholine
DQ	Double-Quantum
DRESS	Depth RESolved Spectroscopy
ds	Double-stranded
DSS	2,2-Dimethyl-2-Silapentane-5-Sulfonate sodium salt
DTT	1,4-DiThioThreitol
DWI	Diffusion-Weighted Imaging
DYANA	DYnamics Algorithm for NMR Applications
ECG	Electro-Cardiogram
ECOSY	Exclusive COrrelation SpectroscopY
EDTA	Ethylene Diamine Tetraacetic Acid
EMBOSS	The European Molecular Biology Open Software Suite
EMR	Electron Magnetic Resonance
ENDOR	Electron Nuclear Double Resonance
EPI	Echo Planar Imaging
EPR	Electron Paramagnetic Resonance
ESR	Electron Spin Resonance
ESRA	Endonuclease Sensitive Repeat Amplification
EVI	Echo Volume Imaging
F6P	Fructose 6-Phosphate
FAD	Flavin Adenine Dinucleotide
FBP	Far-upstream element-Binding Protein
FID	Free-Induction Decay
FISP	Fast Imaging with Steady Precession
FLAIR	FLuid Attenuation by Inversion Recovery
fMRI	functional MRI
FPLC	Fast Protein Liquid Chromatography
FROG	Fast Rotating Gradient Spectroscopy
FS	Frequency Selective
FSE	Fast-Spin-Echo

FT	Fourier Transform
FTNMR	Fourier Transform Nuclear Magnatic Resonance
Fuc	D-fucose
G	Guanine
G6P	Glucose 6–Phosphate
GABA	Gamma-AminoButyric Acid
Gal	D-Galactose
GFT	G-matrix Fourier Transformation
Glc	Glucose
GlcNAc	N-AcetylGlucosamine
GPC	GlyceroPhosphorylCholine
GRASP	Graphical Representation And analysis of Structural Properties
GRE	Gradient Recalled Echo
GROMOS	General-purpose molecular dynamics
GST	Glutathione-S-Transferase
GTP	Guanosine Tri-Phosphate
HASTE	Half-Fourier Acquired Single shoT spin-Echo
Hb	Haemoglobins
HB	Hydrogen Bonded
HDL	High Density Lipoprotein
HETCOR	HETeronuclear COrrelation
HFI	Half Fourier Transform Imaging
HIV-1	Human Immunodeficiency Virus type-1
HMBC	Heteronuclear Multiple Bond Correlation
HMQC	Heteronuclear Multiple–Quantum Correlation spectroscopy
HPLC	High-Performance Liquid Chromatography
HRP	Horse Radish Peroxidase
HSQC	Heteronuclear Single–Quantum Correlation spectroscopy
HTS	High Through-put in silico based Screenings
INEPT	Insensitive Nuclei Enhanced by Polarization Transfer
IPAP	In Phase Anti Phase approach
IR	Inversion Recovery experiment

ISIS	Image Selected *In vivo* Spectroscopy
IUBMB	International Union of Biochemistry and Molecular Biology
IUPAB	International Union of Pure and Applied Biophysics
IUPAC	International Union of Pure and Applied Chemistry
JDV	Jembrana Disease Virus
JR	Jump and Return
Lac	Lactate
LC	Liquid Chromatography
LDL	Low Density Lipoproteins
LG	Lee-Goldburg sequence
LPO	LactoPerOxidase
M9	Minimal medium
MALDI-TOF	Matrix Assisted Laser Desorption Ionization Time-Of-Flight mass spectrometry
Man	D-Mannose
MARDIGRAS	Matrix Analysis of Relaxation for DIscerning the GeometRy of an Aqueous Structure
MAS	Magic Angle Spinning
Mb	Myoglobin
MD	Multi-Dimensional
MD	Molecular Dynamics
MEF2	Myocyte Enhancer Factor 2
MEF2A	Myocyte Enhancer Factor 2A
MEMP	Multi-Echo Multi-planar
MIDAS	Monitoring, Intrusion, Detection, Administration System
MM	Molecular Mechanics
MOLMOL	MOLecule analysis and MOLecule display
MON	MONetite
MORASS	Multiple Overhauser Relaxation AnalysiS and Simulation
MQ	Multiple Quantum
MQF-COSY	MQ-Filtered COrrelation SpectroscopY
MRA	Magnetic Resonance Angiography

MRCP	Magnetic Resonance CholangioPancreatography
MRI	Magnetic Resonance Imaging
m-RNA	Messenger Ribonucleic Acid
MRS	Magnetic Resonance Spectroscopy
MRSI	Magnetic Resonance Spectroscopic Imaging
MS	Mass Spectroscopy
msec	milliseconds
MSG	Maleate Synthase G
MT	Magnetization Transfer
MV	Multi-Voxel
MVSI	Multiple Volume Spectroscopy
NAA	N-Acetyl Aspartate
NAAG	N-Acetyl Aspartyl Glutamate
NAD	Nicotinamide Adenine Dinucleotide
NADH	Nicotinamide Adenine Dinucleotide
NANA	N-Acetyl Neuraminic Acid
NIDDM	Non-Insulin-Dependent Diabetes Mellitus
NMR	Nuclear Magnetic Resonance
NOE	Nuclear Overhauser Effect
nsec	nanoseconds
OCP	Octacalcium Phosphate
ORD	Optical Rotatory Dispersion
PAPS	Phase Alternating Pulse Sequence
PAS	Principal Axis System
PC	PhosphoCholine
PCr	PhosphoCreatine
PD	Proton Density
PDB	Protein Data Bank
PDE	Phospho-DiEsters
PET	Positron Emission Tomography
PFGs	Pulsed Field Gradients
PFK	PhosphoFructoKinase

photo-CIDNP	photoChemically Induced Dynamic Nuclear Polarisation
PISA	Polarity Index Slant Angle wheels
PISEMA	Polarization Inversion Spin Exchange at the Magic Angle
PO	Product Operators
POF	Product Operator Formalism
POPC	1,Palmitoyl-2-Oleoyl-sn-glycero-3-PhosphoCholine
ppm	parts per million
PRESS	Point REsolved SpectroScopy
PROCHECK	PROgrams to CHECK stereochemical quality of protein structure
ps	parallel-stranded
PS I	PhotoSystem I
PS II	PhotoSystem II
ps-DNA	Parallel stranded DNA
psec	picosecond
PW	Pulse-Width
QSAR	Quantitative Structure Activity Relationships
R&D	Research and Development
RARE	Rapid Acquisition with Relaxation Enhancement
RC	Reaction Centers
RD	Relaxation Delay
RDC	Residual Dipolar Couplings
REDOR	Rotational-Echo Double Resonance
RF	Radio Frequency
RMSD	Root Mean Square Deviation
RNase	RiboNucleases
ROESY	Rotational Nuclear Overhauser Effect SpecroscopY
ROI	Region Of Interest
ROPE	Respiratory-Ordered Phase-Encoding
RR	Rotational Resonance
RRE	Rev-Response Element
r-RNA	ribosomal RNA
RuBP	Ribulose-1,5-BisPhosphate
S:N	Signal-to-Noise Ratio

SAIL	Stereo-Array Isotope Labeling
SC	Scalar Coupling
SDS	Sodium Dodecyl Sulphate
SDS-PAGE	Sodium Dodecyl Sulphate- Poly Acrylamide Gel Electrophoresis
SE	Spin-Echo
SF1	Splicing Factor 1
SI	Spectroscopic Imaging
SPECT	Single Photon Emission Computed Tomography
SPIO	Super Paramagnetic Iron Oxide
SPM	Statistical Parametric Mapping
SQ	Single-Quantum
SQC	Single-Quantum Coherences
SR	Spin Rotation
ssr	sequence specific resonance
STAR	Signal Transduction and Activation of RNA
STEAM	STimulated Echo Acquisition Mode
STIR	Short T_1 Inversion Recovery
SV	Single Volume
T	Thymine
TAD	Torsion Angle Dynamics
TATAPRO	Tracked AuTomated Assignments in PROteins
TBFE	1,1,2,2 TetraBromoFluoroEthane
TCA	Tri-Carboxylic Acid
TE	Time to Echo
TI	Inversion Time
TIM	Triosephosphate IsoMerase
TMA	TriMethylAmine
TMAO	TriMethyl-Amine-N-Oxide
TMR	Topical Magnetic Resonance
TMS	Tetra Methyl Silane
TOCSY	Total COrrelation SpectroscopY
TOSS	TOtal Suppression of Side bands

TPPM	Two-Pulse Phase Modulation
TR	Repetition Time
t-RNA	transfer RiboNucleic Acid
TROSY	Transverse Relaxation Optimized SpectroscopY
TRUS	TRansrectal Ultra Sound
TSE	Turbo Spin Echo
TSP	3-(TrimethylSilyl)-Propionic acid-D4 sodium salt
U	Uracil
UDP	Uridine DiphosPhoglucose
VLDL	Very Low Density Lipoproteins
VOI	Volume Of Interest
WAHUHA	WAugh, HUber and HAeberlen
WATERGATE	WATER suppression by GrAdient Tailored Excitation
WC	Watson-Crick
ZQC	Zero-Quantum Coherences

CHAPTER 1

BASIC CONCEPTS IN NMR SPECTROSCOPY

1. HISTORICAL PERSPECTIVES

The phenomenon of Nuclear Magnetic Resonance (NMR) is based on the property of nuclear magnetism. This property is possessed by several atomic nuclei having either odd number of protons or neutrons or both. Rabi and co-workers in 1937 made the first experimental observation of NMR using molecular beams. Rabi was awarded Nobel Prize in 1944 for this work. In 1945, Purcell, Torrey and Pound observed proton (^1H) magnetic resonance in paraffin wax. Independently, Bloch, Hansen and Packard observed the phenomenon in liquid water. In 1952, Bloch and Purcell were jointly awarded the Nobel Prize for their discoveries. Soon after this, NMR emerged as a tool for accurate measurements of nuclear magnetic moments and earth's magnetic field.

During the two decades that followed the discovery of NMR in bulk materials, the concepts and basic theories underlying NMR parameters, such as chemical shifts, nuclear spin-spin couplings, dipolar interactions, nuclear quadrupole interactions, relaxation rates, effect of chemical exchange on line-shapes and nuclear Overhauser effect (NOE), were established. Several technological developments during this period improved the power of NMR in chemical research. These include: introduction of signal averaging to improve signal to noise (S/N) ratio; field/frequency lock to stabilize magnetic fields; electronic shims to improve field homogeneity; development of magnets with higher and more homogeneous magnetic fields; better probe designs; and use of double resonance techniques. One of the major discoveries during this period was that of *spin-echo* (SE) by Hahn. This technique plays a major role both in Magnetic Resonance Spectroscopy (MRS) and Magnetic Resonance Imaging (MRI). Developments in electronics and computer technology greatly enhanced the power of NMR spectroscopy. During this period, NMR became an indispensable tool in chemical and physical sciences. Good sources for information available by the middle of 1950's are the classical books by Pople, Bernstein and Schneider and by Abragam.

A major break-through occurred in 1966, when Ernst and Anderson developed the technique of Fourier Transform (FT) spectroscopy. They showed that the use of short and intense radiofrequency (RF) pulse for excitation, acquisition of the signal in time-domain, followed by FT of the acquired signal, improves the sensitivity of NMR experiment several fold. The FT technique soon replaced the previously used slow passage continuous wave (CW) method which involved changing the magnetic field to reach resonance condition (The Nobel prize was awarded to Ernst in 1991). This was followed by the development of several multiple-pulse experiments and introduction of two-dimensional (2D) FTNMR experiments. These discoveries led to new frontiers in NMR. Developments in computer hardware and software, magnet technologies and better design of the RF probes collectively opened new frontiers of NMR applications in life sciences.

In 1971, Damadian showed that the ^{1}H relaxation rates of water, present in normal and malignant cells, are different. This opened up the possibility of using this property for medical diagnosis. Paul Lauterbur used magnetic field gradients to localize NMR signal information in space. He showed that this can be used to generate images of objects. This gave birth to the technique called *Magnetic Resonance Imaging* (MRI). Richard Ernst used phase and frequency encoding and FT for obtaining MR images, which formed the basis of commercial MRI instruments. Peter Mansfield developed the technique of echo-planar imaging (EPI), which provides images much more rapidly than the earlier methods. Lauterbur and Mansfield received the Nobel Prize in 2003.

In this Chapter, the basic concepts of NMR spectroscopy are described. These may be particularly useful for those who do not have previous knowledge of NMR. Some readers may prefer to skip the mathematically oriented sections in the book and straight-away turn to applications. Those who want to go deeper into the theoretical and experimental aspects of NMR may consult the more advanced text books suggested at the end of each Chapter.

1.1 Quantum Mechanical Model for Spin-1/2 Nuclei

The basis of NMR spectroscopy lies in the fact that certain atomic nuclei possess spin (I; discovered by Goudsmit and Black in 1926) (and spin angular momentum) and consequently, a magnetic moment (μ). According to the quantum theory, angular momentum is *quantized*. The interaction of the magnetic moment of such nuclei with an external magnetic field (B_0) results in an energy manifold (Figure 1.1). There are 2I+1 *stationary* or *energy eigenstates* (also called *spin states*) for a nucleus having spin I, corresponding to the direction of the angular momentum relative to an arbitrary axis z. For the simplest case of nuclei with spin ½, there are two spin states, corresponding to the magnetic quantum number m_I =½ and -½, respectively. The two spin states referred above are identified by the *eigenfunctions*, |α> (m_I = ½) and |β> (m_I = -½).

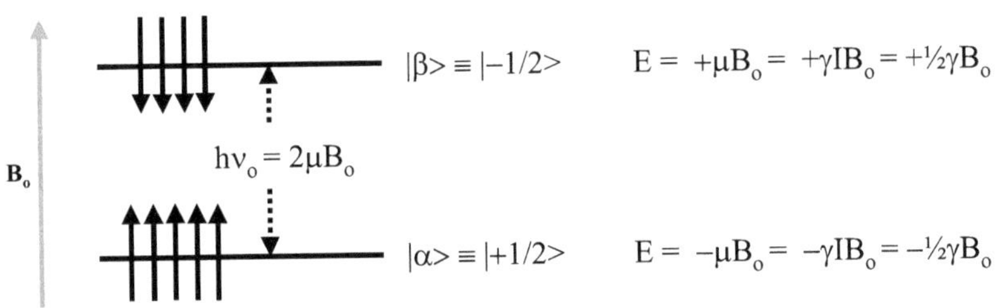

Figure 1.1: Energy levels for a nucleus with spin ½, in the presence of a magnetic field (B_0). The lower level (|α>) has the nuclear spins aligned parallel (m_I=½), while the upper energy level (|β>) has the spins aligned anti-parallel (m_I=-½), relative to the applied magnetic field B_0. Note that $\gamma = \mu/I(h/2\pi)$.

In the absence of a magnetic field (B_0), the two levels have equal energy. This degeneracy is lifted in the presence of B_0. The energy of the two states corresponds to $-\frac{1}{2}\gamma B_0$ (for the state α; $m_I = \frac{1}{2}$, parallel to B_0) and $+\frac{1}{2}\gamma B_0$ (state β; $m_I = -\frac{1}{2}$, antiparallel to B_0). Here, γ is called the *magnetogyric* or the gyromagnetic ratio ($\gamma=\mu/I(h/2\pi)$), and is a property of the nucleus under study. The energy separation (ΔE) between the two levels is γB_0, and is thus proportional to B_0. In a macroscopic sample, containing a large number of such nuclei, the distribution of the spin populations between the two states at a temperature T (expressed in Kelvin) is given by the Boltzmann relation:

$$N_\beta/N_\alpha = exp\ (-\Delta E/kT) = exp\ (-h\nu_0/kT) \approx 1 - h\nu_0/kT \qquad 1.1$$

Here, k is the Boltzmann constant and the lower state α is slightly more populated. The value of $\Delta E/kT$ is about 8×10^{-5} for ^{1}H at room temperature (T = 300 K) and at a Larmor frequency of 500 MHz. Therefore, it is sufficient to retain only the first term in the Taylor series expansion of the exponential term in equation 1.1. This is known as the *high temperature approximation*. One may note that the approximation holds good even at temperatures as low as 4K, where $\Delta E/kT$ *is* $\sim 10^{-3}$.

According to Bohr's resonance condition ($\Delta E = h\nu_0 = (h/2\pi)\omega_0$; where h is the Planck constant), one needs an energy quantum $(h/2\pi)\omega_0$ to induce a transition between the states α and β, where,

$$(h/2\pi)\omega_0 = 2\mu B_0 \ \ or\ \ \omega_0 = 2\mu B_0/(h/2\pi)\ or\ \omega_0 = \gamma B_0 \qquad 1.2a$$
$$where,\ \gamma = \mu/I(h/2\pi) \qquad 1.2b$$

Note that in Equation 1.1, the frequency ν_0 is called the Larmor frequency or the precessional frequency of the nucleus, expressed in Hertz (Hz) or cycles per second, while the frequency ω_0 ($2\pi\nu_o$) is called the angular precessional frequency of the nucleus (expressed in radians per second; 1 radian = $360^\circ/2\pi$ or approximately 57°).

Both representations have been used in NMR literature. These frequencies depend on the gyromagnetic ratio γ of the nucleus, and are proportional to the applied magnetic field B_0. One can excite the system by flipping the nuclear spins from the lower energy state α to the upper energy state β using radiation of frequency ν_0. A net absorption of energy is possible so long as the population of nuclear spins in the higher energy level is smaller than that in the lower one. However, in the presence of electromagnetic field B_1 containing photons of an energy $h\nu_0$, the population of the two levels will become equal and further absorption of energy will not occur. Later, we will discuss processes by which the spin system can *relax* back to the equilibrium state.

As is evident from Equation 1.2, the resonance frequency (ω_0) is different for different nuclear spins since γ is different for each spin. The resonance frequency is proportional to the applied field B_o. For example, at a B_o of 10 Tesla (T; 1 T=10000 Gauss), the resonance frequency for protons (^{1}H) is 425.53 mega-Hertz (MHz). This frequency falls in the radio frequency (RF) region and lies in the low frequency region of the electromagnetic spectrum. At this magnetic field, the difference of population between the states α and β is only about 0.001%. This population difference determines the number of photons absorbed during the upward transition and thus the strength of the NMR signal. Therefore, NMR is intrinsically a very insensitive technique.

Modern techniques use pulse NMR, which is discussed later. The positive z–axis (longitudinal axis) is conventionally chosen as the direction of B_0. However, some books use the convention of placing B_0 along –z-axis. This does not make any difference, except to change the sign of energies corresponding to the two spin states. NMR signal is usually detected by placing a receiver coil in the x-y plane (transverse plane) which is perpendicular to the z-axis.

1.2 Classical Model

Both classical and quantum mechanical viewpoints are integral parts of concepts in NMR spectroscopy. In the classical approach, the phenomenon of NMR is discussed in terms of net magnetization (**M**), in a macroscopic sample containing a large number of nuclei under study, rather than the behaviour of individual spins. In the presence of B_0, the angular momentum of the nucleus causes a precessional motion around the z-axis (Figure 1.2). The angular velocity of such a motion is given by $\omega_0 = -\gamma B_0$.

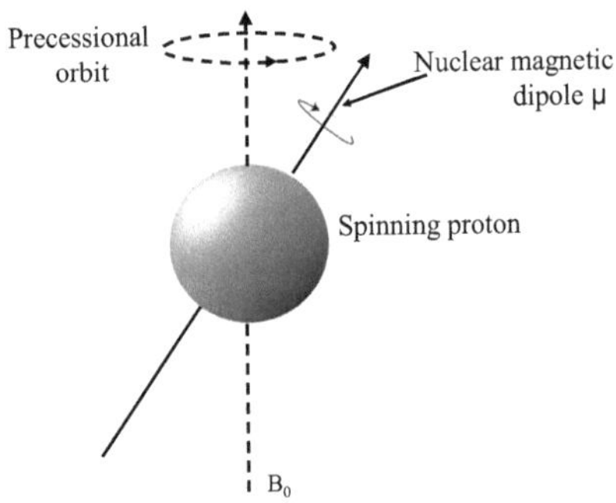

Figure 1.2: Larmor precession of a spin in magnetic field.

In the presence of B_0, more nuclei align in the parallel direction resulting in a net magnetization in the z direction (M_z). If an oscillating radiofrequency field (B_1) with a frequency equal to ω_0 is now applied in y-direction for a short time (t), then the net magnetization *nutates* (flips) from the z-direction towards x direction. The *flip-angle* due to such an RF pulse, is given by $\theta = \gamma B_1 t$ (radians), where B_1 and t are the strength and the time duration respectively, of the applied RF field. To start with, all spins precess with the same phase, after the application of RF field. This results in a transverse magnetization ($M_{x/y}$). In other words, the RF pulse creates a *phase-coherence* among different precessing spins.

1.2.1 Rotating Frame of Reference

It is essential to understand the concept of rotating frame of reference, in which it is easier to follow the evolution of components of transverse magnetization (M_x and M_y). The rotating frame of reference is a coordinate system which rotates at the spectrometer frequency and is usually close to the individual Larmor frequencies of nuclear spins. In the rotating frame, M_x and M_y appear to process with a frequency $\Omega_i = \omega_o(i) - \omega_{rf}$, where $\omega_o(i)$ are the individual Larmor frequencies and ω_{rf} is the applied radio frequency (distinguished by the corresponding offset from the Larmor precession frequency).

It may be noted that Ω_i is of the order of kHz, as compared to ω_o, which is several MHz. When $\omega_o(i) = \omega_{rf}$, $\Omega_i = 0$, the corresponding transverse magnetization appears stationary in the rotating frame. Such a reference frame helps to simplify several concepts in pulsed NMR. We will extensively use rotating frames. Hereafter, when the precessional frequencies are expressed by the symbol Ω, it is assumed that one is using the rotating frame of reference.

1.2.2 Strength of RF Pulses

In modern NMR techniques, RF pulses are characterized by their respective flip angles and pulse widths. For example, a $\pi/2$ (90°) pulse flips the net magnetization from the z-axis to the x-y plane and creates a transverse magnetization (Figure 1.3). Small flip angles rotate only a part of **M** on to the transverse plane. A π pulse inverts the magnetization to the negative z direction.

It may be pointed out that a $\pi/2$ RF pulse equalizes the population between the states α and β and hence there is no net z-magnetization. The transverse magnetization arises because of the *phase coherence* of the spins, which is created by the RF pulse. This coherence decays with time as spins go out of phase due to relaxation processes discussed later.

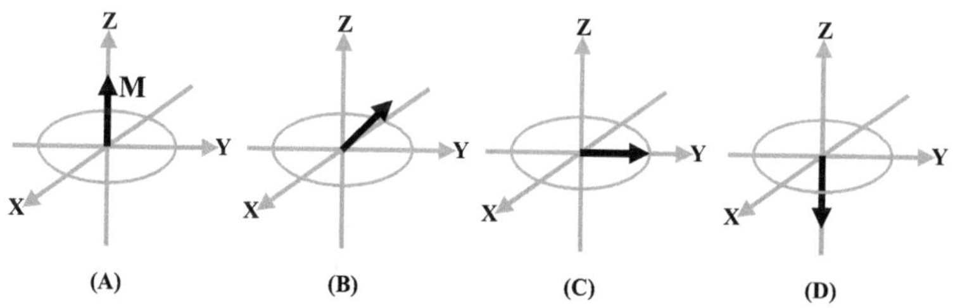

*Figure 1.3: A RF pulse applied to the bulk sample creates a net magnetization in the transverse (xy) plane. In the rotating frame, the effect of the pulse is to tilt the net magnetization **M** which appears stationary. The flip angle depends on the strength and duration of the pulse. (A) The equilibrium magnetization. The net magnetization after the application of: (B) $\pi/4$; (C) $\pi/2$; and (D) π pulse along -x-axis. In the laboratory frame, the net magnetization vector rotates with an angular velocity ω_0 around the longitudinal axis.*

1.3 Basic Design of a NMR Spectrometer

The basic elements of a modern NMR spectrometer are shown in Figure 1.4. A large and highly homogeneous static magnetic field (B_o) is required to remove degeneracy among the nuclear-spin energy levels. The gap between the ground and the excited states increases with higher magnetic fields. Hence, the difference of populations of the spins between the two states is higher. Stronger magnetic fields help in achieving higher sensitivity. As discussed later, higher fields also help in a better dispersion of spectral lines. Thus, for studies on biological molecules, it is desirable to use the highest possible magnetic field strength. Spectrometers using field strengths as high as 21.15 T are currently available. This field corresponds to a ^1H resonance frequency of 900 MHz. It is also essential that the magnetic fields are homogeneous. Such high and homogeneous field strengths are produced using superconducting magnets. One often uses the ^1H resonance frequency to define the magnetic field used in a particular NMR spectrometer.

Spectrometers are also distinguished by the magnet bore size. Magnets with narrow bores (typically a few cm), are used mostly for high-resolution solution work. The sample used in a high-resolution NMR experiment is generally around 600 μl and is taken typically in a sample tube of 5 mm in diameter. Wider bore magnets are used for studies on solid-state materials, which require additional hardware within the magnet bore such as the magic-angle spinning assembly and equipment for work over a wide temperature range.

When working with humans and animals, lower magnetic field strengths are used because of safety considerations. RF penetration and heating issues and their associated neurological effects are important. The magnets should be able to accommodate humans or live animals. For animal studies, magnetic field strengths up to 7 T have been used. However, for humans, the optimum field strength for both MRI and MRS so far has been 1.5 T. In special cases 3 T magnets have been used.

One of the important requirements of NMR experiments is a high homogeneity of the magnetic field, in the region of the sample (typically 1 in 10^9 over the entire sample volume). This is achieved through the use of shim coils, which provide additional fields to compensate for any variations. The shim coils are placed both in the cold part of the magnet and around the probe. A spinner assembly containing a small turbine system is used to spin the NMR sample tube to achieve better homogeneity.

The heart of the spectrometer is the probe. This has arrangement to hold the sample. Probe consists of various resonant circuits, and a set of RF coils and capacitors, needed for excitation and detection of the NMR free induction decay (FID) signal. At least three to four RF frequencies are needed to provide the ^1H frequency and one or two X-nucleus frequencies (e.g. for ^{13}C, ^{15}N and ^{31}P). The magnetic field is locked to a master frequency called the deuterium lock.

For molecular studies, the RF probe head is usually introduced into the magnet from the bottom. In addition, there is provision to maintain sample at a constant temperature within ±0.1 °C. Most commercial spectrometers are equipped for variable temperature studies in the range –100 to +150 °C.

Magnetic field gradient coils are used to alter the static field. Such coils are incorporated either within the magnet assembly or surrounding the RF coil. Data accumulation over long time intervals requires a good field/frequency stability. An internal lock using ^2H of the solvent (^2H_2O) helps in compensating the drift in static magnetic field.

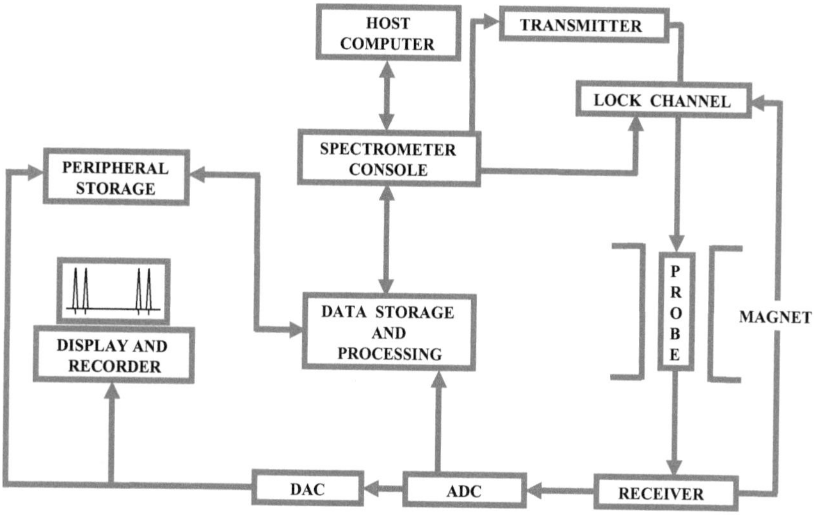

Figure 1.4: A schematic view of NMR spectrometer.

The console part of NMR equipment has a host computer which serves as a controller of the whole system and coordinates all the activities of the spectrometer. It has an arrangement for generating and transmitting RF at desired frequencies and power to the sample. Experiments are performed using RF fields (B_1) in the form of pulses. For multinuclear studies, RF fields at several frequencies are generated simultaneously. They are controlled, amplified, pulsed and transmitted to the RF-probe head. The RF field strengths are in the range of 0.01 to 0.2 mT, which correspond to several hundred kilowatts of power. RF pulses are programmed for diverse applications and are applied in a very precise time frame. Generation of different RF frequencies is achieved from a common master frequency derived from an oven packed crystal oscillator. Though RF pulses are applied for short durations, they can cause significant heating of the sample.

The detected NMR signal is an audio signal, which is digitized by an analogue-to-digital converter (ADC). ADC is characterized by its sampling rate (which is at least twice the frequency of the highest frequency signal that is to be detected; this is commonly known as *Nyquist criterion*) and the dynamic range (the ability of ADC to accurately measure signals having wide-ranging amplitudes). For optimum signal-to-noise (S/N) ratio, proper matching is provided between the excitation and the detection hardware. The detected signal which is a function of time (*time domain signal*) is fed into the computer memory. Later it is Fourier transformed (FT) and phase-corrected to get the frequency domain spectrum using either the host computer or an additional work-station. Software is available for acquiring, processing, displaying, plotting and basic analysis of the data on the host computer. This forms an important component of the spectrometer console.

1.4 A Simple Pulse Sequence for NMR Excitation and Detection

The simplest pulse sequence used in NMR is a single RF pulse for creating transverse magnetization. This is followed by detection of the decay of the magnetization in the transverse plane as illustrated in Figure 1.5. Generally, a $\pi/2$ pulse is used to excite the sample, which tips the orientation of the net magnetization (M_z) from the z-axis to the transverse plane. Immediately after this pulse, all spins start precessing together around the z-axis in the transverse plane with the same phase. A strong pulse, thus establish a *phase-coherence* among the spins. For example, a $\pi/2$ pulse applied along the y-axis rotates the M_z towards the +x-axis, which starts precessing around B_o in the x-y plane (Figure 1.5). This gives rise to detectable M_x and M_y magnetizations which are given by the following equation:

$$M_x(i) = M_o(i) \cos \omega_i t \; ; \; M_y(i) = M_o(i) \sin \omega_i t \qquad 1.3$$

Here ω_i are the individual precessional frequencies, which are close to the spectrometer frequency. $M_o(i)$ are the corresponding equilibrium magnetizations.

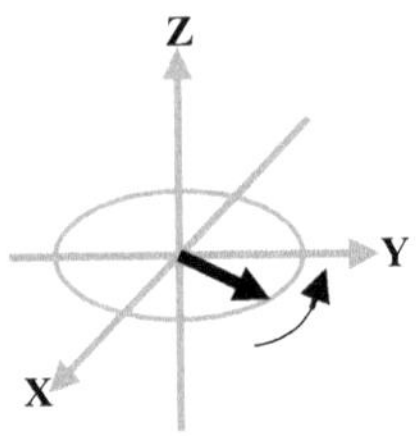

Figure 1.5: Free precession of magnetization vector in the x,y plane.

As spins precess with their respective Larmor frequencies, a signal is induced in the receiver coil. This is called the *free-induction decay* (FID). This signal decays with a time constant called the transverse relaxation time (T_2). During this time all the nuclear spins go out of phase due to nuclear spin-spin interactions. This subject is discussed later in this Chapter. The FID signal thus measures transverse magnetization (M_{xy}) as a function of time. Since the transmitter pulse is several orders of magnitude stronger than the detected signal, the collection of the signal is delayed by 0.1 to 5 milliseconds (called the *acquisition delay* or the dead time) after turning off the transmitter. The detected FID signal S(t), is measured as a function of time (Figure 1.6).

The pulse-width (PW) of the $\pi/2$ pulse is chosen so as to cover the entire spectral range of the spins under study (wider spectral widths require shorter pulses). For ^1H at 600 MHz, the typical value of the PW is around 10 μsec. The time domain

FID signal thus acquired has information about various frequencies with the corresponding amplitudes and phases. In fact, the two modes of signal in Figure 1.6 are representation of the same data set; one is in the time domain and the other in the frequency domain.

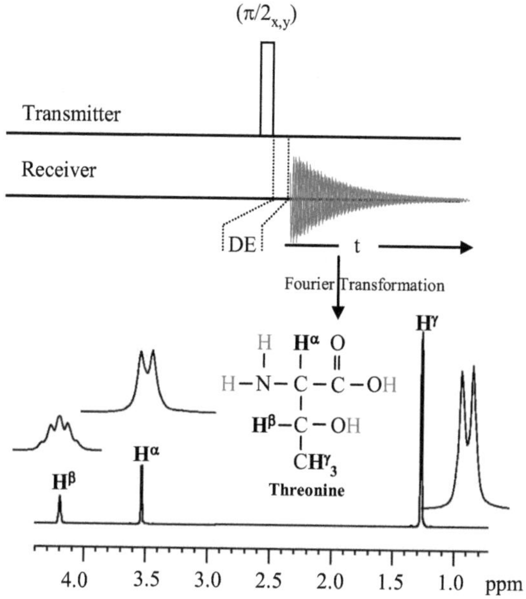

Figure 1.6: NMR excitation and detection of the FID, using a single RF pulse for threonine dissolved in ^2H_2O (D_2O). DE is the acquisition delay. The labile hydroxyl and amide protons are not observed in ^2H_2O, because of their exchange with ^2H. H^α and H^γ protons appear as doublets at 3.52 and 1.26 ppm, respectively; H^β appears as a complex multiplet (doublet of a quartet) at 4.2 ppm. The behaviour of individual signals has been discussed later.

1.5 Fourier Transform

Though the FID contains all the desired information, it is not amenable to a simple interpretation. A mathematical operation called Fourier transformation (FT) on S(t) yields the frequency domain signal F(ω) and *vice-versa*. The two representations are related by:

$$F(\omega) = \int_{-\infty}^{\infty} S(t)\, e^{-i\omega t}\, dt \qquad 1.4$$

$$S(t) = \frac{1}{2\pi} \int_{-\infty}^{\infty} F(\omega)\, e^{i\omega t}\, dt \qquad 1.5$$

These transformations are conveniently carried out using a computer. The FID of a single spin can be represented by a time function as given in Equation 1.6a:

$$S(t) = M_{xy}(t)exp(-t/T_2) \quad 1.6a$$

It may be noted that the right hand side of Equation 1.6a has two terms. The first is a sinusoidal function represented by Equation 1.3. The second term is a damping factor arising from the transverse relaxation time T_2, which is discussed later. The FT of S(t) is a complex quantity, whose real part is given by:

$$F(\omega) = M_0T_2/\{1 + (\omega-\omega_0)^2T_2^2\} \quad 1.6b$$

Here ω_0 is the Larmor frequency of the spin. In the absence of the second term, the FT is a Dirac function [represented by $\delta(\omega-\omega_0)$]. In presence of both the terms, the frequency domain spectrum for a single spin has Lorentzian shape and is centred at ω_0. For a given molecule, there may be several resonances which may contribute to the FID. The FT of the complex FID is carried out numerically, using a computer.

1.6 Line Widths

As indicated by Equation 1.6b, the FT of the FID has a Lorentzian line-shape. The line-width of the resonance signal at half-height is equal to $2/\pi T_2$. However, the decay of transverse magnetization can also be caused by in-homogeneities in the B_0. This leads to an additional broadening of the NMR signal. One therefore defines another relaxation time (T_2^*).

$$1/T_2^* = 1/T_2 + \gamma\Delta B_0/2 \quad 1.6c$$

If the field inhomogeneity is significant, then the line-width is given by $2/\pi T_2^*$.

1.7 Properties of Spin Operators and Pauli Matrices

We have seen that a spin ½ system can be described by two independent states $|\alpha>$ and $|\beta>$, which characterize the orientations of the nuclear magnetic moment with respect to magnetic field (i.e. the z-direction). To obtain further insights on the quantum mechanical description of NMR, one needs to calculate expectation values of the spin operator **I**. One can divide **I** into components I_x, I_y and I_z in the Cartesian space. The expectation values of I_x, I_y and I_z are given by Pauli 2×2 matrices, which have elements $<r|I_v|s>$, where *r* and *s* are either α or β and v is *x*, *y* or *z*. In these two states, the *z* component of the spin (I_z) has definite value of ±½. In other words, α and β are eigenfunctions of I_z having the eigenvalues ±½. The wave function $|\alpha>$ is associated with the state for which $I_z = ½$, while the state $|\beta>$ is associated with $I_z = -½$ (Figure 1.1). With $|\alpha>$ and $|\beta>$ as a basis set, expectation values $<r|I_v|s>$ of the three components of I are given by:

$$I_x = \frac{1}{2}\begin{pmatrix} 0 & 1 \\ 1 & 0 \end{pmatrix};\ I_y = \frac{1}{2}\begin{pmatrix} 0 & -i \\ i & 0 \end{pmatrix}\ and\ I_z = \frac{1}{2}\begin{pmatrix} 1 & 0 \\ 0 & -1 \end{pmatrix};\ E = \begin{pmatrix} 1 & 0 \\ 0 & 1 \end{pmatrix} \qquad 1.7$$

The factor ½ in front of these matrices is for normalization. Notice that without the factor ½, Tr $(I_x^{\dagger} I_x)$ = Tr $(I_y^{\dagger} I_y)$ = Tr $(I_z^{\dagger} I_z) = 2$. Pauli matrices are orthogonal (e.g. $I_x.I_y = I_y.I_z = I_z.I_x = 0$) and can not be expressed as a linear combination of others. Together with a unit matrix E, these four matrices form a complete set for a single spin ½ system.

Pauli matrices play an important role in deriving properties of spin operators and their evolution under pulse sequences. For example, using matrix multiplication, the following cyclic commutation rules for spin-1/2 operators can be derived.

$$[I_x, I_y] = -[I_y, I_x] = iI_z;\ [I_y, I_z] = -[I_z, I_y] = iI_x;\ [I_z, I_x] = -[I_x, I_z] = iI_y \qquad 1.8$$

$$I_xI_y=-I_yI_x=(i/2)I_z;\ I_yI_z=-I_zI_y=(i/2)I_x;\ I_zI_x=-I_xI_z=(i/2)I_y;\ and\ I_xI_x=I_yI_y=I_zI_z=½E \qquad 1.9$$

One may note here that, some of these rules do not apply for spin operators in general, but only for the spin-1/2 matrix representations.

Two more spin operators (I^+ and I^-) are called the *raising* (I^+) and *lowering* (I^-) operators, respectively. These are defined by:

$$I^+ = I_x + iI_y;\ I^- = I_x - iI_y \qquad 1.10$$

It can be shown that these operators have the following properties:

$$I^+|\beta\rangle = |\alpha\rangle\ and\ I^-|\alpha\rangle = |\beta\rangle \qquad 1.11$$

2. NMR SPINS USED IN LIFE SCIENCES

All living systems have five major elements; carbon (C), hydrogen (H), nitrogen (N), oxygen (O) and phosphorous (P). Of these, the naturally abundant isotopes of C (^{12}C) and O (^{16}O) do not have nuclear magnetic moments. The more abundant isotope of N (^{14}N) has spin 1. Nuclei with spin > ½ have quadrupole moments. As discussed later, such nuclei generally give rise to relatively broad signals with poor S/N ratio. Thus, one generally uses four nuclei in life sciences: ^1H, ^{13}C, ^{15}N and ^{31}P, each of which has spin ½. These nuclear spins serve as reporter groups and provide complementary information on biological systems.

The individual NMR properties of above nuclei are listed in Table 1.2. Both hydrogen and carbon are present in all biological molecules and ^1H and ^{13}C can provide information on all classes of biological functions. The high sensitivity of ^1H NMR coupled with its high natural occurrence makes it the most widely studied nucleus. Even ^1H resonance in a simple molecule like water (which constitutes almost 75% of all living systems) provides a wealth of information on biological systems. Nitrogen is an essential component of nucleic acids and proteins and ^{15}N

provides useful information on such macromolecules. To avoid problems of sensitivity due to low natural abundance of ^{13}C or ^{15}N, one can use isotopically enriched (labelled) molecules. Like-wise, one can use ^{13}C or ^{15}N labelled substrates to trace metabolic pathways. ^{31}P with its 100% natural abundance is another useful NMR probe. Though present in relatively smaller number of biological molecules, ^{31}P has been extensively used in the study of structure of nucleic acids and biological membranes. It has found wide applications for monitoring metabolism through molecules such as adenosine triphosphate (ATP) and phosphocreatine (PCr), which are involved in providing energy for essential processes of life.

Table 1.1: Important NMR active nuclear spins used in Life Sciences and their properties.

Nucleus§	Natural Abundance	γ (x 10^6) Rads^{-1}T^{-1}	ν_0 in MHz at 11.743 T	Relative Sensitivity	Chemical Shift Range (ppm)
^{1}H	99.985	267.522	500.000	100.0	15
^{2}H	0.015	41.066	76.753	.00965	20
^{13}C	1.11	67.283	125.725	0.02	250
^{15}N	0.37	-27.126	50.684	0.0004	1000
^{31}P	100.00	108.394	202.606	6.6	430

§All the listed nuclei possess spin (I) = ½, except the ^{2}H whose I = 1.

While the above mentioned nuclei are those which are the most widely studied, use has also been made of isotopes such as ^{2}H (I=1), ^{7}Li (I=3/2), ^{14}N (I=1), ^{19}F (I=1/2), ^{23}Na (I=3/2), for studying biological systems.

3. INTERACTION OF NUCLEAR SPINS AND NMR PARAMETERS

It may appear that each nuclear spin shall give rise to a single line in NMR spectrum. However, what makes NMR so powerful is the fact that nuclear spins are not isolated. The electronic clouds and other magnetic nuclei in the environment of such spins modify their NMR spectrum. This section describes these properties.

3.1 Chemical Shift (δ)

Figure 1.7 shows the ^{1}H and ^{31}P spectra of a small DNA duplex containing 12 base pairs. It is obvious that the ^{1}H spectrum is complex and has several lines. The ^{31}P spectrum likewise has several lines.

These spectral features arise because different ^{1}H and ^{31}P are present in different electronic environments. The electrons modify the magnetic field seen by the nucleus. The field at the nuclear site (B_n) is slightly different from the one applied (B_0). The differential screening due to electrons and other local fields in the vicinity of the nucleus under study results in a property called the *chemical shift*. This is the origin for the different positions of the signals for the H^{α}, H^{β} and H^{γ} protons in threonine shown in Figure 1.6.

One of the major factors contributing to the chemical shift is due to the fact that the applied magnetic field induces circulation of the electron cloud, such that the local field at the site of the nucleus (B_n) is slightly different from the applied field B_0. This is called the local diamagnetic effect (σ_d^{loc}) and results in the magnetic shielding of the nucleus by an amount $\boldsymbol{\sigma} B_0$.

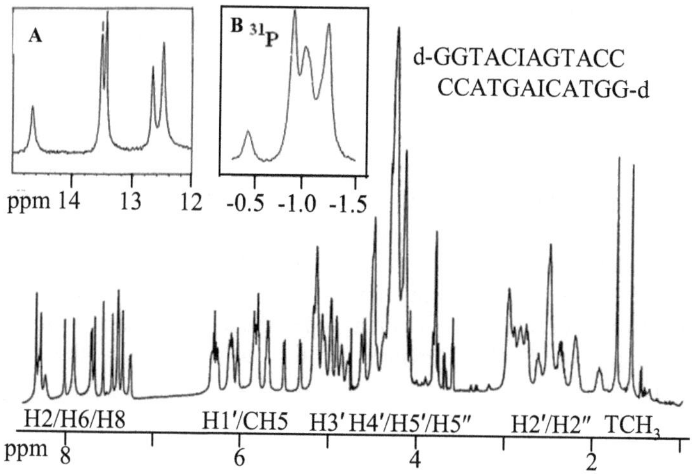

Figure 1.7: ^1H and ^{31}P NMR spectra of a self-complementary DNA duplex in H_2O. The positions of different ^1H are marked relative to TSP. The imino proton region is shown as an inset A. ^{31}P chemical shifts (inset B) are relative to 85% phosphoric acid.

$$B_n = B_0\,(1 - \sigma) \qquad 1.12$$

$\boldsymbol{\sigma}$ is called the *shielding* or the *screening constant.* It is in fact a tensor. Elements of σ can be represented by a 3×3 matrix of real numbers. One can use a suitable coordinate transformation, such that $\boldsymbol{\sigma}$ can be brought in a diagonal form, with the elements σ_{xx}, σ_{yy}, and σ_{zz}. In solutions, one measures the trace of this tensor, namely $1/3(\sigma_{xx} + \sigma_{yy} + \sigma_{zz})$. In solids, one gets a complex line-shape, which is dependent on the direction of the molecular axis relative to magnetic field B_o (Chapter 9). Equation 1.12 implies that at a constant applied magnetic field, the frequency of resonance is different for nuclear spins present in different chemical environment, due to differential electronic screenings.

Several short and long-range factors in molecules contribute to $\boldsymbol{\sigma}$. Therefore, one has to consider other contributions to the shielding effect. A theoretical calculation or experimental determination of the shielding tensor is not easy. However, one can break down contributions from various factors and estimate changes in the screening constant ($\Delta\sigma$) rather than its absolute value. The perturbation of the spherical symmetry of the electron distributions due to other atoms in a molecule, results in a paramagnetic term ($\Delta\sigma_p^{loc}$). This term is particularly important for heavy atoms, where low-lying atomic orbitals are present. In the presence of magnetic field, a mixing of the ground and excited state wave-functions can take place. The secondary field $\Delta\sigma^{loc}$ is influenced by substituents in a molecule, as the local electron density changes due to inductive and mesomeric effects of the substituents. In addition, neighbouring atoms and groups may give rise to local magnetic fields at the nuclear site ($\Delta\sigma^{m}$). Electric fields ($\Delta\sigma^{e}$), van der Waals interactions ($\Delta\sigma^{vW}$) and solvent effects ($\Delta\sigma^{med}$) also contribute to the chemical shifts. Thus, one can write:

$$\Delta\sigma = \Delta\sigma^{loc} + \Delta\sigma^{m} + \Delta\sigma^{e} + \Delta\sigma^{vW} + \Delta\sigma^{med} \qquad 1.13$$

In general, differences in chemical shifts are easier to interpret as compared to absolute values of σ. Quantum chemical theories have been used for calculating such changes or empirical relationships can be established. These are very helpful in predicting and comparing chemical shifts. For example, one may compare the chemical shift value of a specific ^1H in an ordered state of a macromolecule and its value in a random coil. Such differences can then be correlated with changes in its conformation.

Therefore, instead of worrying about absolute values of σ, one uses differences (δ) from a reference compound ($\delta = \sigma_{ref} - \sigma$), which is given by:

$$\delta_i = \frac{\omega_i - \omega_{ref}}{\omega_0} = \frac{\nu_i - \nu_{ref}}{\nu_0} \qquad 1.14$$

where, ν_0 is the spectrometer frequency.

The most common standard for ^1H in organic compounds is the methyl protons in tetra-methyl silane (TMS). Since TMS is insoluble in aqueous solvents, other silane derivatives such as 3-(trimethylsilyl)-propionic acid-D4 sodium salt (TSP) or 2,2-dimethyl-2-silapentane-5-sulfonate sodium salt (DSS) are used as references in biological systems.

The values of $\Delta\sigma$ are of the order of 10^{-6}. Therefore, values of δ are expressed as parts per million (ppm). In this form, δ is a dimensionless quantity. However, one can also express chemical shifts as differences in frequencies in Hz ($\Delta\nu = \nu_0\Delta\sigma$). When expressed in this form, the shifts are proportional to the applied field. For example, 1 ppm on a 100 MHz spectrometer corresponds to 100 Hz, while on a 900 MHz, this is equivalent to 900 Hz.

For ^1H, contributions to σ come almost entirely from the 1s electron of the hydrogen involved in chemical bonding. The chemical shift range for ^1H is around 15 ppm. For other atoms, many more orbitals contribute to the shifts. In some cases such as nitrogen, atoms may be involved in multiple valence states. Thus the chemical shifts for nuclear spins other than ^1H are much larger. The chemical shift ranges are around 250 ppm for ^{13}C, 1000 ppm for ^{15}N and 430 ppm for ^{31}P (Table 1.1). Since chemical shift is a tensor quantity, it may have different values in different directions of the molecular axis. Thus, the chemical shift can be anisotropic. However, in solutions we observe an average behaviour.

The total NMR spectrum of a molecule can be expressed as:

$$S = \sum n_i \nu_i \qquad 1.15$$

Here, n_i is the number of nuclei giving rise to a particular signal at the frequency ν_i. It is obvious that resonances within each set of *like spins* are separated by less than a few kHz, while separations among different sets of *unlike spins* are of the order of several MHz. In a normal NMR experiment, one directly detects only one family of nuclei at a time (referred to as *like-spins* or *homonuclear spins*).

Looking at the spectrum in Figure 1.7 more carefully, one finds that different protons have different chemical shifts. These are characteristic of the chemical groups, to which the protons belong. Thymine (T) CH_3 protons resonate around 1.5 ppm and are at the highest field (*upfield shifted*, a jargon which comes from the era of CW NMR and still persists in literature). The sugar protons resonate in the range 2 to 6.5 ppm, cytosine H5 are in the 5.5 and 6.5 ppm, while other aromatic base protons are in 7 to 8.5 ppm range. The signals in the 12-15 ppm range (inset A) are due to hydrogen bonded imino protons involved in Watson and Crick base pairs. The hydrogen-bonded protons invariably appear in the low-field region of the spectrum compared to the corresponding signals from non-hydrogen bonded protons. If the hydrogen bonds are broken (for example, during melting of an ordered structure into a random coil), the same protons move upfield by 2-8 ppm. Even protons in similar chemical groups such as those in the CH_3 group belonging to the two thymines, show different chemical shifts because the two thymines have different magnetic environments.

3.2 Peak Intensities

The intensities (integrals of area under the signal) in NMR spectra are directly proportional to the number of nuclei (n_i), giving rise to individual signals. Interpreted intensities provide relative number of such nuclei in the chemical groups. In the spectrum shown in Figure 1.6, the signal due to $C^{\gamma}H_3$ protons has three times larger intensity than those due to H^{α} and H^{β}. In practice, the amplitudes may appear to be different if signals have different line-widths and hyperfine structures. Hence, NMR spectrometers have a built-in subroutine to obtain integrated intensities of various lines. This provides area under each peak, which in turn provides precise information on the relative number of nuclei in each chemical group. In a mixture of compounds, integrated intensities help in the measurement of relative molar concentration of each compound.

3.3 Nuclear Spin-Spin (Scalar) Coupling (J)

The H^{α}, H^{β} and H^{γ} resonances in threonine (Figure 1.6) show further fine structures. The splittings arise because each spin interacts with its neighbour through a mechanism called scalar coupling. The scalar interaction between two nuclear spins i and j is given by coupling Hamiltonian $\mathcal{H}_J$ where,

$$\mathcal{H}_J = 2\pi J_{ij} \boldsymbol{I}_i . \boldsymbol{I}_j \tag{1.16}$$

The spin-spin interaction J_{ij} is called scalar coupling constant. It is a scalar quantity in an *isotropic* situation (otherwise it is a *tensor* of rank 2). It is independent of the magnetic field and can be either positive or negative. The nuclear spin-spin interaction is transmitted through the electronic cloud joining the two nuclear spins i and j. The sign of the coupling constants relates to the contribution of the scalar interaction to the interaction energy. In some situations, the spectral properties are influenced by the relative signs of J.

The magnitudes of J_{ij} are important from a structural view-point. The values of J-couplings depend on a number of factors, including the number and the nature of chemical bonds intervening the atoms i and j. A superscript, which is usually shown to the left of J indicates the number of chemical bonds (N) between the spins i and j. ^{N}J has a relatively large magnitude for the directly bonded nuclei (i.e. ^{1}J with N=1).

Values of coupling constants also depend on a number of other stereochemical factors involving the two spins. Since J coupling requires an interaction of nuclear spins and electrons and only *s* orbitals have a finite density at the nuclear site, J values depend on the hybridisation of the atomic orbitals participating in chemical bonding. For example, the ^{1}J value for directly bonded ^{13}C and ^{1}H is around 125 Hz for saturated carbon atoms (sp^3 hybridization), 170 Hz for ethylenic C (sp^2 hybridization), and 250 Hz for acetylenic C (sp hybridization). There is thus a direct relation between the s orbital contribution and the ^{1}J values. The other one-bond couplings of primary interest are ^{15}N-^{1}H, ^{13}C-^{13}C and ^{13}C-^{15}N.

The two-bond ^{1}H-^{1}H coupling (^{2}J) between *geminal* protons (H-C-H) range between -23 and +42 Hz. The usual range for *geminal* protons attached to a tetrahedral carbon atom (for example, the two protons attached to C^{α} of Gly or to the C2′ in deoxyribose), is -12 to -16 Hz. For ethylenic system the *geminal* coupling is of the order of -0.5 to -3 Hz. In the case of cytosine, the interaction between H5 and H6 spins occurs via the H5-C5-C6-H6 fragment and the corresponding ^{3}J is about 7 Hz. In an ethylenic system, ^{3}J (H-C-C-H) is around 12 Hz for *cis* and 18 Hz for *trans* configuration. The values of scalar couplings are relatively small when the number of bonds between i and j exceed three. The coupling between the T(CH_3) and the T(H6) proton is a four bond interaction and its magnitude is around 1 Hz.

There are two properties of J couplings, which differentiate them from chemical shifts. First, the values are independent of the magnetic field at which the spectrum is obtained. Secondly, if the spectrum of one family of spin *i* is observed (e.g. ^{1}H), then it will show manifestation of the coupling to another family *j* (e.g. ^{13}C or ^{15}N). Heteronuclear scalar couplings (J_{ij}) therefore provide an effective way to correlate *unlike spins*, a property which forms the basis of several heteronuclear experiments in NMR spectroscopy.

3.4 Hamiltonian for a 2-spin System

Note that the fine structure in NMR spectra in solutions arises from both chemical shifts and coupling constants. The Hamiltonian operator that defines the energy levels for multi-spin systems in solutions has the form:

$$\mathcal{H} = \mathcal{H}_\delta + \mathcal{H}_J = -\sum_i 2\pi\nu_i I_z(i) + \sum_{i<j} 2\pi J_{ij} \boldsymbol{I}_i \cdot \boldsymbol{I}_j \qquad 1.17$$

Here, $I_z(i)$ are the z-components of spin angular momentum operators for individual spins and J_{ij} are the couplings between the spins i and j. The first term ($\mathcal{H}_\delta$) describes Larmor frequencies for the chemically shifted spins. The second term $\mathcal{H}_J$ arises because of the spin-spin interaction $I_i.I_j$ (or $I_{ix}I_{jx} + I_{iy}I_{jy}+ I_{iz}I_{jz}$). The summation over i<j has been introduced so that these terms are counted only once. Note that in Equation 1.17, the Hamiltonian has been expressed in units of radians. In case one uses frequencies in Hertz, then the right hand side is divided by 2π.

A molecular system consisting of p nuclear spins, each with spin ½, is characterised by p chemical shifts and p(p-1)/2 coupling constants. For example, a three-spin system has three chemical shifts and three coupling constants. The spectral pattern depends on the relative values of chemical shifts and coupling constants. A simple nomenclature is used to describe various types of J-coupled spin systems, which may be present in a molecule. For a spin system having two nuclei with widely different resonance frequencies ($\nu_1 - \nu_2 >> J_{12}$), the spins are denoted by alphabets such as A and X, which are far apart. These are called weakly coupled spin-systems. In case, $\nu_1-\nu_2 \sim J_{12}$, the spin systems are strongly coupled and corresponding spins are denoted by alphabets close to each other, such as A and B. AX spin system arises if the two spins belong to unlike nuclei (e.g. one of them may be ^1H and the other may be ^{13}C, ^{15}N, ^{31}P or ^{19}F), or when within the same family of spins when the chemical shift differences are large.

A group of spins are called chemically equivalent if they are related by a symmetry operation of the molecule and have the same chemical shifts. If they also have identical J couplings, then they are called magnetically equivalent (e.g. the three protons in the CH_3 group of threonine). In such cases, the basic functions are written as a linear combination of the product functions, taking into account the molecular symmetry. A suffix is used to denote spins having identical shifts due to molecular symmetry.

For a homonuclear system, an AX spectrum at higher field may approach an AB system at lower fields, because of lower chemical shift (in Hz). For example, the ^1H spin system in threonine can show AKX_3, ABX_3, or ABC_3 type of pattern depending on the magnetic field used.

The notation α and β introduced earlier for a single spin can be extended to spins of all interacting nuclei in a molecule. This is done by introducing product functions. If the system contains p spins each with I= ½, then the simplest set of functions describing the p-spin system are the 2^p product functions of the type α(1)β(2)β(3)…α(p). For a two spin system one can write the product functions, as αα (ϕ_1), αβ (ϕ_2), βα((ϕ_3) and ββ (ϕ_4). The total m_T (m_A+m_X) for these four states are 1, 0, 0 and -1, respectively.

The Schrödinger equation for the Hamiltonian in equation 1.17 can be solved by setting the Secular determinant ($|H_{ij} - E\delta_{ij}| = 0$). The Secular determinant is of the order 2^p, but can be factorised into equations of lower order. For this purpose, the properties of spin operators are used. The Hamiltonian (expressed in Hertz) in the present case can be expanded as follows:

$$\mathcal{H} = -\nu_A I_z(A) - \nu_X I_z(X) + J_{AX} I_{Az} I_{Xz} + \frac{1}{2} J_{AX} [I_A^+ I_X^+ + I_A^- I_X^-] \qquad 1.18$$

The first three terms in the Hamiltonian contribute only to the diagonal terms. For example,

$$<\alpha\alpha| -\nu_A I_z(A) - \nu_X I_z(X) + J_{AX} I_{Az} I_{Xz} |\alpha\alpha> = -\frac{1}{2}(\nu_A + \nu_X) + \frac{1}{4} J_{AX} \qquad 1.19$$

$I_A^+ I_X^+ + I_A^- I_X^-$ in the above equation are called *zero-quantum* operators. The matrix elements connect only the states |αβ> → |βα> and |βα> → |αβ>. These are called flip-flop terms and contribute to off-diagonal terms ($\mathcal{H}_{23}$ and $\mathcal{H}_{32}$) as follows:

$$<\alpha\beta| \frac{1}{2} J_{AX} [I_A^+ I_X^+ + I_A^- I_X^-] |\beta\alpha> = <\beta\alpha| \frac{1}{2} J_{AX} [I_A^+ I_X^+ + I_A^- I_X^-] |\alpha\beta> = \frac{1}{2} J_{AX} \qquad 1.20$$

Thus all the elements $\mathcal{H}_{ij}$ can be put in a matrix form as follows:

$$\begin{pmatrix} -½(\nu_A+\nu_X)+J/4 & 0 & 0 & 0 \\ 0 & ½(-\nu_A+\nu_X)-J/4 & ½J & 0 \\ 0 & ½J & ½(\nu_A-\nu_X)-J/4 & 0 \\ 0 & 0 & 0 & ½(\nu_A+\nu_X)+J/4 \end{pmatrix}$$

When ν_A-ν_X >> J_{AX}, the off-diagonal terms are small as compared to the differences between the diagonal terms and may be neglected. In such a situation, the product functions are directly the *eigenvectors* and the diagonal terms give the *eigenvalues*. These are called *weakly coupled* spin system. Thus the AX spin-system can be treated in a straight forward manner. In such a situation, the Hamiltonian is defined without the flip-flop terms. Thus:

$$\mathcal{H} = -\nu_A I_z(A) - \nu_X I_z(X) + J_{AX} I_{Az} I_{Xz} \qquad 1.21$$

Let us now consider the complication introduced in *strongly coupled* (AB type) spin systems, where ν_A-ν_B ~ J_{AB}. The off-diagonal terms, $\mathcal{H}_{23}$ and $\mathcal{H}_{32}$ cannot be neglected in this case. These terms lead to mixing of the states corresponding to the basic product functions $\alpha\beta$ and $\beta\alpha$. Thus, in the AB case, the eigenvectors corresponding to $m_T = 0$ will be a linear combination of ϕ_2 and ϕ_3. This involves solving a 2×2 secular determinant. The eigenvectors and eigenvalues in an AB spin system are given in Table 1.2.

Table 1.2: Eigenvectors and eigenvalues for a strongly coupled two spin system

Eigenfunctions	Eigenvalues
$\psi_1 = \alpha\alpha$	$E_1 = -\frac{1}{2}(\nu_A+\nu_B) + 1/4J$
$\psi_2 = \cos\theta\,(\alpha\beta) - \sin\theta\,(\beta\alpha)$	$E_2 = -\frac{1}{2}\,[J^2 + (\nu_A-\nu_B)^2]^{1/2} - 1/4J$
$\psi_3 = \sin\theta\,(\alpha\beta) + \cos\theta\,(\beta\alpha)$	$E_3 = \frac{1}{2}\,[J^2 + (\nu_A-\nu_B)^2]^{1/2} - 1/4J$
$\psi_4 = \beta\beta$	$E_4 = \frac{1}{2}(\nu_A+\nu_B) + 1/4J$

Where $\tan(2\theta) = J/(\nu_B-\nu_A)$

The frequencies of allowed transitions in the NMR spectrum can be calculated in terms of the chemical shifts and coupling constants. The relative intensities require consideration of the selection rules. The allowed transitions are such that $\Delta m_T = \pm 1$, and there is a flip of a single nuclear spin. There are four such transitions both in AX and AB systems. These are shown in the Figure 1.8. Note that all the four transitions have equal intensities in the AX case. However, the inner lines have larger intensities in the AB case. The transition $\psi_4 \rightarrow \psi_1$ or $\psi_1 \rightarrow \psi_4$ involves flips of the spins of both nuclei. Such a *double-quantum* transition ($\Delta m_T = \pm 2$) is not allowed. Same is the case for $\psi_2 \rightarrow \psi_3$ or $\psi_3 \rightarrow \psi_2$ transitions ($\Delta m_T = \pm 0$). These are called *zero-quantum* or flip-flop transitions.

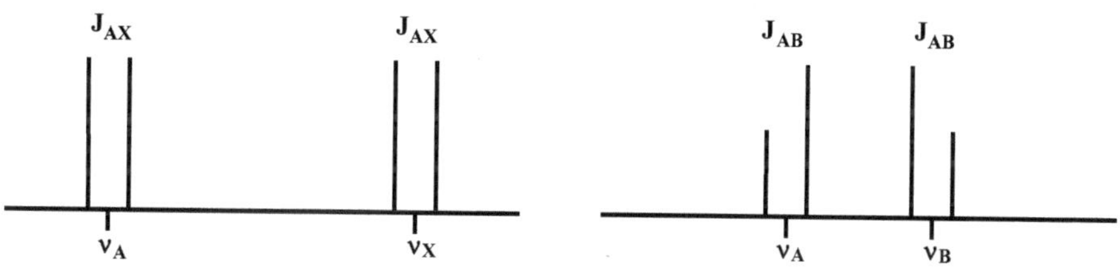

Figure 1.8: NMR spectra for AX and AB spin systems. In the second case, the chemical shifts of A and B correspond to the centre of gravity of the two lines which are separated by J_{AB}.

In a three spin system, one has eight basic product functions. The spin systems may belong to the types ABC, ABX or AKX, depending on relative values of the chemical shifts and coupling constants. In addition to the *zero-, double-, and triple-quantum* transitions, the transition $\alpha\beta\alpha \rightarrow \beta\alpha\beta$ is not allowed by the selection rules, since the later involves simultaneous flipping of all the three spins. Similar equations can be written for more complex spin systems.

In short, it is possible to solve the secular determinant for an NMR spectrum arising from multi-spin systems and to obtain the spectrum in terms of chemical shifts and coupling constants. The two sets of parameters (ν_i and J_{ij}) can thus be obtained from an analysis of the observed spectrum. It may be pointed out that one encounters matrices larger than 2×2 and in such cases an analytical solution to the eigenvector-eigenvalue problem is feasible only upto 4×4 matrices. However, numerical solutions can be obtained. Computer programs for analysis of NMR spectra of strongly coupled systems are available.

To avoid complications due to strongly coupled spin-systems, it is desirable to deal with weakly coupled systems. Besides the simplification of the spectra, the theory and experimental results of the multinuclear 2D and 3D NMR experiments are easier to treat in the case of weakly coupled spin systems. The results are amenable to simple interpretation. The fact that higher magnetic field can lead to weakly coupled systems by higher dispersion of the chemical shifts is another advantage for using higher magnetic fields in NMR spectroscopy.

3.5 Dipolar Coupling (D_{ij})

Nuclear spins also interact directly due to their proximity in space. Each NMR nucleus acts as a tiny magnet. The magnetic field B_D generated by spin *i* at the site of spin *j* is different depending on whether it is in the state α or β. Further, the local field B_D is orientation dependent and is given as:

$$B_D^j = \gamma_i (3 \cos^2 \theta_{ij})/r_{ij}^3 \qquad 1.22$$

Where, r_{ij} is the distance between the two interacting nuclei *i* and *j* and θ is the angle between the inter-nuclear vector (r_{ij}) and B_0. This is called dipole-dipole (DD) interaction or dipolar spin-spin coupling and is represented by the Hamiltonian in equation 1.23. One may note that this Hamiltonian applies only in high field in the context of spectral calculations. It cannot be used for calculation of relaxation rates.

$$\mathcal{H}_{i,j} = d\,[I_i.I_j - 3\, I_{zi}I_{zj}], \text{ where } d = \gamma_i\, \gamma_j\, \hbar\, (r_{i,j})^{-3} \qquad 1.23$$

In rigid solids, this interaction results in the splitting of individual resonances belonging to the interacting spins. For like spins such as protons, the separation between the two signals is given by:

$$\Delta\omega = 3\mu(3 \cos^2 \theta_{ij} - 1)\, r_{ij}^{-3}\, (\mu_0/4\pi) \qquad 1.24$$

Here, μ_0 is the permeability of the free space (its value is $4\pi\times10^7$ NA^{-2}).

Thus, dipolar interaction contains geometrical information about the molecule. This information is used in high-resolution solid-state NMR studies. In solutions, because of the rapid motion of the molecules, the angle θ varies with time. The time average of the term ($3\cos^2\theta_{ij}$-1) is zero and therefore dipole-dipole interactions are not observed directly. When molecules are dissolved in an anisotropic liquid crystalline solvent, the value of this term does not average to zero due to anisotropic molecular motions. There is a partial ordering of the molecules. In such cases, the geometrical information can be retrieved through the so called *residual dipolar couplings* (RDC). The dipolar interaction is also a major contributor to individual relaxation rates, which are manifested in the form of nuclear Overhauser effects (NOE), as discussed later.

3.6 Quadrupolar Interactions

So far, discussion has been restricted to the simplest of the NMR active nuclei, namely those with spin ½, which provide maximum information in biological systems. The majority of nuclei in the periodic table have spin > ½. Such nuclear spins have more than two energy levels. For example, for a nucleus with I=1, there are three spin states (Figure 1.9). Secondly, for nuclei with spin quantum number > ½ the nuclear charge distribution is not spherically symmetric. These nuclei therefore have an additional property called the *quadrupole moment* (Q). When the electric field gradient at the site of the nucleus produced by electronic clouds is asymmetric, such an interaction manifests directly in the NMR spectrum. For an electronic cloud having an axial symmetry, the magnitude of such an interaction is given by (eq)(eQ) or e^2qQ to the first order, where e is the electronic charge and q is the z-component of the electric field gradient at the nuclear site. This interaction is called quadrupole coupling constant.

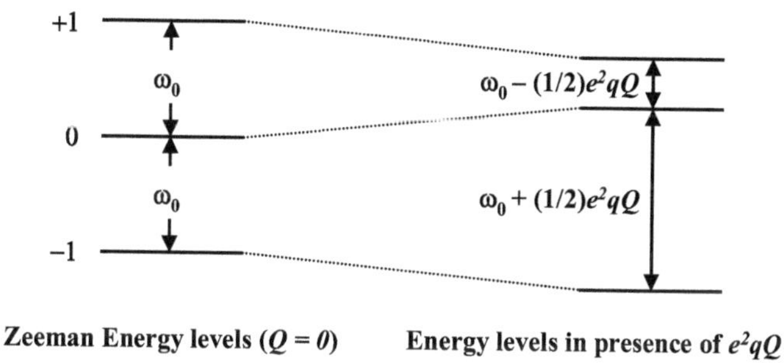

Figure 1.9: Energy levels of a nucleus with I=1. The spin states are characterised by three orientations corresponding to m_I = 1, 0 and −1. Allowed transitions are such that $\Delta m_I = \pm 1$. In the presence of quadrupolar coupling the two transitions occur at different frequencies, with a spacing of e^2qQ.

Compared to other interactions discussed so far, nuclear quadrupolar interactions are the strongest and can have magnitudes as large as a few MHz. One consequence of such an interaction is that the nuclear relaxation in such cases is dominated by quadrupolar interactions (leading to broad resonances). Secondly, the first order quadrupolar interactions lead to a lowering of energies associated with m_I = +1 and -1, and an increase in the energy level with m_I = 0. This leads to a splitting of spectral lines, as the two allowed transitions have different frequencies (Figure 1.9).

Deuterium (^2H) has spin 1. Therefore, in a magnetic field, its nuclear Zeeman levels correspond to $m_I = 1$, 0 and –1 (Figure 1.9). The value of Q for ^2H is relatively small ($Q=0.111$ e $\times 10^{-28}$ m^2) and therefore, the NMR signals are relatively sharp. Since ^2H generally appears in the form of C-D bonds, one can assume an axial symmetry for the electric field gradient in such cases. The signal splits into a doublet according to the equation:

$$\nu_Q = 3e^2qQ(3\cos^2\theta-1)/4h \qquad 1.25$$

For molecules undergoing isotropic motion, the quadrupolar interaction is averaged to zero because of the ($3\cos^2\theta-1$) term, just as in the case of dipolar interaction. In an anisotropic medium, such as molecules embedded in biological membranes, there is a net orientation of the molecule. The time average $<(3\cos^2\theta-1)>$ term in such cases can be replaced by an order parameter S_{CD}. In fact, the quadrupole coupling constant (e^2qQ) for ^2H has been used extensively for obtaining useful information in certain biological problems.

The value of quadrupole coupling constant (e^2qQ) is around 100-300 kHz for ^2H, 5 MHz for ^{14}N and 80 MHz for ^{35}Cl. Due to higher relaxation rates arising from quadrupolar relaxation, the signals for most nuclear spins with I > ½ are broad and generally elude detection. However, there are two components in the quadrupole coupling constants. Even when Q is large, the coupling may be lower if the electrical field gradient (eq) at the nuclear site is small. Thus, spins in a symmetrical chemical environment give relatively sharp signals. This is the case for Na^+ and other quadrupolar ions present in solutions as hydrated ions, or in NaCl and NH_4Cl in solid-state.

3.7 Electron-Nuclear Interaction

Till now the discussion has been focussed on diamagnetic molecules in which the electron spins are paired. The electron also has a spin ½ (discovered by Goudsmit and Uhlenbeck in 1925) and these are not paired in paramagnetic molecules. Electron spin behaves very much like nuclear spins. However, the nuclear magnetic moments are much smaller than the electronic spin magnetic moment. Thus, analogous to NMR we can observe electron spin resonance (ESR; also called electron magnetic resonance EMR or electron paramagnetic resonance, EPR). In fact, ESR was discovered a year before NMR was observed.

Our interest in this book is not in the direct detection of ESR but on aspects, where the electron spin influences NMR properties. The principles of electron-nuclear spin-spin interactions are similar to nuclear-nuclear interactions. In fact, in paramagnetic molecules, both the chemical shifts and the relaxation rates are dominated by the electronic moments. The chemical shifts of nuclear spins can change due to a direct delocalization of the electron spins at various nuclear sites (presence of finite spin densities), referred as *contact term*, or due to the anisotropy of the *g* tensor of the electron spin (*pseudo-contact term*). The nuclear spins in paramagnetic systems relax mainly through electron-nuclear interactions. Endogenous paramagnetic reagents in the medium such as metal ions can influence the relaxation of spins present in diamagnetic molecules.

The energy difference between the two levels arising from the electron-spin is larger and so is the population difference. This property has been utilized in a technique called Electron Nuclear Double Resonance (ENDOR). ENDOR allows detection of nuclear resonances in paramagnetic molecules with a high sensitivity. This has been used in biological systems.

4. NMR RELAXATION

Relaxation rates play an important role in NMR. Isolated nuclear spins do not have a mechanism to interact with the surroundings. In the absence of suitable molecular interaction, it will take a very long time for the magnetization to relax back to their equilibrium values. Some of the methods by which nuclear spins can transfer their magnetization and dissipate energy to the surroundings are discussed below. These processes play an important role in NMR.

4.1 Relaxation Rates

To start with, when a sample containing spin ½ nuclei is subjected to magnetic field, the populations in the two levels α and β are equal. The net longitudinal magnetization $M_z = 0$ and the system is in a non-equilibrium state. To attain the equilibrium magnetization M_o, the system has to exchange energy with the surroundings. The build up of equilibrium magnetization follows a first order differential equation (called the Bloch equation) with a typical rate constant R_1:

$$\frac{dM_z}{dt} = R_1(M_0 - M_z) \qquad 1.26$$

The build-up or the decay of the longitudinal magnetization requires exchange of energy between the nuclear spins and the surroundings (*lattice*). This process is called longitudinal relaxation. The inverse of R_1 is called spin-lattice relaxation time or the longitudinal relaxation time (T_1). The rate at which the system attains equilibrium is governed by the mechanisms through which such an exchange of energy takes place.

Similarly, after excitation by a π/2 pulse there is no net magnetization in the z-direction. The recovery of the longitudinal magnetization as a function of time (t) is mathematically described by:

$$M_z(t) = M_0\,[1 - exp\,(-t/T_1)] \qquad 1.27$$

Once the spins have been excited by a π/2 RF pulse, (i.e. the net magnetization has been nutated to the transverse plane), the coherent, precessing spins produce a detectable signal. The phase coherence or M_{xy} decays with a different time constant, T_2. This process is called spin-spin relaxation, because the energy for this process is exchanged through interaction between neighbouring spins in the sample. The return to equilibrium is again governed by a simple first order differential equation with a transverse relaxation rate $R_2(=1/T_2)$

$$\frac{dM_{xy}(t)}{dt} = -R_2 M_{xy}(t) \qquad 1.28$$

The solution to this differential equation is given by:

$$M_{xy}(t) = M_{xy}(0)\ exp(-t/T_2) \qquad 1.29$$

For non-viscous liquids such as pure water, T_1 and T_2 are nearly equal. However, in general, the transverse magnetization decays faster than the longitudinal magnetization.

4.2 Molecular Mechanisms leading to Relaxation

There are several mechanisms, whereby nuclear spins interact with electrons and other surrounding nuclear spins. Molecules constitute tiny magnets embedded in the form of nuclear magnetic moments and in some cases as unpaired electrons. In a condensed state, the molecules undergo rotational and translational motions. This leads to fluctuations in the local magnetic fields at nuclear sites. Those frequency components in such fields, which exactly match the transition frequencies for the spin system under consideration, act as effective mechanisms for exchange of energy between the nuclear system and its surroundings. Thus, relaxation depends upon two factors:

(i) a magnetic interaction, which is responsible for creation of the desired magnetic fields, and
(ii) molecular motions, which cause fluctuations in these fields.

Both intra- and inter-molecular interactions lead to relaxation processes. Except for paramagnetic systems and quadrupolar nuclei, the major mechanism for energy exchange for spin ½ nuclei is the magnetic dipole-dipole interaction (DD). For ^1H in molecules dissolved in aqueous solutions, a dominant relaxation mechanism is intermolecular dipole-dipole interactions, since most of these nuclei are exposed to the solvent. On the other hand, ^{13}C spins are normally in the interior of the molecule and the relaxation is dominated by intramolecular ^1H-^{13}C DD interactions. Other mechanisms for relaxation in diamagnetic samples include chemical shift anisotropy (CSA), scalar coupling (SC) and spin rotation (SR). In paramagnetic systems, interaction between the unpaired electron (e) and nuclear spins dominate relaxation mechanism rates. The magnetic moment of electron is several times larger than those of nuclei. Even a small amount of paramagnetic material (such as deoxyhemoglobin, dissolved oxygen or paramagnetic metal ions) can strongly enhance relaxation rates. Nuclei with spin > ½ relax mainly through electric quadrupole interaction (Q). Such nuclei also influence the relaxation of surrounding spins. It is possible to quantitatively define each of these interactions and estimate their relative contributions in relaxation processes. Thus, the relaxation rates (R) can be written as a sum of several contributions:

$$R = R(DD) + R(SR) + R(CSA) + R(SC) + R(Q) + R(e). \qquad 1.30$$

4.3 Theoretical Treatment of Relaxation Rates

Very early in the history of NMR spectroscopy, researchers set out to study various aspects of NMR relaxation. Pioneering work in this direction was done by Bloch, Bloembergen, Overhauser and Slichter during the period 1945-60. Since then, the theoretical methodologies have been developed at levels ranging from fully classical to fully quantum mechanical treatment.

The fluctuating magnetic field due to molecular motions $\mathcal{H}(t)$ can be treated as perturbation to the time independent Hamiltonian $(\mathcal{H}_\delta + \mathcal{H}_J)$. The density matrix ρ, which represents the ensemble averaged state of the spin system is then given by:

$$d\rho/dt = -i/\hbar\,[\mathcal{H}, \rho] \text{ with } \mathcal{H} = \mathcal{H}_0 + \mathcal{H}_1(t) \qquad 1.31$$

Normally, it is not possible to solve this equation analytically. Several approximation treatments have been introduced to treat relaxation processes. The situation is complicated by the fact that a number of interactions discussed above contribute to R_1 $(1/T_1)$ and R_2 $(1/T_2)$ relaxation. The relative extent to which they do so depends on the strength of the interaction and on the rate of fluctuations of the secondary magnetic fields.

The fluctuating magnetic fields present in molecules in motion (which have *x, y* or *z* components), provide the source for spin transitions. In particular, the *z*-component of the fluctuating magnetic fields can add or subtract from the static magnetic field B_o. This results in differences in the precessional frequency and hence in dephasing of the spin motions. The components of fluctuating magnetic fields corresponding to the Larmor frequency can induce transitions between the states of the nuclear spin system. The energy received by the system is then transformed into thermal energy. An additional mechanism for the T_2 relaxation is the energy transfer within the spin system. Transition in the spin state of a nucleus changes the magnetic field at a nearby nucleus and contributes to the shortening of the life-time of the spin state.

Rotational and translational motions of a molecule in a liquid are responsible for fluctuations in the magnetic field, arising from interactions discussed in the previous section. A small molecule such as water or chloroform tumbles at a fast rate and produces rapid fluctuations. Large molecules, such as protein or nucleic acid, move slowly and produce fluctuations at a relatively lower rate. Molecular assemblies such as membranes move even more slowly and result in anisotropic motions. This leads to even slower fluctuations. Finally, solid-like materials such as tissues and bones generally have restricted molecular motions.

Consider the energy levels for a two spin-system (Figure 1.10). The coupling here is through one of the mechanisms responsible for relaxation, discussed above. We first discuss T_1 relaxation mediated by dipolar coupling, which involves change in the total magnetization. To make a transition from $\beta\beta \rightarrow \beta\alpha$ or $\alpha\beta$, an exchange of energy quantum with energy corresponding to the Larmor frequency is required. Relaxation can also occur through a double quantum transition $\beta\beta \rightarrow \alpha\alpha$. This requires energy equal to twice the Larmor frequency for *like spins* and $(\omega_A+\omega_X)$ for *unlike spins*. Such spin-flips can occur when molecular motions can provide frequencies, which exactly match the energy differences involved in such transitions. The fluctuating magnetic fields arising from molecules in motion, have *x* or *y* components at these frequencies, and provide the source for such spin transitions.

In the case of T_2 relaxation, any transition in Figure 1.10, can disrupt the phase coherence of the system. Thus, not only the transitions discussed above for T_1, but also zero quantum transition $\alpha\beta \rightarrow \beta\alpha$ (i.e. low frequency fluctuations for like spins $[(\omega_A-\omega_X)$ for unlike spins] can lead to loss of phase coherence or spin-spin relaxation. For like-spins (e.g. both spins are ^{1}H), relaxation occurs through components of the spectral density at ω, 2ω and ω_A-ω_X (i.e. frequencies close to 0 for like spins). Thus, T_2 relaxation is generally faster than T_1, though in special cases T_2 may be as large $2T_1$.

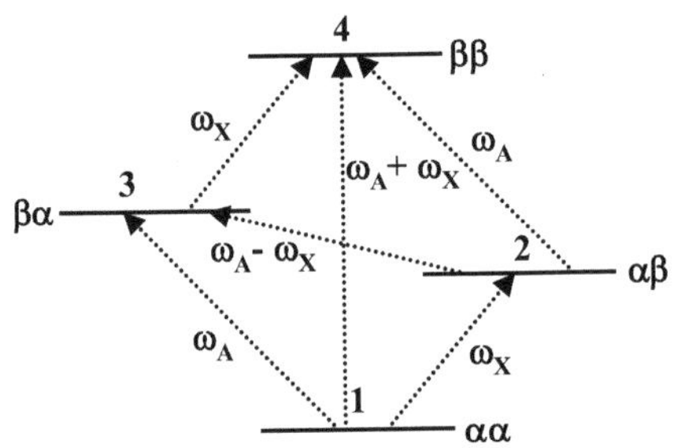

Figure 1.10: Energy levels and transitions for a system consisting of two spins A and X. Transitions involving flip of a single spin (W_1) require a quantum of energy at Larmor frequencies (ω_A or ω_X). Relaxation can also occur through double- and zero-quantum transition (W_2 and W_0).

4.4 Correlation times

For a more quantitative understanding of relaxation, it is necessary to introduce the concept of *correlation time* (τ_c), which represents the time scale at which molecular fluctuations occur. τ_c is the average time that molecules take to rotate through one radian. In principle, for solutions, τ_c can be calculated from the Debye equation:

$$\tau_c = 4\pi\eta r^3/3kT \qquad 1.32$$

Here η is the viscosity of the medium and r is the hydrodynamic radius of the molecule. Correlation times are usually of the order of nanoseconds for fast tumbling molecules. Slow motions associated with macromolecules result in longer τ_c, while faster motions associated with smaller molecules are characterised by shorter τ_c. In solutions, the value of τ_c ranges between 10^{-7} and 10^{-12} seconds.

To understand the frequencies of motion associated with the fluctuating fields for a molecule in solution, one introduces the concept of a correlation function ($g(\tau)$). Let us treat the molecule as a rigid-body undergoing Brownian rotational motion and focus on an internuclear vector which, to start with is oriented at an angle θ with respect to B_0. The function $g(\tau)$ is defined as:

$$g(\tau) = \langle f(t)\, f(t+\tau)\rangle \approx exp\ (-\ t/\tau_c) \quad 1.33$$

Here $f(t)$ and $f(t+\tau)$ describe the position or orientation of the internuclear vector at time t and $t+\tau$, respectively, and the < > indicates an ensemble average.

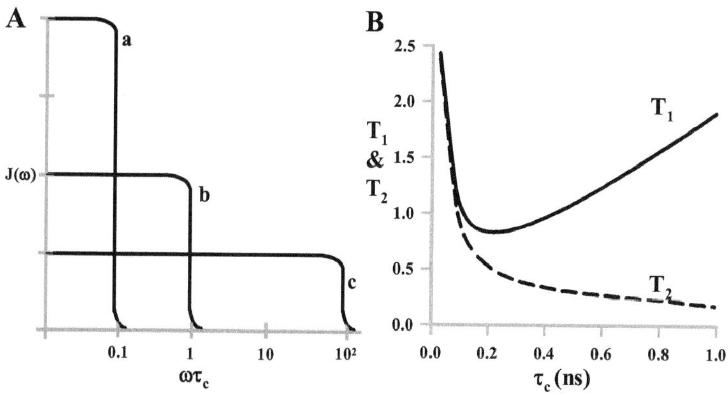

Figure 1.11: (A) Spectral density (energy) available for relaxation for two-like spin system as a function of $\omega\tau_c$. (a) In the fast tumbling range ($\omega\tau_c \ll 1$), only a small part of the motions contributes to relaxation and therefore both T_1 and T_2 are large; (b) In the intermediate region ($\omega\tau_c \sim 1$), relaxation rate T_1 is minimum. (c) For slowly tumbling molecules ($\omega\tau_c \gg 1$), the spectrum has components at lower frequencies only. (B) Corresponding relaxation times as a function of τ_c at 500 MHz spectrometer frequency.

The Fourier transformation of $g(\tau)$ gives the spectral density function $J(\omega)$, which provides density of various frequency components:

$$J(\omega) = \frac{2\tau_c}{1+\omega^2\tau_c^2} \quad 1.34$$

$J(\omega)$ has the information about the entire range of frequencies associated with the motion of the system. The total area under the spectral density depends on the thermal energy of the system and is constant. In other words, the total energy associated with molecular motions is constant [$\int J(\omega)d\omega = \pi/2$]. However, τ_c changes the distribution of $J(\omega)$.

One distinguishes three different ranges of $\omega\tau_c$. Slow ($\omega\tau_c >> 1$), fast ($\omega\tau_c << 1$) and medium range motions ($\omega\tau_c \sim 1$). Figure 1.11A shows the plot of $J(\omega)$ for these correlation times. As mentioned, the spectral density function $J(\omega)$ of the fluctuating magnetic fields depends on the correlation time (τ_c). It has a Lorentzian line-shape centred at a zero frequency with half-maximal line-width of $1/\tau_c$. The uppermost frequency limit for correlation time τ_c is $1/(2\pi\tau_c)$. Thus a molecule tumbling with τ_c of around 1.6×10^{-8} second would have an upper limit of 10 MHz in its field fluctuation. The density of energy contributing to relaxation depends on the rate of fluctuations ($\omega\tau_c$) arising from molecular motions. The following situations may be encountered.

(i) The rate of fluctuations is much higher than the resonance frequency ($\omega\tau_c << 1$). This condition is also called *extreme narrowing*. Small molecules dissolved in non-viscous solvents fall in this category. Both relaxation rates are small. The values of T_1 and T_2 are almost equal in this range. The values of relaxation rates for spin *i* due to a spin *j*, are given by:

$$R_1 = R_2 = \tau_c d^2 \tag{1.35}$$

(ii) The fluctuation rates are in the intermediate region ($\omega\tau_c \sim 1$) such that there are large contributions to the relaxation through single, double- and zero-quantum transitions. Such a situation exists for medium size molecules or viscous solutions. This leads to the most efficient relaxation for both T_1 and T_2 as the spectral densities at both ν and 2ν are large.

(iii) The rate of fluctuation is low ($\omega\tau_c >> 1$) such as in viscous medium, solid-like materials or large macromolecules, which tumble very slowly. There are no components for relaxation near the Larmor precession frequencies. Therefore, the T_1 relaxation time is higher. However, T_2 relaxation can still occur through low molecular components in the spectral density map. In this region we have $R_2 > R_1$. Even though there are components in the spectral density at ω, 2ω or ($\omega_A + \omega_X$) for unlike spins, the spectral density is spread thinly and the amplitude of spectral density at any particular frequency is small.

The longitudinal relaxation rate, R_1 ($1/T_1$) for dipolar coupled *like spins* depends on $J(\omega)$ and $J(2\omega)$, while the transverse relaxation rate R_2, depends on $J(0)$, $J(\omega)$ and $J(2\omega)$. For *unlike spins*, the rates depend on $J(\omega_i)$, $J(\omega_j)$ and $J(\omega_i \pm \omega_j)$. More explicitly expressions have been derived for R_1 and R_2. These are given by the following equations for dipolar coupled relaxation in case of *unlike spins*.

$$R_1 = 1/T_1 = (1/8)d^2 (\mu_0/4\pi)^2 \{6J(\omega_i) + 2\,J(\omega_i - \omega_j) + 12J(\omega_i + \omega_j)\} \tag{1.36}$$

$$R_2=1/T_2=(1/8)d^2(\mu_0/4\pi)^2\{4J(0)+J(\omega_i-\omega_j)+3J(\omega_i)+6J(\omega_j)+6J(\omega_i+\omega_j)\} \quad 1.37$$

For the purpose of nuclear Overhauser effect (NOE) discussed in next section, we also need to calculate a cross relaxation term σ_{ij}, which is given by:

$$\sigma_{ij} = (1/8)d^2(\mu_0/4\pi)^2\{12J(\omega_i+\omega_j)-2J(\omega_i-\omega_j)\} \quad 1.38$$

Similar equations have been derived for *like spins.* Thus, R_1, R_2 and NOE are dependent on the gyromagnetic ratios of the two interacting spins, the correlation time for the vector joining the two spins and the distance between the spins. Knowledge of R_1, R_2 and NOE provides information on the structure (through inter-nuclear distances) and about dynamical parameters such as the overall tumbling correlation time of the molecule in solutions and the internal motions in molecules (through the spectral densities). It is therefore important to measure these rates.

4.5 Dipolar Relaxation due to Several Interacting Spins

Above equations can be generalized to account for relaxation due to several interacting spins. For example, in the extreme narrowing limit, the relaxation rates R_1 and R_2 of a nucleus *i* relaxed by several *j* spins (with spin 1/2), for intra-molecular dipolar relaxation is given by:

$$(R)_{DD} = \gamma_i^2\gamma_j^2\hbar^2\tau_c\Sigma r_{ij}^{-6} \quad 1.39$$

Here, r_{ij} is the inter-nuclear separation and τ_c is the rotational correlation time.

4.6 Field Dependence of Relaxation Rates

Since the Larmor frequencies depend on the magnetic field, the relaxation rates are field dependent. Secondly, within the same molecule interactions leading to relaxation are different for different spins. Therefore their respective relaxation times may differ. Finally, even a small molecule such as water may have relaxation times which may lie in all the three ranges (a, b and c) discussed earlier, depending on its environment in biological systems. It may show fast motions when present in biological fluids such as blood, intermediate motion when present in cells and slow motion when present in tissues and bones. In fact, such differential relaxation rates forms the basis for Magnetic Resonance Imaging (MRI) of whole organs.

5. CHEMICAL EXCHANGE: DYNAMIC EFFECTS IN NMR SPECTROSCOPY

Chemical exchange or reactions affects NMR parameters and the spectral behaviour depends on the rate of chemical reactions. This forms the basis of the so called *dynamic NMR*.

5.1 Conformational Equilibrium

Let us consider chemical exchange between two species A_1 and A_2,

$$A_1 \rightleftharpoons A_2 \qquad 1.40$$

As a result of the chemical exchange, the information on magnetization of each spin is transferred from the state A_1 to A_2 and *vice-versa.* A simple case is internal rotation around the C-C bond in a substituted ethane, 1,1,2,2 tetrabromofluoroethane ($CHBr_2$-$CFBr_2$, abbreviated as TBFE) (Figure 1.12). It is known that the internal rotation around C-C bond in ethane leads to three low energy conformations, such that the substituents are in staggered position. The stable conformations for TBFE are shown in Figure 1.12. In most substituted ethanes, the interconversion between such conformations occurs at a very fast rate. In TBFE, the substitution of H by Br atoms increases the barrier to internal rotation, and the reaction rate is slower. Since the two conformers where the H and F are in *gauche* conformation are identical, there are two low energy conformations: *gauche* (g) and *trans* (t).

The two spin ½ nuclei (^1H and ^{19}F) in TBFE form an AX spin system. A two-line pattern both in the ^1H and ^{19}F spectra, arising from the $^3J(^1H$-$^{19}F)$ is expected. For each of these conformers, the ^1H and ^{19}F spectra are doublet, separated by J_g and J_t, respectively. At temperatures below 171 K one observes two doublets. The one due to *g* isomer appears in the centre (J_g = 1.2 Hz) of the ^1H spectrum (Figure 1.12). This doublet is flanked by the component of *t* isomer (J_t= 22.2 Hz). As temperature is raised, the spectrum goes through a complex pattern finally giving rise to only one set of doublet above 228 K. At higher temperatures, the separation between the two lines corresponds to an intermediate J value of about 7 Hz.

This type of temperature dependent behaviour is linked to the ability of a particular physical technique to identify signals from chemical species, which are in dynamic equilibrium. Separate signals from each molecular state are observed if the reaction rate is slower than the time resolution of the technique. In case of a fast process, an average behaviour is observed. This property is common to all measuring techniques, but the time-scale at which individual states can be distinguished, depends on the time-resolution of the method. For example, vibrational techniques can resolve much faster reacting species.

Consider a NMR property such as the ^1H chemical shifts, denoted by ν_{A1} and ν_{A2} for the two conformers with $\delta\nu = \nu_{A1} - \nu_{A2}$. Three types of situations, which depend on the energy of activation for the reaction, can be envisaged.

(i) The rate of exchange (k_{ex}) for transition between A_1 and A_2 is slow, such that the life times of the two states τ_1 and τ_2 are long compared to $1/\delta\nu$. This is called *slow exchange* on the NMR time-scale. Distinct signals from A_1 and A_2 are observed with intensities equal to their relative populations (p_{A1} and p_{A2}). The equilibrium constant for the reaction ($K= p_{A1}/p_{A2}$) and therefore the free energy difference (ΔG) can be measured from the integrated intensities of signals from A_1 and A_2.

(ii) The rate of exchange is fast ($k_{ex} > 1/\delta\nu$). In such a situation, NMR is unable to resolve the two states and the observed spectrum is a time-average of the two states. For example, the chemical shift will average as follows:

$$\nu_{obs} = p_{A1}\nu_{A1} + p_{A2}\nu_{A2} \qquad 1.41$$

In a more general case of several states, $\nu_{obs} = \Sigma p_i \nu_i$ *1.42*

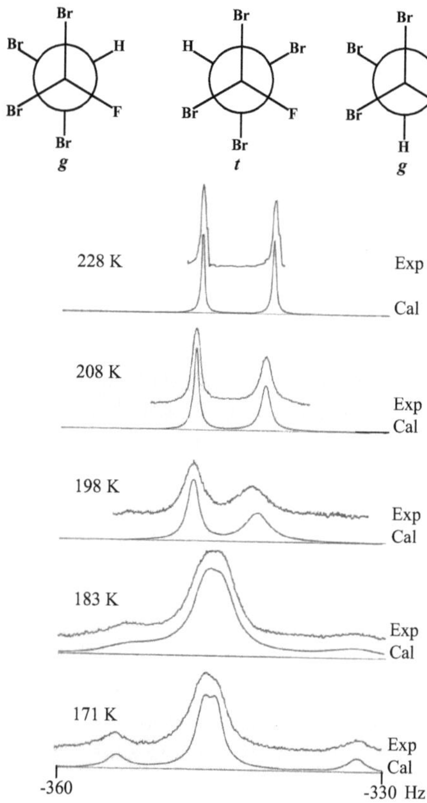

Figure 1.12: ^1H spectrum (at 60 MHz) of 1,1,2,2 tetrabromofluoroethane (TBFE) at various temperatures. At low temperatures, spectra of each conformer appear as distinct signals. At higher temperatures only an average behaviour is observed. The line shapes at intermediate temperatures have been calculated and matched with the experimental spectra using ΔG^*= 42kJ/mole. Stable conformers of TBFE are shown on top.

If the values of ν_{A1} and ν_{A2} can be estimated from by other means or from the low temperature data, then the relative populations can be estimated.

(iii) The value of $k_{ex} \cong 1/\delta\nu$. This is called the coalescence region. Separate signals in the slow exchange regime coalesce before giving a sharp line corresponding to the fast exchange signature. The signal in this situation is broad. However, the line-shape can be calculated theoretically from the NMR parameters and the rate of the reaction (k_{ex}). The life-times of the conformational states and the free energy of activation (ΔG^*) can be estimated, using analysis of the line-shape in this range of temperature. For example, the barrier to internal rotation in TBFE has been estimated to be 42 kJ/mole. This value may be compared with the energy barrier of 12 kJ/mole, observed for ethane. The theoretical line shapes in Figure 1.12 have been simulated using this value.

Thus, NMR is a useful tool for measurement of thermodynamic and kinetic parameters. In particular, intensities are useful for studying thermodynamics and line-shapes for the dynamics of the exchange processes. Spectra for reactions for which the free energy is such that $k_{ex} \sim \delta\nu$, are amenable for studies such as the one discussed. Since the chemical shifts are in the range 1 to 1000 Hz, line shapes are sensitive to rates in the range 1 to 10^3 sec^{-1}. Using standard rate equations for first-order reactions, it can be estimated that reactions having free energy of activation in the range 40 to 80 kJ/mole, can show temperature dependent spectra in the range of -100 to +150 °C, which is generally used in NMR. Several biomolecular reactions fall in this category. Slower reaction rates can be studied by looking at the change in intensities of signals of A_1 and A_2 as a function of time or by using magnetization transfer experiments.

Examples of such reactions in biomolecules are internal rotation (conformational transitions), inversion, ring flips and tautomerism. From the view-point of biological systems, conformational changes in biological molecules are the most important reactions. Such reactions be studied using the dynamic effects in NMR spectroscopy.

5.2 Solvent Exchange

One of the extensively used techniques in the study of biological molecules is the exchange rate between ^2H and ^1H. Hydrogen atoms such as those in O-H and N-H bonds called *labile* protons exchange freely with the solvent water. Therefore, when an amino acid, sugar or a nucleotide is dissolved in ^2H_2O, the signals from the labile ^1H vanish. However, in ordered proteins, nucleic acids and carbohydrates, several such groups are locked in hydrogen bonds and/or buried in the core of these molecules. The H-D exchange rates in such cases are substantially slower. Signals from labile ^1H may be observed for several hours after the sample is dissolved in ^2H_2O in such cases. Such slow exchanging protons help in identifying hydrogen bonded networks.

6. NUCLEAR MAGNETIC DOUBLE RESONANCE

Double resonance experiments play an important role in NMR spectroscopy. Such experiments require two RF fields (B_1 and B_2). The frequencies of the two RF fields are chosen depending on the type of experiment. For example, for homonuclear double resonance, the frequencies of B_1 and B_2 differ by a few kHz, and can be derived from a single master oscillator. For heteronuclear applications the two frequencies are widely different and one needs two RF transmitters. The experiments can be designed by choosing the strength and frequencies of the two fields. As is the case with other areas of NMR, the field of nuclear magnetic double resonance was developed in early 60s. However, the use of pulsed NMR has changed the current methodologies.

6.1 Spin Decoupling

J-couplings help in spin-system identification, sequence-specific resonance assignments and provide information about local stereochemistry in a molecule. However, at times, the splitting caused by such couplings may become undesirable. A resonance signal split into a doublet results in two-fold loss in its S/N. The problem is particularly acute while observing a rare spin such as ^{13}C. Further, J-splitting results in spectral overlaps. In such situations, there is a need for *decoupling* the effect of J-coupling. We note from Equation 1.18 that in an AX system, the coupling term is given by $J_{AX}I_{Az}I_{Xz}$. Thus, by manipulating this term, one can remove the effect of J-coupling in the NMR spectrum.

One of the early applications of double resonance was the spin-decoupling in homonuclear systems. This is illustrated in Figure 1.13, for threonine. In this experiment the frequency of the field B_2 has been placed at the γ-protons, while the spectrum of the remaining spectral region is observed. If the field B_2 has an amplitude > $2\pi J$ and is applied at the resonance frequency of one of the spin, then it results in the removal of the spin coupling of that spin. This results in the simplification of spectra. In an FT experiment, the receiver and the decoupler channel are usually synchronised in such a way that the decoupling is applied between the excitation and the data collection.

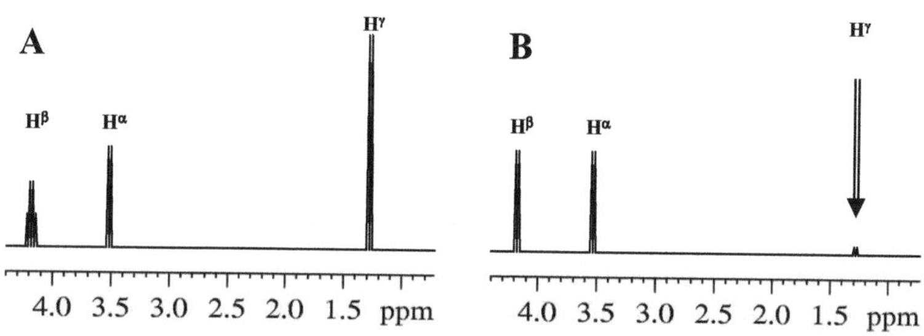

Figure 1.13: ^1H spectrum of threonine (A) without and (B) with the decoupling field at CH_3 protons.

6.2 Heteronuclear and Broad-band Decoupling

One can also apply the field B_2 such that it leads to irradiation of ^1H, while the resonances of a heterospin such as ^{13}C, ^{15}N or ^{31}P are observed. For example, during the acquisition of a ^{13}C spectrum one may like to remove all ^1H couplings. This requires irradiation of a broad range of ^1H frequencies (spread over ~15 ppm). Such techniques are called *broad band* decoupling. A variation of this experiment is off-resonance decoupling, such that the spectrum still contains residual J-couplings. This is particularly useful for distinguishing CH_3 (quartet), CH_2 (triplet), CH (doublet) and quaternary carbons (singlet) in the ^{13}C spectra.

The band-width of decoupling region increases with an increase in the spectrometer frequency. Thus, the decoupling power required for the second RF field increases with the magnetic field. Hence, removal of *spin-spin coupling* becomes more challenging at higher magnetic fields. The power dissipation within the sample results in sample heating, which in turn has significant effect on the spectral quality. This problem is more acute in the case of spins with higher γ, where one needs to use larger values B_{dec}, as the decoupling field strength is proportional to γB_{dec}.

The criteria for an efficient proton decoupling for observing a good signal from the rare spin require the following considerations:

(i) Uniform irradiation of entire ^1H spectrum, such that the residual splittings are smaller compared to the line-widths.

(ii) Low power dissipation and temperature control such that the strong decoupler RF does not heat the sample.

(iii) Insensitivity to the RF pulse-widths and their phase-shifts.

(iv) Small side-band signals.

A number of pulse experiments have been suggested as alternatives to the *broad-band* decoupling. In these, decoupling is achieved by irradiating the sample with a train of composite π pulses such as MLEV, WALTZ-16 and GARP (discussed later), applied at proton frequency, while simultaneously detecting the X-nuclei ($^{13}C/^{15}N/^{31}P$). This keeps the decoupled spin (^1H) states α and β equally populated and spins are flipped rapidly between the two states, such that the X-spins do not feel their presence.

6.3 Nuclear Overhauser Effect (NOE)

One of the most important double resonance experiment used in conformational studies of biological molecules is the nuclear Overhauser effect (NOE). Its origin goes back to the observation of enhancement of nuclear signals when the magnetic levels of unpaired electron are perturbed. The theory and concepts were later extended to nuclear-nuclear double resonance.

NOE (η_{AX}) is defined as fractional enhancement in the signal intensity of a nucleus X, when a spin in its close spatial proximity (A) is saturated.

$$\eta_{AX} = [(\textit{Intensity of X, when A is saturated})/(\textit{Equilibrium intensity of X})] - 1 \qquad 1.43$$

NOE is caused by differential relaxation rates between transitions in a multi-spin system. This leads to population distributions of spins in the various states, which deviate from the Boltzmann distribution and results in what is called a *polarization* of nuclear spin distribution.

Let us again restrict to a two spin AX system. For simplicity, assume that $J_{AX} = 0$ and that the two spins A and X are coupled only through intra-molecular dipole-dipole interaction. It has been discussed that the relaxation between the four energy levels can be brought about by single- (W_1), double- (W_2) and zero- (W_0) quantum transition probabilities. Suppose that the transitions corresponding to the nucleus A are saturated by applying a strong RF power at the frequency ω_A (Figure 1.14). This will tend to equalize the populations in levels 1 and 3 as well as between 2 and 4. The system tends to return to thermal equilibrium through transitions between states which are not saturated. W_2 tends to populate level 1 by transferring the population from 3 via level 4. Like-wise, W_0 will equalize populations of levels 2 and 3.

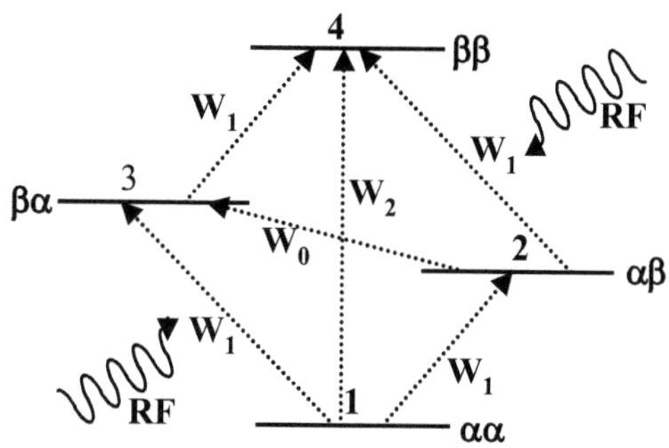

Figure 1.14: The principle of NOE. Saturation of levels corresponding to spin A results in changes in the populations of the energy levels compared to their equilibrium values. This results in a change in the intensity of transitions corresponding to spin X.

The intensities of the transitions of the X nucleus, 1→2 and 3→4 will consequently be perturbed. The result is a change in intensity of the signal for X, when the transitions of spin A are saturated. If W_2 dominates over W_0 ($W_2 >> W_0$), then the result is higher populations in levels 1 and 3, and one gets a positive NOE. In the reverse case, one gets a negative NOE. Mathematically one can write:

$$NOE = \eta_{AX} = \Delta I_A\,(X) = (W_2 - W_0)/(2W_1 + W_0 + W_2)(\gamma_X/\gamma_A) \qquad 1.44$$

W_2-W_0 is called the *cross-relaxation rate* (σ_{AX}), while $2W_1+W_2+W_0$ is called the *auto-relaxation rate* (ρ). For homonuclear spins and under continuous irradiation of spin X (M_{AX}=0), NOE can be written as:

$$NOE = \eta_{AX} = \Delta M_{Az}/M_{A0} = \sigma_{AX}/\rho \qquad 1.45$$

The *single-* (W_1), *zero-* (W_0), and *double-quantum* (W_2) transition probabilities are given by:

$$W_1 = (3/20)\, d^2 \{\tau_c/1+\omega_A \tau_c^2\} \qquad 1.46$$

$$W_0 = (1/10)\, d^2 \{\tau_c /1+(\omega_A-\omega_X)^2 \tau_c^{\,2}\} \qquad 1.47$$

$$W_2 = (3/5)\, d^2 \{\tau_c /1+(\omega_A+\omega_X)^2 \tau_c^{\,2}\} \qquad 1.48$$

Thus, NOE depends on the terms involved in T_1 relaxation. It therefore depends on the correlation time (τ_c) and on the interaction leading to relaxation. If the relaxation is dominated by intramolecular dipole-dipole interaction and one is in fast molecular motion regime, then:

$$W_2\text{-}W_0 = \sigma_{AX} = 1/2(\mu_0/4\pi)^2\, (\hbar)^2\, \gamma_A^2\, \gamma_X^2\, \tau_c\, r_{AX}^{-6} \qquad 1.49$$

It is possible to use this relation for measurements of relative internuclear distances. If it is assumed that τ_c is the same for all protons in the molecule then one can write:

$$\eta_{AX}/\eta_{AY} = (r_{AY}/r_{AX})^6 \qquad 1.50$$

If one of the two distances (say r_{AX}) is known, then the other distance (r_{AY}) can be estimated. Several distances in biological molecules of interest are rigidly fixed. These can be used as reference. Some examples of such distances in molecules of interest are the H5-H6 distance in cytosine or the H1′-H2″ distance in methylene groups. Since NOE depends on the inverse of the sixth power of distance, the effect goes down rapidly with inter-nuclear distances. Therefore, NOE effect is limited to distances of about 5 Å. NOE may be quenched if relaxation mechanisms other than the intra-molecular dipole-dipole interactions are present.

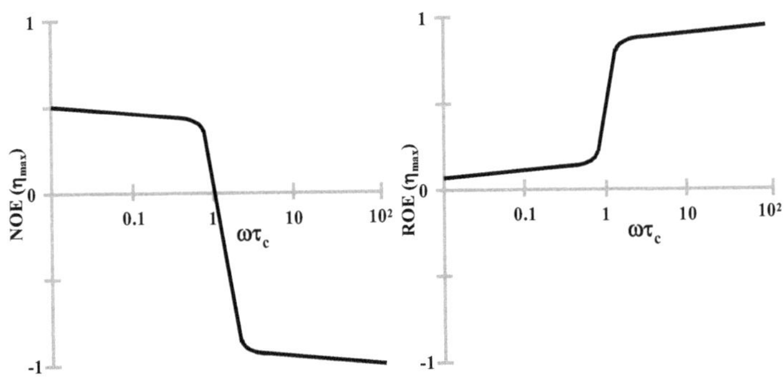

Figure 1.15: The observed NOE goes from +ve to –ve values, depending on $\omega\tau_c$. One can also use enhancement in rotating frame, which has positive value in the entire range of $\omega\tau_c$.

As discussed earlier, relaxation rates are strongly dependent on molecular motions, which affect the correlation time τ_c and through it, the values of W_1, W_0 and W_2. The effectiveness of the relaxation processes W_0 and W_2 is determined by the rate of the molecular reorientation. For small molecules undergoing fast motions ($\omega\tau_c < 1$), the z magnetization of spin X will increase to a value higher than its Boltzmann equilibrium value when the resonance of A is saturated. This results in a positive NOE. The relaxation through W_2 is only possible when the molecular reorientation is fast, which corresponds to short τ_c. The opposite is true for large molecules undergoing slow motions ($\omega\tau_c > 1$). Here, the cross-relaxation occurs mainly through the low-frequency transitions W_0. Depending on whether W_0 or W_2 dominates, the cross-relaxation rate W_2-W_0 may be either negative or positive and hence the sign of the corresponding NOE effect (Figure 1.15). Positive cross-relaxation rate gives rise to positive NOEs and negative cross-relaxation rate gives negative NOEs. For W_2-W_0=0, a special situation arises, when no NOE is seen. This is the case for molecules with $M_r \sim 3000$ Da (more specifically when $\omega_0\tau_c$=1).

For a two spin homonuclear case the limiting values of NOE is 0.5 for fast molecular motions and -1.0 for slow motions. In a heteronuclear case, the maximum value of η is given by γ_A/γ_X. For example, for a ^1H, ^{13}C system a nuclear Overhauser enhancement of 4 can be achieved by broad band decoupling of ^1H while observing resonances of ^{13}C.

7. LINE SHAPES IN NMR

7.1 Absorptive and Dispersive Signals

In the classical description of the behaviour of an ensemble of the magnetic moments in presence of the magnetic and RF fields (B_0 and B_1, respectively), Felix Bloch used a set of macroscopic or phenomenological treatment based equations to describe the time dependence of the net magnetization. These equations are popularly known as Bloch equations. The treatment of Bloch's equations is given in most text books on NMR and will not be discussed here. However, one important aspect of the Bloch equations is the prediction of the line shapes of nuclear induction signals in a slow passage experiment, which depend on the position of the receiver coil in the transverse plane. These are called absorptive and dispersive signals and are related to the so called *Bloch susceptibilities.* For weak RF fields ($\gamma B_1 \ll (T_1T_2)^{-1/2}$), the transverse magnetization and hence the line shape $g(\omega)$ in the rotating frame, as a function of ω_0- ω in the x′ and y′ directions is proportional to:

$$g\,(absorption) = -M_0\,\gamma B_1\,T_2\,/\{1+ T_2^2\,(\omega_0-\omega)^2\} \qquad 1.51$$

$$g\,(dispersion) = M_0\gamma B_1 T_2^2\,(\omega_0-\omega)\,/\{1+ T_2^2\,(\omega_0-\omega)^2\} \qquad 1.52$$

These line shapes of the absorption and dispersion signals as predicted by Bloch equation are shown in Figure 1.16.

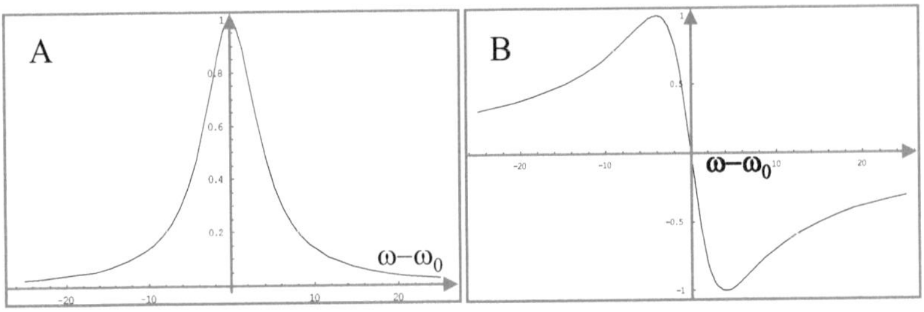

Figure 1.16: (A) Absorption and (B) dispersion line shapes of a resonance line.

7.2 Characteristics of FT-NMR Signals

A similar situation exists in FT spectroscopy. The previous treatment of FT of the FID signal was in fact, a simplification. The resonances have components of absorptive and dispersive character. Each signal may have a unique phase. Modern FT-NMR spectrometers can detect transverse magnetization in any direction of the *xy* plane. The orientation of the vector M_{xy} is defined by the phase of the signal. The receiver-phase dictates the shape of the detected signal. Following situations may arise:

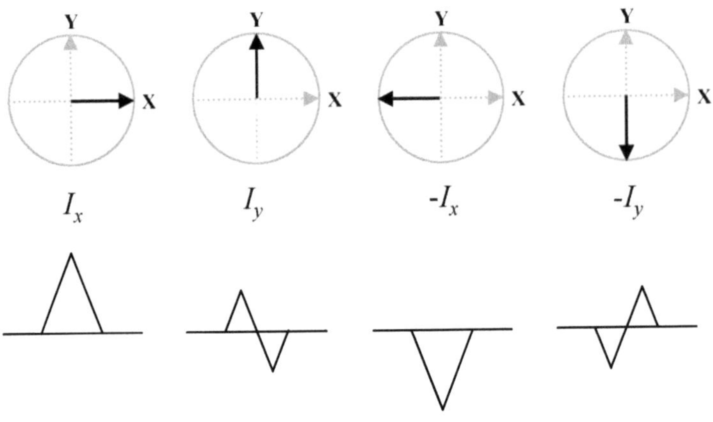

Figure 1.17: Line-shapes of NMR signals detected with the receiver aligned along the x-axis and the position of the magnetization vector along different axes at the start of the FID.

(i) The receiver is aligned along the x-axis and the magnetization is along the x-axis (i.e. x-magnetization) is detected, or the receiver is aligned along y-axis to detect y-magnetization. In such situations, one gets a pure absorption mode signal.

(ii) Receiver is aligned along the x-axis to detect y-magnetization or it is aligned along the y-axis to detect x-magnetization. In such situations, the result is a dispersion mode spectrum.

The shape of NMR signals is therefore complex. They can have absorptive or dispersive line-shapes. Figure 1.17 shows the line-shapes of the detected NMR signal, with the receiver aligned along the x-axis and the magnetization vector aligned along different axes at the start of the FID. It may be noted that the shapes of the detected signal can be controlled by phase correction after Fourier transformation. In the rest of the book, we shall use a convention in which I_x gives rise to pure absorptive signal while I_y gives pure dispersive signal.

Because of the acquisition delay *(dead time)*, introduced for recovery of the receiver, spins having different Larmor frequencies acquire different initial phases at the start of the acquisition of FID. This results in frequency dependent phase errors in the acquired signals. These can be corrected after the acquisition of the data.

7.3 Quadrature Phase Detection

One can now analyse the results of FT of FID in greater detail. If the FID is given by $S(t) = \sin(\omega_0 t)$, then its Fourier transform is:

$$F(\omega) = \int_{-\infty}^{\infty} \sin(\omega_0 t)\, e^{-i\omega t}\, dt = \int_{-\infty}^{\infty} \left\{ \frac{e^{i\omega_0 t} - e^{-i\omega_0 t}}{2i} \right\} e^{-i\omega t}\, dt \qquad 1.53$$

$$F(\omega) = \frac{1}{2i} \int_{-\infty}^{\infty} \left\{ e^{-i(\omega-\omega_0)t} - e^{-i(\omega+\omega_0)t} \right\} dt = \frac{i}{2}\left[\delta(\omega+\omega_0) - \delta(\omega-\omega_0)\right] \qquad 1.54$$

The resultant spectrum has two components (at ω_0 and $-\omega_0$) and has dispersive anti-phase character (see Figure 1.18 A).

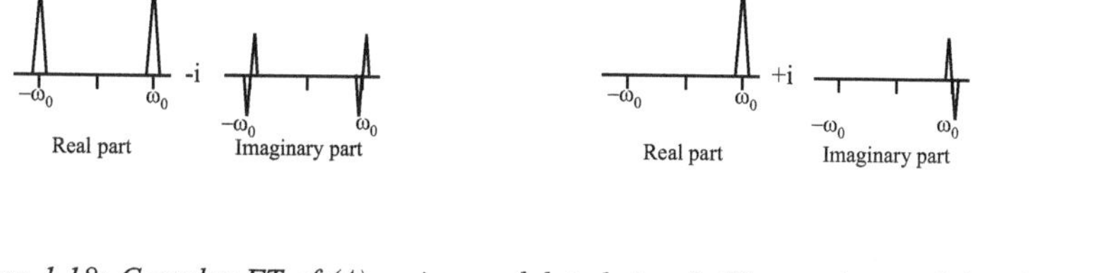

Figure 1.18: Complex FT of (A) a sine modulated signal, (B) a cosine modulated signal and (D) a complex signal ($M(t) = M_x(t) + iM_y(t)$). (C) The principle of quadrature detection.

On the other hand, if $S(t) = \cos(\omega_0 t)$, the Fourier transform is given by:

$$F(\omega)=\int_{-\infty}^{\infty} \cos(\omega_0 t)\, e^{-i\omega t}\, dt = \frac{1}{2}\left[\delta(\omega+\omega_0)+\delta(\omega-\omega_0)\right] \qquad 1.55$$

The resultant spectrum has absorptive in-phase character (Figure 1.18 B).

Thus in both cases, whether the detected signal is *sine* or *cosine* modulated, the resultant spectrum contains components at $+\omega_0$ and $-\omega_0$. Hence, there is a redundancy of information. A distinction between the two signals is not possible unless one sets the offset at one end of the spectrum. This implies that one has to use double the required spectral-width. This results in two-fold reduction in the spectral resolution. The method of quadrature detection, in which both x and y magnetizations are detected, enables a distinction between the two signals. While the *cosine*-modulated signal is detected in the x-channel (x-magnetization), simultaneously the *sine*-modulated signal is detected in the y-channel (y-magnetization). This results in a complex signal $M_x(t)+iM_y(t)$. Experimentally, a complex signal is generated by splitting the detected signal into two and shifting the phase of one component by $\pi/4$ and the other by $-\pi/4$ (Figure 1.18 C). The FT of such a complex signal $[M(t) = M_x(t) + iM_y(t)$ or $\exp(i\omega_0 t)\exp(-t/T_2)]$ gives rise to a complex spectrum $F(\omega) = F_x + iF_y$, which has a single line at $+\omega_0$ as shown in Figure 1.18 C. The quadrature detection allows the offset to be kept at the center of the spectrum and results in resolution enhancement by a factor of two.

8. REFERENCES

The NMR literature is vast and is expanding at a fast rate. There are also frequent changes in formalism from new developments in technology. Following books are suggested for further reading and are categorized according to their emphasis.

8.1 History and Early Developments

J.A. Pople, H.G. Schneider and H.J. Bernstein, *High-Resolution Nuclear Magnetic Resonance*, McGraw Hill Publications (1959).

A. Abragam, *Principles of Nuclear Magnetism*, Clarendon Press, Oxford (1961). Historical account of the developments of NMR is covered in 200 articles in *Encyclopedia of Nuclear Magnetic Resonance*, Vol 1, D.M. Grant and R.K. Harris (eds), John Wiley and Sons, England (1996).

8.2 Theoretical and Physical Aspects of NMR

T.C Farrar and E.D. Becker, *Pulse and Fourier Transform NMR*, Introduction to Theory and Methods, Academic Press (1971).

R.K. Harris and B.E. Mann, *NMR and the Periodic Table*, Academic Press (1978).

A. Carrington and A.D. McLachlan, *Introduction to Magnetic Resonance*, Chapman Hall, London (1979).

G.C. Levy and R.L. Lichter, *Nitrogen-15 Nuclear Magnetic Resonance Spectroscopy*, John Wiley & Sons (1979).

J. Sandstrom, *Dynamic NMR Spectroscopy*, Academic Press, New York (1982).

D. Shaw, *Fourier Transform NMR Spectroscopy*, Elsevier, Amsterdam (1984).

R.R. Ernst, G. Bodenhausen and A. Wokaun, *Principles of Nuclear Magnetic Resonance in One and Two Dimensions*, Oxford University Press, New York (1985).

N. Chandrakumar and S. Subramanian, *Modern Techniques in High-Resolution FT-NMR*, Springer-Verlag, New York (1987).

R. Freeman, *A Handbook of Nuclear Magnetic Resonance*, Longman, Harlow (1987).

R. Freeman, *Magnetic Resonance in Chemistry and Medicine*, Oxford University Press, New York (1987).

R. Freeman, *Spin Choreography: Basic Steps in High-Resolution NMR*, Oxford University Press, New York (1987).

J. Mason, *Multinuclear NMR*, Plenum Press, New York (1987).

H.O. Kalinowski, S. Berger and S. Braun, *Carbon-13 NMR Spectroscopy*, John Wiley & Sons (1988).

M. Munowitz, *Coherence and NMR*, John Wiley, New York (1988).

C.P. Slichter, *Principles of Magnetic Resonance*, 3rd Edition , Springer-Verlag, Berlin (1990).

M. Goldman, *Quantum Description of High-Resolution NMR in Liquids*, Oxford Science Publications, Clarendon Press, Oxford (1992).

E.D. Becker, *High-Resolution NMR: Theory and Chemical Applications*, 3rd Edition, Academic Press (1999).

D. Neuhaus and M.P. Williamson, *The Nuclear Overhauser Effect in Structural and Conformational Analysis*, 2nd Edition, Wiley-VCH (2000).

M.H. Levitt, *Spin Dynamics. Basics of Nuclear Magnetic Resonance*, John Wiley & Sons, Singapore (2001).

8.3 Chemistry oriented books

A.E. Derome, *Modern NMR Techniques for Chemistry Research*, Pergamon Press, Oxford (1987).

J.K.M. Sanders and B. K. Hunter, *Modern NMR Spectroscopy - A Guide for Chemists*, 2nd Edition, Oxford University Press, Oxford (1993).

H. Gunter, *NMR Spectroscopy: Basic Principles, Concepts, and Applications in Chemistry*, 2nd Edition, John Wiley & Sons, Chichester (1995).

G.A. Webb (senior reporter), *Oriented Molecules: Specialist Periodical Reports*: Nuclear Magnetic Resonance 33, The Royal Society of Chemistry, England (2004).

8.4 Specialized Topics

P.L. Corio, *Structure of High-Resolution NMR Spectra*, Academic Press, New York (1966).

G. Govil and H.J. Bernstein, J. Chem. Phys. **48**, 285–291 (1968).

J.H. Noggle and R.H. Shirmer, *The Nuclear Overhauser Effect, Chemical Applications*, Academic Press, New York (1971).

M.V. Phillipsborn, *Methods and Applications of Nuclear Magnetic Double Resonance*. Angew. Chemie Int Ed. **10**, 472–490 (1971).

J.I. Kaplan and G. Fraenkel, *NMR of Chemically Exchanging Systems*, Academic Press, New York (1980).

S.W. Homans, *A Dictionary of Concepts in NMR*, Clarendon Press, Oxford (1992).

J.N.S. Evans, *Biomolecular NMR Spectroscopy*, Oxford University Press (1995).

M.W.F. Fisher, A. Mazumdar and E.R.P. Zuiderweg, *Proton NMR Relaxation: Theory, applications and outlook*, Prog. Nucl. Magn. Reson. Spect. **33**, 207–272 (1998).

CHAPTER 2

INTRODUCTION TO BIOLOGICAL NMR

Even though NMR had become an indispensable tool in chemical studies in early 50s, its applications to biology evolved at a much slower pace. However, today NMR techniques span the whole range of biological systems. These include: structure determination of large biological molecules; understanding biochemistry and functional behaviour of proteins and nucleic acids; cell metabolism; structure and function of tissues: imaging; metabolism and functions of whole organs; pharmacology; plants and agricultural sciences and several other areas of life sciences. The applications of NMR range from a wide variety of biological molecules and micro-structures, human health and physiology to transgenic plants. NMR has wide applications in medicine, human physiology, development of new pharmacological compounds, material science and agriculture.

It is difficult to summarize all the path-breaking discoveries, which have made NMR such a powerful tool in biological systems. A summary of some of the milestones in the development of biological NMR is listed in Table 2.1.

Table 2.1: Frontiers of NMR in Life Sciences

Year	*Research areas opened*	*Major researchers**
1946	NMR in bulk materials	Bloch & Purcell et al.
1950-60	NMR as a tool in Chemistry	Gutowsky & others
	Multiple-pulse NMR	Hahn
1966	First FT NMR experiment	Ernst & Anderson
1970-1980	T_1, T_2 in tissues	Damadian
	Magnetic Resonance Imaging (MRI)	Lauterbur & Ernst
	Echo Planar Imaging	Mansfield & Edelstein
	2D NMR	Jeener, Ernst & others
1980-90	NMR Angiography	Dumoulin
	MRI as a clinical tool	Several workers
	Macromolecular structure, dynamics	Ernst, Wüthrich & others
	3D NMR	Several workers
	High-Resolution solid-state NMR	Waugh & Pines
1990-	f MRI: BOLD	Ogawa & Ugurbil
	MRS in cells and organs	Shulman & others

*Many researchers have been responsible for the overall development of NMR spectroscopy. Only a few selected names are given in the last column.

In the last sixteen years, four Nobel prizes have been awarded to NMR spectroscopists, whose work has revolutionised the area of life sciences (Figure 2.1). There is no sign of abatement in further advances. The basic principles of biological NMR are introduced in this chapter. The subsequent chapters deal with these advances in greater detail.

Figure 2.1: Awardees of Nobel prizes in the field of NMR (Courtesy: http://nobelprize.org).

1. LEVELS OF BIOLOGICAL STRUCTURES

The basic foundations of modern biology were laid during the period 1940-1970. It was established that deoxyribonucleic acid (DNA) is the carrier of genetic information. The information which is coded in the form of sequences of DNA is transcribed into m-RNA. The information is then translated into specific amino acid residues in the proteins synthesized by living systems.

Proteins are responsible for different functions of cells. Cellular functions in turn are responsible for the chemistry and functioning of whole organs in living systems, which include humans, animals and plants. Understanding protein-function and metabolism under *in-vitro* and *in-vivo* conditions is an important component of biochemistry.

Information transmission in living systems proceeds from the 1D information in nucleic acids in the form of sequence of nucleotides, which is transcribed onto m-RNA molecules. This information is translated through the 3D structures of t-RNA and ribosomal RNA (rRNA) to sequence of amino acid residues in proteins

(Figure 2.2). The 3D structures of proteins in turn, are responsible for metabolic activities of cells.

This picture is more complicated for evolved living systems such as mammals, which have organs with different functions. All cells originate from the same genomic material (paternal genome from spermatozoa and maternal genome from egg) and are derived from a single fertilized cell. The initial fertilized cell, after the first few rounds of division, starts differentiating into cells of different organs. This leads to differential expression of the genes in different organs. Each organ has more than one type of cells. Further, cells from the same organ may have different expression and altered functional behaviour when subjected to different kinds of stress such as disease, diet, chemical and environmental changes and other factors. For example, upon ingestion of a toxin, the protein level may not change, but the toxin may inhibit a critical enzyme and reduce its activity to a lower level.

During the last 40 years there has been significant progress in the decoding of gene sequences of various organisms, sequencing and structure determination of macromolecules such as proteins, carbohydrates and nucleic acids, understanding of macromolecular assemblies such as biological membranes and viruses, cellular metabolism, biological functions of tissues and understanding the functioning of organs.

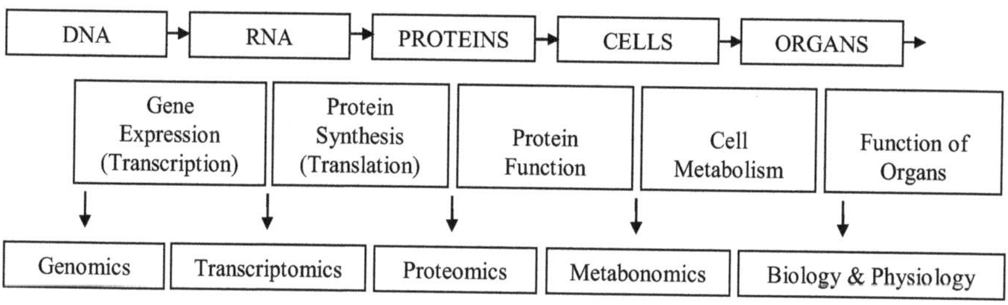

Figure 2.2: Flow of information and controls in biological systems. The function is controlled at several levels. It is important to understand chemistry and biology of living systems at all levels shown in the above diagram.

The knowledge of gene sequences of a large number of organisms including human has emerged at a relatively faster rate. In fact, genomics is the first step towards understanding metabolic behaviour of living systems at a molecular level. However, as yet one does not know all the DNA sequences in humans and higher animals, which are expressed as proteins.

The next step in understanding biological systems is proteomics. Proteins are the molecules responsible for most of the body functions and serve as 50% of the building materials. At this stage, one does not know the total number of proteins responsible for functional behaviour of humans and animals. The current estimate is that about 30000 genes may be involved in humans. An antibody based portrait of human proteome (www.proteinatlas.org) lists the location and quantity of only about 700 proteins in humans. The future information is being generated at a rapid pace.

2. NMR AND BIOLOGICAL STRUCTURES

Applications of NMR to biological systems can be discussed at different levels of the complexity of biological structures. The essential molecules involved in biological systems range from small molecules including water to complex and large macromolecules such as proteins and nucleic acids. Several of these molecules form multi-molecular assemblies, such as protein-nucleic acid complexes, biological membranes, viruses, cell organelles. At the same time, a cell can be thought as the smallest unit of life. The cells are ultimately responsible for properties of organs in human, animals and plants. NMR has been applied to the whole range of biological complexity.

The knowledge of the 3D structures of proteins, t-RNA, ribosome and their structure-function relationship is the next step in understanding various biological processes. Today, the 3D structures of a large number of such molecules are known. However, the available number is much smaller than what is needed for understanding the biochemistry of a human body. Several proteins with similar functions have similar 3D structures (examples are kinases, dehydrogenases.). Therefore, knowledge of the structure of a particular class of protein from one organism may help to understand a particular biochemical reaction in a cell or organ of another organism. Thus, it is important to investigate all the unique protein folds that are involved in the function of living systems. This number of unique folds is estimated to be around 1000. The structure determination of proteins and nucleic acids and their *in-vitro* function thus forms an important area of biophysical research. NMR and X-ray crystallography are two major techniques used to unravel 3D structure of biological molecules.

The next level at which NMR has helped us to understand living systems is the macromolecular organization. An interesting application in this regard is the study of biological membranes, which show liquid crystalline behaviour. Membrane bound proteins are often insoluble in water and are difficult to crystallize. Recent developments in high-resolution solid-state NMR techniques have made it possible to study such proteins.

Cell function is a result of an intricate network of functioning proteins, small molecules and ions, which may activate or control enzyme functions. Even though attempts have been made to study enzyme function *in-vitro,* one can understand *in-vivo* biological functions by experimenting at cellular level. NMR has proved to be an invaluable tool to understand cell metabolism and in exploring factors which control and modify such actions.

Finally, in plants, animals and humans, different tissues or organs perform distinct functions. NMR is the technique of choice to understand such complex functions. Future strides in the methodologies of high-throughput NMR will further enhance our investigative capabilities to unravel complex biological structures and functions. Such studies are important for human health and understanding human physiology. A relatively new area is the development of NMR in plant and agricultural sciences.

3. DIFFICULTIES IN STUDYING BIOLOGICAL SYSTEMS BY NMR

It took almost 30 years for NMR to advance from chemical to biological systems. This is primarily because of several problems which one encounters when dealing with living systems. The solutions to these problems have led to major advances in applications of NMR to biology.

3.1 Sensitivity

By its very nature, NMR is a relatively insensitive technique. The intrinsic sensitivity of a nucleus is proportional to the cube of the nuclear magnetic moment. Since ^1H has the highest magnetic moment (μ), it is the most sensitive nucleus. This is fortunate because of the high abundance of ^1H in biological systems. The next most sensitive NMR spin is ^{19}F, which is less useful in biology. The natural abundance of the nucleus directly determines the number of nuclei available for detection of NMR signal. For example, natural abundance of ^{13}C is 1 in 100, while that of ^{15}N is 1 in 300. The sensitivity of less abundant nuclei however, can be increased by using labelled compounds.

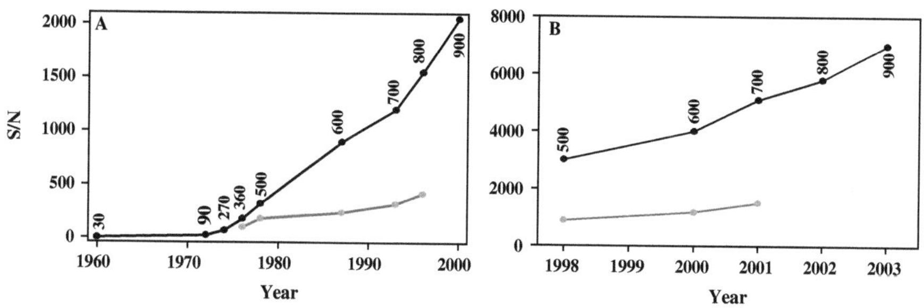

Figure 2.3: Historical development of the improvement in ^1H (black) and ^{13}C (grey) NMR sensitivity, with (A) normal probe and (B) cryogenically cooled probe.

The population difference between the ground and the excited site depends on the value of B_0. Use of higher fields therefore results in higher S/N. In the early days of NMR, solutions of 100 millimolar concentration of samples were required to obtain a decent ^1H NMR spectrum. It is almost impossible to obtain such high concentrations of biological macromolecules. With the advent of FT technology and advances in better RF probe designs, magnet and computer technology, this limit has now come down to sub-mM concentrations. One of the recent advances is the use of super-cooled RF-probes, with which one can enhance the S/N by an additional 3-4 times. In early 1960s, the S/N ratio recorded with a 100 mM sample of ethyl benzene was around 0.04 on a 30 MHz NMR spectrometer. Today, on a 900 MHz NMR spectrometer equipped with cryogenically cooled RF-probe one can achieve S/N of 8000, with 0.1% solution of ethyl benzene (Figure 2.3). This reflects a net gain of 200,000. However, in spite of these developments, the sensitivity of NMR experiments is much lower than that achieved by other spectroscopic techniques. In fact, the improvements in sensitivity are closely linked to the wider range of problems that can be handled by NMR.

3.2 Resolution

As seen in the example discussed in Chapter 1, even a short oligonucleotide (12 nucleotide long) gives a large number of ^1H resonances. The ^1H signals lie in a narrow range of 0-15 ppm. Thus, it is difficult to resolve all the signals for larger molecules. This is particularly true for spins in similar chemical environment, for example the amide protons in proteins or the imino protons in nucleic acids. In the above mentioned example, one notices severe crowding of resonances for sugar ^1H resonances.

One of the ways to handle such problem is to use higher fields for biomolecular studies. Higher fields also simplify the spectrum as they lead to better dispersion and convert strongly coupled spin systems to weakly coupled ones. The advent of multi-dimensional FT techniques has made further revolutionary advances in this regard, by expanding the information in more than one dimension

3.3 Assignments

To make meaningful conclusions, one needs to know which resonance is associated with what nuclear spin in a molecule. In the case of the oligonucleotide discussed in Chapter 1, there are two thymines and hence two CH_3 resonances are observed in the expected ^1H region (Figure 1.6). However, it is not straight forward to decide, which resonance corresponds to which of the two thymines (T3 or T9). Hence, sequence specific resonance assignment is a crucial step in the study of biological systems. Even when complete resonance assignments are not possible, partial assignments help in addressing certain specific problems.

Multi-nuclear and multi-dimensional NMR have made it easier to resolve the problem of sequence specific resonance assignment.

3.4 Water Signal

With two protons, the molar concentration of water protons is almost 110 M in aqueous solutions. The signals from mM concentrations of molecules of interest have to be picked up against this background signal. In most chemical applications, NMR studies are conducted using deuterated solvents. However, in biological applications studies are performed in H_2O, which is the milieu of living system. One can use ^2H_2O (D_2O) instead of H_2O to avoid strong water resonance for molecular studies. However, ^2H_2O should be avoided in animal studies. Even in studies on macromolecules, the use of ^2H_2O leads to H/D exchange for the labile NH and OH protons, which are important in studying hydrogen bonding.

At the same time, the abundant presence of water in biological tissues, its strong signal, and sensitivity to its relaxation behaviour due to its interaction with other molecules and its fluidity have been exploited in the development of powerful techniques such as magnetic resonance imaging (MRI).

3.5 Line Widths

The line-widths of ^{1}H resonances in solutions of most organic compounds are less than 0.3 Hz. The lines in biological macromolecules are broader, because of restricted motion and small values of T_2. Line-width increases with the molecular weight. Therefore studies on very large proteins are complicated not only by problems of resolution and assignments but also by the higher line widths. In cells and organs, the resonances are even broader (Figure 2.4). In recent years, methods to overcome such problems have been developed.

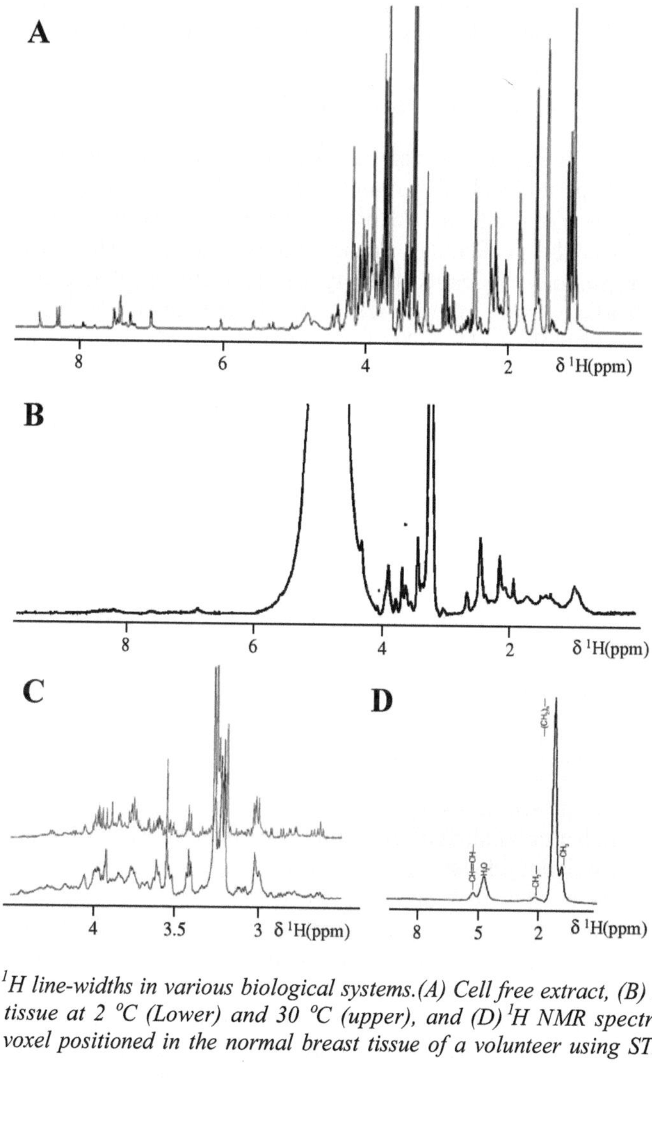

Figure 2.4: ^{1}H line-widths in various biological systems.(A) Cell free extract, (B) Intact Cell, (C) Kidney tissue at 2 °C (Lower) and 30 °C (upper), and (D) ^{1}H NMR spectrum from a $2\times2\times2$ cm^3 voxel positioned in the normal breast tissue of a volunteer using STEAM pulse sequence.

3.6 Quantification

The integrated area of the NMR signal is directly proportional to the concentration of the spin(s) giving rise to the signal. The relaxation rates and therefore the line-widths of NMR signals can be quite different depending on the molecular flexibility. Such differences are quite pronounced for the same molecule present in different parts of the cell or a living system. Line widths of even a simple molecule such as water, may be different depending on its location in the living system. In many instances the resonances may broaden below the noise level and may not be detected.

Even for the same molecule such as a biopolymer dissolved in aqueous solutions, different parts of the molecule may exhibit different motion and the line-widths may not be comparable. This can create serious problem in quantification.

4. BIOLOGICAL MACROMOLECULES

The molecules of living systems can be divided into three categories: water, small molecules and large molecules (macromolecules). Almost 75% of the body weight of living beings is water. About 10% of the body mass consists of small molecules, such as amino acids, nucleotides, sugars, lipids and ions. The remaining mass consists of macromolecules called biopolymers [proteins, nucleic acids (DNA and RNA) and carbohydrates or polysaccharides] and organized multi-molecular assemblies such as biological membranes and protein nucleic acid assemblies.

Macromolecules, such as proteins, nucleic acids, carbohydrates and their assemblies, generally adopt specific shapes both in their active and passive states. Knowledge of the three dimensional (3D) structure, dynamics and interaction of these biological macromolecules and their assemblies helps in understanding molecular mechanisms of biochemical reactions. The information about the molecular architecture and the forces responsible for multi-molecular assemblies such as membranes, chromosomes and cell organelles provides a key to understand various sequential and linked molecular processes in biology, which occur in a highly orchestrated fashion. The stereochemical details of biological macromolecules and their interactions with endogenous and exogenous molecules also form the basis for drug-design.

4.1 Building Blocks of Biological Molecules

Several molecules present in living systems are highly complex. However, nature has used a small number of building blocks with a specific stereochemistry, to design such molecules. The building blocks are amino acids, nucleotides, sugars, fatty acids and lipids. Each one of them has one or more chiral atoms. Therefore two or more configurations (stereoisomers) can exist. Changing from one configuration to another (e.g. from D- to L-form of amino acids) requires breaking and making of chemical bonds. Therefore, transition from one configuration to another requires a fairly high energy of activation. In reality, such changes do not occur in biological systems.

Nature has thus, carefully adopted only one of the several chiral forms, while constructing biological macromolecules (Figure 2.5). For example, there are two configurations of amino acids. In living systems only α–L-amino acids are selected for building proteins. Similarly, deoxyribose, the central moiety in DNA can exist in 8 configurations; only D-ribose is used. Configurations of hexoses and lipids are also specific. Such a stereo-selectivity implies that during metabolic pathways, scrambling between different configurations is not allowed.

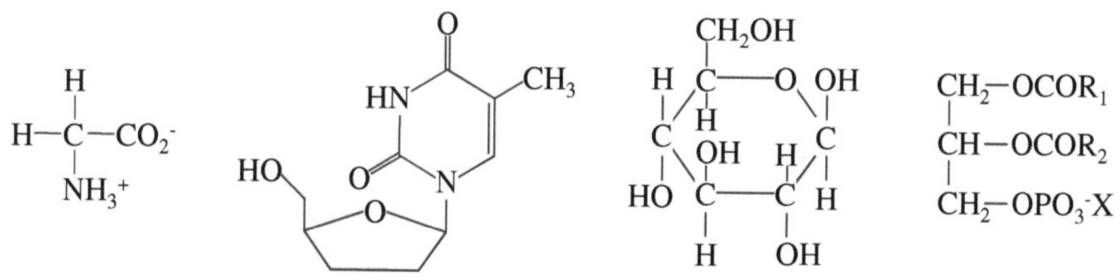

Amino-acid Nucleoside Glucose Phospholipids

Figure 2.5: Basic building blocks of biomolecules have a specific stereochemistry.

With such restrictions and selectivity, macromolecules in living systems are built from a few simple motifs. L-amino acids are primary precursors for peptides and proteins. Ribose or deoxyribose (five member sugar rings), purine and pyrimidine bases, and phosphate groups are the fundamental units of nucleotides, which are the building blocks of nucleic acids. D-Glucose (Glc) and certain other sugars are the main building blocks for carbohydrates.

Lipids are essential components of cell membranes. A large fraction of these molecules are phospholipids, which are built from fatty acids, glycerol and a phosphate ester. Lipids usually have well defined regions in their chemical structures, which are either hydrophilic or hydrophobic. This property of phospholipids endows membranes with ability to solubilize non-polar molecules in their matrix. They act as scaffolds for the location and assembly of biological molecules which are insoluble in water.

4.2 Biopolymers

Each biological macromolecule has a specific role to play in the biochemistry of living systems. Some molecules may have more than one role. The 3D structures provide insights into the functional aspects of these molecules in fundamental biological processes and how their functions differ in healthy and diseased organisms.

The 3D structures of macromolecules, which have been determined using either NMR or X-ray crystallography, can be down-loaded from the protein data bank (PDB). Currently even personal computers (PCs) can produce stunning coloured pictures of such structures and provide an insight into their detailed architecture. However, this often creates an illusion that biopolymers have a rigid structure, which is not true. In fact, they exhibit a fair amount of flexibility. Such a dynamic property of biopolymers plays an important role in their interaction with other molecules and hence in their function. Therefore, the study of structure, dynamics, interaction and function of biological molecules has aroused tremendous interest in recent years.

A large number of techniques have been used for characterization, determination of 3D structures, interaction and dynamics of biological molecules. Fluorescence, vibrational and electronic spectroscopy, Optical Rotatory Dispersion (ORD), Circular Dichroism (CD), Atomic Force Microscopy (AFM) and Dynamic Light Scattering (DLS) all have a high degree of sensitivity and spectra can be obtained with μg quantities. However, such techniques generally provide only a gross picture of large molecules. In recent years, mass spectroscopy (MS), using ion-spray (ESI-MS) and matrix assisted laser desorption/ionization time-of-flight (MALDI-TOF) have become indispensable tools to determine molecular weights (M_r) of large molecules. An accuracy better than one Dalton can be easily obtained. More recently, cryo-electron microscopy has also been used for structure determination of biological macromolecules.

Theoretical methods using both quantum chemical and semi-empirical approaches have gained importance in the study of macromolecular structure. With increasing computational powers, the limitations in applying such theoretical studies to large macromolecules are fast receding. Though the treatment of the role of water in stabilizing molecular structures using theoretical methods is difficult, some strategies have been developed for this purpose. The theoretical methods also provide information on the dynamic nature of macromolecules.

4.3 3D Structures of Biological Molecules

NMR and X-ray crystallography are the two major techniques, which provide detailed 3D structural information on large biological molecules. Both techniques have been used extensively, often in a complementary fashion and in combination with other methods.

For several years, X-ray diffraction served as the only source of information on the 3D arrangements of atoms in biomolecules. Initially, fibres of long chain polymers, which show ordering along the axis of the fibre, were subject of interest. Using periodicities in the X-ray fibre-diffraction patterns, models of the structures of DNA double helix, collagen triple-helix, α-helix and β-sheet structures in proteins and structures of certain polysaccharides were proposed. Thereafter, X-ray diffraction from single-crystals developed as an effective tool in mid 1950s. X-ray diffraction provides detailed information on the electron-density map of the molecule. The positions of heavy atoms (contributing to large X-ray scattering), can be located with a high degree of precision and accuracy. Determination of positions of lighter atoms such as hydrogen are less precise and are usually deduced by model building.

During the last two decades, NMR has emerged as a potential tool for structure determination, following developments in multinuclear and multidimensional NMR spectroscopy. Simultaneously, with the availability of high speed computers, theoretical methods have provided valuable aid for the analysis of data from both X-ray diffraction and NMR.

4.4 Comparison of 3D Structures Obtained from NMR and X-ray

NMR has evolved as a powerful alternative to X-ray crystallography for determination of 3D structures of biological macromolecules. However, X-ray crystallography is still the dominant technique and accounts for more than 80% of the structures in the PDB files. NMR on the other hand, has contributed to fewer structures though the numbers are growing with recent advancements. In many cases NMR and crystallography act as complementary techniques. It is therefore desirable that one understands the relative advantages and disadvantages of the two methods.

The main requirement for X-ray crystallography is the availability of single crystals and ability to produce heavy atom derivatives. Not all molecules crystallise well. For example, membrane bound proteins, which form almost 30% of the proteome do not crystallise easily. In favourable cases, X-ray scattering from different atoms provides electron density maps, from which the positions of all atoms (with the exception of hydrogens) can be deduced with high precision. The coordinates of hydrogen atoms can be obtained by model building. Proteins, nucleic acids, protein-nucleic acid complexes and multi-molecular complexes with large molecular weights (M_r) have been solved by this technique. This includes molecular assemblies such as ribosomes containing more than 100,000 atoms. Structures of protein-DNA complexes, viruses and molecular assemblies such as nucleosides, and molecular assemblies such as ATP synthase and ribosomes are some of the other major achievements of X-ray crystallography. Most of these supramolecular structures are still beyond the reach of NMR.

Recent advances such as the availability of synchrotron sources and high speed detection devices have drastically cut down on the time of data collection. Fairly small crystals can be investigated. Using cryo-techniques, snap shots of intermediate stages in biochemical reactions can be obtained using time-resolved crystallography.

The success of NMR experiments likewise depends on our ability to produce relatively large quantities of macromolecules. For proteins, this is often achieved by gene cloning techniques. Unlike crystallography, the structure is determined by compiling structural clues in the form of NMR parameters. These are then used as constraints in a structure determination protocol. The conformational geometry of the residues, where experimental information is lacking, may be poorly defined. The success of structures of higher M_r, depends on the ability to produce ^{13}C and ^{15}N labelled molecules. Even then, the structures solved by NMR are generally limited to less than 50 kDa. Several proteins have a tendency to aggregate under NMR concentrations.

NMR investigations are made in aqueous solutions, under conditions which can be controlled to mimic *in-vivo* situations of pH and ionic concentrations. The structures are not dictated by packing forces present in crystals. The NMR structures therefore show much more flexibility compared to crystal structures. In fact, one of the major advantages of NMR is the ability to study dynamics of macromolecules, which plays an important role in biological functions. Changes in structural characteristics as a result of binding by a ligand, metal ion, cofactors or exogenous compounds such as drugs and toxins, can be handled much more conveniently by NMR. Conformational changes can be probed and thermodynamic parameters can be determined in favourable cases.

Protein folding and enzymology are some of the recent trends in biochemical research. NMR has proved to be a powerful technique in such studies. A recent advance is the development of high-resolution solid-state NMR, which allows structures of membrane bound proteins, large DNA molecules and other solid materials.

Recent advances in molecular biology have led to a new appreciation of the complexity of the living state. On the other hand, recent developments for determination of biomolecular structures using multi-dimensional NMR methods have enabled the structures to be obtained with greater precision. Also, much larger structures are amenable to studies today as compared to a decade back. The developments in NMR technology and structural algorithms for structure simulation have developed side by side. NMR has been extensively used to study the structure and dynamics of proteins, nucleic acids and organized supra-molecular assemblies. It thereby enables to understand associated biochemical activities of various genes identified by molecular genetics. The success of such ventures often depends on our ability to have labelled molecules. These are mostly synthesized through microbiological route. This approach has provided structures of several key proteins and other biomolecules, and in turn, to understand their function at an individual level. This provides physico-chemical explanations for several properties, through a reductionist approach, where complex reactions are broken down into individual components.

5. NMR IN CELLS AND TISSUES

As discussed above, small molecules, such as amino acids, nucleotides, sugars and lipids are the building blocks of larger molecular systems and assemblies. They act as substrates and ligands or may be the final products in biochemical reactions. Small molecules are also involved in energy transduction and act as cofactors in enzymatic reactions. Such molecules have relatively simple and often flexible structures. This enables them to interact relatively freely with larger molecules. They can readily adapt themselves to the shape of macromolecules to achieve optimum interaction during binding.

However, even in the most basic biological processes, hundreds of reactions and enzymes may be involved. Understanding of biological phenomenon such as bioenergetics, biosynthesis, molecular transport, changes arising from disease, behaviour, evolution, development and stress, require a bottom-up approach. How does one utilize information discussed above, to understand complex but highly coordinated processes involved in the biochemistry of living systems?

Some of these questions can be answered using NMR studies on cells, tissues and body fluids such as blood, bile and urine.

5.1 Cellular Metabolism is the Bridge between Proteomics and Function

Metabolic pathways consist of a series of coupled chemical reactions, catalysed by enzymes. The cells control these pathways through regulation of genetic expression, feed-back mechanisms and signalling pathways. The demand on metabolism changes as a result of variation in the environmental factors. The rates of reactions in metabolic pathways may change to meet the requirements of organisms.

Metabonomics is a recent term coined, for an area devoted to the understanding of the behaviour of cellular systems and organs at a molecular level, and aberrations in their function as a result of physiological or genetic modification, diet, gender, stress, toxicity, and changes during cell development and differentiation. NMR of cells, tissues, body fluids and whole organs, has therefore become an important area of research in biological and medical sciences.

The NMR approaches on cells thus bridge the gap between the properties of molecules on one hand, and the function and behaviour of living systems on the other. It can be used to monitor changes in metabolism as a result of stresses, both in plants and in animals. Several small molecules are detected with a fair degree of sensitivity. Cellular NMR can be used in conjunction with microscopic techniques. In fact, the NMR methodology is particularly useful in studies of genetically modified plants.

6. NMR IN STUDIES OF ORGANS

While working with NMR of biological macromolecules and cellular systems, one determines properties, which are time and space averages of the macroscopic systems. The observed spectrum is an average over sample volume within a region excited by RF pulses and over the time of measurement. The sample in such cases is typically 600 μl of solution, placed in a 5 mm sample tube. To obtain meaningful information on an organ of human body such as brain, kidney, heart, breast, and joints, the volume or region of interest (abbreviated as VOI or ROI, respectively) needs to be selected for NMR investigation. Naturally, such volume is much larger than the one used in molecular or cellular studies. In this sense, the techniques used in *in-vivo* NMR differ substantially from those discussed earlier. However, the basic principle is the same.

The NMR techniques used in animal studies are often divided into two categories (i) (Nuclear) Magnetic Resonance Imaging (MRI; notice the missing *N*) and (ii) (*in-vivo* Nuclear) Magnetic Resonance Spectroscopy (MRS). The word *nuclear* has been dropped because of two reasons. Firstly, a number of other nuclear techniques are extensively used in medical diagnosis; secondly, NMR does not use ionising radiations (in most cases stable isotopes are used).

MRI investigations in hospitals are often conducted side by side with X-ray based computed tomography (CT) and techniques based on radioactive isotopes, which provide complementary information.

6.1 Historical Development of MRI

Way back in 1946, Felix Bloch placed his finger inside the probe of NMR spectrometer and observed a strong signal. In a sense, this was the first biological application of NMR. Clinical MRI developments primarily originate from the work of Damadian (1971) who showed that the ^{1}H relaxation rates for water are different for normal and malignant tissues, and that this information can be used for diagnostic purposes. Paul Lauterbur (1973) showed that the ^{1}H signal of a single molecule such as H_2O, can be separated in space by using magnetic field gradients and that an image of an object containing water can be constructed. The work of Richard Ernst and co-workers (1975) based on the FT technique opened the way for fast signal acquisition for imaging by using phase and frequency encoding methodology. With in few years, human *in-vivo* images were published, and MRI of human organs such as finger, hand and wrist were reported. Around the same time, Peter Mansfield (1977) developed the technique of echo-planar imaging, which can provide images much more rapidly. Both Paul Lauterbur and Peter Mansfield received Nobel Prize for their contributions in 2003.

6.2 Basis of MRI

Most MRI studies concentrate on the use of proton signal from water. Water is the elixir of the whole biological world. It influences properties of other molecules in its surroundings through electrostatic interactions and hydrogen bonding. It is a highly polar molecule which is simple in terms of its structure in isolation, but highly complex when viewed in the context of its overall architecture and properties in biological systems. With two protons that can act as donors in hydrogen bonding and two lone pairs of electrons of the oxygen atom, which can act as acceptors, this triatomic molecule forms a variety of hydrogen bonded networks with itself and with other molecules. Water is present in the hydration spheres of ions and other biological molecules, and forms an integral part of crystals of biological molecules used in structural studies. It has a high dielectric-constant, which considerably weakens electrostatic interactions between charged groups and salt bridges in solute molecules. Water plays a key role in the molecular assembly of membranes through hydrophobic and hydrophilic interactions. It is either a product or a substrate in most biochemical reactions. In fact, this universal biological solvent is not inert but radically changes the properties of all other molecules and molecular assemblies in living cells. These properties of water vary from one region of an organ to the other and form the foundation of imaging of body organs using NMR.

Today, MRI has become a tool of choice for clinical studies and diagnostics. The advances have been led by the developments in technologies in design of magnets, electronics and computational power. The presence of water in large amounts in most living systems coupled with the high sensitivity in detection of signals from ^1H is a major advantage in water based MRI. It enables water images to be obtained relatively easily. The differences in property of water in different tissues allow one to obtain biologically relevant information.

Recent advances have made it feasible to obtain images of other molecules and nuclear spins (such as ^{31}P). MRI provides a spatial map of the distribution of a specific NMR active nucleus in a particular compound (mostly ^1H in water) within the volume of interest (VOI). It is one of the latest modalities in clinical imaging and represents a continuing revolution in medical technology for human health. MRI provides detailed images of human body with unprecedented soft tissue contrast and sub-millimetre resolution and that too in few seconds to minutes. Images of any desired orthogonal (perpendicular to B_o) or oblique planes can be obtained by manipulation of the magnetic field gradients and RF pulses. Tissue anatomy, pathology, metabolism, flow and diffusion are all amenable to non-invasive evaluation by MRI.

6.3 Comparison of Images from NMR and other Techniques

In view of its importance to human health, a wide range of techniques have been developed or are under development for clinical imaging. In fact, the entire electromagnetic spectrum has been used for imaging of human organs (Table 2.2). In addition, ultrasonography based on sound and other techniques are routinely used in hospitals.

Table 2.2: Some Techniques used for Molecular Imaging

Radiation	Technique	Target	Application
γ rays	PET*	Glucose	Metabolism in brain and heart
γ rays	SPECT*	Radio isotopes	Imaging
X-rays	CT* scan	Absorption	Hard tissues; bones
Visible/IR	Optical methods	Several molecules	Soft tissues; flow; diffusion; still in development stage
Microwave	ESR*	Exogenous spin probes and oxygen	In development stage
Radiowave	NMR	H_2O, fat, other molecules	Soft tissues; flow; diffusion; metabolism
Ultrasound		Rigid matter, fat, collagen.	Wide variety of pathological conditions

* PET: Positron Emission Tomography; SPECT: Single Photon Emission Computed Tomography; CT: Computer Tomography; ESR: Electron Spin Resonance.

One of the major advantages of NMR as an imaging technology is the ease with which sections of the organ to be imaged can be selected. For example, simply by selecting the direction of the field gradients, one can select axial, orthogonal, sagittal or an oblique plane in human body. Since one usually obtains water images, it is an ideal tool for imaging soft tissues with good contrast. The NMR sensitivity for water images is particularly good and even small structures can be seen with good contrast in a reasonable time.

A variety of NMR parameters can be used for contrast and this aspect can be used to obtain images in more than one way and to obtain diverse types of information. For imaging of the soft tissues, MRI is the best among all the existing medical imaging techniques. The current resolution is better than sub-mm and this is improving with better design of the electronics and pulse sequences. It is a non-invasive technique and is relatively safe since it does not use ionizing radiations. Magnets have been developed so that imaging can be performed during the course of surgery. Unlike other techniques, the limit of spatial resolution is not limited by the principles of wave optics. MR Spectroscopy has been used for study of metabolism in organs.

Computed Tomography (CT) imaging is dependent solely on differences in X-ray attenuation. There is essentially a linear relation between image brightness and the X-ray linear attenuation coefficient (μ). The image interpretation is largely understood in terms of this single parameter. Usually μ is expressed in Hounsfield units (H) with water having a value zero and air a value of 1000. Most soft tissues have values in the range 20-60 H. Calcification usually increases value of μ, while oedema in brain decreases it. From changes in μ in tissues, it is possible to study anatomy, physiology, biochemistry and pathology of the tissue, particularly whether it is normal or abnormal. In particular, CT gives good pictures for hard tissues and bone structures. It uses ionizing radiations. Extensive and repeated use of such radiations may be injurious to the patient.

Ultrasound imaging depends primarily on the rigidity of tissues and gives good maps of collagen and fat. However, it is not possible to image micro-calcifications. There is a significant overlap between benign and malignant lesions.

In spite of potential injury to body, imaging using radio-isotopes is fairly common. Positron Emission Tomography (PET) provides good pictures of glucose metabolism. PET uses radio isotopes with high energies. Single Photon Emission Computed Tomography (SPECT) scans radiotracers introduced in the organ in the form of complexes of Technitium or Thallium, which travel to an intended organ of the body. The γ ray photons are then picked up by gamma cameras and image of the organ is constructed. PET scan is more quantitative in terms of obtaining metabolic fluxes than functional MRI. Also these techniques are able to locate cancer in soft tissues and bones at a very early stage. However, they use radioisotopes which are injected into the body.

In recent years, several other ranges of electromagnetic radiations have been explored for imaging such as optical and microwave (electron spin resonance). However, most of them are in developmental stages. Though different techniques may provide complementary information, the advantages of NMR techniques are unique and unsurpassed.

6.4 Magnetic Resonance Spectroscopy (MRS)

We have seen that water images provide anatomical features. Certain properties of water are used to measure flow, perfusion, and blood oxygen related properties for studying biological activities of organs. The *in-vivo* MRS of organs in humans and animals is an extension of studies of metabolism in cells and tissues. It is used to observe levels of different compounds in specific volumes of interest (VOI) in an organ. The VOI is selected by an appropriate use of pulse sequences and magnetic field gradients. Determination of changes of levels of concentration of metabolite provides information on the state of normal and abnormal tissues, changes in metabolic processes with age and development, response of tumours and other diseases to different treatments, drug development and a number of other applications related to physiology and metabolism of living systems.

The dual techniques of MRI and MRS have found wide applications in clinical, pharmaceutical and biological research. Though the two techniques have evolved independently, there is a strong link between the two.

7. BASIC MULTI-PULSE NMR EXPERIMENTS IN BIOLOGICAL SYSTEMS

Some of the basic pulse experiments that form the backbone of biological NMR are discussed below. More advanced multi-dimensional techniques are discussed in Chapter 3.

In a single pulse excitation and detection experiment (Chapter 1), the FID has the signatures of all *like* spins. Fourier transformation of this complex FID provides a spectrum amenable for direct interpretation. Several of the problems discussed in the previous section can be tackled using multi-pulse experiments. Such applications use several RF pulses applied in a well defined sequence and at very precise time intervals. The band-widths and the frequencies of the excitation pulses, duration of pulse(s), their shapes and phases, and inter-pulse delays all play important roles in the signal finally acquired.

The train of pulses thus applied is called a *pulse sequence*. Designing of pulse sequences is an important area of NMR spectroscopy and is often referred to as *pulse engineering*. Novel sequences have opened up several new areas of NMR spectroscopy. Some of the basic multiple-pulse experiments, which form basic building blocks of more advanced multi-dimensional and multi-nuclear pulse sequences, are discussed below. More advanced 2- and 3-D FT-NMR sequences are discussed in Chapter 3.

7.1 Signal Averaging and Partial Saturation

A simple method for excitation and detection of NMR signal using a $\pi/2$ pulse has been described earlier. For the purpose of obtaining better S/N ratio, the experiment is repeated several times and the collected FIDs are added coherently using a computer. As the signals add coherently while noise is random, accumulation of n_s transients results in an enhancement of S/N by $\sqrt{n_s}$. This is the principle behind signal averaging.

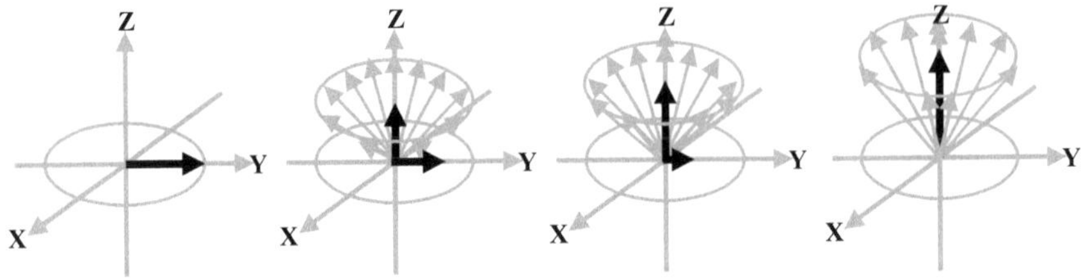

Figure 2.6: The decay of transverse magnetization and the build up of longitudinal magnetization as a function of relaxation times.

Since T_2 is usually shorter than T_1, the rate of decay of transverse magnetization dictates the acquisition time (sometimes called the *read-out* time) of the FID signal (Figure 2.6). This is usually of the order of 10-100 milliseconds. In fact the decay of the transverse magnetization is governed not only by *true* T_2 but also by the magnetic field inhomogeneity. The T_2^*, which includes the relaxation contribution arising from magnetic field inhomogeneity, is shorter than the true T_2. Thus, while the FID may have decayed due to the dephasing of the transverse magnetization, the longitudinal magnetization may not reach the equilibrium value (Figure 2.6).

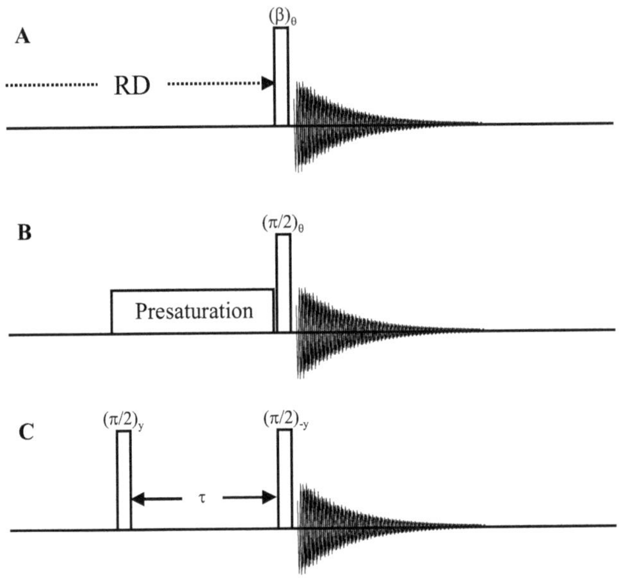

Figure 2.7: Some basic pulse sequences used in NMR (A) 1D ^1H acquisition by the method of partial saturation (β represents the Ernst angle; RD is the relaxation delay) (B) 1D ^1H acquisition with presaturation (C) Jump and Return scheme for water suppression.

Typically, one waits for a time less than or equal to $5T_1$ (called the relaxation delay; abbreviated as RD) to re-establish the equilibrium z-magnetization before repeating the process of excitation and detection again (Figure 2.7 A). This is called the repeat time (TR) of the experiment. In many cases, a faster repetition time (TR) is used and therefore, the observed signal is weaker. In such cases, after few transients (scans or repetition of the experiment), the system reaches a steady state value for initial z-magnetization, given by:

$$M_z = M_z(0)\,[1-\exp(-TR/T_1)] \qquad 2.1$$

The FID signal is maximum when a $\pi/2$ excitation pulse is used. However, one has to use higher RD in order to allow the longitudinal magnetization to relax. To make efficient use of time, lower flip angles are used in experiments involving data accumulation. For a pulse with a flip angle θ, the transverse magnetization induced is $M_0 \sin\theta$, while the z-magnetization is reduced to a value $(1- M_0 \cos\theta)$. For small flip angles ($\theta < 30\text{-}50^{\circ}$), we have $\sin\theta > 1- M_0 \cos\theta$. Thus the loss in signal intensity due to the use of a weaker pulse, is more than compensated by the shorter relaxation delay, which has to be used to allow z-magnetization to recover. The optimum pulse angle (θ) is given by $\cos\theta = \exp(-TR/T_1)$ and is called the *Ernst angle.*

7.2 Presaturation

As mentioned earlier, while recording ^1H spectra of a biomolecule, signals from mM concentrations of molecules of interest have to be picked up against the background of the large water signal. There are several water suppression schemes. One of the ways to overcome this problem is to use pre-saturation technique. Water signal is selectively saturated with low RF power, prior to the application of the $\pi/2$ acquisition pulse. The pulse sequence for such a scheme is shown in Figure 2.7 B. Though this sequence suppresses the water substantially, it also reduces the S/N of exchangeable protons to a large extent, because of the magnetization transfer from the solvent. Further, the information under the water resonance is lost because of the ridge caused by the pre-saturation.

7.3 Jump and Return (JR) Sequence

The Jump and Return (JR) pulse sequence for water-suppression uses two $\pi/2$ pulses separated by suitable time delay, $(\pi/2)_y$-τ-$(\pi/2)_{-y}$-FID (Figure 2.7 C). After the first $(\pi/2)_y$ excitation pulse, the net magnetization arising from both the sample of interest and that of the solvent is along the *x*-axis in the rotating frame. During the following time delay τ, all the spins fan out in the *xy*-plane except the solvent resonance. This is chosen at resonance at the transmitter-offset. It remains along the *x*-axis. The second $(\pi/2)_{-y}$ pulse brings back the solvent to the *z*-axis and the rest of the spins from the *x/y*-plane to positions in the *z/y*-plane. Immediately after the 2nd pulse, FID is recorded. Transverse magnetizations mainly arising from the sample of interest start along *y* on one side of the solvent peak and along *-y* on the other. The spectrum recorded with JR sequence exhibits the characteristic 180° phase difference between the two sides of the solvent peak. This experiment provides superior water suppression compared to pre-saturation experiment and is devoid of problems associated with the chemical exchange.

7.4 Spin-echo (SE)

An ingenious method was proposed by Hahn in early 1950s, to counteract the effect of inhomogeneous magnetic fields. This is called *spin-echo* (SE). SE uses two pulses separated by suitable time delays (τ). The SE pulse sequence is $(\pi/2)_y$-τ-π_x-τ- or sometimes written as $(\pi/2)_y$-TE/2-π_x-TE/2 (Figure 2.8 A). This experiment is one of the most important building blocks of 2D pulse sequences and also forms the backbone of MRI experiments. A number of aspects of multi-pulse experiments can be understood from SE sequence. In particular, the behaviour of the transverse magnetization M_{xy} as a function of transverse relaxation, inhomogeneity of the magnetic field, Larmor precession and scalar couplings are very well illustrated by SE experiments.

7.4.1 Effect of Inhomogeneity of B_0

In considering the effect of inhomogenous magnetic field (B_0), it may be recalled that T_2^* is responsible for the line-width of NMR signals. The magnetic field inhomogeneity leads to an avoidable signal loss. Consider an ensemble of a single uncoupled spin I, such that the different spins at different points in the sample experience slightly different magnetic fields. After the first $(\pi/2)_y$ excitation pulse in the SE pulse sequence, the net magnetization is brought along the *x*-axis, which evolves with time in the transverse plane. During the time τ, it moves by an angle $\Omega\tau$ away from the positive x-axis, where Ω is the offset frequency for the spin. If the magnetic field is not uniform over the sample, then some spins will move faster (offset frequency Ω_f, represented by vector *b* in Figure 2.9) and others slower (offset frequency Ω_s, vector *a*). This leads to dephasing of the coherence and hence the decay of transverse magnetization (M_{xy}). The effect of -τ-π_x-τ- sequence is primarily to avoid such de-phasing due to magnetic field inhomogenity.

Application of a π_x pulse tilts the magnetization vectors with respect to *x*-axis as shown in Figure 2.9. The vectors *b* and *a* continue to precess with frequencies Ω_f and Ω_s, respectively. After the second time period of τ, these magnetization vectors collectively end up in-phase along the *x*-axis. In the event of any field inhomogenity, all spins refocus along the *x*-axis at the end of the -τ-π_x-τ- sequence. Such build up of phase coherence in the *xy* plane leads to a signal, which is called spin-echo. The period 2τ is referred as time to echo (TE). Thus, SE pulse sequence removes the dephasing caused by the field inhomogenity. However, other processes which contribute to the decay of transverse magnetization continue to exert their influence.

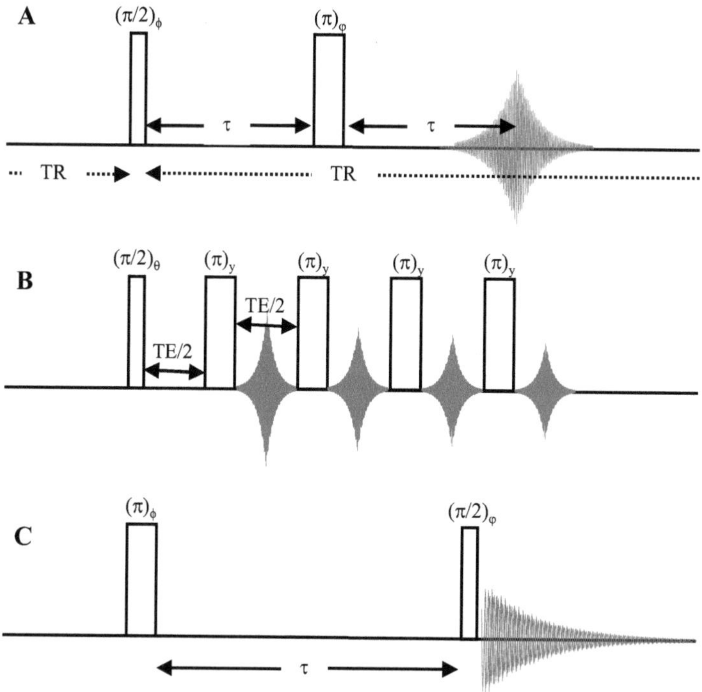

Figure 2.8: Some important multi-pulse experiments: (A) Spin-echo (SE) sequence, which can be repeated after time intervals of TR. (B) CPMG sequence; following the single π/2 excitation pulse, refocussing pulses are applied at several time intervals of TE/2 which are followed by detection of echo signals. (C) Inversion Recovery (IR) sequence.

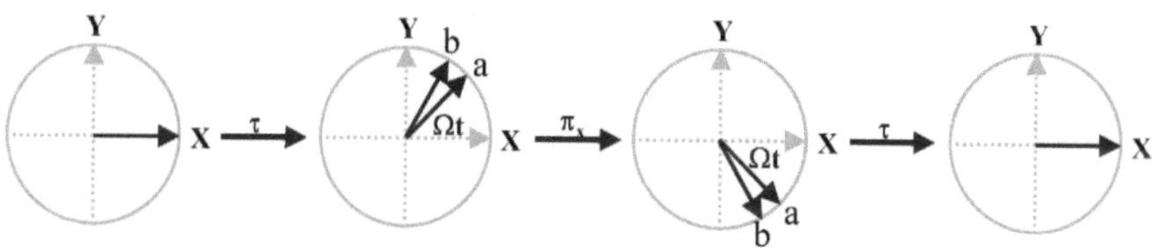

Figure 2.9: Effect of a SE sequence on an uncoupled spin with precessional frequency Ω.

7.4.2 Behaviour of J-Coupled Systems

The effect of SE sequence on chemical shifts and J-couplings is different. It refocuses chemical shifts even in the case of a coupled spin system. In fact, SE can be used to correct for frequency dependent phases of chemically shifted resonances.

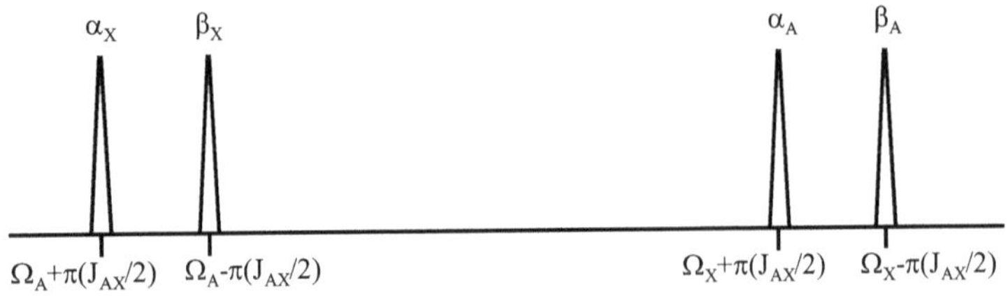

Figure 2.10: NMR spectrum of an AX spin-system. Offset frequencies are depicted below the lines and the spin states of the associated coupled partner are indicated on top.

For J coupled spin systems, SE phase modulates the J-couplings. This is an important feature of the SE sequence and the behaviour is different for homonuclear and heteronuclear cases. To understand this aspect, let us consider a two-spin system (AX), where the spin-spin coupling constant $J_{AX} \neq 0$. In the case of like spins, the spectrum of each spin is a doublet (Figure 2.11). Each line of the doublet is associated with a specific spin state of the coupled partner.

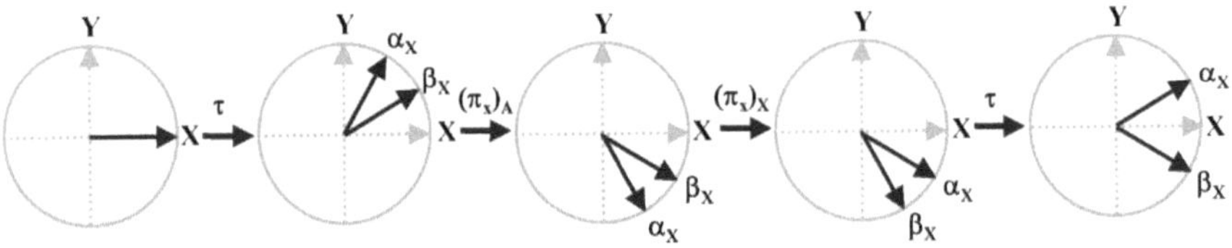

Figure 2.11: The effect of a SE sequence $(\pi/2)_y$-τ-π_x-τ on a J-coupled spin A. α_x and β_x represent the spin states of the coupling partner X.

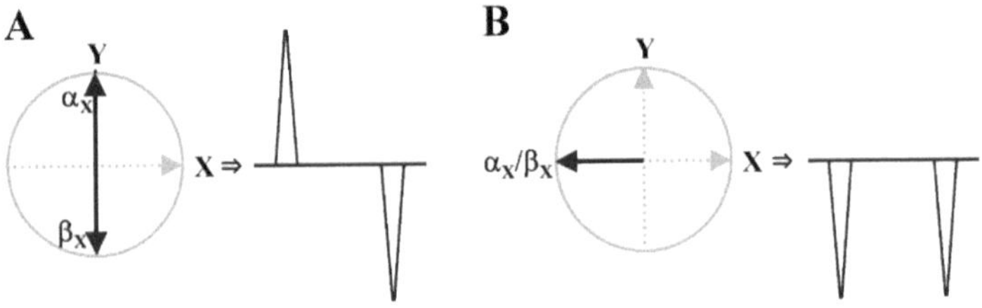

Figure 2.12: The effect of a SE sequence $(\pi/2)_y$-τ-π_x-τ on a J-coupled spin A doublet; (A) when $\tau = 1/4J_{AX}$, and (B) when $\tau = 1/2J_{AX}$. α_x and β_x represent the states of the coupling partner.

The $(\pi/2)_y$ pulse tilt the individual magnetization vectors of both A and X to x-axis for a homonuclear system. The magnetization vectors evolve during time τ in the transverse plane. For example, the vector corresponding to $\Omega_A+2\pi(J_{AX}/2)$ moves faster than the one corresponding to $\Omega_A-2\pi(J_{AX}/2)$ (Figure 2.11). At the end of the first delay time τ, application of $(\pi)_x$ pulse results in tilting the magnetization vectors with respect to x-axis. This results in the swapping of the associated spin states of the coupled partner. Hence, during the second delay time, the vectors precess with exchanged frequencies, Ω_A-$2\pi(J_{AX}/2)$ and Ω_A+$2\pi(J_{AX}/2)$. The vector associated with the spin state α_x, which was precessing with a frequency $\Omega_A+2\pi(J_{AX}/2)$ during the first time interval, acquires a frequency $\Omega_A-2\pi(J_{AX}/2)$ after the $(\pi)_x$ pulse. At end of the -τ-(π_x)-τ- sequence, the vector effectively spans an angle $[-\{\Omega_A+2\pi(J_{AX}/2)\tau\} + \{\Omega_A-2\pi(J_{AX}/2)\}\tau]=-2\pi J_{AX}\tau$. Its partner spans $[-\{\Omega_A-2\pi(J_{AX}/2)t\} + \{\Omega_A+2\pi(J_{AX}/2)\}t] = 2\pi J_{AX}\tau$.

Thus, the SE sequence in the case of homo-nuclear spin system refocuses the chemical shifts, leaving behind the phase modulation due to the J-couplings (Figure 2.11). Note the influence of the delay, when the value of τ is set to $1/4J_{AX}$ or $1/2J_{AX}$. When τ is $1/4J_{AX}$, the two frequencies align along the ±y-axes (Figure 2.12) giving rise to an *anti-phase* doublet. When τ is set to $1/2J_{AX}$, the two frequencies align along the –x-axis giving rise to an inverted *in-phase* doublet.

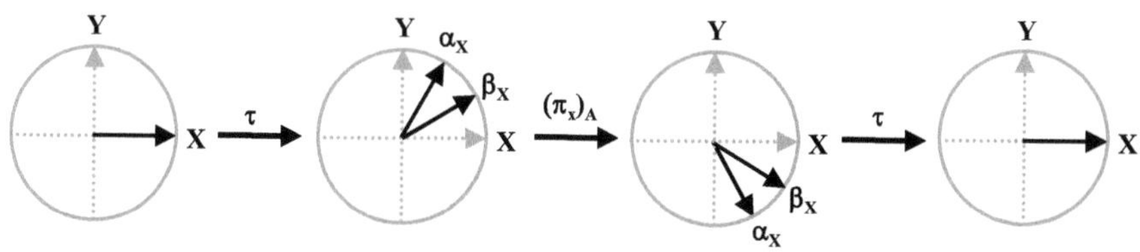

Figure 2.13: The effect of a SE sequence $((\pi/2)_y)_A$-τ-$(\pi_x)_A$-τ on a J-coupled spin A. The α_x and β_x are the spin states of the coupling partner spin X. Note that π_x is applied only on the spin A.

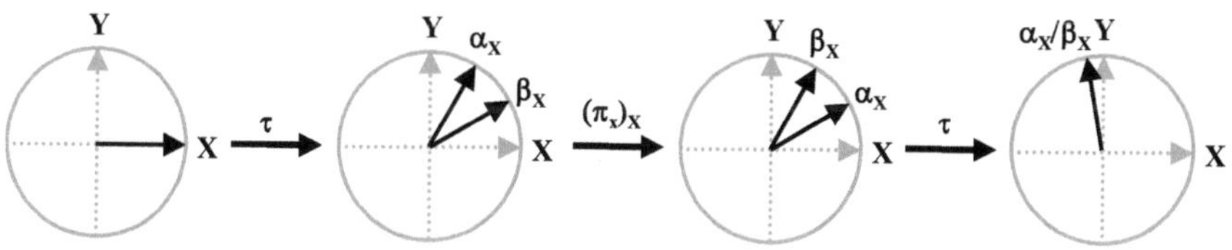

Figure 2.14: The effect of a SE sequence $((\pi/2)_y)_A$-τ-$(\pi_x)_X$-τ on a J-coupled spin A. The α_x and β_x are the spin states of the coupling partner spin X. Note that π_x is applied only on the spin X.

In the case of *unlike spins*, the effect of the SE sequence is different depending on whether the $(\pi)_x$ pulse is applied on both the spins or only on one. When the $(\pi)_x$ pulse is applied on both the spins, one ends up observing the same phase modulation as in the case of *like spins*. However, when it is applied only along A spin, then there is no phase modulation of the hetero-nuclear J_{AX} (Figure 2.13). Further, when the $(\pi)_x$ pulse is applied only to X spin, the A spin gets frequency labelled and there is no phase modulation due to the hetero-nuclear J_{AX} (Figure 2.14).

7.4.3 Applications of SE

The differences in the evolution of spin magnetization under Larmor precession and scalar couplings have been used to separate the two terms in the form of (J, δ) spectroscopy. SE sequences are extensively used in magnetic resonance imaging (MRI). In MRI experiments, the time between the excitation pulse and the echo is called time to echo (TE) and is equal to twice the delay time. TE can be chosen by choosing the time τ and thus determines the time for which the transverse magnetization is allowed to decay due to T_2. For the purpose of signal averaging, the entire sequence can be repeated after a time-interval TR (repetition time). The signal intensity S of the echo signal is given by:

$$S = C\,\rho\,M_z\,(0)\;[1 - 2\exp\{-(TR-TE/2)/T_1\} + \exp(-TR/T_2)]\;\exp(-TE/T_2) \qquad 2.2$$

C is a constant related to the electronics of the experiment and ρ is the proton density. For short echo times and $T_1 < T_2$ (which is the case for slowly tumbling molecules), this equation simplifies to:

$$S = C\,\rho\,M_z\,(0)\;[1 - \exp(-TR/T_1)\,]\;\exp(-TE/T_2) \qquad 2.3$$

There are three terms in the above equation, which depend on the properties of the sample, namely ρ, T_1 and T_2. By choosing the experimental parameters, TR and TE, one can obtain signals which are dominated by one of the three parmeters.

7.5 Carr-Purcell-Meiboom-Gill (CPMG) Sequence

CPMG is a sequence which is particularly useful for measuring T_2. In the SE experiment, as the spins are re-phased after a time period TE_1, application of a second re-phasing π pulse after a time $TE_2/2$ following the first spin-echo, is equivalent to starting the SE experiment once again. This results in another echo signal after time $TE_2/2$. The time to the second echo is therefore TE_2 (Figure 2.9 B). The signal intensity of the second echo signal is lower due to the loss of magnetization due to T_2 relaxation during the time elapsed from the original excitation. This process may be repeated several times and echo signals acquired after each delay. After n such repetitions, the echo signal will be similar to that achieved if a single experiment is performed with a time of echo TE_t, which is the sum of the TE_i used ($TE_t = \sum TE_i$). In this way, T_2 information can be obtained with several values of TE using a single excitation. CPMG sequence is used extensively in MRI as well as in studies on biological molecules.

7.6 Inversion Recovery Experiment (IR)

One often uses a pulse sequence such as $\tilde{RD}-\pi-\tau-(\pi/2)_y$ (Figure 2.9 C) to measure longitudinal (spin-lattice) relaxation time T_1, where RD is the relaxation delay. This is called the inversion-recovery pulse sequence. As the name suggests, the magnetization is first inverted with a π pulse down to the –z-axis. In the process of re-establishing the equilibrium magnetization M_z, the inverted magnetization undergoes a longitudinal relaxation, during a variable delay period τ. The $(\pi/2)_y$ read pulse rotates the longitudinal magnetization into the x-y plane, giving rise to a detectable FID. The detected signal is a function of a variable time τ. The signal intensity in an IR experiment is given by:

$$S = k\,\rho\,(1 - 2\,exp(-\tau/T_1)) \qquad 2.4$$

The signal therefore goes from negative values to positive value as the system relaxes due to T_1 during the time τ. A null signal occurs for $\tau = T_1\,ln2$. Thus, one can directly measure the value of T_1. For efficient results, several factors should be considered. One should uniformly excite all the resonances. The number of points used to collect the data should be large enough to accurately integrate the peaks. Use of long relaxation delay (RD) is strongly recommended to take care of the slowly relaxing nuclei and is usually kept at $5T_1$. IR pulse sequence has found wide applications in MRI.

8. COMPARISION OF NMR AND OTHER PHYSICAL TECHNIQUES

Several strong points in favour of NMR spectroscopy can be realized from above discussions. There are several nuclear spins in living systems, which can provide NMR signals. Each type of spin reports distinct and complementary information. There are several properties of resonance signals, which can be utilised to obtain biological information.

The RF radiations involved are low energy photons, which do not lead to ionisation or rupture of chemical bonds. In fact, NMR is a *non-invasive* technique, harmless to living systems. In clinical medicine, repeated use of the technique can be made without injury or long term harmful affects to the patient. Samples used in *in-vitro* studies can be retrieved for further studies.

NMR provides a rich arsenal of techniques, which can probe structure, dynamics, interaction, biochemical reactions *in-vivo and in-vitro* and functional aspects of molecules, cells and organs. In fact NMR covers the entire range of problems in living systems - from molecules to humans. Studies can be made in aqueous medium, which is the milieu of living systems. When used as an analytical tool, the analysis is not geared to detect a particular compound. All molecules are detected simultaneously.

9. REFERENCES

9.1 Theoretical and Physical Aspects of NMR

F. Bloch and I.I. Rabi, *Atoms in variable magnetic fields*, Rev. Mod. Phys. **17**, 237– 244 (1945).
F. Bloch, W.W. Hansen and M. Packard, *Nuclear induction*, Phys. Rev. **69**, 127–127 (1946).
F. Bloch, *Nuclear induction*, Phys. Rev. **70**, 460–474 (1946).
F. Bloch, W.W. Hansen and M. Packard, *The nuclear induction experiment*, Phys. Rev. **70**, 474–485 (1946).
E.M. Purcell, H.C. Torrey and R.V. Pound, *Resonance absorption by nuclear magnetic moments in a solid*, Phys. Rev. **69**, 37–38 (1946).
N. Bloembergen, E.M. Purcell and R.V. Pound, *Relaxation effects in NMR absorption*, Phys. Rev. **73**, 679–712 (1948).
E.L. Hahn, *Spin echoes*, Phys. Rev. **80**, 580–594 (1950).
I. Solomon, *Relaxation processes in a system of two spins*, Phys. Rev. **99**, 559–565 (1955).
R.R. Ernst and W.A. Anderson, *Application of Fourier transform spectroscopy to magnetic resonance*, Rev. Sci. Instru. **37**, 93–102 (1966).

9.2 Biological Molecules

G. Govil and R.V. Hosur, *Conformation of Biological Molecules: New Results from NMR*, Springer-Verlag, Berlin (1982).
K. Wüthrich, *NMR of Proteins and Nucleic Acids*, John Wiley & Sons, New York (1986).
B. Lewin, *Genes V*, Oxford University Press, Oxford and New York (1994).
R.W. Old and S.B. Primos, *Principles of Gene Manipulation: An Introduction to Genetic Engineering*, Blackwell Science, Inc. (1994).
J. Cavanagh, W.J. Fairbrother, A.G. Palmer and N.J. Skelton, *Protein NMR Spectroscopy: Principles & Practice*, Academic Press Inc., San Diego (1996).

9.3 Biomedical Systems

R. Damadian, *Tumor detection by nucelar magentic resonance*, Science **171**, 1151–1153 (1971).
P.G. Lauterbur, *Image formation of induced local interactions: Examples employing NMR*, Nature **242**, 190–191 (1973).
A. Kumar, D. Welti, R.R. Ernst, *NMR Fourier zeugmatography*, J. Magn. Reson. **18**, 69–83 (1975).
D.G. Gadian, *Nuclear magnetic resonance and its applications to living systems*, Oxford University Press, Oxford (1982).
P. Mansfield and P.G. Morris, *NMR Imaging in Biomedicine*, Academic Press, New York (1982).
C.E. Hayse, W.A. Edelstein, J.F. Schenck, O.M. Mueller, M. Eash, *An Efficient Highly Homogeneous Radiofrequency Coil for Whole-Body NMR Imaging at 1.5T.*, J. Magn. Reson. **63**, 622–628 (1985).
M.R. Bendall, *Surface Coil Technology*, in *Magnetic Resonance Imaging*, C.L. Partain, R.R. Price, J.A. Patton, M.V. Kulkarni, A.E. James (eds), Saunders, Philadelphia (1988).
S.R. Thomas, L.J. Busse, J.F. Schenck, *Gradient Coil Technology*, in *Magnetic Resonance Imaging*, C.L. Partain, R.R. Price, J.A. Patton, M.V. Kulkarni, A.E. James (eds), Saunders, Philadelphia (1988).
C.L. Dumoulin, S.P. Souza, M.F. Walker, W. Wagle, Magn. Reson. Med. **9**, 139–149 (1989).
P.T. Callaghan, *Principles of Nuclear Magnetic Resonance Microscopy*, Oxford University Press, Oxford (1991).
M.S. Albert, G.D. Cates, B. Driehuys, W. Happer, B. Saam, C.S. Springer Jr., A. Wishnia, *Biological magnetic resonance imaging using laser-polarized 129Xe*, Nature **370**, 199–201 (1994).
D. Le Bihan (ed.), *Diffusion and Perfusion Magnetic Resonance Imaging: Application to Functional MRI*, Raven Press, New York (1995).
R. de Graaf, *In-vivo NMR Spectroscopy: Principles and Techniques*, John Wiley & Sons (1999).
V. Kupermann, *Magnetic Resonance Imaging; Physical Principles and Applications*, Academic Press (2000).
N.R. Jagannathan (ed.), *Recent Advances in MR Imaging and Spectroscopy*, Jaypee (2005).

CHAPTER 3

MULTI-DIMENSIONAL NMR

1. INTRODUCTION

The multi-pulse experiments introduced in Chapter 2 form the basic building blocks for multi-dimensional (MD) and multi-nuclear NMR spectroscopy. The basic sequences illustrate how the evolution of the spin coherence of uncoupled and coupled spins can be controlled. Chemical shifts can be refocussed and heteronuclear couplings can be removed during their evolution by using different pulses sequences. Several such trains of pulses can be combined to obtain biological information which may not be amenable by a simple 1D NMR.

1.1 Multi-dimensional (MD) NMR

The concept of two-dimensional (2D) NMR was proposed by Jean Jeener in 1971 at a summer school. The significance of this great idea was realized when Richard Ernst experimentally showed its utility and demonstrated the potential of this method. Notable contributions later emerged from the laboratories of Ray Freeman and Kurt Wüthrich. This concept has been extended into the third dimension in late 1980s, and later to the fourth and fifth dimensions, for which substantial contributions came from several other laboratories. These techniques are collectively called as multi-dimensional (MD) NMR spectroscopy.

The MD NMR spectroscopy has provided a powerful arsenal of versatile tools for yielding information on spectral parameters, such as, δ, J, NOE, chemical exchange and dipolar couplings. Today, it is possible to conduct such experiments even by a non-expert to obtain desired information. These techniques form important tools for studying the structure, dynamics and interaction of biological macromolecules and for studies for other biological systems.

1.2 General Scheme for 2D NMR

A general 2D experiment can be characterised by four time-periods (Figure 3.1). During each time-period, one uses RF pulses, time-delay or a combination of pulses with time delay(s) depending on the need of the experiment. The complete framework of pulse sequences and time delays is collectively called the pulse sequence.

The first time-period in a pulse sequence is called the preparation period. Typically, a $\pi/2$ RF pulse may be applied during this period to create transverse magnetization. When a pulse sequence is applied repeatedly for achieving time averaging, a relaxation delay (RD) precedes the $\pi/2$ RF pulse to allow the system to regain thermal equilibrium from the perturbations caused to the spin-system by the previous pulse sequence. A decoupler pulse may be applied to saturate the solvent signal during this period.

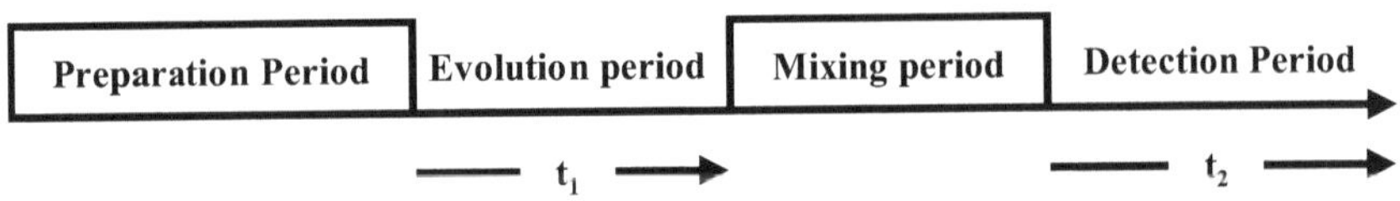

Figure 3.1: Schematic representation of various time-periods in a typical 2D NMR experiment.

During the evolution period (t_1), the transverse magnetization is allowed to precess under the free Hamiltonian ($\mathcal{H}$ in the absence of RF; comprising of the terms $\mathcal{H}_\delta$ and $\mathcal{H}_J$ in solution state). Additional perturbation may be imposed by the spectroscopist. During this period, each spin is characterized by its specific precessional frequency and acquires a specific phase. The evolution time t_1 is incremented in steps of Δt_1. Such incremental delays constitute the second time-dimension (this is the new one; though it is represented by t_1), often called the indirect dimension.

During the mixing period, transfer of magnetization (coherence transfer) takes place between the spins which are coupled to each other (through J coupling, dipole-dipole interaction or chemical exchange). In other words, spins are made to interact or mix during the mixing time.

For each t_1 incremental period, a separate FID is acquired, as a function of time t_2, during the detection period. Each FID thus collected, reflects the history of the spin system as a function of t_1. If there are *n-1* incremental periods (Δt_1) in the period t_1, then *n* FIDs are acquired as a function of t_2 (Figure 3.2 A) during the detection period. The value of the maximum t_1 (t_{1max}) should be less than T_2, since the transverse magnetization decays with a time-constant T_2.

Thus, one acquires a 2D dataset $S(t_1, t_2)$ consisting of detected FID signal as a function of two time variables. One of them is the direct dimension t_2, and the other one is a function of an indirect dimension t_1. A double-Fourier transformation of such a data set provides the 2D NMR spectrum with two frequency domains Ω_1 and Ω_2. One first obtains FT with respect to t_2 to obtain $S(t_1, \Omega_2)$. A second FT with respect to t_1 then gives the spectrum in two frequency dimensions $S(\Omega_1, \Omega_2)$. Thus, the signals in a 2D spectrum are characterized by a pair of frequencies (Ω_1, Ω_2). Depending on the pulse sequence used, the two frequencies may correspond to *like* spins (*homonuclear* 2D) or *unlike* spins (*heteronuclear* 2D). In a homonuclear 2D spectrum, the diagonal peaks ($\Omega_1 = \Omega_2$) correspond to the 1D NMR experiment. The off-diagonal peaks (commonly called the *cross peaks*) have important information about interactions between different spins (Figure 3.2 B and C).

A wide range of 2D methods have been developed. These are based on the physical mechanisms mediating interactions between nuclear spins: e.g. J couplings, dipole-dipole interactions or chemical exchange. Thus, different interactions between the spins can be selectively picked up in a 2D spectrum by a careful choice of using appropriate pulse sequences.

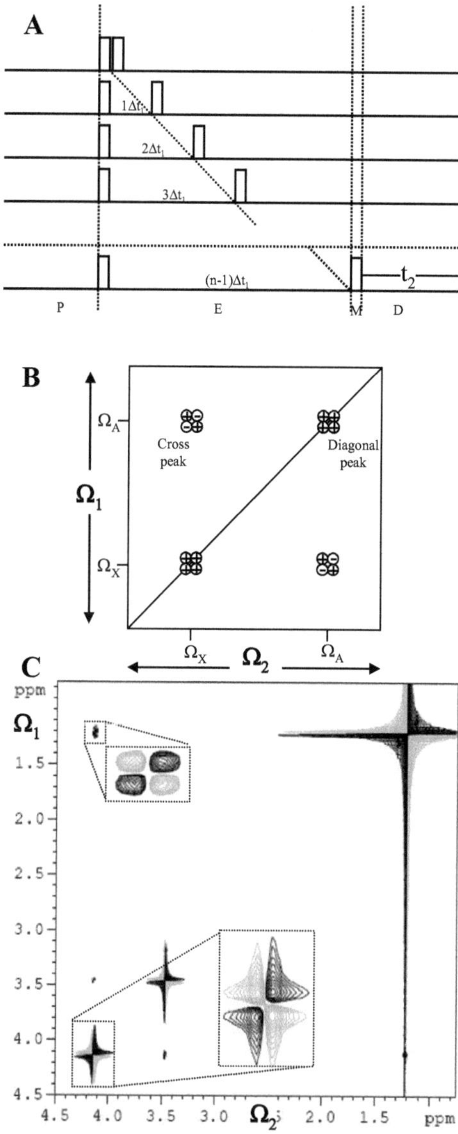

Figure 3.2: (A) The concept of a simple 2D FT NMR spectroscopy. (B) Schematic 2D COSY spectrum in the case of an AX spin system. (C) Contour plots of the 2D COSY spectrum of threonine taken in ^2H_2O. Negative and positive components are shown in grey and black.

While recording homonuclear 2D NMR spectra of macromolecules, one typically uses 512 incremental steps (n_1) during the evolution period. If each FID involves an acquisition of 2048 points (n_2) in the t_2 domain, then $S(t_1, t_2)$ is a data matrix of 512×2048 points. Even though each experiment may take only a few msec, generally a relaxation delay of about 1 sec is required between each scan. Therefore, if 4 scans are used for signal averaging, then the total time taken for a 2D homonuclear experiment such as a 2D [^{1}H-^{1}H] COSY is 4×512 sec or about half an hour. For heteronuclear experiments such as [^{15}N-^{1}H] correlations n_1 is usually limited to 64 to 128 steps. Therefore such 2D experiment can be acquired in a few minutes.

The outcome of a 2D experiment is a 3D stacked plot, where the n_1 spectra are stacked on one top of another. A more convenient method to represent a 2D spectrum is through contour plots, where a section of the 3D stacked plot at conveniently chosen amplitude is chosen (Figure 3.2 C). This cut-off amplitude is selected such that one does not miss cross peaks of interest and all the spectral features are above the noise level.

1.3 COrrelated SpectroscopY (COSY)

The simplest form of a 2D NMR experiment consists of a sequence of two $\pi/2$ pulses which are separated by the evolution time t_1 (Figure 3.3). The frequency axes contain chemical shifts and the cross peaks arise due to *J-couplings*. This experiment is popularly known as COSY or the Jeener experiment. The COSY has been discussed in some detail as it forms the most basic form of more complex MD NMR spectra.

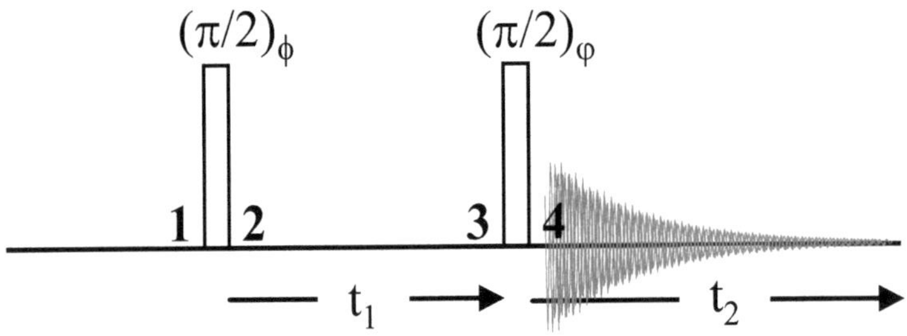

Figure 3.3: Pulse sequence for the 2D COSY experiment.

Let us look at the phase sensitive COSY spectrum for a 2-spin homonuclear AX system (Figure 3.2 B-C). The diagonal peaks are centred at (Ω_A, Ω_A) and (Ω_X, Ω_X). The off-diagonal or the cross peaks are at (Ω_A, Ω_X) and (Ω_X, Ω_A). Each of these peaks has four components arising from J_{AX}. The individual components of diagonal peaks are pure dispersive while the cross peaks are pure absorptive. The reasons for this kind of phase behavior are discussed later in this Chapter.

The information provided by COSY is similar to a 1D decoupling experiment where A and X are decoupled. However, the beauty of COSY spectrum lies in the fact that in a single experiment, the whole network of coupled spins can be analysed.

The COSY spectrum has several problems. The most common artefact in any multi-dimensional spectrum is the appearance of the so called axial peaks. These signals arise from the transfer of *z*-magnetization by the mixing sequence into observable *x-/y*-magnetization. Such a transfer gives rise to peaks parallel to Ω_2 axis at $\Omega_1 = 0$. These peaks can be removed using *phase cycling* in the pulse sequences, which are discussed in Section 3.2. Another drawback is the appearance of what is called t_1 *noise*. Instrumental instabilities in RF generation and field-frequency lock during the incremental periods in the t_1 time-domain cause a noise-like modulation of strong signals such as that from the solvent. Special precautions are taken in modern instruments to minimize such instabilities.

It will be shown later that the cross peak intensity depends on the magnitude of J_{AX} and is proportional to $\sin(\pi J_{AX} t_1)\sin(\pi J_{AX} t_2)$. The dispersive and in-phase nature of individual components of diagonal peaks leads to long tails near the diagonal and masks cross peaks which are close to the diagonal (Figure 3.2 C). Further, it results in very large ratio for the intensities of diagonal to cross peak. The spectra show an asymmetry with respect to the diagonal, since the number of data points in the two dimensions are not the same. Such problems can be overcome simply by editing techniques such as symmetrization.

Some of these major drawbacks of the most basic form of 2D spectrum, are also common to several other MD spectra.

1.4 Advantages of 2D NMR

With the above introduction, the advantages of 2D spectroscopy can be readily conceived. In fact, the advent of 2D spectroscopy has resolved several problems associated with resolution, solvent suppression and assignments, which one encounters while dealing with large molecules. The advantages are summarized below and will become more obvious as one proceeds further.

(i) Since the information is spread in two dimensions, the chemical shifts get resolved by the cross peaks, even when one of the two chemical shifts is in a crowded region suffering from spectral overlaps.

(ii) Correlation parameters between all the nuclear spins such as J couplings, NOE and chemical exchange can be obtained in a single experiment. In turn, such correlations provide information on the structure and dynamics of biological molecules.

(iii) Resonances from large biopolymers can be grouped into separate networks of coupled spin systems, belonging to individual monomeric units. This serves as a prelude to sequence specific resonance assignments.

(iv) MD NMR spectroscopy enables an indirect detection of the forbidden fruits of spectroscopy, namely the *multiple-quantum (MQ)* transitions. The *MQ* coherences are usually excited by RF pulses or specific combinations of RF pulses and pulse delays. Such coherences are frequency labelled during the indirect detection time

and then transformed into observable *single-quantum (SQ) coherences* for detecting the signal, during the detection period.
(v) Heteronuclear correlations between ^{1}H and other spins ($^{13}C/^{15}N/^{31}P$) can be established for indirect detection of the later nuclei. This helps in correlating sequential monomeric units in biopolymers based on J-couplings and provides an efficient method for sequence specific resonance assignments.
(vi) Solvent resonance, such as that from water and unwanted signals can be filtered, suppressed or tamed.
(vii) Spin decoupling can be achieved efficiently to simplify spectra with the state of the art probes.
(viii) One can carry out *protonless* homo- and hetero-nuclear correlation experiments, which form an alternate method for assignments in biopolymers.
(ix) In presence of more than one chemical species, each one can be characterized individually.
(x) It has helped in the development of techniques such as magnetic resonance imaging and spectroscopy.

In this chapter, the basic principles of some of the important and most commonly used MD techniques have been discussed. The principles behind the MD NMR spectroscopy are difficult to understand. The theoretical treatment is complicated. This is particularly so, for strongly coupled spin systems.

The theoretical analysis of such spectroscopy has been presented through *Product Operator Formalism* (POF). The principles have been illustrated through practical examples. The mathematical treatment of pulse sequences has been minimized to what is essential for understanding how each magnetization evolves under the effect of multi-pulse sequences. Readers whose intentions are to understand the physics behind each pulse sequence, may work out the details of the evolution of spin-coherences using the formalism described.

2. PRODUCT OPERATOR FORMALISM (POF)

There are three ways of analyzing the outcome of pulse experiments. The simplest is the vector model introduced for the SE sequence in Chapter 2. The vector model fails to provide understanding in the case of experiments on molecules having more than two spins.

The second approach is based on the density matrix. The density matrix approach is a powerful formalism. However, it is more suitable for use on computers. To know the extent of transverse magnetization, at any particular time during a pulse sequence, one needs to take the trace of the result obtained by matrix multiplication with the density matrix, ρ. However, this is not trivial when the number of spins is more than two and the pulse sequence involves more than 3-4 pulses. Note that the number of interacting NMR spins (active spins; N) essentially dictates the number of elements ($2^N \times 2^N$) in the corresponding density matrix. For example, in the case of a two spin-½ system, the density matrix size is $2^2 \times 2^2$, while the matrix size is $2^3 \times 2^3$ for a three-spin system.

The product operator formalism (POF), developed by Ernst and his group, is widely used to explain the outcome of MD spectroscopy. It is based on the algebra of spin matrices described below. Using this formalism, one can get an insight into a NMR experiment, irrespective of the length and the number of pulses and the delays involved. However, the formalism is limited to weakly coupled spin systems. This formalism is first described and then used it to analyze some of the 2D and 3D NMR experiments, which are used more frequently in studies of biological molecules.

2.1 Product Operators (PO)

The physical significance of the various single-spin representations used in POF can be explained by considering a set of normalized product operators appropriate for the spin system under consideration. The density operator may be expressed as a linear combination of base operators, represented by B_S. For a non-interacting spin-½ system, the set $\{B_S\}$ is essentially the Cartesian operators I_x, I_y and I_z together with an identity matrix, ½E (half the unit operator). The operators I_x and I_y represent the transverse (x- and y-) magnetization of the spin *I* while I_z represents its longitudinal magnetization. For a spin system with two or more interacting spin-½ nuclei, the set $\{B_S\}$ is more complex. One needs to consider all products of such operators. For N number of interacting spins, the set {Bs} contains 4^N product operators.

For a two spin-½ nuclei AX spin system (each of which is identified by symbols k and l), there are 16 product operators: ½E, I_{kx}, I_{ky}, I_{kz}, I_{lx}, I_{ly}, I_{lz}, $2I_{kx}I_{lx}$, $2I_{kx}I_{ly}$, $2I_{kx}I_{lz}$, $2I_{ky}I_{lx}$, $2I_{ky}I_{ly}$, $2I_{ky}I_{lz}$, $2I_{kz}I_{lx}$, $2I_{kz}I_{ly}$, $2I_{kz}I_{lz}$. Each of these operators can be written in a matrix form as given in Appendix 3.1.

It is important to note that two-spin matrices are created from the single spin matrices by performing a cross-product (⊗). The outcome of a pulse experiment can be calculated simply by matrix maipulations. For example,

$$I_{kx}=2\, I_{kx} \otimes \tfrac{1}{2} E = 2 \times \frac{1}{2}\begin{pmatrix} 0 & 1 \\ 1 & 0 \end{pmatrix} \otimes \frac{1}{2}\begin{pmatrix} 1 & 0 \\ 0 & 1 \end{pmatrix} = \frac{1}{2}\begin{pmatrix} 0 & 0 & 1 & 0 \\ 0 & 0 & 0 & 1 \\ 1 & 0 & 0 & 0 \\ 0 & 1 & 0 & 0 \end{pmatrix} \qquad 3.1$$

The product operators I_{kz} and I_{lz} represent z-magnetizations of spins k and l, respectively, while $2I_{kz}I_{lz}$ represents longitudinal two-spin order. I_{kx} and I_{ky} are called in-phase x- and y-magnetizations of spin k, respectively, and represent transverse magnetizations. While I_{kx} and I_{lx} give rise to absorptive in-phase doublets, I_{ky} and I_{ly} give rise to dispersive in-phase doublets. These are schematically shown in Figure 3.4.

2.2 Coherences

The concept of *coherence* plays an important role in pulse NMR. The term describes all possible mechanisms, whereby spin populations between different states can be exchanged. Generally, in any NMR experiment (1D, 2D, 3D, ...) one observes transitions only between states with $\Delta m = \pm 1$. These are called *single-quantum (SQ) coherences.*

In general, coherences are classified according to the difference in the *magnetic-quantum number* (m_I) associated with the two connected states. Coherences connecting states for which $\Delta m = \pm 1$ are called *single-quantum coherences* (with the associated *coherence order,* $p=\pm 1$). Coherences which connect states for which $\Delta m = 0, \pm 2, \pm 3, \ldots \pm p$ are called *multiple-quantum coherences (MQC)* or more explicitly, *zero-quantum* (p=0), *2-quantum* (p=2), *3-quantum* (p=3) and *p-quantum coherence.*

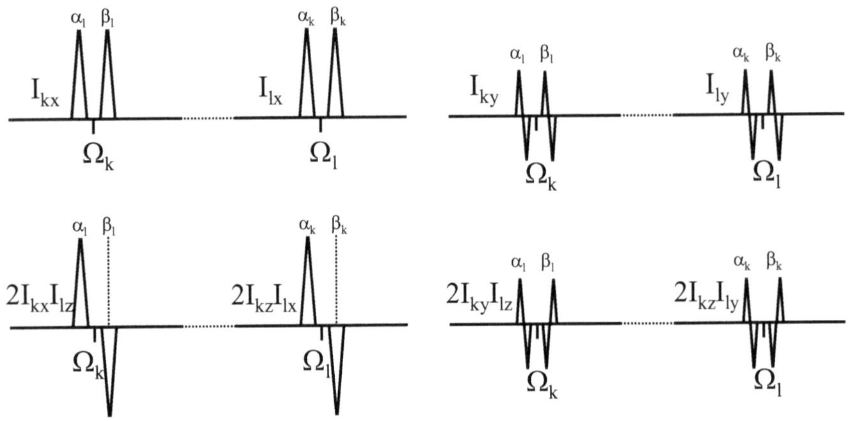

Figure 3.4: Line shapes of single-quantum ($\Delta m = 1$) coherences for a J-coupled AX system.

The number of coherences associated with various spin systems is listed in Table 3.1. For example, $I_{kx/lx}$ and $I_{ky/ly}$ are *single-quantum coherences* ($\Delta m = \pm 1$), and represent off-diagonal elements in the density matrix.

Table 3.1: Number of coherences associated with various spin systems.

Spin System	Zero-QC	Single-QC	2-QC	3-QC	4-QC
Single	0	1	0	0	0
Two	1	4	1	0	0
Three	6	12	6	1	0

Coming back to the product operators, terms such as $2I_{kx}I_{lz}$, $2I_{ky}I_{lz}$, $2I_{kz}I_{lx}$ and $2I_{kz}I_{ly}$ represent anti-phase magnetization. For example, $2I_{kx}I_{lz}$ is called as the x magnetization of the spin k that is anti-phase with respect to the spin l. They also represent transverse magnetizations. The terms $2I_{kx}I_{lz}$ and $2I_{kz}I_{lx}$ give rise to absorptive anti-phase doublets, while $2I_{ky}I_{lz}$ and $2I_{kz}I_{ly}$ give rise to dispersive anti-phase doublets (Figure 3.4). Anti-phase magnetization can occur if the two spins are coupled by scalar (J) coupling. The other four product operators ($2I_{kx}I_{lx}$, $2I_{kx}I_{ly}$, $2I_{ky}I_{lx}$ and $2I_{ky}I_{ly}$) represent superposition of *zero-* and *double-quantum* coherences. The terms ½($2I_{kx}I_{lx}+2I_{ky}I_{ly}$) and ½($2I_{ky}I_{lx}-2I_{kx}I_{ly}$) represent *x-* and *y-* components of pure *zero-quantum* coherences, respectively, while ½($2I_{kx}I_{lx}-2I_{ky}I_{ly}$) and ½($2I_{ky}I_{lx}+2I_{kx}I_{ly}$) represent *x-* and *y-* components of pure *double-quantum* coherences, respectively.

2.3 Observable Coherences

Observable coherences are those which are detected by the receiver in the t_2 time-domain. The density matrix can be written as a combination of a set of operators:

$$\rho(\tau) = a(t)I_x + b(t)I_y + c(t)I_z \qquad 3.2$$

where the coefficients *a, b* and *c* depend on time. The expectation value of an operator A is obtained by using the relation:

$$<A> = Tr[\rho A] = Tr[(a(t)I_x + b(t)I_y + c(t)I_z)A] = Tr[a(t)I_xA + b(t)I_yA + c(t)I_zA] \qquad 3.3$$

It turns out that $Tr[I_pI_q]$ is zero unless p = q, when the trace is = ½.

For example, $Tr[I_xI_x] = ½$; *while* $Tr[I_yI_x] = Tr[I_zI_x] = 0$

The following rules can be derived from such considerations.
(i) POs containing a single transverse component I_x or I_y lead to observable signals. (for example, one spin operators I_{kx} and I_{ky})
(ii) POs with more than one transverse component (for example $2I_{ky}I_{lx}$) lead to multiple quantum coherences and are not observed.
(iii) POs with one transverse component and rest of the components as z-terms, lead to signals with anti-phase character (for example, $2I_{kx}I_{lz}$ and $3I_{kx}I_{lz}I_{mz}$).

2.4 Effect of RF pulses on Product Operators

During a pulse experiment, the spin operators evolve under the influence of pulses and the free Hamiltonian. The effect of a RF pulse with a flip angle β, applied along the *x* and *y* axes, on various product operators are governed by the following rules.

$$I_{kz} \xrightarrow{(\beta)x} I_{kz}\cos\beta - I_{ky}\sin\beta;\ I_{ky} \xrightarrow{(\beta)x} I_{ky}\cos\beta + I_{kz}\sin\beta;\ I_{kx} \xrightarrow{(\beta)x} I_{kx} \qquad 3.4$$

$$I_{kz} \xrightarrow{(\beta)y} I_{kz}\cos\beta + I_{kx}\sin\beta;\ I_{ky} \xrightarrow{(\beta)y} I_{ky};\ I_{kx} \xrightarrow{(\beta)y} I_{kx}\cos\beta - I_{kz}\sin\beta \qquad 3.5$$

2.5 Evolution of Product Operators under free Hamiltonian

For weakly-coupled spin systems in solutions, various product operators evolve during a time τ under the time-independent Hamiltonian $\mathcal{H}$ consisting of $\mathcal{H}_\delta + \mathcal{H}_J$, where $\mathcal{H}_\delta = \Omega_k I_{kz} + \Omega_l I_{lz}$ and $\mathcal{H}_J = 2\pi J_{kl} I_{kz} I_{lz}$. Note that the terms in the Hamiltonian commute with one another. Hence, one can follow the evolution caused by $\mathcal{H}_\delta$ and $\mathcal{H}_J$ separately and in any order.

For a two-spin system, the evolution under $\mathcal{H}_\delta$ is given by:

$$I_{kz} \rightarrow I_{kz};\ I_{ky} \rightarrow I_{ky} \cos \Omega_k \tau - I_{kx} \sin \Omega_k \tau;\ I_{kx} \rightarrow I_{kx} \cos \Omega_k \tau + I_{ky} \sin \Omega_k \tau \qquad 3.6$$

$$2I_{ky}I_{lz} \rightarrow 2I_{ky}I_{lz} \cos \Omega_k \tau - 2I_{kx}I_{lz} \sin \Omega_k \tau;\ 2I_{kx}I_{lz} \rightarrow 2I_{kx}I_{lz} \cos \Omega_k \tau + 2I_{ky}I_{lz} \sin \Omega_k \tau \qquad 3.7$$

For evolution under $\mathcal{H}_J$, one can write,

$$I_{kx} \rightarrow I_{kx} \cos \pi J_{kl} \tau + 2I_{ky}I_{lz} \sin \pi J_{kl} \tau;\ I_{ky} \rightarrow I_{ky} \cos \pi J_{kl} \tau - 2I_{kx}I_{lz} \sin \pi J_{kl} \tau \qquad 3.8$$

$$2I_{kx}I_{lz} \rightarrow 2I_{kx}I_{lz} \cos \pi J_{kl} \tau + I_{ky} \sin \pi J_{kl} \tau;\ 2I_{ky}I_{lz} \rightarrow 2I_{ky}I_{lz} \cos \pi J_{kl} \tau - I_{kx} \sin \pi J_{kl} \tau \qquad 3.9$$

2.6 Effect of Composite π pulses on Product Operators

A sequence of composite π pulses in sequences such as MLEV, WALTZ and GARP, is used to achieve *broadband decoupling* (see Section 5.5 below), primarily to overcome the problems of π pulse imperfections and resonance-offset effects. The first composite π pulse, $(\pi/2)_x$-$(\pi)_y$-$(\pi/2)_x$ (which is usually represented as R), is used for spin inversion and needs relatively much lower RF power. This shows better offset independence than a simple π pulse. This can be appreciated by looking at the effect of such a composite pulse R on the longitudinal magnetization I_{kz}:

$$I_{kz} \xrightarrow{(\pi/2)_x} -I_{ky} \xrightarrow{(\pi)_y} -I_{ky} \xrightarrow{(\pi/2)_x} -I_{kz} \qquad 3.10$$

Thus, a sequence of R pulses causes the longitudinal magnetization to oscillate between I_z and $-I_z$, causing spin inversion. Such sequence of pulses keep the decoupled spin (^{1}H) states α and β equally populated. The spins are flipped rapidly between the two states, such that the X-spins do not feel the presence of the spin H. Thus, the entire X-spectrum is decoupled from ^{1}H. Such decoupling is called *broadband decoupling*.

The *broadband decoupling* requires repeated application of composite pulses. A small imperfection can lead to an overshoot or undershoot with respect to the expected ideal total rotation. Such cumulative effects are more pronounced as one move away from the centre of the spectrum. These are called resonance-offset effects. Such offset effects are minimized by using the pulse sequence known as MLEV-4, which consist of a sequence of 4 composite pulses in the following manner: $R\bar{R}\bar{R}R$. Here $\bar{R}$ represents the same composite pulse as R with an inverted phase $[(\pi/2)_{-x}\text{-}(\pi)_{-y}\text{-}(\pi/2)_{-x}]$. Such a sequence causes the same rotation as that of R but in the opposite sense. Cyclic permutation of one of the elements of MLEV-4 does not affect its performance. Hence, one can use other MLEV-4 sequences such as $\bar{R}\bar{R}RR$, $\bar{R}RR\bar{R}$ and $RR\bar{R}\bar{R}$. Sequences $\bar{R}\bar{R}RR$ and $RR\bar{R}\bar{R}$ are complementary and so are $R\bar{R}\bar{R}R$ and $\bar{R}RR\bar{R}$. This helps in constructing *supercycles* such as MELV-8 ($RR\bar{R}\bar{R}$ $\bar{R}\bar{R}RR$ and $R\bar{R}\bar{R}R$ $\bar{R}RR\bar{R}$) and MLEV-16 ($R\bar{R}\bar{R}R$ $\bar{R}\bar{R}RR$ $\bar{R}RR\bar{R}$ $RR\bar{R}\bar{R}$). One can even conceive of sequences such as MLEV-64, following similar cyclic permutation rules.

2.7 Evolution of MQ Coherences

While discussing product operators in the case of a 2-spin system, $2I_{kx}I_{lx}$, $2I_{kx}I_{ly}$, $2I_{ky}I_{lx}$ and $2I_{ky}I_{ly}$ were identified as superposition of *zero-* and *double-quantum* coherences. The terms $\frac{1}{2}(2I_{kx}I_{lx}+2I_{ky}I_{ly})$ and $\frac{1}{2}(2I_{ky}I_{lx}-2I_{kx}I_{ly})$ represent *x-* and *y-* components of pure *zero-quantum* coherence, respectively, while $\frac{1}{2}(2I_{kx}I_{lx}-2I_{ky}I_{ly})$ and $\frac{1}{2}(2I_{ky}I_{lx}+2I_{kx}I_{ly})$ represent *x-* and *y-* components of pure *double-quantum* coherence, respectively. Using the table in Appendix 3.1 one can show that

$$\tfrac{1}{2}(2I_{kx}I_{lx}-2I_{ky}I_{ly})=\frac{1}{2}\begin{pmatrix}0&0&0&1\\0&0&0&0\\0&0&0&0\\1&0&0&0\end{pmatrix} \text{ and } \tfrac{1}{2}(2I_{ky}I_{lx}+2I_{kx}I_{ly})=\frac{1}{2}\begin{pmatrix}0&0&0&-i\\0&0&0&0\\0&0&0&0\\i&0&0&0\end{pmatrix} \qquad 3.11$$

These represent *x-* and *y*-components of pure *double-quantum* coherence, respectively.

2.8 Evolution of Zero Quantum (ZQ) Coherences

The evolution of *ZQ coherence* under $\mathcal{H}_\delta$ is given by:

$$\{ZQC\}_x \rightarrow \{ZQC\}_x \cos(\Omega_k+\Omega_l)t_1 + \{ZQC\}_y \sin(\Omega_k+\Omega_l)t_1 \qquad 3.12$$

$$\{ZQC\}_y \rightarrow \{ZQC\}_y \cos(\Omega_k+\Omega_l)t_1 - \{ZQC\}_x \sin(\Omega_k+\Omega_l)t_1 \qquad 3.13$$

Where, $\{ZQC\}_x = \frac{1}{2}(2I_{kx}I_{lx}+2I_{ky}I_{ly})$ and $\{ZQC\}_y = \frac{1}{2}(2I_{ky}I_{lx}-2I_{kx}I_{ly})$ 3.14

2.9 Evolution of Two Quantum Coherences

The evolution of *2Q coherence* under $\mathcal{H}_\delta$ is given by:

$$\{2QC\}_x \rightarrow \{2QC\}_x \cos(\Omega_k+\Omega_l)t_1 + \{2QC\}_y \sin(\Omega_k+\Omega_l)t_1 \qquad 3.15$$
$$\{2QC\}_y \rightarrow \{2QC\}_y \cos(\Omega_k+\Omega_l)t_1 - \{2QC\}_x \sin(\Omega_k+\Omega_l)t_1 \qquad 3.16$$

Where, $\{2QC\}_x = ½(2I_{kx}I_{lx}\text{-}2I_{ky}I_{ly})$ and $\{2QC\}_y = ½(2I_{ky}I_{lx}+2I_{kx}I_{ly})$

For evolution under $\mathcal{H}_J$, $\{2QC\}_{x/y}$ remain as $\{2QC\}_{x/y}$, implying that the *MQ coherences* are invariant under the coupling Hamiltonian $\mathcal{H}_J$.

3. HOMONUCLEAR CORRELATION SPECTROSCOPY

In this section some of the common homonuclear experiments have been described. The result of each experiment has been analysed using POF.

3.1 Correlation Spectroscopy (COSY)

This experiment has been described in some detail in Section 1.1. One can use POF to calculate the outcome of the COSY experiment. Consider a *homonuclear* two-spin AX system (spins labelled as *k* and *l*) and calculate the product operator term (ρ) at various time-points (*1, 2, 3* and *4* in Figure 3.3) during the COSY pulse sequence. Prior to the application of the first $(\pi/2)_x$ pulse, ρ_1 at time-point *1* in Figure 3.3 is represented by:

$$\rho_1 = I_{kz} + I_{lz} \qquad 3.17$$

The $(\pi/2)_y$ pulse tilts the magnetization of both spins *k* and *l* to x-axis. Consequently,

$$\rho_2 = I_{kx}+I_{lx} \qquad 3.18$$

During the evolution time t_1, ρ_2 evolves in the transverse plane under $\mathcal{H}_\delta$ and $\mathcal{H}_J$. One may consider only the term I_{kx} since the evolution of both I_{kx} and I_{lx} is identical. Under $\mathcal{H}_\delta$, ρ_2 evolves as:

$$I_{kx}\cos(\Omega_k t_1) + I_{ky}\sin(\Omega_k t_1) \qquad 3.19$$

Further, because of evolution under $\mathcal{H}_J$ one ends up with ρ_3, which is given by:

$$\rho_3 = [I_{kx}\cos(\pi J_{kl}t_1)+2I_{ky}I_{lz}\sin(\pi J_{kl}t_1)]\cos(\Omega_k t_1)+[I_{ky}\cos(\pi J_{kl}t_1)\mathbf{-2I_{kx}I_{lz}}\sin(\pi J_{kl}t_1)]\sin(\Omega_k t_1) \qquad 3.20$$

The second non-selective $(\pi/2)_y$ pulse transforms ρ_3 into ρ_4, which is given by:

$$\rho_4 = [-I_{kz}\cos(\pi J_{kl}t_1)+2I_{ky}I_{lx}\sin(\pi J_{kl}t_1)]\cos(\Omega_k t_1)+[\mathbf{I_{ky}\cos(\pi J_{kl}t_1)+2I_{kz}I_{lx}\sin(\pi J_{kl}t_1)}]\sin(\Omega_k t_1) \qquad 3.21$$

The transformation which is of interest in detecting the signal is shown in bold. This is the transformation of the anti-phase *x*-magnetization of spin *k* ($-2I_{kx}I_{lz}$) into anti-phase *x*-magnetization of the spin *l* ($-2I_{kz}I_{lx}$). The term implies that the magnetization of spin *k* has been transferred to spin *l*. This process is called *magnetization* or *coherence transfer*.

The first term represents in-phase *y*-magnetization of spin *k*. This term, $I_{ky}\cos(\pi J_{kl}t_1)\sin(\Omega_k t_1)$ is modulated with Ω_k frequency during the period t_1 and continues to precess with the same frequency during t_2. Hence, it gives rise to a diagonal peak. The term corresponds to an in-phase *k* spin dispersive-doublet centered at $\omega_2 = \Omega_k$. Its modulation coefficient $\cos(\pi J_{kl}t_1)\sin(\Omega_k t_1)$, can be expanded as $[\sin(\Omega_k+\pi J_{kl})t_1 + \sin(\Omega_k-\pi J_{kl})t_1]$. This describes the nature of the multiplicity along the ω_1 axis. This again results in an in-phase dispersive doublet, centered at $\omega_1 = \pm\Omega_k$.

The second term, $2I_{kz}I_{lx}\sin(\pi J_{kl}t_1)\sin(\Omega_k t_1)$ is also modulated with Ω_k during the time-period t_1. This term represents *x*-magnetization of spin *l* anti-phase with respect to *k*. It starts precessing with Ω_l frequency during t_2, and gives rise to a cross peak. $2I_{kz}I_{lx}$ corresponds to an anti-phase absorptive-doublet centered at $\omega_2 = \Omega_l$. Its modulation coefficient $\sin(\pi J_{kl}t_1)\sin(\Omega_k t_1)$, can be expanded as $[-\cos(\Omega_k+\pi J_{kl})t_1+\cos(\Omega_k-\pi J_{kl})t_1]$. This results as an anti-phase absorptive-doublet centered at $\omega_1 = \pm\Omega_k$. Thus, the individual components of diagonal peaks are pure dispersive and in-phase, while those of the cross peaks are pure absorptive and anti-phase along both the dimensions of the 2D spectrum.

Because of the $\sin(\pi J_{kl}t_1)$ dependence, the intensity of the cross-peak depends on the magnitude of the scalar coupling (J_{kl}). Larger coupling leads to better cross-peak intensity.

3.2 Phase cycling

Phase cycling plays an important role in MD NMR. Besides suppressing artefacts, it helps to choose signals with the desired order of coherences. In modern spectrometers, the transmitter and receiver phases can be selected according to the needs of the experiment. It is therefore possible to choose signals of interest by a proper choice of the receiver phase.

Phase cycling in a pulse sequence is used to retain magnetization of interest and simultaneously suppress undesirable magnetizations or artefacts. Undesirable signals arising from imbalances in spectrometer hardware, coherent noise and other artefacts generated during the multi-pulse experiments can be suppressed by phase-cycling. In general, a pulse sequence is repeated by changing (or incrementing in specific steps) the phases of a specific pulse or a set of pulses and that of the receiver, while keeping the delays and flip-angles unaltered.

Phase alternating pulse sequence (PAPS) is one of the first examples of phase cycling. The experiment consists of two transients during which the RF pulse is alternated along the opposite axes. For example, if for the 1st transient, the RF pulse is applied along Y axis, then for the 2nd transient it is applied along -Y. In such a sequence, the magnetization results on opposite axes +X and –X, respectively. For signal averaging the second signal is subtracted from the first (Table 3.2). Such a manipulation results in the co-addition of the signals of interest. Artefacts whose phases have not been modified between the two transients, cancel out.

Table 3.2: The PAPS phase cycling.

Transient	*Transmitter (Pulse) Phase*	*Receiver Phase*
1	*Y*	*X*
2	*-Y*	*-X*

3.2.1 CYCLOPS

CYCLOPS is one of the phase cycling procedures used in quadrature detection. During the quadrature detection the two signals, which are $\pi/2$ out of phase, are sampled. If there is an imbalance between the two channels, the acquired data gives rise to undesirable image signals. These are called quadrature images. To remove such artefacts, the simplest procedure is to acquire 4 transients where the phase of the pulse is varied in steps of $\pi/2$. The pulse phase cycling is thus written as X, Y, -X, -Y which tilts the magnetization along the -Y, X, Y, -X respectively. The receiver phase is appropriately chosen as shown in Table 3.3.

Table 3.3: The CYCLOPS phase cycling.

Transient	*Transmitter (Pulse) Phase*	*Receiver Phase*
1	*X*	*-Y*
2	*Y*	*X*
3	*-X*	*Y*
4	*-Y*	*-X*

3.2.2 Axial peak suppression

Axial peaks are the most common artefacts in MD spectra. These arise from the transfer of *z*-magnetization by the mixing sequence into *x*-/*y*-magnetization. Such a transfer gives rise to undesirable peaks at $\Omega_1 = 0$. For example, in the case of COSY, if one has additional *z*-magnetization (I_{kz}) at time 3, then the effect of the second $(\pi/2)_y$ pulse is given by:

$$\rho_4' = [-I_{kz}\cos(\pi J_{kl}t_1)+2I_{ky}I_{lx}\sin(\pi J_{kl}t_1)]\cos(\Omega_k t_1) + [\mathbf{I_{ky}\cos(\pi J_{kl}t_1)+2I_{kz}I_{lx}\sin(\pi J_{kl}t_1)}]\sin(\Omega_k t_1)+I_{kx} \quad 3.22$$

The last term (I_{kx}) gives rise to a peak at $\Omega_1 = 0$. This needs to be suppressed. If the next transient is collected by changing the phase of the second $(\pi/2)$ pulse from *Y* to *–Y*, then the result is:

$$\rho_4'' = [+I_{kz}\cos(\pi J_{kl}t_1)-2I_{ky}I_{lx}\sin(\pi J_{kl}t_1)]\cos(\Omega_k t_1) + [\mathbf{I_{ky}\cos(\pi J_{kl}t_1)+2I_{kz}I_{lx}\sin(\pi J_{kl}t_1)}]\sin(\Omega_k t_1)-I_{kx} \quad 3.23$$

Addition of ρ_4' and ρ_4'' without changing the phase of the receiver, results in the suppression of axial peaks. Thus, a two-step *phase-cycle* shown in Table 3.4 achieves the axial-peak suppression.

Table 3.4: Axial-peak suppression using a two-step phase cycling in COSY.

ϕ	ϕ'	Receiver
X	Y	X
X	-Y	X

3.3 Relayed COSY

Several modifications of the basic COSY sequence have been reported. Each has been designed so as to provide specific information in a biological system. Some of the modified sequences which are useful in biological systems have been discussed below. One of these is relayed COSY, which helps in resolving ambiguities in case of resonance overlaps. For example, a three spin AMX system where the J_{AM} and $J_{MX} \neq$ zero and J_{AX} = 0, can not be distinguished from two independent two-spin systems (AM and M′X), when the chemical shifts for spins M and M′ are not resolved (Figure 3.5 A). Relayed-COSY is useful in such cases. It gives rise to a relayed cross peak (Figure 3.5 B) between spins A and X, which are not directly coupled. In the later case no relay peak is seen (Figure 3.5 A).

In another situation, two independent networks of coupled spin systems in the same molecule may have degenerate chemical shifts. For example, in a protein, two threonine residues may have degenerate H^β chemical shifts, leading to ambiguity in assignments. Such an ambiguity is resolved by monitoring the relay peaks from individual H^α to their respective H^γ via H^β (Figure 3.5 C).

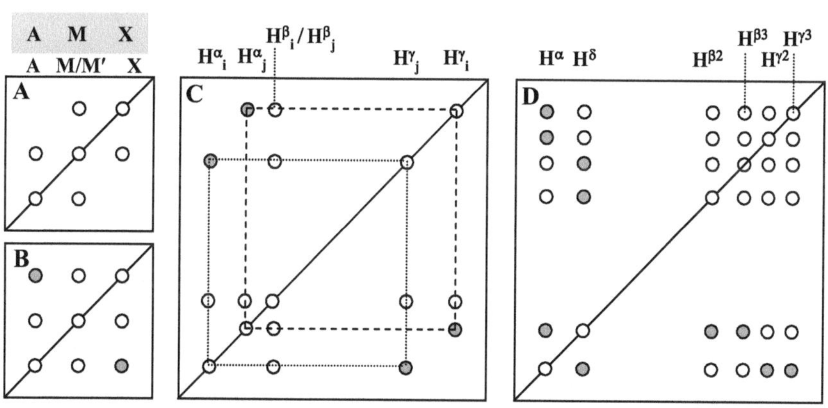

Figure 3.5: (A) Schematic representation of COSY in the case of a three spin system (AMX) or two independent 2-spin systems (AM and M′X) with degenerate chemical shifts for M and M′. (B and C) Schematic representation of a Relayed-COSY. The relay peaks are shown in grey. (D) Schematic representation of a TOCSY. TOCSY peaks are shown in grey.

Consider a three-spin system of spins *k*, *l* and *m*, where J_{kl} and J_{lm} are non-zero and $J_{km} = 0$. The magnetization acquired by *l* from *k* can be relayed to spin *m*, if $J_{lm} \neq 0$. The requirement for the relay transfer from spin *l* to *m* is an additional non-selective $(\pi/2)_x$ pulse. Prior to this, the magnetization acquired by spin *l* from *k* needs to be evolved under $\mathcal{H}_{J(lm)}$, such that one has *x/y*-magnetization of spin *l* anti-phase with respect to *m* (terms such as $2I_{mz}I_{lx}$ or $2I_{mz}I_{ly}$). To achieve this, a delay τ_M is introduced with a π pulse in the middle that prevents evolution under $\mathcal{H}_S$ (Figure 3.6).

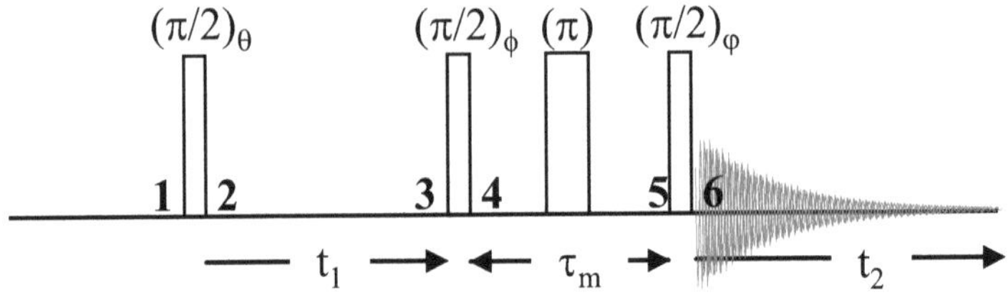

Figure 3.6: Relayed COSY pulse sequence.

3.4 TOtal COrrelation SpectroscopY (TOCSY)

TOCSY is an important and widely used technique. It provides information about all the networks of coupled spin systems. An important advantage of TOCSY over COSY is that it provides complete in-phase diagonal and cross peaks.

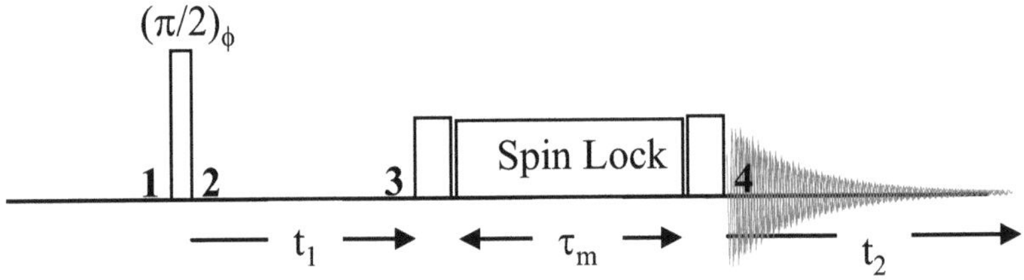

Figure 3.7: 2D-TOCSY pulse sequence.

The pulse sequence for TOCSY is shown in Figure 3.7. The mixing period (*Spin Lock* period) consists of a set of composite π pulses which are sandwitched between two trim pulses. The composite pulse cycle is essentially MLEV-16 discussed above. This is followed by a $(\pi)_x$ pulse. The total cycle is collectively identified as MLEV-17. This MLEV-17 cycle is repeated several times depending on the total required mixing time. The π_x pulse at the end of MLEV-16 is used to remove the overall effect of pulse imperfections that may occur during the composite pulse cycle. The two trim pulses (usually set to 2-3 msec each) at the beginning and end of the mixing period serve to defocus any magnetization that is not parallel to the *x*-axis and ensures proper phasing of 2D spectrum into absorption mode. The magnetization transfer is primarily from $I_{kx} \rightarrow I_{lx} \rightarrow I_{mx} \rightarrow I_{nx} \rightarrow \ldots\ldots$, in one go. The mixing time (τ_m) is about 30-100 msec and is equivalent to an integral multiple of MLEV-17 sequence. Use of shorter τ_m, however, results in cross peaks only between the directly coupled spins.

TOCSY helps in overcoming the problems of peak overlap during sequence specific resonance assignments (discussed more in detail in Chapters 5-9), and in identifying the complete networks of coupled spin systems. For example, TOCSY spectrum of an arginine residue shows cross peaks not only between the directly coupled spins, but also exhibits multiple-relay peaks such as $H^\alpha \rightarrow H^\gamma$ and H^δ, $H^\beta \rightarrow H^\delta$ and *vice-versa* (Figure 3.5 D). In-phase character of the TOCSY spectrum makes it particularly useful in larger M_r molecules, where the individual resonance linewidths are of the order of J-couplings.

3.5 MQ-Filtered COSY

As discussed, in the COSY experiment the dispersive and in-phase nature of individual components of diagonal peaks lead to long tails near the diagonal. This masks cross peaks, particularly those which are close to the diagonal. COSY also suffers from a large ratio between the intensities of diagonal and cross peaks.

MQ-filtered correlation spectroscopy (MQF-COSY) helps to overcome these drawbacks. MQF-COSY offers several other advantages. Resonances from uncoupled spins (particularly those from solvent) are suppressed. The peaks are purely absorptive and anti-phase. Cross-peak multiplet structures and their symmetry properties provide additional information about the various couplings involved.

3.5.1 Multiple-Quantum (MQ) Filtering

While discussing COSY, one encountered terms which are superposition of *zero-quantum* and *double-quantum* coherences. These terms have been ignored so far. Such terms can be transformed into observable *single-quantum* coherences by application of $(\pi/2)_{x/y}$ pulse. For example,

$$2I_{ky}I_{lx}+2I_{kx}I_{ly} \xrightarrow{(\pi/2)_x} 2I_{kz}I_{lx}+2I_{kx}I_{lz} \text{ and } 2I_{ky}I_{lx}+2I_{kx}I_{ly} \xrightarrow{(\pi/2)_y} -2I_{ky}I_{lz}-2I_{kz}I_{ly} \qquad 3.24$$

Thus, one can generate *multiple-quantum* coherences. Coherence of a specific order *p* can be filtered and transformed into observable *single-quantum* coherence by application of a $(\pi/2)_{x/y}$ pulse. Such experiments are called *multiple-quantum filtered* experiments.

With a *p-quantum* filter, it is possible to suppress resonances of spin systems with (*p-1*) coupled spins. For example, in a 2QF-COSY singlets are suppressed. In 3QF COSY, singlets, resonances from AB and AX spin systems are suppressed.

Sensitivity goes down rapidly with higher order of filters. The effects of MQ filtering on COSY spectra have been rationalized by the following coherence transfer rules, which hold for weakly coupled systems:
(i) A diagonal peak in pQF-COSY spectrum appears only when the *active-spin* involved in the magnetization transfer has resolved couplings to at least p-1 spins.
(ii) A pQF-COSY cross peak between two *active spins* can appear only when the two *active spins* have (p-2) additional common coupling partners.

3.5.2 Double-Quantum Filtered COSY (2QF-COSY)

The pulse sequence for a 2QF-COSY experiment is shown in Figure 3.8. It differs from the COSY sequence in the detection pulse. This is replaced by a pair of ($\pi/2$) pulses, separated by a short delay of the order of few μsec. It is used primarily for phase switching i.e., changing the phase of the pulses as desired. The phase cycling of the first two pulses enables the required filtering of the appropriate coherences in MQ-filtered COSY experiments.

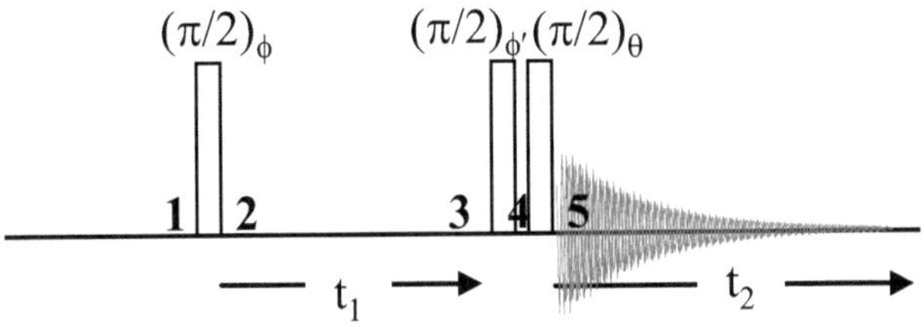

Figure 3.8: 2QF-COSY pulse sequence. The basic four step phase cycling for the first two pulses and the receiver are the following: ϕ *=x, y, -x, -y;* ϕ' *=x, y, -x, -y; Rec = x, -x, x, -x.*

The outcome of such an experiment can be calculated using POF. Starting with equilibrium magnetization I_{kz} prior to the application of the first ($\pi/2$) pulse, the density operator after the application of the second ($\pi/2$) pulse (at time *4*) can be derived. Addition of $\rho_4(1)$ and $\rho_4(3)$, and $\rho_4(2)$ and $\rho_4(4)$ for which the receiver phase is same, results in:

$$\rho_4(1+3) = 2[-I_{kz}\cos(\pi J_{kl}t_1) - 2I_{kx}I_{ly}\sin(\pi J_{kl}t_1)]\cos(\Omega_k t_1) \qquad 3.25$$

$$\rho_4(2+4) = 2[-I_{kz}\cos(\pi J_{kl}t_1) + 2I_{ky}I_{lx}\sin(\pi J_{kl}t_1)]\cos(\Omega_k t_1) \qquad 3.26$$

$$\rho_{resultant} = \rho_4(1+3) - \rho_4(2+4) = -[2I_{kx}I_{ly} + 2I_{ky}I_{lx}]\sin(\pi J_{kl}t_1)\cos(\Omega_k t_1) \qquad 3.27$$

In this expression, the terms individually represent superposition of zero-quantum and *double-quantum* coherences and together lead to pure *double-quantum* coherence. These terms get transformed into observable *single-quantum* coherences by the application of the final $(\pi/2)_x$ pulse as shown below.

$$\rho_{detected} = [2I_{kx}I_{lz} + 2I_{kz}I_{lx}]\sin(\pi J_{kl}t_1)\cos(\Omega_k t_1) \qquad 3.28$$

The term $2I_{kx}I_{lz}$ corresponds to an anti-phase *k* spin doublet centered at $\omega_2 = \Omega_k$, while $2I_{kz}I_{lx}$ corresponds to an anti-phase *l* spin doublet centered at $\omega_2 = \Omega_l$. Both the terms are anti-phase coherences and have the same modulation coefficient $sin(\pi J_{kl}t_1)$ $cos(\Omega_k t_1)$. The two terms originate from I_{kz}. The first term, $2I_{kx}I_{lz}\sin(\pi J_{kl}t_1)\cos(\Omega_k t_1)$, is modulated with Ω_k frequency during t_1. It continues to precess with the same frequency during t_2 and gives rise to a diagonal peak. The second term $2I_{kz}I_{lx}$ $\sin(\pi J_{kl}t_1)\cos(\Omega_k t_1)$ is modulated with Ω_k during t_1. This precesses with Ω_l frequency during t_2 and gives rise to a cross peak.

The modulation coefficient $sin(\pi J_{kl}t_1)\ cos(\Omega_k t_1)$ of both the cross and diagonal peaks can be expanded as $[\sin(\Omega_k+\pi J_{kl})t_1 - \sin(\Omega_k-\pi J_{kl})t_1]$. This describes the nature of the peak multiplicity along the Ω_1 axis. The peaks are anti-phase dispersive doublets centered at Ω_k. Since the peaks have dispersive nature along the Ω_1 axis, a $\pi/2$ phase correction can simultaneously transform both to pure-absorptive line shapes. Thus, both the diagonal and cross peaks in 2QF-COSY are characterized by pure absorption phase for individual components (Figure 3.9 A) and possess anti-symmetric character with respect to chemical-shift axes. The net integral volumes of individual diagonal peaks are zero in a *MQ*-filtered COSY. Hence, the problem of large diagonal to cross peak intensity ratio is averted. It may be noted that in 2QF-COSY, there is a loss in S/N by a factor 2, as compared to COSY.

In the presence of a *passive spin* the cross peak shows a complex multiplet structure, but still possess anti-symmetric character with respect to chemical-shift axes (Figure 3.9 B).

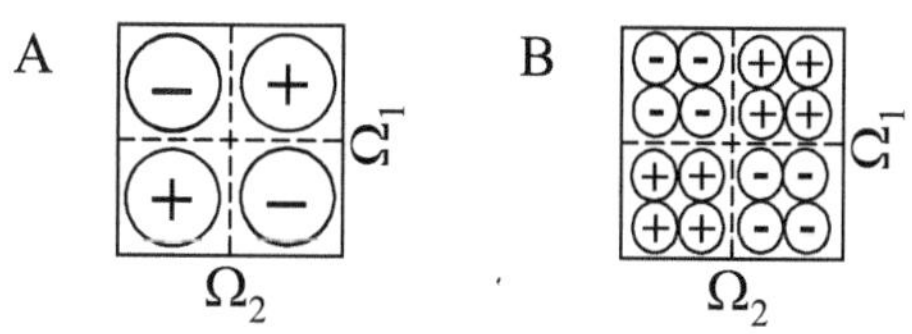

Figure 3.9: A) Schematic structures of cross and diagonal peaks for a two spin system in 2QF-COSY. B) Cross peak (k to l) multiplet structure for a three spin system, where k and l are active spins coupled to a passive partner m in 2QF-COSY. This is characteristic of an even quantum filtered COSY.

3.5.3 Three-Quantum Filtered COSY (3QF-COSY)

The sequence for a 3QF-COSY experiment is similar to the one for 2QF-COSY except that it differs in the phase cycling of the first two pulses and that of the receiver. For filtering of 3Q coherence, the phase of the first two pulses ϕ and ϕ' is incremented in steps of $\pi/3$, leading to a 6 $(2\pi/(\pi/3))$ step phase cycle. This leads to a $p(\pi/3)$ phase shift (where p=0,1,2,3, ….) of each of the p-quantum coherence as shown in Table 3.5 . Simultaneously, the receiver phase is alternated between x and –x. Note the suppression of unwanted *zero-quantum* (ZQC), *single-quantum* (SQC) and *two-quantum* (2QC) coherences.

Table 3.5: Basic phase cycling of the first two pulses and the receiver needed in the 3QF-COSY experiment.

Phase ϕ and ϕ'	Phase of ZQC	Phase of SQC	Phase of 2QC	Phase of 3QC	Phase of Receiver
0	0	0	0	0	x
$\pi/3$	0	$\pi/3$	$2(\pi/3)$	$3(\pi/3)$	-x
$2(\pi/3)$	0	$2(\pi/3)$	$4(\pi/3)$	$6(\pi/3)$	x
$3(\pi/3)$	0	$3(\pi/3)$	$6(\pi/3)$	$9(\pi/3)$	-x
$4(\pi/3)$	0	$4(\pi/3)$	$8(\pi/3)$	$12(\pi/3)$	x
$5(\pi/3)$	0	$5(\pi/3)$	$10(\pi/3)$	$15(\pi/3)$	-x

In 3QF-COSY, cross peaks between two resonances appear only when the two active spins have at least one additional common coupling partner. For example, in Thr and Ile, the J-coupled H^{α} and H^{β} do not have a common coupling partner. Hence, one does not observe a cross peak between H^{α} and H^{β}. However, these spin systems are identified in 2QF-COSY from H^{α}-H^{β} and H^{β}-H^{γ} cross peaks. Absence of H^{α}-H^{β} cross peak in 3 QF COSY confirm these observations.

The final detectable magnetization is given by

$$\sigma_{\mathrm{detectable}} = [4I_{ky}I_{lz}I_{mz} + 4I_{kz}I_{ly}I_{mz} + 4I_{kz}I_{lz}I_{my}]\ \sin(\pi J_{km}t_1)\ \sin(\pi J_{kl}t_1)\ \sin(\Omega_k t_1) \qquad 3.29$$

$4I_{ky}I_{lz}I_{mz}$ is the y-magnetization of spin *k* anti-phase with respect to spins *l* and *m*. $4I_{kz}I_{ly}I_{mz}$ is the y-magnetization of spin *l* anti-phase with respect to spins *k* and *m*. $4I_{kz}I_{lz}I_{my}$ is the y-magnetization of spin *m* anti-phase with respect to spins *k* and *l*. All the three terms are modulated with Ω_k frequency during the t_1. The first term continues to precess with same frequency during t_2 giving rise to a diagonal peak. The second and the third terms precess with Ω_l and Ω_m frequencies during t_2 giving rise to two cross peaks. There is a loss by a factor of 4 in 3QF COSY spectrum compared to the unfiltered COSY experiment. The modulation coefficient of the three terms, $\sin(\pi J_{km}t_1)\ \sin(\pi J_{kl}t_1)\ \sin(\Omega_k t_1)$ can be expanded as

$$\sin(\Omega_k+\pi J_{km}/2+\pi J_{kl}/2)t_1 - \sin(\Omega_k+\pi J_{km}/2-\pi J_{kl}/2)t_1$$
$$- \sin(\Omega_k-\pi J_{km}/2+\pi J_{kl}/2)t_1 + \sin(\Omega_k-\pi J_{km}/2-\pi J_{kl}/2)t_1$$

3QF-COSY gives rise to a multiplet structure of individual peaks symmetric with respect to chemical-shift axes (Figure 3.10).

The *odd-* and *even-quantum* filtered COSY give rise to symmetric and anti-symmetric multiplet structures for their individual peaks, respectively. Further, 3QF-COSY cross peaks have either positive or negative multiplet components near their centers, where as *even-quantum* filtered COSY cross peaks have zero intensity in the center. The sign of individual components depends on the sign of the coupling constants J_{kl} and J_{km}. This can be exploited to distinguish between the cross peaks arising from *vicinal* and *geminal* J-couplings, which are negative and positive, respectively. In describing the MQF-COSY, our discussion has been restricted to 2QF- and 3QF-COSY. One can even record 4QF and 5QF COSY spectra, at an expense of S/N ratio, which reduces by a factor of 2^{p-1}, where p is the coherence order.

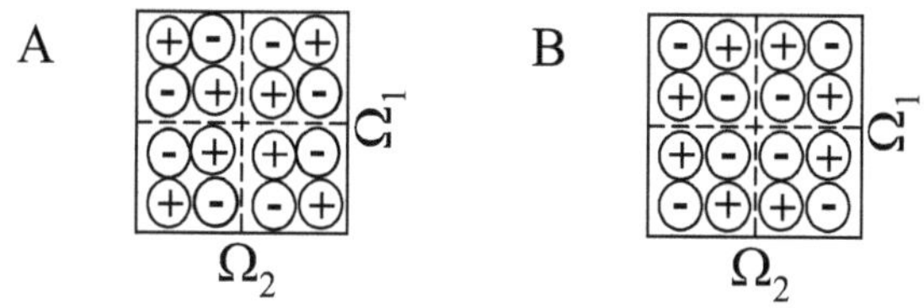

Figure 3.10: Multiplet structure of cross peaks in 3QF COSY symmetric with respect to chemical-shift axes. A) geminal and B) vicinal cross peaks.

3.5.4 Exclusive COSY

Exclusive COSY (E.COSY) is obtained, from a weighted addition or subtraction of 2QF- and 3QF-COSY spectra. The pulse sequence for E.COSY is identical to the one used for 2QF-COSY, except in the phase cycling of the first two pulses and that of the receiver. Without working on the phase cycle and the theoretical outcome of such an experiment, one can appreciate that a weighted summation of cross peak multiplet structures of 2QF- and 3QF-COSY spectra simplifies individual multiplet structures (Figure 3.11). The features of the individual cross peaks in an E.COSY spectrum are influenced by the linewidths and magnitude of J-couplings. Such simplified multiplet structure makes E.COSY amenable to measure both the *active* and *passive* J-couplings.

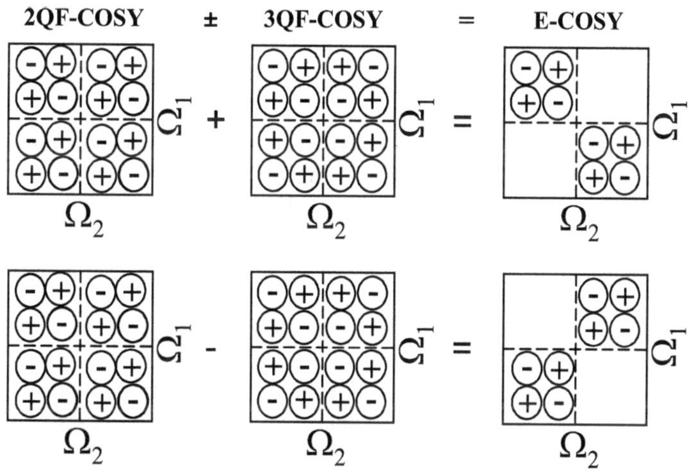

Figure 3.11: Weighted summation and subtraction of multiplet structures of cross peaks observed in a 2QF- and 3QF COSY spectra in the case of a three spin system.

3.6 EXSY and NOESY

The other two mechanisms for magnetization transfer in solution-NMR are chemical exchange and nuclear Overhauser effect (NOE). The pulse sequence for picking such magnetization transfers is the same (Figure 3.12). The difference from the COSY sequence is in the detection pulse. This consists of two ($\pi/2$) pulses separated by a fixed delay of the order of around few tens of milli-seconds 50-500 msec. This delay is called the mixing time (τ_m). The phase cycling of the first two pulses generates the required $-I_z$ magnetization.

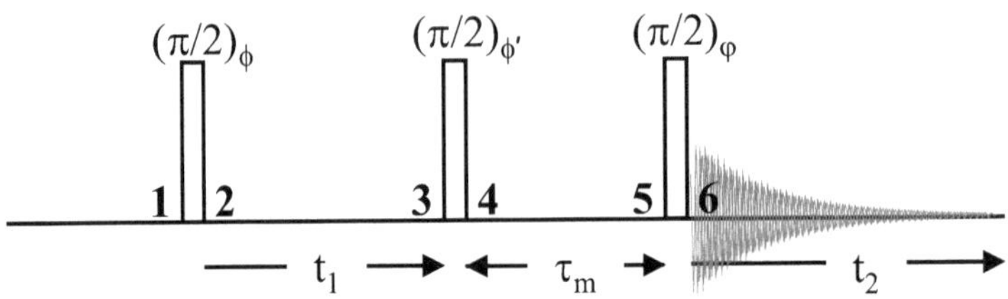

Figure 3.12: 2D-NOESY pulse sequence. The basic eight step phase cycling for the three pulses and the receiver are the following: ϕ =x, -x; ϕ'=x, x, x, x, x, x, x, x; φ =x, x, -x, -x, y, y, -y, -y; Rec = x, -x, -x, x, y, -y, -y, y, y.

Starting from an equilibrium magnetization I_{kz} prior to the application of the first ($\pi/2$) pulse, the density operator after the application of the second ($\pi/2$) pulse (at of time *4*) can be derived:

For, $\phi = \phi' = x$

$$\rho_4 = [-I_{kz}\cos(\pi J_{kl}t_1) - 2I_{kx}I_{ly}\sin(\pi J_{kl}t_1)]\cos(\Omega_k t_1) + [I_{kx}\cos(\pi J_{kl}t_1) - 2I_{kz}I_{ly}\sin(\pi J_{kl}t_1)]\sin(\Omega_k t_1) \quad 3.30$$

So far we have ignored the first term ($-I_{kz}$). The z-magnetization is non-observable. However, during chemical exchange or dipolar interaction, such a term allows magnetization transfer from spin *k* to *l*. Hence, the phase cycling is such that one retains this term and suppress almost all the remaining terms. Concentrating on this term, one obtains

$$\rho_4 = -I_{kz}\cos(\pi J_{kl}t_1)\cos(\Omega_k t_1) \quad 3.31$$

Assume that a fraction (represented by *b*) of the z-magnetization of spin *k* is passed to spin *l* due to chemical exchange or dipolar interaction (called cross-relaxation as discussed in Chapter 1). Then at the end of the mixing time one has:

$$\rho_5 = \{-aI_{kz}-bI_{lz}\} \cos(\pi J_{kl}t_1) \cos(\Omega_k t_1) \qquad 3.32$$

The final $(\pi/2)_x$ pulse transforms this into detectable magnetization I_{ky}, I_{ly}, etc,

$$\rho_6 = \{+aI_{ky}+bI_{ly}\} \cos(\pi J_{kl}t_1) \cos(\Omega_k t_1) \qquad 3.33$$

The coefficients (*a* and *b*) are dictated by the transfer efficiency and depend on the cross relaxation rate and the rate of chemical exchange. Thus, the first term continues to precess at a frequency Ω_k giving rise to a diagonal peak. The second term precesses with a frequency Ω_l and gives rise to a cross peak. Thus, it is the cross-relaxation or chemical exchange process during the mixing time that gives rise to the cross peak. In the absence of such interactions between nuclear spins, such a peak will not be observed.

As discussed in Chapter 1, positive cross-relaxation rate (W_2-W_0) gives rise to positive NOEs and $-I_{kz}$ evolves into I_{lz}, giving rise to diagonal and cross peaks which are opposite in sign. Conversely, negative W_2-W_0 gives negative nOes and $-I_{kz}$ evolves into $-I_{lz}$, giving rise to same sign for both the diagonal and cross peaks. EXSY is characterized by same sign for both the peaks. Further, larger molecules generally give better NOEs at higher fields and smaller molecules give better nOes at lower fields.

Some coherences (specifically *zero-quantum coherences*) generated by J-coupling are not completely removed by phase cycling. In such situations, one can vary τ_m and minimize their contribution as they oscillate sinusoidally, whereas the NOE signals increase with τ_m.

3.7 ROtational nuclear Overhauser Effect SpectroscopY (ROESY)

For W_2-W_0=0, a special situation arises, when NOE is not observed. Such a situation arises when $\omega_0\tau_c$=1.

A different technique called ROESY has been used to observe dipolar interactions in such cases. ROESY provides information about internuclear distances and chemical exchange process that may be present in a given sample. In NOESY, the dipolar coupled spins result in mutual cross-relaxation of z-magnetization. In ROESY, they cause mutual cross-relaxation of transverse magnetization. ROESY is particularly useful for cases where NOESY signals are weak or zero. ROESY experiment also provides cross peaks from chemical exchange. Exchange peaks and TOCSY type artefacts are always positive so they can be distinguished from ROE peaks. To maximize sensitivity, the mixing time should be chosen so as to be half to one times the longitudinal relaxation time (T_1).

The pulse sequence for ROESY is shown in Figure 3.13. The difference between the NOESY and ROESY pulse sequences is in the mixing process. The mixing time in ROESY consists of a spin-lock. The spin-lock is generated by a low-power CW irradiation (γB_1 is set to approximately one-half of the spectral width) or by a series of small angle pulses. During the spin-lock time, the spins are prevented from precessing around the z-axis by the applied RF field. The associated cross-relaxation rate (W_2-W_o) is given by:

$$W_2 - W_0 = (1/10)\,(\gamma_I^2 \gamma_S^2 \hbar^2 \tau_c / r^6)\,\{2+3/(1+\omega^2\tau_c^2)\} \qquad 3.34$$

Thus, unlike NOE, ROE which is proportional to W_2-W_0 is always positive. This means that the ROEs are opposite in their sign with respect to the diagonal peaks irrespective of the value of $\omega\tau_c$.

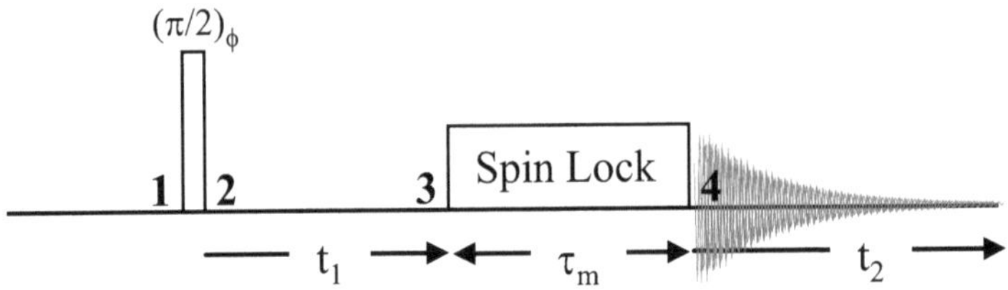

Figure 3.13: 2D-ROESY pulse sequence.

4. MULTIPLE-QUANTUM (MQ) SPECTROSCOPY

A special feature of MD NMR is in the indirect observation of MQ coherences (MQC). MQC coherences can be generated and filtered to convert them into observable single-quantum (SQ) coherences. The MQC are evolved and frequency labelled during an indirect dimension, namely, the evolution period of 2D experiment. This enables one to detect such forbidden fruits of spectroscopy. The scheme for such an experiment involves the following (Figure 3.14):

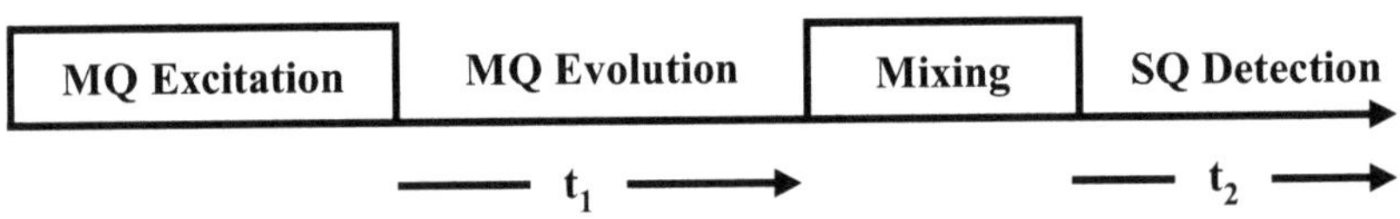

Figure 3.14: Schematic representation of various time-periods in a 2D MQ spectroscopy experiment.

Starting from initial magnetization, at least two pulses are required to create MQC. One needs to set the phases ϕ and ϕ'of the two $\pi/2$ pulses that generate the MQC. For creating even-QC ϕ' is kept equal to ϕ, while $\phi' = \phi+\pi/2$ for creating odd-QC.

4.1 2D Double-Quantum (DQ) experiment

In the case of a 2-spin system, one can generate a *two-quantum coherence* (2QC) with two $\pi/2$ pulses which are phase cycled in four steps as discussed in (Section 3.4.2):

$$\rho = -[2I_{kx}I_{ly} + 2I_{ky}I_{lx}]\, sin(\pi J_{kl}\tau)\, cos(\Omega_k\tau)] \qquad 3.35$$

The undesirable part of such 2QC is the chemical-shift modulation term $\cos(\Omega_k\tau)$. This is avoided by inserting a π_x pulse between the two $\pi/2$ pulses (Figure 3.15). Therefore, one can write

$$\rho = -[2I_{kx}I_{ly} + 2I_{ky}I_{lx}]\, sin(\pi J_{kl}\tau)] \qquad 3.36$$

Such 2QC is frequency labelled along the indirect dimension t_1 and then converted to an observable SQC. Such 2D *double-quantum* experiments help in the identification of individual networks of coupled spin systems, in the identification of magnetically equivalent spins and in the characterization of linear networks of coupled spin systems of AMX type with $J_{AX} = 0$ (in the form of *Remote Connectivity*), which are not possible with COSY spectra. Further, the spectra do not have diagonal peaks.

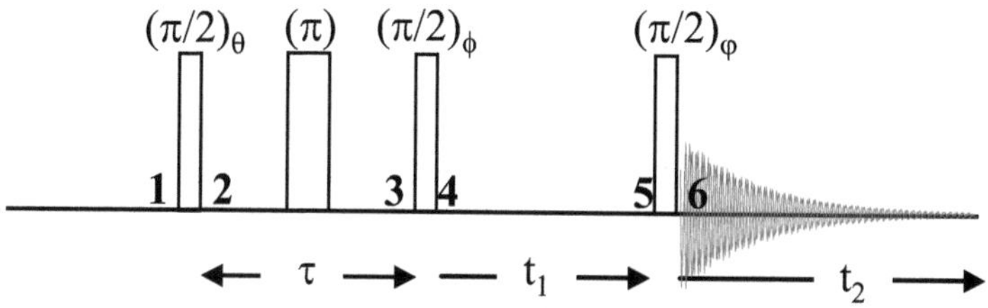

Figure 3.15: Pulse sequence for a 2D DQ experiment.

The pulse sequence for the 2D double-quantum (DQ) experiment is shown in Figure 3.15. The difference from the COSY sequence is in the preparation pulse, which is replaced by a pair of ($\pi/2$) pulses, with a π_x pulse placed in between the two. The phase cycling of the first two $\pi/2$ pulses enables generation of the required 2QC. The following magnetization is present at time *4:*

$$\rho_4 = -[2I_{kx}I_{ly} + 2I_{ky}I_{lx}]\, sin(\pi J_{kl}\tau) \text{ or } \rho_4 = -2[\{2QC\}_y]\, sin(\pi J_{kl}\tau) \qquad 3.37$$

The 2QC thus generated evolves under $\mathcal{H}_\delta$ during the indirect dimension t_1 and gets frequency labelled. Note that this coherence does not evolve under $\mathcal{H}_J$. At the end of the evolution period one has:

$$\rho_5 = -[2\{2QC\}_y\, cos\,(\Omega_k+\Omega_l)t_1 - 2\{2QC\}_x\, sin\,(\Omega_k+\Omega_l)t_1]\, sin(\pi J_{kl}\tau) \qquad 3.38$$

$$\rho_6 = [\{-I_{kx}I_{lz}-I_{kz}I_{lx}\}\, cos\,(\Omega_k+\Omega_l)t_1 - \{I_{kx}I_{lx}+I_{kz}I_{lz}\}\, sin\,(\Omega_k+\Omega_l)t_1]\, sin(\pi J_{kl}\tau) \qquad 3.39$$

$$\boldsymbol{\rho_{detectable} = \{-2I_{kx}I_{lz}\, cos\,(\Omega_k+\Omega_l)t_1 - 2I_{kz}I_{lx}\, cos\,(\Omega_k+\Omega_l)t_1\}\, sin(\pi J_{kl}\tau)} \qquad 3.40$$

The term $2I_{kx}I_{lz}$ corresponds to a spin *k* doublet centered at $\omega_2 = \Omega_k$. The term $2I_{kz}I_{lx}$ corresponds to a spin *l* doublet at $\omega_2 = \Omega_l$. The projection along Ω_1 provides 1D MQ spectrum, and along Ω_2 gives 1D SQ spectrum. All singlets are suppressed. Both the above terms are anti-phase coherences and have the same modulation coefficient $\cos(\Omega_k+\Omega_l)t_1 \sin(\pi J_{kl}\tau)$ (Figure 3.16). Both the terms originate from I_{kz}. The terms are modulated with a 2Q frequency, $\Omega_{k+}\Omega_l$, during the t_1 without any splitting since there is no J modulation. The transfer function has a constant term $\sin(\pi J_{kl}\tau)$. If J_{kl} is small , then the corresponding peaks tend to be weak.

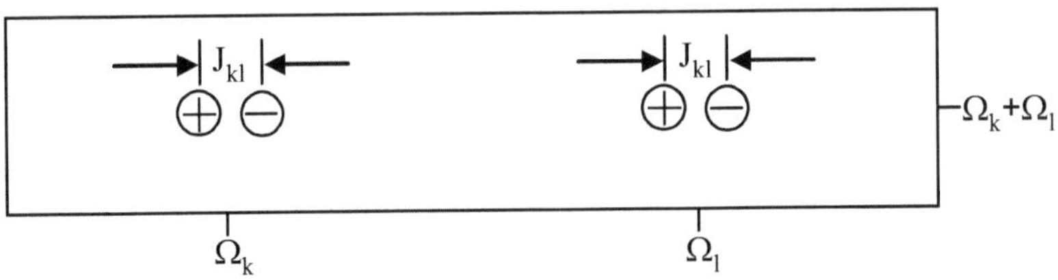

Figure 3.16: Schematic view of the 2D DQ experiment in the case of an AX spin system.

The 2D double-quantum experiment has been extensively used in correlating neighbouring ^{13}C spins under natural abundance. Therefore, the pulse sequence has been given the acronym *Incredible Natural Abundance Double-Quantum Transfer Experiment* (INADEQUATE). The technique establishes ^{13}C-^{13}C coupled networks and aids in ^{13}C resonance assignments. Further, one can measure ^{13}C-^{13}C couplings.

5. HETERONUCLEAR CORRELATION SPECTROSCOPY

So far the discussion has been centered on 2D techniques involving homonuclear spins. Correlation techniques have been used for correlation of other like spins such as ^{13}C-^{13}C and ^{31}P-^{31}P. One can also observe correlations between unlike spins, such as ^{13}C-^{1}H, ^{15}N-^{1}H and ^{31}P-^{1}H. Experiments involving such correlations are called heteronuclear correlations. For conducting such experiments, one needs a spectrometer with two or more RF channels. For example, the experiment for X-^{1}H correlation requires a ^{1}H and an X RF channel. Because of its high abundance and sensitivity, most experiments use ^{1}H as one of the partner in correlation schemes. The one bond couplings (^{1}J(H-X)) which have a large magnitude are utilized for magnetization transfer.

5.1 Insensitive Nuclei Enhanced by Polarization Transfer (INEPT)

The aim of the INEPT experiment is to transfer ^{1}H magnetization to a J-coupled hetero-spin. The process is called polarization transfer. Since the value of γ for ^{1}H is four times that of ^{13}C and ten times that of ^{15}N, the ^{1}H magnetization is much larger than that of the X-spin. By transferring this magnetization to the less sensitive spin, the signal from the spin X can be detected with a higher sensitivity. Figure 3.17 A shows the pulse sequence for an INEPT experiment.

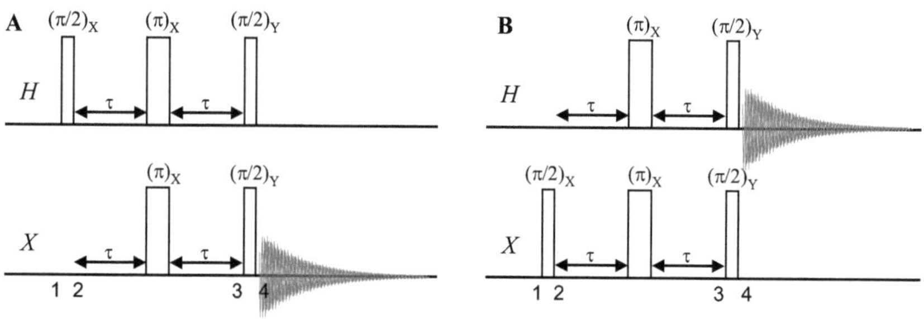

Figure 3.17: Pulse sequences for (A) INEPT and (B) Reverse INEPT.

Consider a directly coupled two-spin H-X system. The equilibrium ^1H magnetization prior to the application of the first $(\pi/2)_x$ pulse on the ^1H channel is $\rho_1 = \gamma_H H_z$. The $(\pi/2)_x$ pulse tilts the magnetization of spin H to –y axis, i.e. $\rho_2 = -\gamma_H H_y$. The effect of $-\tau-(\pi_x)_{I,X}-\tau-$ is similar to that of a spin-echo experiment, which refocuses the chemical shifts of both H and X. However, it phase modulates the heteronuclear $^1J_{HX}$ coupling. It may be noted that the π_x pulses are applied simultaneously on both the channels. One obtains evolution only under the $\mathcal{H}_J$ term arising from the one-bond heteronuclear coupling since $^1J_{HX}$ is much large than $^2J_{HX}$ and $^3J_{HX}$ and all the ^1H-^1H couplings. Therefore, $-H_y$ evolves under $\mathcal{H}_J$ as follows (the effect of π_x pulse is to change the sign):

$$\rho_{3'} = \gamma_H \{H_y \cos(2\pi J_{HX}\tau) - 2H_xX_z \sin(2\pi J_{HX}\tau)\} \qquad 3.41$$

If the value of τ is set to $1/4J_{HX}$, then $\cos(2\pi J_{HX}\tau) = 0$ and $\sin(2\pi J_{HX}\tau)=1$. Thus,

$$\rho_3 = -\gamma_H 2H_xX_z \qquad 3.42$$

The simultaneous application of $(\pi/2)_y$ pulses on both ^1H and X transforms ρ_3 into a detectable coherence which is given by: $\rho_4=+\gamma_H 2H_zX_x$. The term $2H_zX_x$ results in an absorptive-antiphase X spin doublet centered at Ω_X (Figure 3.18 A). In contrast, if a $(\pi/2)_x$ pulse had been applied to the equilibrium X spin magnetization ($\gamma_X X_z$) the result will be $-\gamma_X X_y$. Thus, the intensity ratio of intensities between the INEPT and a direct observation is γ_H/γ_X. As evident, the advantage of INEPT is that the signal intensities are larger for lower values of γ_X. An additional advantage is that the repetition rate of the experiment is set by the relaxation times of the high γ spins, which are generally shorter than those of the spins with low γ. It may be noted that in the INEPT pulse sequence, the $(\pi/2)_y$ pulse on X spin creates an additional term $\gamma_C X_x$. The detectable magnetization is given by,

$$\rho_{eff}(detectable) = +\gamma_H(2H_zX_x) + \gamma_X X_x \qquad 3.43$$

Thus, one ends up with an anti-phase asymmetric doublet (Figure 3.18 A). This is an inherent feature of the INEPT pulse sequence, which can be overcome by a two-step phase cycling.

The INEPT experiment provides a sensitivity enhancement of γ_H/γ_X. The corresponding enhancement by NOE is lower and is given by $1+\gamma_H/2\gamma_X$. Note that for ^{15}N, which has a negative γ, the difference of sensitivity enhancement is much higher.

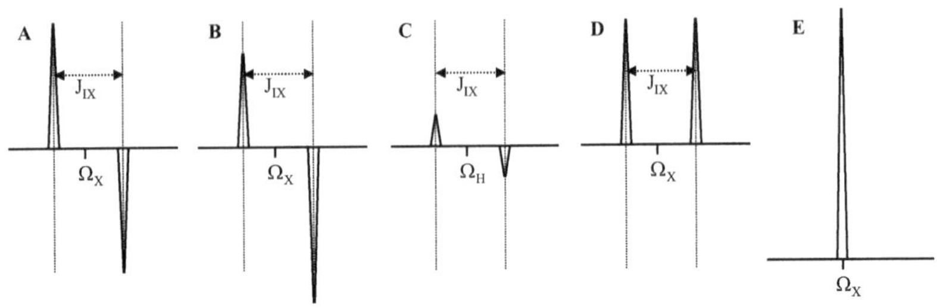

Figure 3.18: Anti-phase asymmetric doublet seen in the case of INEPT (A) with and (B) without two-step phase cycling, (C) in the case of reverse INEPT, (D) in the case of refocussed INEPT without decoupling and (E) with decoupling on.

5.2 Reverse INEPT

In the reverse INEPT experiment, magnetization is transferred from the spin *X* to *H*. Though this sequence in itself is not of practical utility, it serves the purpose of transferring back the magnetization of spin *X* to *H* in several heteronuclear correlation experiments discussed later. Figure 3.17 B shows the pulse sequence for the reverse INEPT experiment. The density operator at various time-points is given by

$$\rho_1 = \gamma_X X_z;\ \rho_2 = -\gamma_X X_y;\ \rho_3 = -\gamma_X 2H_zX_x;\ \rho_4(detectable) = +\boldsymbol{\gamma_X\, 2H_xX_z} \qquad 3.44$$

The term $2H_xX_z$ corresponds to an absorptive-antiphase *H* spin doublet centered at Ω_H (Figure 3.18 C), through magnetization transfer from spin *X* to spin *H*. In this pulse sequence too, the $(\pi/2)_y$ pulse on *H* spin creates an additional term $+\gamma_H H_x$. This term can be suppressed by a two-step phase cycling.

5.3 Refocused INEPT

Figure 3.19 shows the pulse sequence for a refocused INEPT experiment. The ^1H magnetization is transferred to its directly bonded X-spin and is refocussed into an in-phase magnetization. Because of the in-phase character, we can *decouple* ^1H during the acquisition period. This helps to enhance S/N almost two-fold compared to the INEPT experiment (Figure 3.18 D). The density operator at various time-points is given as follows:

$$\rho_1 = \gamma_H H_z;\ \rho_2 = -\gamma_H H_y;\ \rho_3 = \gamma_H 2H_xX_z;\ \rho_4 = -\gamma_H 2H_zX_x \qquad 3.45$$

$$\boldsymbol{\rho_5(detectable) = +\gamma_H X_y} \qquad 3.46$$

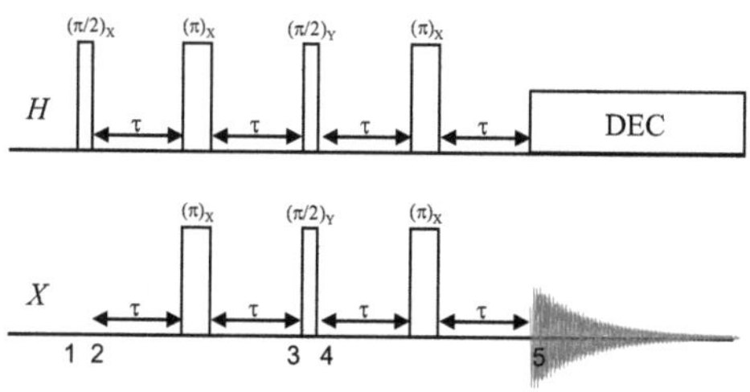

Figure 3.19: Refocused INEPT pulse sequence.

The term X_y corresponds to an in-phase *X* spin doublet centered at Ω_X (Figure 3.18 D), arising from the magnetization transferred from spin *H* to spin *X*. Heteronuclear ^{1}H decoupling incorporated during the detection period removes the splitting and results in a single line (Figure 3.18 E) providing a 2-fold enhancement in S/N.

5.4 Distortionless Enhanced by Polarization Transfer (DEPT)

The DEPT pulse sequence is similar to that for refocused INEPT and provides signals of the insensitive nucleus. All the signals are in-phase at the start of the acquisition period. Thus, decoupled spectra can be acquired without the need for an extra refocusing delay. The coupled spectra retain their familiar multiplet structures, without any distortion. The technique has some spectral editing advantages. Therefore, DEPT is the method of choice for direct identification and assignment of resonances particularly in small molecules.

5.5 Broadband Decoupling

As discussed in Chapter 1, the splitting caused by J-couplings becomes undesirable when only the chemical shift information is desired. For example, during the acquisition of a ^{13}C spectrum one may wish to remove all the proton couplings. This is achieved by irradiating the sample with a sequence of RF pulses (generally π pulses) applied at proton frequency, while simultaneously detecting the X-nuclei (^{13}C/^{15}N). The decoupled spin states α and β are kept equally populated and spins in these states are flipped rapidly between the states, such that the X-nuclei do not feel their presence. Thus, the entire ^{13}C spectrum can be decoupled from ^{1}H. Such decoupling is called *broadband* decoupling.

The bandwidth of decoupled region increases with increase in the spectrometer working frequency. Hence, removal of spin-spin coupling becomes more challenging at higher spectrometer frequencies. The power dissipation within the sample results in sample heating. This in turn has significant effect on the spectral quality. The problem is more acute for low γ nuclei, where one needs to use larger values of B_{dec}, since the decoupling field strength is γB_{dec}. The situation is even worse in the case of biomolecules which are generally sensitive to small changes in temperature. Thus, the criteria for efficient decoupling are:

(i) uniform irradiation of entire ^{1}H spectrum with a given decoupler RF power level;
(ii) higher ratio for decoupling bandwidth to decoupling power;
(iii) the decoupler RF power should not heat the sample.
(iv) insensitivity of signals with regard to the RF pulse-widths and their phase shifts;
(v) absence of residual splittings of decoupled nuclei.

Earlier, a sequence of π pulses was used to achieve *broadband* decoupling. Later, better methods were described that adopt the use of composite π pulses in pulse sequences such as MLEV-4, MLEV-16 and MLEV-64 (Section 2.6). The first composite π pulse $(\pi/2)_x$-$(\pi)_y$-$(\pi/2)_x$, represented as by R is used for spin inversion. This needs relatively much lower RF power and show better offset independence than a simple π pulse. The pulse sequence MLEV-4 has 4 such composite pulses in the following manner: $R\bar{R}\bar{R}R$, while MLEV-16 has 16 such composite pulses in the following manner: $R\bar{R}\bar{R}R\ \bar{R}\bar{R}RR\ \bar{R}RR\bar{R}\ RR\bar{R}\bar{R}$, where $\bar{R}$ is inverse of R ($(\pi/2)_{-x}$-$(\pi)_{-y}$-$(\pi/2)_{-x}$). An alternate composite π pulse, $(\pi/2)_x$-$(\pi)_{-x}$-$(3\pi/2)_x$, has superior off-set dependence (Figure 3.20). The corresponding pulse sequences are identified as WALTZ-4, WALTZ-8 or WALTZ-16. WALTZ-4 sequence is $RR\bar{R}\bar{R}$, where in R is $(\pi/2)_x$-$(\pi)_{-x}$-$(3\pi/2)_x$ and $\bar{R}$ is $(\pi/2)_{-x}$-$(\pi)_x$-$(3\pi/2)_{-x}$. Other sequences identified as *supercycles,* such as WALTZ-8 and WALTZ-16, are designed by permuting or phase inverting the WALTZ-4 sequence. WALTZ-16 is the most commonly used decoupling sequence for heteronuclear correlation experiments.

GARP and DIPSI are two other alternate decoupling pulse trains where in the composite pulse is simulated by numerical optimization. The GARP type composite pulse (R), when used as a *supercycle*, such as $RR\bar{R}\bar{R}$ provides uniform decoupling profile over a two-fold bandwidth achieved with the use of WALTZ-4. On the other hand, a similar *supercycle* generated by DIPSI type composite pulse, is found to be better compared to other decoupling sequences. It results in better S/N ratio and resolution in the ^{13}C spectra of strongly (^{1}H-^{1}H) coupled homonuclear systems.

Broad-band decoupling may not be needed in all cases. At times, decoupling can be frequency selective. In such situations, one achieves selective decoupling by using relatively low decoupler power.

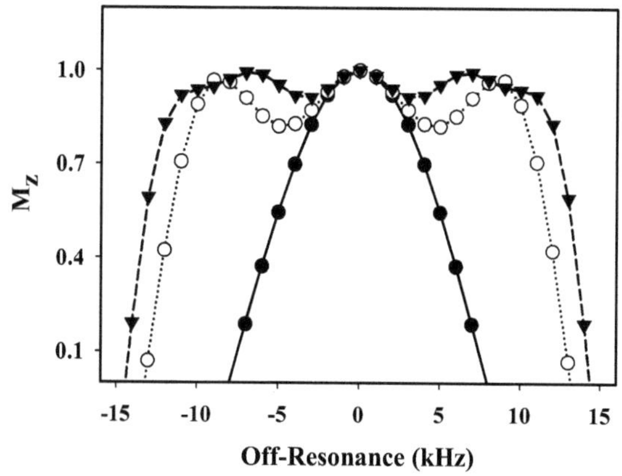

Figure 3.20: Off-set dependence of decoupling by a sequence of π pulses (filled circles) and composite pulses $(\pi/2)_{-x}$-$(\pi)_{-y}$-$(\pi/2)_{-x}$ (open circles) and $(\pi/2)_x$-$(\pi)_{-x}$-$(3\pi/2)_x$ (filled triangles). (Thanks to P.K. Madhu for simulating this figure).

5.6 Heteronuclear Single–Quantum Correlation Spectroscopy (HSQC)

2D heteronuclear correlation spectroscopy is widely used in biomolecular NMR. Figure 3.20 shows the pulse sequence for an HSQC experiment. The ^{1}H magnetization is transferred to a directly bonded X spin using an INEPT sequence, which is frequency labelled during the indirect dimension t_1. The proton π pulse in the middle of t_1 decouples spins H and X. Thereafter, magnetization is transferred back to ^{1}H using a refocussed INEPT pulse sequence and converted into in-phase ^{1}H magnetization, which is detected during the time-period t_2 (Figure 3.21). The delays τ are set to $1/4J_{HX}$. The typical values of τ are ~2.7 msec corresponding to $^1J(^{15}N\text{-}^1H)$ value of 90 Hz and ~1.8 msec for $^1J(^{13}C\text{-}^1H)$, taken as 140 Hz.

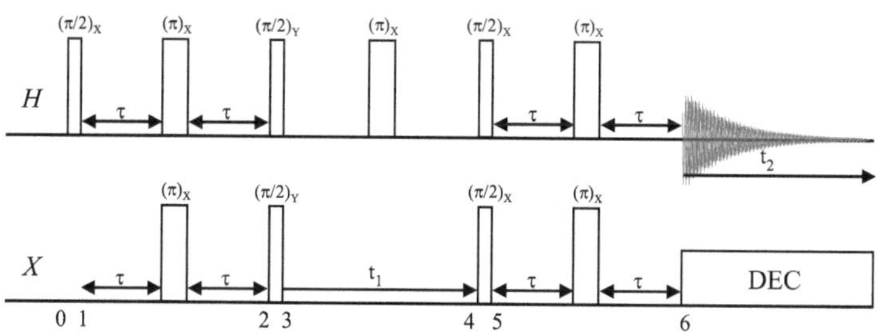

Figure 3.21: 2D HSQC pulse sequence.

The density operator at the final stage in HSQC experiment is given by:

$$\rho_6 = \boldsymbol{H_x \cos(\Omega_X t_1)} + 2H_yX_y \sin(\Omega_X t_1) \qquad 3.47$$

The first term H_x leads to an in-phase H spin absorptive doublet centered at Ω_H. Its modulation coefficient is $\cos(\Omega_X t_1)$, where Ω_X is the offset frequency of the spin X. The second term in the above equation is an non-observable MQ term.

Thus the detectable magnetization corresponds to a cross peak with two components at $(\Omega_H \pm J_{HX}/2, \Omega_X)$. The heteronuclear decoupling incorporated during the ^1H detection period removes splitting along the Ω_2 dimension and results in a peak only at (Ω_H, Ω_X) with in-phase pure-absorptive line-shapes along both the dimensions.

In such a correlation experiment, one is working with the satellites in the ^1H spectrum arising from one-bond $J(X\text{-}^1H)$ couplings. We have ignored so far, protons which are bonded to ^{12}C. When dealing with ^{13}C and ^{15}N spins in natural abundance, there are a large number of ^1H bonded to ^{12}C and ^{14}N, which are also excited by the first $(\pi/2_x)_H$ pulse. Signals arising from these spins need to be suppressed. This is achieved by a two-step phase cycle, in which two or more transients per each t_1 increment are acquired. For example, in a two-step phase cycle used in HSQC, the phase of one of the pulses applied to the X spins is inverted so that the signals arising from the *H→X→H transfer* are inverted, but not those arising from *H→H*. Subtraction of the second transient from the first eliminates undesirable signals. Additional two-step phase cycle is used to eliminate 2Q coherences.

A typical $[^{15}N\text{-}^1H]$-HSQC spectrum is shown in Figure 3.22. One peak is observed for each pair of directly bonded H-X pair, with the corresponding chemical shifts along the two dimensions. Such spectra can be obtained in just a few minutes. They are used extensively in biomolecular systems to detect changes in chemical shifts or peak intensities arising from molecular interactions, conformational transitions and chemical exchanges.

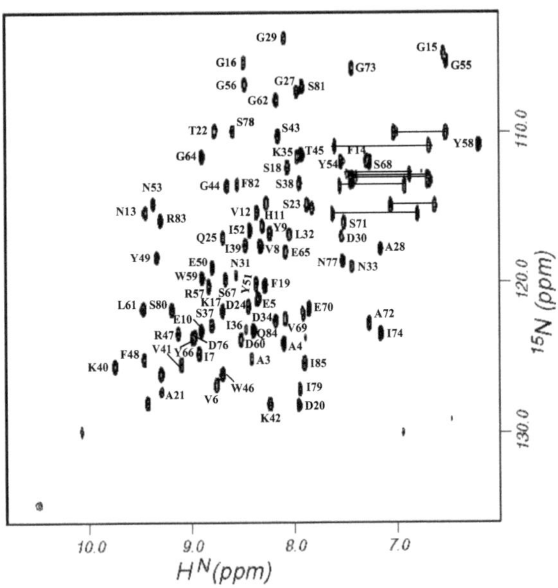

Figure 3.22: Sensitivity-enhanced $[^{15}N\text{-}^1H]$-HSQC spectrum of M-crystallin (M^r=10 kDa) in Ca^{2+} bound form recorded on a 600 MHz NMR spectrometer. Recording time: 10 mins.

5.7 Heteronuclear Multiple–Quantum Correlation Spectroscopy (HMQC)

Figure 3.23 shows the pulse sequence for a HMQC experiment. The in-phase ^{1}H magnetization created by the first $(\pi/2)_x$ ^{1}H pulse is converted into the anti-phase magnetization $-2H_xC_z$ at the end of the first delay τ, which is set to $1/2J_{HX}$. A $(\pi/2)_x$ pulse on the *C* spin converts $-2H_xC_z$ into heteronuclear *multiple-quantum* coherence $(2H_xC_y)$, which evolves only under $\mathcal{H}_\delta$. Thus, it is frequency labelled during the indirect detection time t_1, before it is converted back to an antiphase ^{1}H magnetization $(2H_xC_z)$. Thereafter, it is transformed into an in-phase ^{1}H magnetization using a refocussed INEPT pulse sequence, and is detected during t_2. The delays τ are set equal to $1/2J_{HX}$. The density operator at the final step is given by:

$$\rho_6 = -H_y \cos\Omega_X t_1 + 2H_xC_x \sin\Omega_X t_1; \ \rho_{\text{detectable}} = -H_y \cos\Omega_X t_1 \qquad 3.48$$

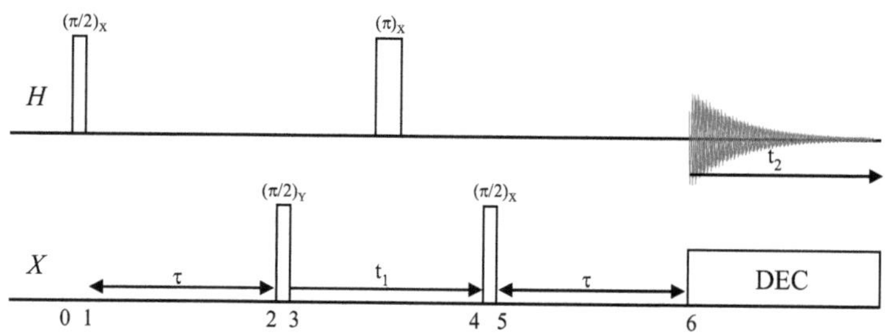

Figure 3.23: The HMQC pulse sequence.

$\rho_{\text{detectable}}$ gives rise to a peak at (Ω_X, Ω_H) with pure in-phase absorptive line shapes along both dimensions. Unwanted coherences are suppressed by a four-step phase-cycling as in the case of the HSQC. As compared to HSQC, HMQC based transfer helps in overcoming signal loses associated with off-resonance effects. In particular, while recording [^{13}C-^{1}H] correlation experiments, HMQC is recommended, since it needs only two ^{13}C pulses.

5.8 Sensitivity enhanced HSQC

In HSQC, ρ_6 has two terms $-H_x \cos(\Omega_X t_1)$ and $-2H_yC_y \sin(\Omega_X t_1)$. As discussed, the second term does not contribute to the detected signal. Thus, one half of the initial *H*-spin polarization is not detected. Sensitivity-enhanced HSQC experiment avoids this problem and enhances the sensitivity of the experiment by a factor of $\sqrt{2}$. A gradient version of the same experiment has been proposed, which has been incorporated in almost all the 3D *triple-resonance* experiments.

The sensitivity enhanced HSQC pulse sequence (Figure 3.24) is similar to the normal HSQC pulse scheme up to the time-point 6. The sensitivity enhancement is achieved by additional pulses and delays (in the form of another refocused-INEPT transfer) used thereafter. As in the case of HSQC, the density operator at the time-point 6 is same. The two terms at the time-point 6 further evolve during the rest of the pulse sequence as follows:

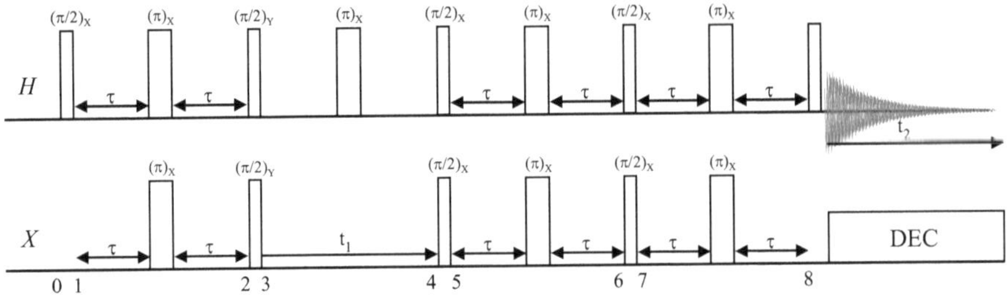

Figure 3.24: Pulse sequence for pure absorption gradient enhanced heteronuclear single quantum coherence (HSQC) spectroscopy.

$$\rho_6 = +H_x cos(\Omega_X t_1) + 2H_yX_y\, sin(\Omega_X t_1);\ \rho_7 = -H_z\, cos\Omega_X t_1 + 2H_yX_z\, sin\Omega_X t_1 \qquad 3.49$$

$$\rho_8 = -H_z\, cos\Omega_X t_1 - H_x sin\Omega_X t_1;\ \rho_{detectable} = +\mathbf{H_y\, cos\Omega_X t_1 - H_x\, sin\Omega_X t_1} \qquad 3.50$$

There are two detectable terms, H_x and H_y which are sine and cosine modulated along the t_1 dimension, respectively. Unlike in the normal HSQC, both terms originate from the initial magnetization H_z. The fact that the two terms have a phase difference of $\pi/2$, their vector sum provides a sensitivity enhancement by a factor of $\sqrt{2}$. The term $\rho_{detectable}$ gives rise to a peak at (Ω_X, Ω_H) with pure in-phase absorptive line shapes along both dimensions. Undesirable coherences are suppressed by a four-step phase-cycling scheme or by using pulse field gradients, discussed below.

6. PULSED FIELD GRADIENTS

Field gradients are routinely used in *Magnetic Resonance Imaging*. In recent years, these have also become popular in molecular studies. Pulsed field gradients (PFGs) are used for solvent suppression and for selection of specific coherence order. The use of PFGs allows minimization of phase cycling steps. This in turn helps in reducing the total instrument time.

In molecular studies, linear field gradients are applied along the z-axis. The gradients change the static magnetic field B_0 across the sample volume linearly for a specific time-period τ (typically $\tau < 10$ msec). Thereafter the original homogeneity is re-established. Hence, these are called pulsed field gradients. When PFG is switched on, it reaches its peak gradient strength with a short rise time (typically < 100 μsec). When switched off it decays within the same time. Such short rise and decay times result in eddy currents in the RF probe. For this reason, PFG coils are self-shielded. During the gradient pulse, the transverse coherences of the order greater than zero are rapidly dephased. The dephased coherence can be restored by the application of a gradient pulse equal in magnitude but opposite in polarity.

Figure 3.25 illustrates the concept of the use of linear PFG. The applied static field, varies along the z-axis with a slope G (G=dB/dz). Thus the magnetic field experienced by the sample is dependent on the value of G, and the z-coordinate. More explicitly, $B_z = B_0+Gz$. In terms of frequency, one can express this relation as $\omega_z = \omega_0+\gamma Gz$. Generally, G is expressed in milli-Tesla cm^{-1}. If the gradient is switched on for a time τ, then the magnetization in the transverse plane at any point z precesses through an angle $\theta_z=\omega_z\tau$. Thus, θ_z is position dependent leading to a net dephasing of *x/y* magnetization over the time-period that the gradients are kept on (Figure 3.25). It may be noted that $\theta_z = -\theta_{-z}$. If another gradient pulse of the same strength (G) but opposite in polarity is applied for the same duration τ, then the dephased magnetization is refocussed. To dephase and then rephrase any magnetization, one may use any combination of the strengths of PFG and their durations, such that the relation $G_1\tau_1=G_2\tau_2$ is satisfied. The duration between dephasing and refocussing gradients should be less than T_2. Diffusion can prevent efficient refocussing.

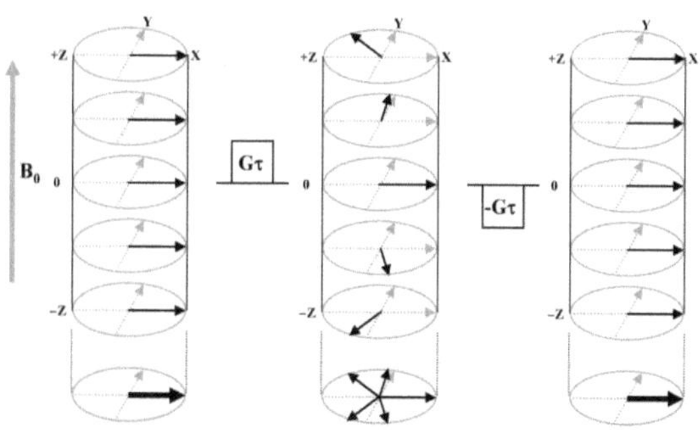

Figure 3.25: Dephasing and refocussing effects on the transverse magnetization of gradient pulses of strength G applied along z-axis for a time-period τ with opposite polarity.

6.1 Coherence Pathway Selection using PFG

In HSQC, magnetization is transferred from ^1H to *X* spin. This is frequency labelled during the indirect detection period, and then transferred back to ^1H. Such a coherence transfer pathway is depicted in Figure 3.26. There is another pathway arising from solvent signal and the protons bonded to the abundant ^{12}C and ^{14}N. These signals need to be suppressed in order to select only the desired pathway (*H→X→H*). As mentioned earlier, such unwanted coherences are suppressed by appropriate phase-cycling, in which two or more transients per each t_1 increment are acquired. For example, in a two-step phase cycle used in HSQC, the phase of one of the pulses applied to the *X* spins is inverted such that the signals arising from the (*H→X→H*) pathway are inverted, but not those arising from (*H→H*). Subtraction of the second transient from the first eliminates the undesirable signals. Problems associated with such a phase-cycling are:

(i) Each transient contains contributions from both the transfer pathways. The dynamic range of ADC is limited by the strong signals from the ($H \rightarrow H$) pathway.
(ii) Even for uniformly labelled samples, the solvent signal is several times stronger than the signals from dissolved molecules.
(iii) Since the method involves subtraction, small changes in pulse widths, phase switching or sample temperature can introduce artefacts.

These problems can be overcome by the use of PFG through a selection of the desired coherence transfer pathway as illustrated in Figure 3.26. The pulse sequence has two gradient pulses; one (G_1) is applied at the end of t_1, and the other (G_2) prior to data-acquisition. The gradient G_1 induces a phase shift of $\gamma_X G_1 \tau_1$ to the X spin transverse magnetization arising from the $H \rightarrow X \rightarrow H$ pathway and a phase shift of $\gamma_H G_1 \tau_1$ to the H transverse magnetization from the $H \rightarrow H$ pathway. After the coherence transfer from X to H, G_2 induces a phase shift of $\gamma_H G_2 \tau_2$ to the H spin transverse magnetization due to $H \rightarrow X \rightarrow H$ pathways and a phase shift of $\gamma_H G_2 \tau_2$ to the H spin transverse magnetization from the $H \rightarrow H$ pathway. Thus, the sum of the phase shifts for the H spin magnetisations arising from the $H \rightarrow X \rightarrow H$ and $H \rightarrow H$ pathways is $\gamma_C G_1 \tau_1 + \gamma_H G_2 \tau_2$ and $\gamma_H G_1 \tau_1 + \gamma_H G_2 \tau_2$, respectively. If the values of G_1, G_2, τ_1 and τ_2 are selected such that $\gamma_C G_1 \tau_1 + \gamma_H G_2 \tau_2 = 0$ (i.e. $G_1 \tau_1 / G_2 \tau_2 = -\gamma_H / \gamma_X$) then the H spin magnetization arising from the $H \rightarrow X \rightarrow H$ pathway is refocussed. The magnetization from the $H \rightarrow H$ pathway acquires a net phase and leads to a net dephasing of the transverse magnetization. Thus, the desired pathway is selected without phase cycling. PFG incorporated pulse sequences thus enhance the dynamic range.

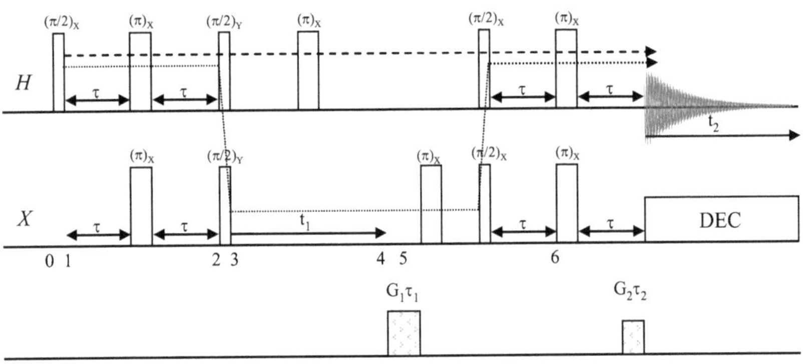

Figure 3.26: Coherence pathway selection using PFGs in HSQC. Rectangular π/2 and π pulses are indicated by thin and thick vertical bars, respectively. Phases are indicated above the pulses. The dotted and dashed lines show the two possible coherence pathways.

6.2 Water Suppression by Gradient Tailored Excitation (WATERGATE)

Efficient water suppression can be achieved using PFG. WATERGATE is one such pulse sequence, where selective pulses and PFGs are used to achieve highly efficient water suppression. The pulse sequence for WATERGATE is shown in Figure 3.27.

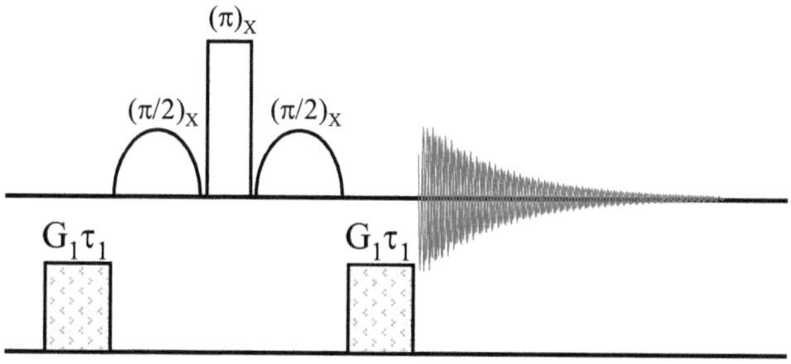

Figure 3.27: WATERGATE pulse sequence. Rectangular π/2 pulse is indicated by thin vertical bar. Selective pulses at water resonance are indicated by dome-like pulses.

The pulse sequence is based on the fact that Gτ–π–Gτ refocusses transverse magnetization. For example, I_x evolves in the following manner under the pulse scheme Gτ–π–Gτ.

$$I_x \rightarrow I_x \cos\theta_z + I_y \sin\theta_z \rightarrow I_x \cos\theta_z - I_y \sin\theta_z \rightarrow (I_x \cos\theta_z + I_y \sin\theta_z)\cos\theta_z - (I_y \cos\theta_z - I_x \sin\theta_z)\sin\theta_z \rightarrow I_x \quad 3.51$$

In the presence of the selective π/2 pulses on the water, one each on either side of the non-selective π pulse, the solvent is subjected to a net nutation of 2π. The effect of the two gradients is given by

$$I_x \rightarrow I_x \cos\theta_z + I_y \sin\theta_z \rightarrow I_x \cos\theta_z + I_y \sin\theta_z \rightarrow (I_x \cos\theta_z + I_y \sin\theta_z)\cos\theta_z + (I_y \cos\theta_z - I_x \sin\theta_z)\sin\theta_z \rightarrow I_x \cos 2\theta_z + I_y \sin 2\theta_z \quad 3.52$$

Thus, the presence of the two selective π/2 pulses applied on water resonance has a dephasing effect on its signal. The solute protons continue to refocus. The success of the technique depends on the selectivity of the pulses, which is achieved by using longer and weaker selective pulses.

7. THREE DIMENSIONAL (3D) EXPERIMENTS

2D experiments can be extended to higher dimensions by concatenating a string of evolution periods, each of which is followed by a mixing period. Figure 3.28 shows the concept of the 3D NMR spectroscopy. It has two evolution and two mixing periods. Depending on the requirement of the experiments, the three frequencies (Ω_1, Ω_2, Ω_3) may correspond to *homo-* or *hetero-nuclear* resonances. Since these experiments involve two incremental time-periods t_1 and t_2, the measurement times are large. Even with 64 data points in each of the two indirect dimensions, 4 transients and 1 sec relaxation delay, the total time works out to more than four hours.

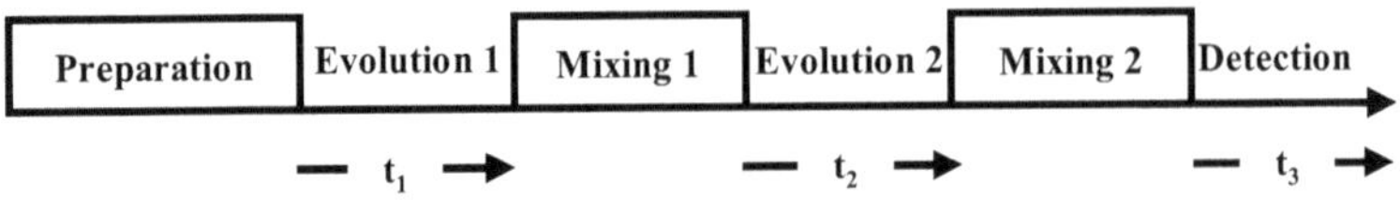

Figure 3.28: Schematic representation of various time-periods in a 3D NMR experiment.

Some examples of *homonuclear* 3D spectroscopy are COSY-COSY, COSY-NOESY, TOCSY-NOESY and NOESY-NOESY. In these experiments, magnetization transfer from each ^1H takes place to more than one coupled partners. The 3D spectra are complex, but provide more information compared to the two individual 2D spectra. The scalar correlations in these spectra rely on small ^1H-^1H couplings. Hence, the spectra have low sensitivity. Such 3D techniques are generally useful for small molecules where the ^1H line-widths are smaller than the J-couplings.

On the other hand, *heteronuclear* 3D experiments have found wide-applications in large biological molecules. Such experiments generally use specific ^{13}C and ^{15}N isotope labelled molecules. One of the two magnetization transfer steps is from a ^1H to its directly attached hetero-nucleus (X). This makes heteronuclear 3D experiments less complex as compared to the homonuclear experiments. In general, the number of peaks in a $^{13}C/^{15}N$-edited-3D experiment corresponds to the number of peaks expected in the corresponding 2D spectrum. Since the *heteronuclear* couplings are large, the corresponding spectra have good sensitivity. The large dispersion in ^{13}C and ^{15}N chemical shifts, which form one of the dimensions in the 3D spectrum is another major advantage.

7.1 3D Double-Resonance Experiments

Concatenating 2D *double-resonance* experiments, such as HSQC and HMQC with COSY, NOESY or TOCSY results in 3D *double-resonance* experiment. For example, one can record COSY-[$^{13}C/^{15}N$-^1H] HSQC, NOESY-[$^{13}C/^{15}N$-^1H] HSQC, TOCSY-[$^{13}C/^{15}N$-^1H] HSQC, TOCSY-[$^{13}C/^{15}N$-^1H] HMQC, NOESY-[$^{13}C/^{15}N$-^1H] HMQC. Such experiments are often used in biomolecular studies.

7.2 3D NOESY-[$^{13}C/^{15}N$-^1H] HSQC

It is difficult to derive complete information on ^1H-^1H distances from the 2D NOESY with large molecules, because of spectral overlaps. This is overcome by recording a $^{13}C/^{15}N$-edited NOESY. The pulse sequence is derived by concatenating NOESY and [$^{13}C/^{15}N$-^1H] HSQC or [$^{13}C/^{15}N$-^1H] HMQC. In the case of ^{13}C-edited NOESY, one uses HMQC for the heteronuclear transfer step as it requires only two ^{13}C pulses. HMQC based transfer helps in overcoming signal loses associated with off-resonance effects. This is particularly important in observing NOEs between the aliphatic and aromatic protons.

The pulse sequence for the 3D NOESY-[^{15}N-^1H] HSQC is shown in Figure 3.29. Figure 3.30 shows a 3D NOESY-[^{15}N-^1H] HSQC spectrum of a 10 kDa protein and a cross section of the same taken at a specific ^{15}N-chemical shift. In the event of spectral overlaps in such 3D resolved spectra, 3D HMQC-NOESY-HMQC or 4D experiments such as [^{13}C-^{1}H]-HMQC-NOESY-[^{15}N-^{1}H]-HSQC spectra or [^{13}C-^{1}H]-HMQC-NOESY-[^{13}C-^{1}H]-HMQC spectra have been used. In all such schemes, the second heteronuclear step results in lower S/N ratio.

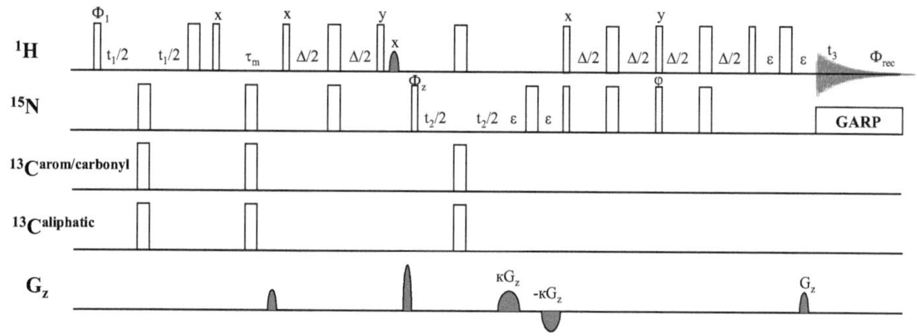

Figure 3.29: The pulse sequence for the 3D NOESY-[^{15}N, ^{1}H]HSQC. Rectangular $\pi/2$ and π pulses are indicated by thin and thick vertical bars, respectively, and phases are indicated above the pulses. $\kappa = \pm 5$; $\psi = \pm y$; $\phi_1 = \pi/4, 5\pi/4 + TPPI(t_1)$; $\phi_2 = x, x, -x, -x$; $\phi_{rec} = x, -x, -x, x$; $\Delta = 5.4$ ms. Where no RF phase is marked, the pulse is applied along x.

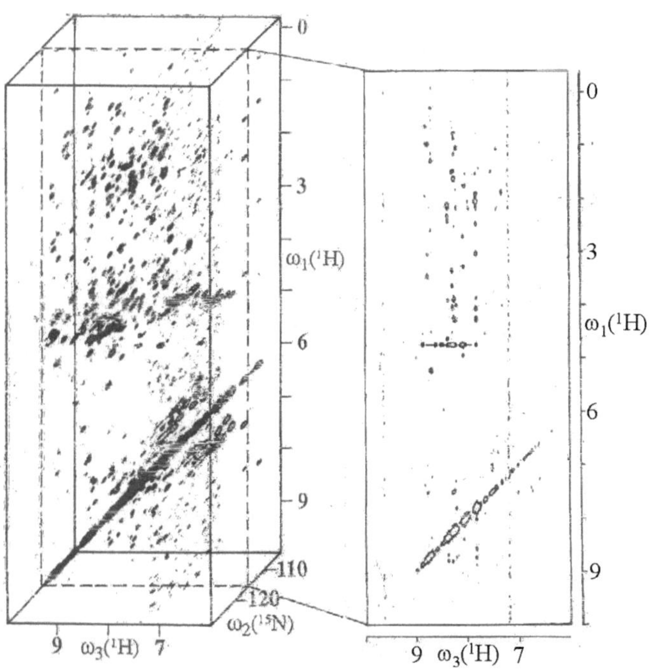

Figure 3.30: 3D NOESY-[^{15}N-^{1}H] HSQC of [u-^{15}N]-labelled glutaredoxin taken in 90% H_2O + 10% $^{2}H_2O$ and a cross section of the same taken at a specific ^{15}N-chemical shift.

7.3 3D TOCSY-[$^{13}C/^{15}N$-^{1}H] HSQC

With larger molecules, it is also difficult to accomplish complete connectivities among various networks of coupled spin systems in the given molecule using 2D-TOCSY. This is again because of spectral overlaps. The $^{13}C/^{15}N$-edited TOCSY helps in overcoming such problems. The pulse sequence is derived by concatenating TOCSY and 2D [$^{13}C/^{15}N$-^{1}H] HSQC or 2D [$^{13}C/^{15}N$-^{1}H] HMQC.

7.4 3D Triple Resonance Experiments

The 3D *triple-resonance* experiments correlate three different kinds of spins namely $^{1}H^{N}$, ^{13}C and ^{15}N spins using one and two-bond J-coupling interactions. These experiments, which are more specific to proteins and nucleic acids are discussed in Chapters 5 and 7, respectively.

8. TECHNIQUES FOR SOLUTION NMR OF VERY LARGE MOLECULES

The *double-* and *triple-resonance* experiments described above, can be used for studies of molecules with M_r of up to ~20-25 kDa. Partial and complete deuteration of the molecular systems may extend this limit to 30 kDa. Beyond this, limitations due to transverse relaxation, restrict the use of these techniques. For work above M_r of 30 kDa, two novel techniques namely, transverse relaxation-optimized spectroscopy (TROSY) and cross-correlated relaxation-enhanced polarization transfer (CRINEPT) have been proposed.

In TROSY, transverse relaxation optimization is achieved by the interference effects between two different types of spin interactions; the chemical-shift anisotropy (CSA) and the dipole-dipole (DD) coupling. TROSY is a heteronuclear-correlation experiment in which ^{1}H magnetization is transferred to X nucleus. This magnetization evolves during t_1 with differential relaxation rate of the X doublet due to CSA of spin X and the DD coupling between spins X and ^{1}H. After the frequency labeling of X, the magnetization is transferred back to ^{1}H, which evolves during t_2 with differential relaxation rate of the ^{1}H doublet due to CSA (of ^{1}H) and DD (^{1}H-X) cross-correlation.

Without decoupling during the time-periods t_1 and t_2, the resulting cross peak has four components. Each of these has different line widths along ω_1 and ω_2 axes (Figure 3.31). As evident from the Figure 3.31 B and C, one of the four components of individual cross peaks experiences line narrowing due to CSA-DD cross-correlation. This is selected by the application of phase-cycling, which simultaneously suppresses the remaining three components. Thus, the use of TROSY enables high-quality spectra for very large molecules.

As an example, Figure 3.31 D shows a TROSY-type ^{15}N-^{1}H correlation spectrum of the uniformly ^{15}N- and ^{2}H-labelled (110 kDa) protein, 7, 8-dihydroneopterin aldolase from *Staphylococcus aureus* in H_2O solution. This protein is a homo-octamer with each subunit consisting of 121 amino acid residues. This is compared with the conventional HSQC spectrum (Figure 3.31 E). Both the spectra have been recorded in an hour with a protein monomer concentration of 0.4 mM (pH

5.5, T 20 °C and ^{1}H frequency 750 MHz). The spectra clearly show that TROSY yields significantly narrower line-widths. The sensitivity improved by 20-50 fold. One may note that the positions of corresponding peaks in these spectra (Figure 3.31 D and E) differ by $^1J(^{15}N\text{-}^1H)/2$ along both the dimensions.

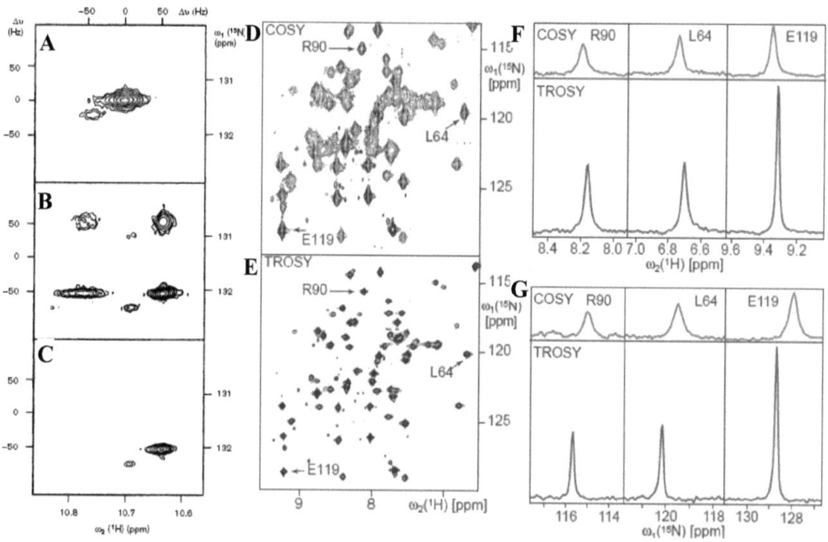

Figure 3.31: Contour plots of a tryptophan indole $^{15}N\text{-}^1H$ cross peak from three different types of $^{15}N\text{-}^1H$ correlation spectra. (A) Conventional broad-band decoupled $[^{15}N\text{-}^1H]$-COSY. (B) Same as (A), without decoupling during the evolution and detection periods. (C) $[^{15}N\text{-}^1H]$-TROSY. G. Wider and K. Wüthrich. Curr. Op. Struct. Biol. 9, 594-601 (1999) (Reproduced with permission from Curr. Opp. Struct. Biol) Comparison of (D) TROSY-type $^{15}N\text{-}^1H$ correlation and (E) conventional HSQC of a 110 kDa protein. (F) shows cross sections along the ^1H frequency axis (ω_2) through three peaks from L64, R90 and E119. (G) shows cross sections along the ^{15}N frequency axis (ω_1)through the same three peaks R. Riek, K. Pervushin and K. Wüthrich. TIBS 25, 462-468 (2000). (Reproduced with permission from TIBS).

Note that TROSY is applied only during the periods t_1 and t_2. Hence, transverse relaxation during the INEPT transfers, tend to become a limiting factor for higher M_r systems. In such situations, CRINEPT becomes a highly efficient transfer mechanism, which overcomes this limitation. It combines INEPT with cross-correlated relaxation-induced polarization transfer. Thus, TROSY and CRINEPT exploit constructive interference between DD coupling and CSA. One can record high resolution spectra of macromolecules with M_r beyond 100 kDa. Combined use of TROSY and CRINEPT enables studies on very large-biomolecular structures. One can even record TROSY-based NOESY experiments which provide structural constraints. These approaches are useful in the studies of large macromolecules intermolecular interactions in supramolecular assemblies.

9. REFERENCES

9.1 Theoretical and Physical Aspects

R.R. Ernst, G. Bodenhausen and A. Wokaun, *Principles of Nuclear Magnetic Resonance in One and Two Dimensions*, Oxford University Press, New York (1985).

R. Freeman, *A Hand Book of Nuclear Magnetic Resonance*, Longman, Harlow (1987).

R. Freeman, *Magnetic Resonance in Chemistry and Medicine*, Oxford University Press, New York (1987).

R. Freeman, *Spin Choreography: Basic Steps in High-Resolution NMR*, Oxford University Press, New York (1987).

M. Munowitz, *Coherence and NMR*, John Wiley, New York (1988).

M. Goldman, *Quantum Description of High-Resolution NMR in Liquids*, Oxford Science Publications, Clarendon Press, Oxford (1992).

J. Cavanagh, W.J. Fairbrother, A.G. Palmer and N.J. Skelton, *Protein NMR Spectroscopy: Principles & Practice*, Academic Press Inc., San Diego (1996).

M.H. Levitt, *Spin Dynamics. Basics of Nuclear Magnetic Resonance*, John Wiley & Sons, Singapore (2001).

9.2 Review Articles

O.W. Sorensen, G.W. Eich, M.H. Levitt, B. Bodenhausen and R.R. Ernst, *Product Operator Formalism for the Description of NMR Pulse Experiments*, Prog. NMR Spect. **16**, 163–192 (1983).

P-K Wang and C.P. Slichter, *A Pictorial Operator Formalism for NMR Coherence Phenomena*, Bull. Magn. Reson. **8**, 3–16 (1986).

9.3 Three Dimensional NMR

C. Griesinger, Ø.W. Sorenson and R.R. Ernst, *A Practical Approach to 3-Dimensional Nmr-Spectroscopy*, J. Magn. Reson. **73**, 574 (1987).

C. Griesinger, Ø.W. Sorenson and R.R. Ernst, *Novel 3-Dimensional Nmr Techniques for Studies of Peptides and Biological Macromolecules*, J. Am. Chem. Soc. **109**, 7227 (1987).

G.W. Vuister and R. Boelens, *Three-Dimensional J-Resolved NMR Spectroscopy*, J. Magn. Reson. **73**, 328 (1987).

S.W. Fesik and E.R.P. Zuiderweg, *Heteronuclear three-dimensional NMR dimensional NMR spectra*, J. Magn. Reson. **78**, 588–593 (1988).

D. Marion, P.C. Driscoll, L.E. Kay, P.T. Wingfield, A. Bax, A.M. Gronenborn, G.M. Clore, Overcoming the overlap problem in the assignment of 'H-NMR spectra of larger proteins by use of three-dimensional heteronuclear 'H-''N Hartrnann-Hahn multiple quantum coherence and nuclear-Overhauser multiple quantum coherence spectroscopy: Application to interleukin 1β, Biochemistry **28**, 6150–6156 (1989).

S.W. Fesik and E.R.P. Zuiderweg, *Heteronuclear three-dimensional NMR spectroscopy of isotopically labeled macromolecules*, Q. Rev. Biophys. **23**, 97–131 (1990).

L.E. Kay, M. Ikura, R. Tschudin, and A. Bax, *Three-dimensional triple resonance NMR spectroscopy of isotopically enriched proteins*, J. Magn. Reson. **89**, 469–514 (1990).

R.T. Clubb, V. Thanabal, G. Wagner, *A constant-time 3-dimensional triple-resonance pulse scheme to correlate intraresidue proton ($^{1}H^{N}$), nitrogen-15 (^{15}N) and carbon-13 (^{13}C) chemical shifts in nitrogen-15, carbon-13 labeled protons*, J. Magn. Reson. **97**, 213–217 (1992).

S. Grzesiek and A. Bax, *Improved 3D triple-resonance NMR techniques applied to a 31 kDa protein*, J. Magn. Reson. **96**, 432–440 (1992).

G. Wider and K. Wüthrich, *NMR spectroscopy of large molecules and multimolecular assemblies in solution*, Curr. Opin. Struct. Biol. **9**, 594–601 (1999).

R. Riek, K. Pervushin and K. Wüthrich, *TROSY and CRINEPT: NMR with large molecular and supramolecular structures in solution*, TIBS **25**, 462–468 (2000).

Appendix 3.1: Matrix Representation of Two-spin Product Operators

$$\tfrac{1}{2}E = \tfrac{1}{2}\begin{pmatrix} 1 & 0 & 0 & 0 \\ 0 & 1 & 0 & 0 \\ 0 & 0 & 1 & 0 \\ 0 & 0 & 0 & 1 \end{pmatrix};\ I_{kx} = I_{lx} = \tfrac{1}{2}\begin{pmatrix} 0 & 0 & 1 & 0 \\ 0 & 0 & 0 & 1 \\ 1 & 0 & 0 & 0 \\ 0 & 1 & 0 & 0 \end{pmatrix};\ I_{ky} = I_{ly} = \tfrac{1}{2}\begin{pmatrix} 0 & 0 & -i & 0 \\ 0 & 0 & 0 & -i \\ i & 0 & 0 & 0 \\ 0 & i & 0 & 0 \end{pmatrix}$$

$$I_{kz} = I_{lz} = \tfrac{1}{2}\begin{pmatrix} 1 & 0 & 0 & 0 \\ 0 & -1 & 0 & 0 \\ 0 & 0 & 1 & 0 \\ 0 & 0 & 0 & -1 \end{pmatrix};\ 2I_{kz}I_{lz} = \tfrac{1}{2}\begin{pmatrix} 1 & 0 & 0 & 0 \\ 0 & -1 & 0 & 0 \\ 0 & 0 & -1 & 0 \\ 0 & 0 & 0 & 1 \end{pmatrix};\ 2I_{kx}I_{lz} = \tfrac{1}{2}\begin{pmatrix} 0 & 0 & 1 & 0 \\ 0 & 0 & 0 & -1 \\ 1 & 0 & 0 & 0 \\ 0 & -1 & 0 & 0 \end{pmatrix}$$

$$2I_{kz}I_{lx} = \tfrac{1}{2}\begin{pmatrix} 0 & 1 & 0 & 0 \\ 1 & 0 & 0 & 0 \\ 0 & 0 & 0 & -1 \\ 0 & 0 & -1 & 0 \end{pmatrix};\ 2I_{ky}I_{lz} = \tfrac{1}{2}\begin{pmatrix} 0 & 0 & -i & 0 \\ 0 & 0 & 0 & i \\ i & 0 & 0 & 0 \\ 0 & -i & 0 & 0 \end{pmatrix};\ 2I_{kz}I_{ly} = \tfrac{1}{2}\begin{pmatrix} 0 & -i & 0 & 0 \\ i & 0 & 0 & 0 \\ 0 & 0 & 0 & i \\ 0 & 0 & -i & 0 \end{pmatrix}$$

$$2I_{kx}I_{lx} = \tfrac{1}{2}\begin{pmatrix} 0 & 0 & 0 & 1 \\ 0 & 0 & 1 & 0 \\ 0 & 1 & 0 & 0 \\ 1 & 0 & 0 & 0 \end{pmatrix};\ 2I_{ky}I_{ly} = \tfrac{1}{2}\begin{pmatrix} 0 & 0 & 0 & -1 \\ 0 & 0 & 1 & 0 \\ 0 & 1 & 0 & 0 \\ -1 & 0 & 0 & 0 \end{pmatrix};\ 2I_{ky}I_{lx} = \tfrac{1}{2}\begin{pmatrix} 0 & 0 & 0 & -i \\ 0 & 0 & -i & 0 \\ 0 & i & 0 & 0 \\ i & 0 & 0 & 0 \end{pmatrix}$$

$$2I_{kx}I_{ly} = \tfrac{1}{2}\begin{pmatrix} 0 & 0 & 0 & -i \\ 0 & 0 & i & 0 \\ 0 & -i & 0 & 0 \\ i & 0 & 0 & 0 \end{pmatrix}$$

CHAPTER 4

BIOMOLECULAR STRUCTURES USING NMR: GENERAL PRINCIPLES

1. INTRODUCTION

Biomolecules such as proteins and nucleic acids are linear polymers formed by condensation reactions between individual monomeric units. The condensation leads to specific linkages, namely, the peptide bond in proteins and the 3′-5′ sugar-phosphate linkage in DNA and RNA (Figure 4.1). Each class of these molecules therefore have a repeating backbone (B_i). However, they have variations in the side chains, R_i .

The building blocks for peptides and proteins are the twenty L-amino acids: Alanine (Ala; A), Arginine (Arg; R), Asparagine (Asn; N), Aspartic acid (Asp; D), Cysteine (Cys; C), Glutamic acid (Glu; E), Glutamine (Gln; Q), Glycine (Gly; G), Histidine (His; H), Isoleucine (Ile; I), Leucine (Leu; L), Lysine (Lys; K), Methionine (Met; M), Phenylalanine (Phe; F), Proline (Pro; P), Serine (Ser; S), Threonine (Thr; T), Tryptophan (Trp; W), Tyrosine (Tyr; Y) and Valine (Val; V).

In the case of nucleotides and nucleic acids, the five-member furanose ring (α-D-ribose for RNA and deoxyribose for DNA) forms the central moiety. These sugars are linked to each other through a 3′-5′ sugar-phosphate backbone. DNA and RNA use four different aromatic bases in their side-chains: Guanine, (Gua; G) Cytosine (Cyt; C), Adenine (Ade; A) and Thymine (Thy; T)/Uracil (Ura; U).

Both proteins and nucleic acids are functional molecules. Generally, each molecule has a specific role to play in living systems. To perform their tasks, each of these polymers selects a unique sequence in which the monomeric units are linked together.

Carbohydrates are condensation products of six member sugar rings (hexoses), such as D-glucose (Glc). Glc is involved in a large number of metabolic processes and both its α- and β-configurations are readily inter-convertible in such reactions. Glc is stored in the form of glycogen, a branched chain polymer. Besides being an important source of energy, sugars and oligo-saccharides play specific roles in intercellular recognition. Their conjugates with proteins and lipids also have specific roles. Sugar molecules such as Glc can be linked to each other in polysaccharides, by condensation through the hydroxyl groups of the two units. As can be seen from the Figure 2.5, there is more than one way to link these units. In polysaccharides, one observes a wider variability in the variety of sugars and their linkages, as compared to proteins and nucleic acids.

In this Chapter, the general principles employed in the studies of biopolymers are outlined. The focus is mainly on proteins and nucleic acids. The subject will be discussed in more detail in Chapters 5 and 6 (for proteins), Chapter 7 (for nucleic acids) and Chapter 8 and 9 (for carbohydrates and biological membranes).

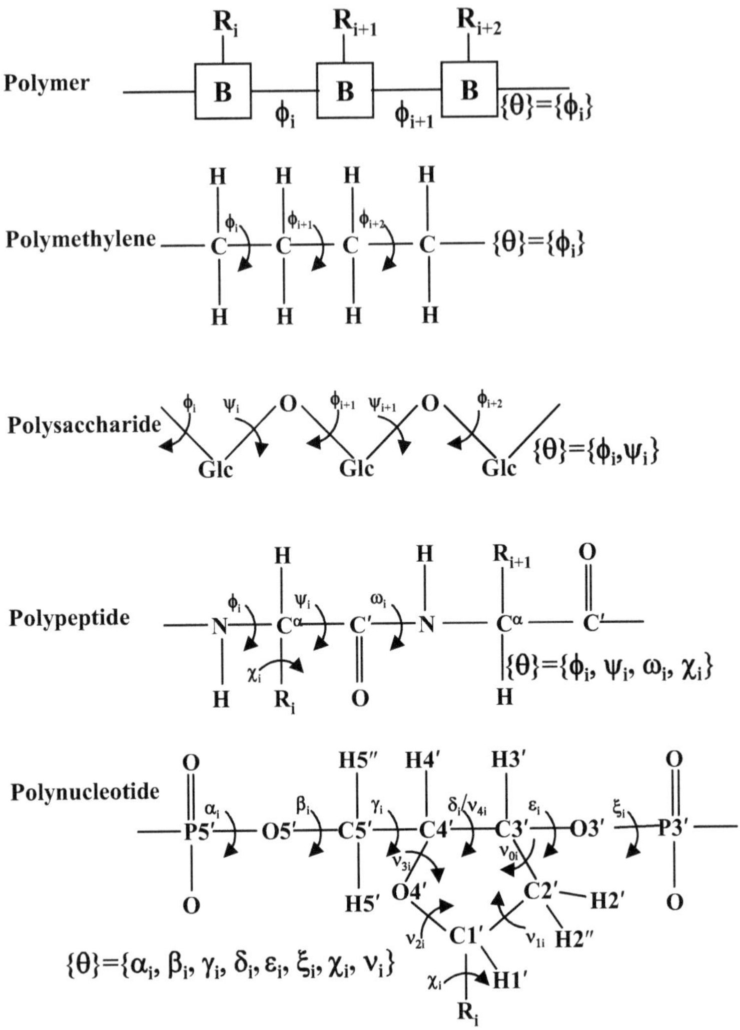

Figure 4.1: Schematic representation of linear polymers. B represents backbone, R_i are the side chains and $\{\theta_i\}$ is a set of torsion angles which determine the conformation of the ith unit.

1.1 Conformation of Biological Molecules

Consider a linear molecule consisting of four atoms A-B-C-D. The 3D structure of such a molecule is determined by the three bond lengths (A-B, B-C and C-D), two bond angles (A-B-C and B-C-D) and one torsional angle (the dihedral angle θ, between the two planes containing atoms A, B and C, and atoms B, C and D). In general, to define the 3D geometry for a linear molecule containing N atoms, one needs N-1 bond-lengths, N-2 bond angles and N-3 torsional angles, a total of (3N-6) structural elements. A more convenient way to define the 3D structure of a molecule is through the x, y and z coordinates of the atoms in a Cartesian space, with 3N structural elements.

Of the 3N-6 independent parameters, the values of bond length and bond angle in similar molecular systems change only slightly. However, potential energies for rotation around single bonds are significantly smaller and are of the order of a few kJ/mole. Therefore, molecules generally have a high degree of flexibility for rotation around chemical bonds. Thus, the problem of obtaining the 3D structure of a molecule can be reduced to determination of N-3 torsional angles, or the conformation of the molecule.

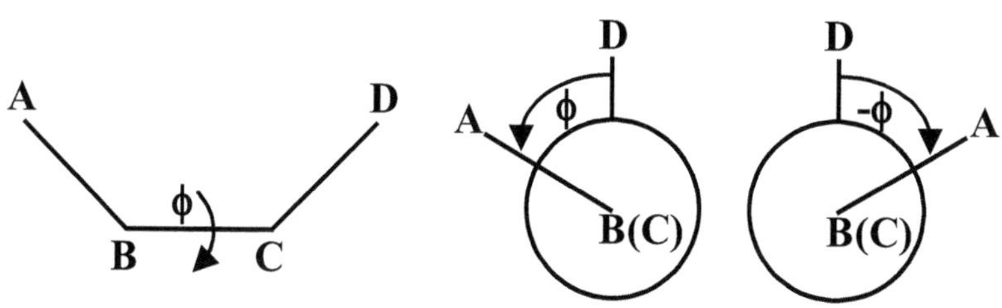

Figure 4.2: The Newman projections and definition of torsional angles for a four-atom molecular segment, A-B-C-D.

In Chapter 1, a simple example of the conformational behaviour of a substituted ethane has already been presented. As a result of rotation around the C-C bond in 1,1,2,2 tetrabromofluoroethane, it adopts two conformations, *gauche (g)* and *trans (t)*, corresponding to the H-C-C-F angles of 60° and 180°, respectively. These two states are called conformers, rotational isomers, rotamers or secondary structures. Such conformers are most conveniently distinguished by their Newman projections (Figure 4.2). The torsional angle ϕ around the bond B-C is usually measured using a right handed rotation: i.e. while viewing along the bond B-C, the front atom A moves in an anticlockwise fashion relative to the far atom D. The zero value for ϕ, corresponds to the situation when the bonds A-B and C-D eclipse each other (*eclipsed* conformation). Since the barrier to internal rotation around single bonds is of the order of only a few kJ/mole, a fair degree of conformational freedom exists around the mean position. It is, therefore, preferable to use a nomenclature, which represents a family of structures covering a range of torsion angles, rather than precise values of ϕ. Table 4.1 lists some of the common conformations.

1.2 Conformational Theory

Certain basic principles of conformational theory will prove helpful in later discussions. These concepts are introduced through the examples of n-butane and n-pentane. Such hydrocarbon chains are important in the conformational behaviour of lipids, glycolipids and lipoproteins.

Table 4.1: Typical conformations in biopolymers along with their ranges of torsional angles.

Classical Name	*Range of torsional angles and areas of application*
gauche	60 ± 20; Side chains in proteins; fatty acids chains
trans	180 ± 20; Protein side chains; fatty acids; backbone of nucleic acids
endo, exo	Ring conformation of furanose in nucleic acids
chair, boat	Conformation of six member rings in sugars
syn, anti	± 40; 180± 40; Orientation of bases in nucleotides

The potential energy of n-butane as a function of torsion angle around the central C-C bond is shown in Figure 4.3. There are three low energy conformations; two *gauche* (g^+ and g^-), and one *trans* (*t*) conformers, which correspond to minima in the potential energy curve. The two *gauche* isomers are equivalent. The difference of free energy between *g* and *t* conformer (ΔG) is related to population distribution (P_g and P_t) between the two isomeric states (Equation 4.1).

$$P_g/P_t = 2\ exp\ (-\Delta G/RT) \qquad 4.1$$

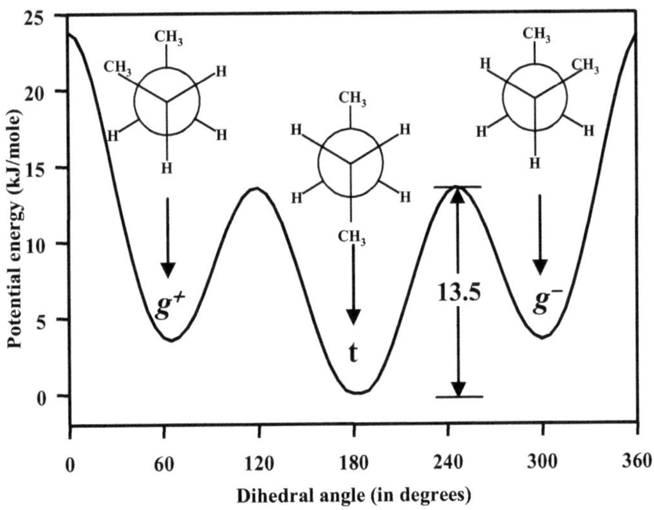

Figure 4.3: Potential energy of n-butane as a function of dihedral angle around the central C-C bond.

The energy of activation for transition from *t* to one of the *g* forms (ΔG^*) is related to the rate of interconversion (k_r) between the two conformers through Equation 4.2:

$$k_r(t \rightarrow g) = (kT/h)\ exp\ (-\Delta G^*/RT) \qquad 4.2$$

At room temperature (T = 300K), the experimental value of ΔG for hydrocarbon chains (around 13.5 kJ/mole) corresponds to a value of 1:4 for P_g/P_t. The value of ΔG^* (around 13.5 kJ/mole) corresponds to a rate constant of 10^9 per second. As seen in Chapter 1, such an interconversion is fast on the NMR time scale. Therefore a time-averaged spectrum is observed.

When dealing with n-pentane, torsional angles along two adjacent C-C bonds (represented by two torsional angles, ϕ and ψ) have to be considered. The conformational energies can be represented by a two dimensional (2D) isoenergy map (Figure 4.5). *A priori*, nine low-energy conformational states are expected. Arising from the molecular symmetry, these states can be divided into four sets with equal energies for each member in the group, namely: [g^+g^+, and g^-g^-]; [g^+t, g^-t, tg^+ and tg^-]; [g^+g^- and g^-g^+]; and [*tt*]. A simple model building shows that in g^+g^- and g^-g^+ states, terminal methyl groups are close in space. Therefore, such conformers have high energies due to steric repulsions. The energies of other states are as indicated in Figure 4.4.

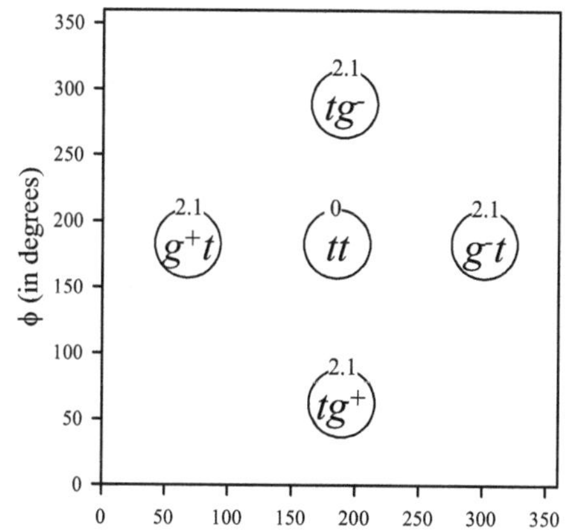

Figure 4.4: 2D iso-energy map of n-pentane showing low energy conformers for a hydrocarbon chain.

Two points emerge from these considerations. The stable structures of larger polymers are more restricted than those of smaller molecules from which they are derived. On the other hand, significantly larger domains of the conformational space need to be probed to get a complete picture of the conformational energy landscape of polymers. Complex structures require determination of energy in a multi-dimensional conformational space. Such energies can be calculated either by Classical Potential Functions (CPF) or by quantum chemical methods. Thus, a complete energy landscape for larger molecules can be obtained through theoretical calculations.

Fortunately, conformational energies are mainly determined by nearest neighbour interactions. The long distant interactions are small, unless chemically distal parts of the molecule fold in such a way that these residues come close to each other (example is the g^+g^- and g^-g^+ conformers in n-pentane). Therefore, isoenergy maps are extremely useful to predict possible conformations of a polymer. These maps provide clues for the starting conformations in energy minimization schemes in simulations of biomolecular structures based on experimental results. The most frequently referred map in this connection is the Ramachandran (ϕ, ψ) plot for proteins, which is used to assess the correctness of a protein structure determined experimentally. Similar maps are available in literature for other classes of biological molecules.

1.3 Conformational Domains of Proteins and Nucleic Acids

Figure 4.1 represents a simplistic view of linear polymers such as proteins and nucleic acids. The backbone B consists of the same repeating unit throughout the sequence. The repeating unit in proteins is the peptide linkage formed by the condensation of the C-terminal of one amino acid with the N-terminal end of the other. In nucleic acids the backbone consists of the 3′-5′ linked deoxyribose-phosphate in DNA and 3′-5′ ribose-phosphate in RNA. Each residue in such a polymer is characterized by a specific side chain, R_i. The 3D structures of linearly linked polymers are therefore described by a set of torsional angles $\{\theta_i\}$, which define the backbone conformation. Side-chain torsional angles define the spatial distribution of atoms in the side-chains relative to the backbone of the biopolymer.

In proteins, the backbone torsion angles are denoted by three angles, ϕ_i, ψ_i and ω_i (Figure 4.1). Arising from a partial double bond character in the peptide bond, the amide group is planar and ω_i acquires a value close to 180° (*trans* conformation). The conformational determinants in proteins are thus generally reduced to a set of angles $\{\phi_i, \psi_i\}$. In addition, the side chain conformations have to be determined, which are usually denoted by a set of conformational angles $\{\chi_i\}$.

The conformations of nucleic acids are determined by six backbone torsion angles, namely, α_i, β_i, γ_i, δ_i, ε_i, and ξ_i (Figure 4.1). Another variant is the conformation of the five member sugar ring, ribose or deoxyribose. Further, the orientation of individual bases with respect to the sugar moiety is described by the glycosidic torsional angle χ_i. The notation and presentation of structures of proteins and nucleic acids that are used for different atoms in these molecules are governed by certain international conventions, which are discussed later.

To uniquely define the conformation of these two classes of biopolymers, each of these conformational angles has to be determined. Since a typical biopolymer may contain hundreds of residues, the 3D structure determination is a Herculean task. Fortunately, most 3D structures of proteins and nucleic acids usually consist of certain repeating motifs, called the secondary structural elements. Identification of such segments greatly simplifies the problem.

2. ELEMENTS OF MACROMOLECULAR STRUCTURES

The structures of proteins and nucleic acids are generally described at four levels: primary, secondary, tertiary and quaternary. Multi-molecular assemblies are a higher level of organization. Such assemblies are formed by interactions between a set of molecules, often belonging to different classes of biopolymers.

2.1 Primary Structure

The sequence of amino acid residues in proteins or nucleotides in nucleic acids is called the primary structure. Powerful chemical and biological techniques are available for the sequence determination of proteins, nucleic acids and polysaccharides. In fact, the success of human genome project is essentially based on the sequence analysis of three billion nucleotide long DNA. Generally, the primary sequence needs to be known before a researcher embarks on determination of the 3D structure using either of the two popular experimental techniques: NMR or X-ray crystallography. However, there have been instances where the sequence determined by biochemical techniques had errors, which have been corrected by NMR or X-ray crystallography.

```
MAEALFKQLDANGDGSVSYEEVKAFVSSKRPIKNEQLLQLIFKAIDIDGNGEIDLAEFTK 60
MAEALFKQLDANGDGSVSYEEVKAFVSSKRPIKNEQLLQLIFKAIDIDGNGEIDLAEFTK 60
MAEALFKEIDVNGDGAVSYEEVKAFVSKKRAIKNEQLLQLIFKSIDADGNGEIDQNEFAK 60
*******::*.****:***********.**.************:** *******  **:*

FAAAVKEQDLSDEKVGLKILYKLMDADGDGKLTKEEVTTFFKKFGYEKVVDQIMKADANG 120
FAAAVKEQDLSDEKVGLKILYKLMDADGDGKLTKEEVTTFFKKFGYEKVVDQIMKADANG 120
FYGSIQGQDLSDDKIGLKVLYKLMDVDGDGKLTKEEVTSFFKKHGIEKVAEQVMKADANG 120
* .::: *****:*:***:******.************:****.* ***.:*:*******

DGYITLEEFLAFNL 134 EhCaBP2
DGYITLEEFLAFNL 134 Calmodulin_Fragment
DGYITLEEFLEFSL 134 EhCaBP1
********** *.*
```

*Figure 4.5: Primary sequences of three Ca^{2+} binding proteins. The sequences have been aligned such that the Ca^{2+} binding domains overlap. * denotes that the residues in the column are identical; : denotes conserved substitutions; .implies semi-conserved substitutions.*

As an example, a set of primary sequences of three medium size proteins is given in Figure 4.5. CLUSTAL W (1.82) (www.ebi.ac.uk/clustalw) has been used in the alignment of the primary sequences of three Ca^{2+} binding proteins. The common functional feature of these proteins is that each one has four binding sites for Ca^{+2}. The four Ca^{2+}-binding loops in the set of primary sequences (underlined in Figure 4.5) are highly homologous. Several additional points may be observed in such an alignment. Firstly, these proteins have similar lengths. Secondly, most of the amino acid residues are similar. Such a similarity is known as sequence homology. It is a good way to probe whether two proteins are likely to have similar biophysical and biochemical properties and also similar 3D structures.

2.2 Secondary Structures

A helical structure is formed when the values of the backbone torsional angles $\{\theta_i\}$ are nearly the same for all residues in a particular molecular stretch, and do not depend significantly on the nature of side chains. The major stabilization forces in helical structure arise from non-covalent interatomic interactions in the backbone. The helical structures are further stabilized by intra- and inter-molecular hydrogen bonds. Each residue in such segments acquires a conformation such that the backbone of each unit is identical to its neighbours. Clearly, in such a situation, the number of independent structural parameters is less. The conformation dependent NMR properties of such helical segments show a high degree of regularity. Examples of some well known helical structures are DNA duplex, triplex, tetraplex structures; triple stranded structures of collagen, α- and 3_{10}-helices, parallel and anti-parallel β-sheet structures in proteins; and structures of certain polysaccharides. These are often referred to as secondary structural elements.

2.3 Tertiary Structures

A more general situation is encountered in globular proteins, enzymes and RNA. Even though α-helix, β-sheets or duplex conformation dominate 3D structures of such molecules, these ordered segments are usually connected by regions such as loops and turns. Some of the regions may have well defined geometry while others may be relatively more flexible. The complete 3D structure of a protein (tertiary structure) is thus the sum-total of all such elements. A similar situation occurs in RNA, where intramolecular duplexes are connected by loops. On the other hand, DNA generally has a linear helical structure involving two strands, held together by inter-strand hydrogen bonds. The side chain interaction with the backbone plays a major role in dictating the conformation or tertiary folding of the molecules.

2.4 Quaternary Structures

Certain proteins are formed from more than one polypeptide chain, which are held together by inter-chain hydrogen bonding and salt bridges. The 3D structures at this level are called quaternary structures. A typical example of such a multimeric structure is haemoglobin, the carrier of oxygen in blood, which contains four polypeptide chains.

The NMR spectra of ordered structures show well resolved resonances from each residue and provide structural details at atomic level. Even though the ordered structures of biopolymers are relatively rigid, NMR evidences suggest that there is a fair amount of flexibility, particularly in the loops. Such a dynamic behaviour plays a key role in biological activity.

2.5 Multi-molecular Assemblies

Biological molecules may bind to each other to form multi-molecular assemblies. In these cases, the complex entity as a whole is involved in a biological function. Most membrane bound molecules fall in this category. Hundreds of lipids and sometimes sugar molecules may bind to a macromolecule. Other examples of multi-molecular assemblies are chromosomes, protein-DNA complexes, photosynthetic systems and viruses.

2.6 Random Coil Structures

Biopolymers can go to disordered states by heating, a process usually called melting or thermal denaturation. Changes in pH, solvent, ionic concentrations or addition of certain compounds such as urea, which weaken hydrogen bonded network, can also cause transition from an ordered to a disordered state. In such a situation, the set of backbone torsion angles $\{\theta_i\}$ are not rigidly fixed, but fluctuate both in time and space. Such polymeric structures are called random-coils. However, a short-range order may be present in such molten structures. Thus, though the random coils may show structural propensities for certain secondary structures, the over all conformation of the molecule is not defined in space. The NMR spectra in such cases show a time-averaged behaviour.

3. SAMPLE PREPARATION FOR NMR: LABELLING TECHNIQUES

For NMR studies, a relatively large amount of material is required. Preparation, isolation and purification of biological molecules in itself, is a vast and multi-faceted subject. The sample preparation involves expression, isolation and purification of one of the four major types of biological molecules: proteins, nucleic acids, carbohydrates and lipids. For structure determination of larger biopolymers, it is desirable to carry out 3D- or 4D- multinuclear NMR. In such cases, it is essential that the molecules be labelled with ^{13}C and ^{15}N. Deuterium (^{2}H) labels are also used either to simplify the spectra or to achieve narrowing of various spin resonances (^{1}H, ^{13}C or ^{15}N) in large molecules. In some cases, ^{19}F label has been incorporated in biological molecules. ^{19}F has a high sensitivity and large chemical shifts. In fact, one uses as many NMR active nuclei (^{1}H, ^{2}H, ^{13}C, ^{15}N, ^{19}F and ^{31}P) as possible.

Genetic engineering has made an enormous impact in our ability to uniformly or selectively label macromolecules by taking advantage of the metabolic pathways in cell cultures, or by employing cell free expression. The isotope labelling procedures have helped in the development of several 2D and 3D double and triple resonance experiments to carry out sequence specific ^{1}H, ^{13}C and ^{15}N resonance assignments in molecules. This is an important and essential step towards complete 3D structural characterization. The cost of labelled compounds, such as ^{15}N-labelled NH_4Cl or ^{13}C-labelled Glc, which are used as the sole sources of nitrogen or carbon respectively, has been decreasing substantially. New advances in biotechnology are leading to new and better methods for protein expression and purification.

Thus, the emergence of NMR as a powerful tool to unravel 3D structures of proteins and nucleic acids and their complexes, has been greatly aided by the simultaneous emergence of new techniques in molecular biology for isotopic enrichment of biomolecules with stable isotopes. This, in turn, has enabled heteronuclear multidimensional NMR experiments, that utilize the favourable properties of ^{13}C and ^{15}N nuclei, such as the large one- and two-bond heteronuclear scalar coupling constants [$^{1}J(^{15}N-^{1}H) \sim 90$ Hz and $^{1}J(^{13}C-^{1}H) \sim 150$ Hz] and the large chemical shift dispersions (amide $^{15}N \sim 40$ ppm and $^{13}C \sim 200$ ppm). The larger ^{1}J values result in an increased sensitivity of double and triple resonance NMR. The large chemical shift dispersions help in overcoming problems of spectral overlaps.

3.1 Preparation of Samples for NMR

High purity of the sample is necessary for NMR experimentation. Even small amount of impurities can seriously hamper interpretation of the spectrum. Tags and end-groups which are used during the protein expression, synthesis and purification should be removed. It is recommended that mass spectroscopy analysis be carried out to check that the sample is pure and the sequence is correct.

Samples are prepared by taking appropriate amounts of the compounds of interest, salts and buffer solutions which are added together in an Eppendorf tube, in proportion to the desired final concentrations. These are lyophilized, and redissolved in an appropriate solvent, and transferred to NMR tube. Commonly, two sets of solutions are prepared. One is with almost fully deuterated solvent (typically 99.99% $^{2}H_2O$), such that the water resonance is weak and resonances of non-labile protons can be detected with ease. Special care is taken in preparing the sample in 99.99% $^{2}H_2O$ to avoid any residual water. For this purpose, one may lyophilize the sample and redissolve it in 99.99% $^{2}H_2O$ and repeat the process two to three times. The final addition of the 99.99% $^{2}H_2O$ is carried out in a nitrogen chamber. A number of protons such as those in hydroxyl, imino, amino and amide groups are exchanged with ^{2}H in such solutions. These signals provide important information on hydrogen bonds. Therefore, a second solution is prepared in H_2O. Typically a mixture of 95% H_2O and 5% $^{2}H_2O$ is used; a small amount of $^{2}H_2O$ is needed for the lock signal.

Conditions of pH, salt concentrations and temperatures are carefully optimised to obtain good spectra amenable for interpretation. Thin pH micro-electrodes that fit into 5 mm NMR tubes help in accurately measuring the pH. It is advisable to check the pH of the sample, before embarking on to long time acquisitions of multidimensional NMR data, and again after the experiment is complete.

Obtaining a stable biological sample in concentrations high enough to acquire reasonably good spectrum, is the central and pivotal step in its structural study using NMR spectroscopy. NMR requires highly concentrated and purified samples of the order of 0.5-1 mM. For spectrometers equipped with cryogenically cooled probes, a concentration of 50 μM may suffice, in special situations. The volume of the sample used in a typical experiment is 600 μl. The need for such small volumes allows use of reasonably expensive purification protocols. Salient features of each aspect, from cloning to purification, involved in acquiring a sample suitable for NMR spectroscopy, are briefly described below:

3.1.1 Cloning and Expression of Proteins

The methods for protein expression and purification by biosynthetic techniques are well established, primarily because of their demand in industrial applications. Solid-state synthesis has been extensively used for preparing small peptides. In most cases, proteins are prepared by gene expression in a microbiological system. (Readers may find the site *www.lsbu.ac.uk/biology/enztech* useful for an introduction to this subject). In such biosynthesis, protein yield depends not only on the specific sequence but also on the vector type, host cell and culture conditions used. The design and PCR amplification of the proper gene is followed by screening of potential clones for insertion into a donor vector. These are then introduced into the recipient vector (*E. coli,* yeast, mammalian cells, etc.) for subsequent protein expression (Figure 4.6).

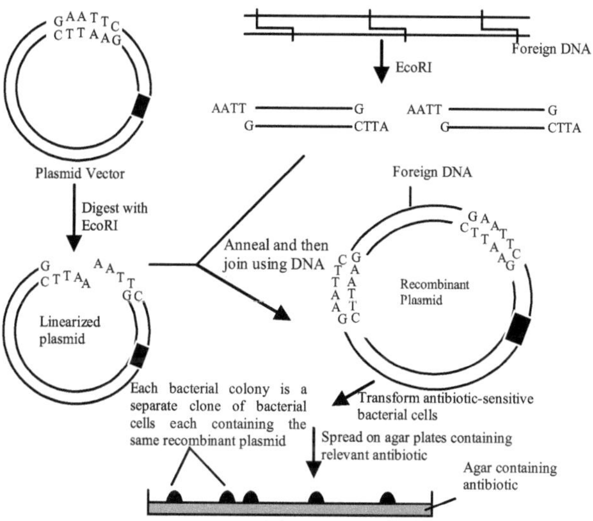

Figure 4.6: Protocol for biosynthetic expression of proteins.

Although both prokaryotic and eukaryotic organisms have been used as hosts for expression, the bacterial systems (especially *E. coli*), are cheaper and faster than eukaryotic systems. In addition, the bacterial systems produce proteins without post-translational modifications and provide high yields allowing quantities of the order of a few milligrams per litre of the culture medium. This is a typical amount needed for NMR studies. Another advantage is that a number of vector/host systems are commercially available. So is the case with the N-terminal expression tags, promotors, regulators, initiation and translation signal elements, and genetically modified strains. A cost-effective *E. coli* expression is achieved using T7 RNA polymerase and T7 promotor. Specialised strains of *E. coli* are being developed on a regular basis. The modified strains can confer certain advantages and provide cost effective systems. For example, protease deficient strains improve the likelihood of isolating full-length recombinant protein.

In addition to the host strain genotype, optimum protein expression depends on several other important variables such as the media conditions, the nature of induction, temperature, induction levels and specific media formulations (such as supplementary salts, cofactors or prosthetic groups). It is important that the proteins do not unfold during preparation and purification phases. The conditions should be controlled so as to result in high yields, preventing proteolysis and denaturation. Optimizing the yield of properly folded protein may require reduction in the induction level. In some cases, optimal expression may require supplementary salt and cofactors or prosthetic groups. When human genes are cloned in bacterial systems, several heterologous proteins lead to insoluble proteins during their expression. These are known as inclusion bodies. Eukaryotic expression system is preferred for proteins that tend to form inclusion bodies, since a post-translational modification is required for their proper folding. Several modifications in expression system have been suggested to minimize inclusion bodies.

A number of proteins cannot be expressed by bacterial systems, since they do not have the machinery for post-transitional modification, such as glycosylation. In such cases, yeast expression systems are preferred. An alternative is to use mammalian cell systems. Hybridoma cells have been used to express antibodies and a number of other proteins.

3.1.2 Taming Proteins

Certain precautions are needed during protein preparation before proceeding to recovery and purification. Problems related to protein solubility can be due to a number of reasons including aggregation, denaturation, instability and degradation. The high concentrations used in NMR may result in aggregation of proteins into dimers or higher oligomers due to a variety of reasons. The most common problem is aggregation due to oxidation of cysteine residues. This can be reduced by addition of 1,4-dithiothreitol (DTT) or β-mercapto-ethanol. Aggregation of proteins with large surface hydrophobicity is another common phenomenon. It is difficult to prevent such aggregation without compensating for the loss of native structure, by adding surfactants like SDS or agents such as urea and thiosulfates.

Denaturation can be caused by factors such as temperature, extreme pH and surface effects. The first two have been studied extensively. The surface effects, are due to spreading out ability of amphiphilic protein molecules at accessible regions like air/water interfaces. Protein degradation due to proteases released during the cell disruption, is also common. It can be avoided by the inclusion of protease inhibitors in the cell suspension during cell disruption. Optimization of these factors is essential for protein stability. Use of low temperatures, suitable buffers with optimum pH and shorter data acquisition times during NMR studies, can minimize the problems of protein denaturation to a certain extent.

As mentioned, several heterologous proteins are over-expressed in *E. coli* and tend to form insoluble protein aggregates known as inclusion bodies. 15-29% of human gene constructs, expressed in *E. coli* are soluble; 20-40% form inclusion bodies; the remaining do not express significantly or are degraded. Inclusion-body formation can be minimized by incorporating modifications into the expression and induction procedures. Periplasmic secretion also influences protein solubility. However, inclusion body formation can be advantageous since these provide a ready source of highly concentrated proteins and are protected from proteolytic degradation. The inclusion bodies can be partially refolded by the use of agents such as urea and guanidine-HCl or by separation processes such as hydrophobic interaction chromatography.

3.1.3 Protein Recovery

Proteins are recovered from the microbial cells after fermentation using centrifugation. If the protein of interest is extracellular, as in the case of mammalian cell culture, the supernatant is subjected to further processing, while the cells are discarded. However, most proteins from *E. coli* are intracellular, which require cells to be retained, subjected to cell disruption followed by the separation of cell debris from protein solution. Cell disruption is usually carried using sonication. A number of mild techniques for cell permeabilization exist, such as osmotic shock and use of surfactants, organic solvents and chelating agents like EDTA. These are generally used in conjunction with sonication. For sonication, the cells are suspended in a suitable buffer. The buffer may contain lysozyme, EDTA for the release of cell wall bound proteins, or non-ionic surfactants such as Triton-X or Tween 80 for retaining the released protein in its folded form, without aggregation. The number of such additives should be kept to a minimum, to avoid complications in further purification.

3.1.4 Protein Purification

Protein purification is a multi-step process. Protein precipitation by ammonium sulphate is a common pre-purification step. Specific methods such as heat treatment for thermophilic proteins and subjecting the protein to extremes of pH are also used. To be able to purify a target protein, knowledge about its physico-chemical nature is essential. This may be inferred on the basis of the protein primary sequence.

The final high-resolution macromolecular purification is usually carried out by chromatography. Chromatographic separations are seldom applied directly to the crude extract without a partial pre-purification step. This is because most of such processes require expensive matrices. Ligands are recycled for cost effectiveness. Increasing the number of purification steps lead to a loss in protein yield. The cost of purification is often ignored in NMR in order to obtain a stable, highly purified sample of protein, in high yield. It is, therefore common to find expensive, high-resolution chromatographic operation being used directly after cell disruption. The major chromatographic operations are described below.

3.1.5 Ion-exchange Chromatography

The protein solution is passed either through a cationic or anionic ion exchanger comprising of charged groups chemically linked to an inert solid. Depending on the surface charge, proteins get linked to different layers of the column. These are eluted out by changing the pH and/or the ionic strength of the eluting solution. This process provides high-resolution, high speed and high capacity purification and is relatively less expensive.

3.1.6 Affinity Chromatography

One of the most powerful purification tools, it provides good resolution, high speed and high capacity of separation. A specific molecule is used as a ligand, which is covalently linked to an inert solid support. The molecule has a high binding affinity to the target protein, which is bound and retained on the chromatographic column. The proteins can be eluted by addition of another competing ligand. Several affinity tags can be incorporated into the protein sequence at the expression stage. These include glutathione-S-transferase (GST), His- and T7-tag. The major drawback of this method is the high cost of the ligand. The necessity of cleaving the tags after protein purification can sometimes complicate the process.

3.1.7 Gel Filtration

The solution of protein is applied to a stationary, porous gel phase consisting of cross-linked dextran, polyacrylamide or agarose. Molecules are separated on the basis of their size. The largest particles move fastest through the gel beads without entrapment. The smaller particles enter the pore of gel of appropriate pore diameter and move slowly. This process does not have high-resolution and high-capacity capabilities. Other methods such as hydrophobic interaction chromatography, reverse phase chromatography and adsorption chromatography can also be used for macromolecular isolation.

3.1.8 Approaches in Structural Genomics

Both NMR and X-ray crystallography have emerged as powerful techniques for studies in structural genomics. A major obstacle in such studies is in high-throughput expression of the desired protein in a soluble form and its purification. The gene constructs should be expressed in a synchronous fashion, and the purification procedures should be similar. These are achieved by using N-terminal expression tags and affinity purification tags, as discussed above.

3.2 Isotope Labelling

Biosynthetic methods are extensively used to label biological molecules, with isotopes such as ^{2}H, ^{13}C and ^{15}N. The incorporation of these isotopes is carefully manipulated such that the labels are introduced either uniformly or selectively, as may be desired. Both the levels of enrichment and enrichment patterns can be manipulated.

E. coli can efficiently incorporate isotopes into the expressed proteins from labelled compounds in the growth medium. The amino acid pathways of the bacteria are well understood. This allows development of strategies for the control of desired enrichment pattern. Uniform ^{13}C and ^{15}N labelling can be achieved using $^{15}NH_4Cl$ and $^{13}C_6$ Glc as the sole source for ^{15}N and ^{13}C, respectively. Alternatively, addition of $^{13}C/^{15}N$ labelled amino acid of interest to the growth medium results in selective labelling of specific amino acid residues. More than one kind of labelled amino acid can be incorporated, by supplying relevant labelled amino acids or a single labelled amino acid, which acts as a precursor for other amino acids of interest. It is also possible to achieve selective un-labelling by using uniformly labelling media, but incorporating selected unlabelled amino acid residues in the medium. The level of enrichment correlates with the level of enrichment in the substrates.

Stereo-selectively labelled amino acids have been used for introducing selective Val and Leu side chain methyl groups. This helps in stereo specific assignments of Pro-R and Pro-S methyl groups. A similar pattern of enrichment is achieved if a mixture of labelled (10%) and unlabelled (90%) Glc is used by taking advantage of the metabolic pathways by which the two groups are introduced, in such pro-chiral side chains.

Use of photosynthetic bacteria has an advantage that these organisms can grow on media consisting of simpler precursors, such as $^{13}CO_2$ and $K^{15}NO_3$. This leads to considerable cost saving. The biomass can be used as a source of labelled amino acid hydrolysate for use in other expression systems.

The demand of enriched proteins for NMR studies is continuously increasing. Introduction of protocols using yeast and mammalian cells has improved the potential for expressing diverse sets of proteins, including glycoproteins.

3.3 Concentration

The eluted protein solution after chromatography has to be brought to high concentrations suitable for NMR studies. Concentration using laboratory scale ultrafiltration is the most common technique used. Water from the solution is removed by passing through a membrane capable of retaining molecules with M_r above the cut-off limit.

3.4 Quality Assessment and Storage

Protein sample should be free from contaminants, denatured species, structural micro-heterogeneities and should be mono-disperse. These aspects can be assessed through the use of SDS-PAGE, MALDI and dynamic light scattering (DLS).

After expressing and purifying biological molecules, proper storage of the sample is of prime importance. The method of storage depends on the stability of the biomolecule, and on its end-use. Addition of compounds, which interfere with the desired applications, should be avoided. For short term storage (~24 h), most proteins can be kept at 4°C. It is desirable to filter and sterilize the protein preparation or add a bacteriostatic agent (such as sodium azide) to avoid bacterial growth for long-time storage. For storing more than a week, it is advisable to rapidly freeze the sample using liquid nitrogen or a dry ice/ethanol mixture. Stabilizing agents, such as glycerol, serum albumin, reducing agents (such as DTT), and ligands (depending on the nature and concentration of the target protein) are helpful. Proteins can be stored for several months at -20°C with the addition of 50% glycerol to avoid freezing.

3.5 Synthesis of Nucleic Acids

In most NMR studies on DNA model compounds, oligomers of 10-30 nucleotides have been used. Such oligonucleotides have been used for understanding DNA polymorphism and DNA-protein interactions. Such samples are usually synthesized chemically. The method of choice is the solid-state synthesis for which, automated synthesiser based on phosphoramidite chemistry are commercially available. The technique uses stepwise incorporation of nucleotides in a sequential manner. The OH group, which is not supposed to participate in the condensation reaction, is protected at each step by suitable chemical modification. Usually, a 10 μmol synthesis is sufficient for NMR studies. Approximate DNA concentrations are determined directly by weight or by absorbance at 260 nm.

For preparing non-self complementary duplexes and other higher order complexes of DNA such as triplexes and tetraplexes, individual strands are mixed in appropriate molar ratios. Higher concentrations of DNA may lead to increased viscosity of the sample and consequently broader NMR line-widths.

3.6 RNA Samples

RNA can also be synthesized using automated synthesizers. However, the technique involves additional steps as two hydroxyl groups have to be protected. It is advantageous to use RNA from natural sources and obtain smaller model fragments using enzyme based cleavage. This is the usual way to obtain t-RNA and fragments of r-RNA. For preparing RNA samples, the dialyzed RNA is freeze-dried and redissolved in appropriate buffer. The buffer with the required pH contains 10-20 mM sodium/potassium phosphate, 100 mM NaCl, and 100 μM deuterated EDTA. Once the RNA samples are transferred to the NMR tube, the tubes are often flame sealed for long-time data acquisition and sample storage.

3.7 Purification of Nucleic Acids

Though automated solid-state synthesisers carry out DNA synthesis with a high degree of accuracy, small errors in fidelity at each stage may add up, to make the final product relatively impure. Thus, purification of the sample becomes an important step. The synthesized product is first deprotected. A typical purification protocol consists of heating the sample in concentrated NH_4OH at 50-60° C. The oligomers are purified by anion exchange chromatography. A 2.5×24.5 cm column of Q-sepharose attached to a FPLC chromatography system is best suited. Tubes containing the leading, middle and trailing portions are pooled separately. Each fraction is concentrated to a small volume in vacuo and dialyzed against NaCl-EDTA, NaCl and water until the dialysate tests negative for chloride. After filtration, each fraction is concentrated to dryness in vacuo and recovered in water.

3.8 ^{13}C and ^{15}N Labelling of Nucleic Acids

Although, heteronuclear NMR techniques have been routinely applied for studying proteins, their application to large RNA and DNA has started only recently. This has been due to lack of efficient methodologies for isotope labelling in DNA and RNA.

Solid-phase phosphoramidite chemistry has been widely used to prepare labelled DNA oligomers. Despite the fact that the chemical synthesis is simple, its utility is restricted due to the requirement of expensive labelled chemicals. Efficient synthetic methods for producing labelled nucleoside phosphoramidite are lacking. Several enzymatic methods have been proposed. These are based on DNA polymerase *fill-in* reaction and PCR techniques. The methods use ^{13}C and/or ^{15}N labelled d-NTPs as precursors, which are prepared from bacteria grown in $^{13}C/^{15}N$ labelled nutrient media. One such methodology is *E*ndonuclease *S*ensitive *R*epeat *A*mplification (ESRA). Oligonucleotides containing two repeats of the target sequence, separated by restriction enzyme recognition site are subjected to PCR amplification. During each cycle of denaturation and annealing, these repeats can align in one of the different frames, giving rise to 5′ overhangs, which are then filled in by *Taq* polymerase. This results in molecules with a larger size distribution, corresponding to varying numbers of repeat units. The technique has been used to generate ^{13}C or/and ^{15}N labelled DNA by feeding the PCR reaction with ^{13}C or ^{15}N labelled dNTPs.

In another approach, PCR products bearing the repeats are cloned in a plasmid and propagated in BL21 (DE3) cells. ^{32}P labelled PCR product has been used to screen the putative plasmid clones for the presence of *insert* by Southern hybridization. A clone that contains relatively large size of the insert is selected. This method enables additional DNA amplification, since the plasmid is maintained as multiple copies inside the cell. The cost is reduced significantly by the use of ^{13}C labelled Glc and $^{15}NH_4Cl$, as the sole source of carbon and nitrogen, respectively, instead of the more expensive labelled-dNTPs.

The yield of the plasmid is optimized when the growth is performed in minimal medium (M9). This is achieved by increasing the plasmid copy number at the cost of chromosomal DNA, by adding chloramphenicol to the growing culture. This procedure inhibits replication of chromosomal DNA, but not of the plasmid DNA. The yield of the DNA from such a culture is typically ~30 mg per litre. Digestion of the plasmid by the corresponding restriction enzyme leads to the release of the insert. It recovered from the gel, by the *crush and soak* method of elution.

For preparing isotopically labelled RNA, bacterial cells are grown in a minimal salts medium containing $^{15}NH_4Cl$ and $^{13}C_6$-Glc or ^{13}C-methanol as the sole source of nitrogen and carbon, respectively. The preferred bacterial cells are *E.coli* for ^{15}N-labelled nucleotides and *methylophilus methylotrophus,* for ^{13}C-labelled or ^{13}C/^{15}N-doubly labelled nucleotides. The deoxyribonucleotides are separated from the ribonucleotides using a boronate affinity column. The ribonucleoside mono-phosphates are converted to the corresponding nucleoside tri-phosphates by enzymatic phosphorylation. Such labelled tri-phosphates are used in *in-vitro* transcription reactions to produce the desired RNA of defined sequence and length.

3.9 In-vitro RNA Transcription and Purification

In-vitro transcription of RNA is carried out by using monomeric RNA polymerases from T7 or similar bacterophages. The reaction mixture contains a double stranded promoter/single stranded template or completely double stranded template. Both templates give fairly good yields. The *in-vivo* transcription product of interest is purified from other transcripts by denaturing gel electrophoresis or ion-exchange chromatography. The purified RNA is dialyzed using DEPC-treated tubing against water containing chelex-100 to remove contaminating metal ions, which influence the relaxation times.

4. NMR APPROACH TO STRUCTURAL STUDIES

There is a major difference between the approaches used in X-ray crystallography and NMR for deriving 3D structures. The NMR structures are based on utilizing bits and pieces of information, which provide clues on the structure of the molecule. These parameters are fed as geometrical constraints in a theoretical algorithm, which sets up a target function, error function or a potential function. This function is minimized to obtain the final structure(s). In general, more than one structure may satisfy the NMR derived structural constraints. The final structures thus derived, are consistent with the NMR properties of the molecule as well as other criteria used as inputs in the program, such as non-bonded interaction and hydrogen bond energies between the atoms in the molecule. The precision of the structure depends on the resolution of the resonances and the structural parameters obtained from the experimental data.

It is obvious that larger the number of NMR constraints for a particular residue, more precise will be the description of its local conformation. For example, consider the conformation of a peptide backbone. The conformation of each residue is defined by two torsional angles (ϕ and ψ), as ω is in a *trans* conformation. Thus, the minimum number of experimental constraints required to define the local conformation of each unit in a peptide backbone is two. The number of available constraints can be 10-15 in favourable cases. However, for certain residues, which are poorly defined in the final spectra, experimental information may not be available. The number of conformational variants for each residue in nucleic acids and oligosaccharides is substantially larger and hence more number of structural constraints are required.

After the structure has been obtained, a critical analysis is required to check if the experimental constraints are spread over the entire structure. Fortunately, with the development of labelling techniques and multi-nuclear NMR, better and more stringent constraints are available. Increasingly, efforts are being made to obtain adequate number of constraints to obtain reliable structures.

4.1 NMR Strategies

The NMR determination of biomolecular structures, primarily involves finding structure(s), which are consistent with the NMR derived geometrical constraints. The approach involves following steps:

(i) Resonance assignments of NMR spins; establishment of chemical shifts.

(ii) Determination of experimental parameters (such as NOE, J-couplings, hydrogen-bonds, long-range interactions) which can be used as structural constraints.

(iii) Choice of a starting conformation. This may be random or based on considerations such as an existing structure of a related molecule, favourable stereochemical torsional angles and secondary structural propensities derived from other experiments.

(iv) Setting the primary structure with correct chirality, covalent distances, metal coordination etc.

(v) Use of stereospecific assignments for prochiral groups, whenever possible.

(vi) Optimization of the structure using experimental constraints and an algorithm, which sets up a suitable function to be minimized. The final acceptable structures are reached by using appropriate conformational drivers.

(vii) Structure refinement through energy minimization.

(viii) Validation by checking the accuracy of the final structure for agreement with experimental data and stereochemical considerations.

(ix) Presentation of the acceptable structures in a graphical form.

Normally these steps are repeated in an iterative fashion, till an acceptable structure is obtained. Generally, more than one structure is satisfied by the observed NMR data. Therefore, one obtains a family of acceptable structures, rather than a single conformer.

4.2 General Approach for the Resonance Assignments in Biomolecules

To make use of the information contained in NMR spectrum, identification of the nuclear spin in the molecule from which a particular signal arises is essential. In other words, a 1:1 correlation is required between the chemical shifts and the atoms, which contribute to such signals. Considering that on an average, each amino acid in a protein contributes about six ^{1}H resonances and each nucleotide in a nucleic acid gives rise to around ten resonances, assignment of 1D ^1H spectrum can be difficult even for a peptide or a small oligonucleotide.

Fortunately, for small macromolecules (M_r less than 10 kDa), homonuclear [^1H-^1H] 2D NMR spectroscopy has largely solved this problem. For larger molecules, heteronuclear multidimensional experiments are necessary. The general approach is illustrated through a simple example (Figure 4.7A and B).

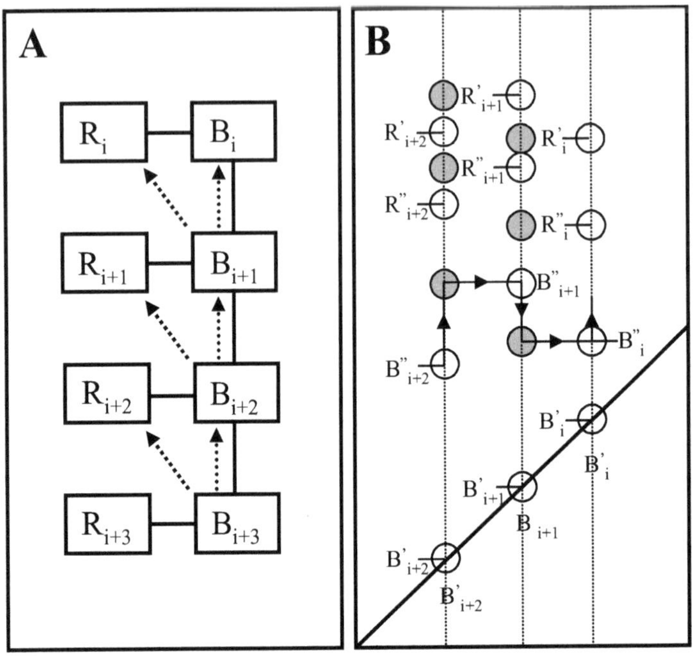

Figure 4.7: (A) Schematic diagram illustrating the principle of sequential assignments, in biopolymers. Dotted arrows indicate NOE connectivities. (B) Sequential walk along the chain in a biopolymer in a 2D spectrum, using self (empty circles) and sequential (grey) NOEs between sets of spins $\{B_i\}$ and $\{R_i\}$.

Within each individual residue, the ^1H spins form a network of coupled spin-system. For example, Thr has 6 protons, NH, H^{α}, H^{β} and CH_3, and forms an $AKMX_3$ type J-coupled spin-system. Thus, it is possible to first identify all ^1H spins which belong to individual side-chains (R_1, R_2, R_3). These can be grouped as networks of coupled spin-systems. One can use COSY and TOCSY correlations for this purpose. Intra-residue ^1H-^1H NOEs can also be used. Following this, each spin-system thus identified is linked to their respective backbone spins again using COSY or TOCSY. It is desirable that these resonances are well resolved. A 600 MHz spectrometer is generally sufficient to overcome overlap problem for molecules with M_r of less than 15 kDa.

What is more difficult is to assign the identified chemical shifts according to the positions of the residues in the sequence. The process to achieve this is called sequence specific resonance assignment. One uses NOEs between the backbone protons belonging to sequential residues (Figure 4.7A). Following this approach, one is able to walk along the biomolecule from one end of the chain to the other as illustrated in Figure 4.7B. Problems may arise because of near degeneracy of chemical shifts, broadened resonances or at times due to certain specific residues, which do not have a backbone proton. There is another problem that concerns the starting point for such assignments. These problems are molecular specific and will be discussed in detail in Chapters 5-9.

For larger molecules, higher dimensional (3D and 4D) experiments using the chemical shifts of ^{13}C and ^{15}N as additional dimensions have emerged as powerful tools. Recall that the chemical shifts for ^{13}C and ^{15}N are much larger than those of ^{1}H. This is helpful in removing overlap problems in the [^{1}H-^{1}H] 2D NMR. However, labelled compounds are needed, since the natural abundance of both ^{13}C and ^{15}N is small. Special techniques such as TROSY along with the use of ^{2}H labels enable studies for still larger molecules. The general principles used for resonance assignments of different classes of biological macromolecules are similar. However, the details vary depending on the methodology adopted, the size of the molecules and their nature.

5. NMR PARAMETERS FOR STRUCTURAL STUDIES

In this section the structural parameters used in NMR and how these are introduced as geometrical constraints to obtain acceptable structures, have been discussed.

5.1 Chemical Shifts: Magnetic Anisotropy of Chemical Groups

Chemical shift is highly sensitive to changes in conformational structure. The NMR spectrum of a random coil polymer (δ_r) is fairly well reproduced by mixing different monomers in the right proportion and then recording the spectrum. However, spectra of ordered structures (δ_o) exhibit large deviations from that of a random coil. The difference in chemical shifts ($\Delta=\delta_o-\delta_r$), show a sequence dependent conformational behaviour. The same residue type at different positions may exhibit significant differences in their conformational dependent chemical shifts.

One of the major factors influencing chemical shifts of ordered polymers is the magnetic anisotropy of groups such as aromatic moieties, the carbonyl group, double bonds and lone pair orbitals. In a rigid molecule, the spatial position of such groups is geometrically fixed with respect to the nuclear spin under investigation.

The concept of variations in chemical shifts arising from chemical groups having magnetic anisotropy can be discussed with the help of the Figure 4.8. Let O be the electric centre of gravity of the chemical bond or group, and χ_{xx}, χ_{yy} and χ_{zz} be the components of the magnetic susceptibility tensor $\boldsymbol{\chi}$ along the three axes x, y and z. The change in the electronic screening (σ_P) arising from the magnetic susceptibility of the group at a point P (defined by $\mathbf{r_P}$), is given by:

$$\sigma_P = [(1\text{-}3\cos^2\theta_x)\,\chi_{xx} + (1\text{-}3\cos^2\theta_y)\,\chi_{yy} + (1\text{-}3\cos^2\theta_z)\,\chi_{zz}]/\,3r_P^{\,3}\,. \qquad 4.3$$

Where θ_x, θ_y and θ_z are the angles made by the position vector **r**, with the three axes.

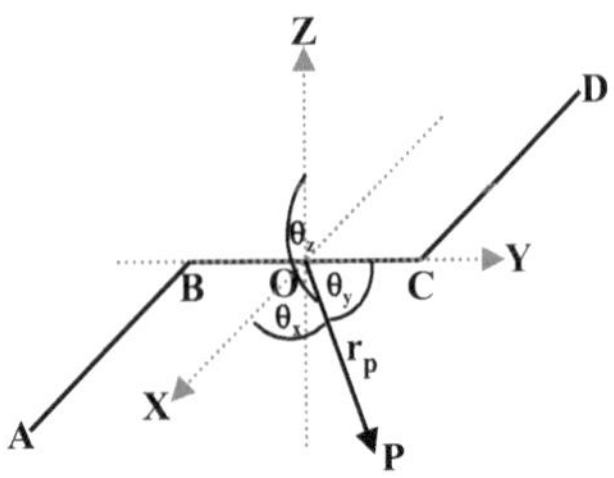

Figure 4.8: Effect of magnetic anisotropic group on the chemical shift of a nucleus at point P.

For groups having axial symmetry, this equation simplifies to:

$$\chi_{yy} = \chi_{zz}\,,\quad \Delta\chi = \chi_{xx} - \chi_{zz} \text{ and } \sigma_P = \Delta\chi\,(1\text{-}3\cos^2\theta_x)\,/3r_P^{\,3} \qquad 4.4$$

5.1.1 Ring Current Effects

The most important examples of magnetic anisotropic groups, which influence the chemical shifts to a major extent, are the aromatic side chains Tyr, Phe, and Trp in proteins and purine and pyrimidine bases in nucleic acids. The magnetic anisotropy due to these groups arises from the induced motions of the π electronic cloud of the ring in the presence of external magnetic field. The changes in the shielding of nuclear spins shift their resonance signals to low or high fields depending on their location with respect to the plane of aromatic rings (Figure 4.9). This so called *ring current effect* moves the resonance of a spin located above or below the ring to a higher field. Resonances from spins in the plane of ring shift to a lower field.

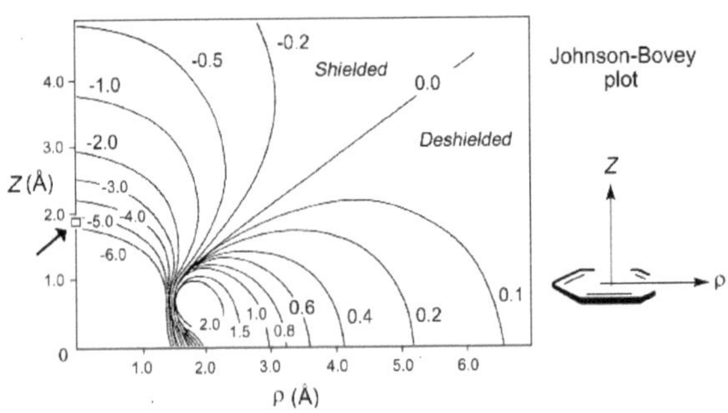

Figure 4.9: Ring current induced shifts ($\Delta\delta$ in ppm) due to a phenyl ring as a function of radial (ρ) and axial (Z) distances from the centre of the ring. C.E. Johnson and F.A. Bovey, J. Chem. Phys. 29, 1012, (1958). (Reprinted with permission from the American Institute of Physics).

The stacking of purine and pyrimidine rings in helical nucleic acids gives rise to large changes in chemical shifts, when molecules go from a random coil (molten) to an ordered state. Such changes have been used extensively to determine the melting profiles of nucleic acids.

The ring current effects can be estimated to a fair degree of accuracy. Thus, in an ordered protein or nucleic acid structure, the shifts of spins near aromatic residues can be calculated and used to check the accuracy of a derived structure.

5.1.2 Changes in Chemical Shifts and Secondary Structures

The chemical shifts are not capable of providing quantitative point-to-point constraints for structural calculations. Therefore, it is not possible to introduce such shifts directly as structural constraints, in the same way as can be done with NOE, hydrogen bonds or coupling constants. However, when several residues in a sequence are involved in the formation of secondary structures, systematic changes occur in the chemical shifts over the entire segment. Such changes in chemical shifts have been characterized for α-helices and β-sheet structures in proteins and are called chemical shift indices (CSI). The observations of CSI serve as good indicators of existence of secondary structural elements in proteins.

Likewise, helical structures in nucleic acids can be characterized by ring current shifts. In a linear helical duplex, protons in each base are influenced by the base pair below and the one above. These shifts depend on the geometry and type of base pairs involved in stacking. Extensive theoretical estimates are available in literature, which correlate well with experimental data. Such calculations have also been used for RNA structures, which show wider range of stacking patterns.

5.2 Chemical Shifts and ^{2}H Exchange Rates of Hydrogen Bonded Protons

The O-H and N-H groups in ordered biological molecules are frequently locked in hydrogen bonds. The most common hydrogen bonds in biopolymers are N-H…N, N-H…O and O-H…O. These strong hydrogen bonds result in significant low field shifts of the corresponding ^{1}H signals. Non-bonded N-H and O-H protons in random coils appear at higher fields and are usually broad to an extent that they may not be observed. When the carbonyl oxygen is involved in bonding, then carbonyl ^{13}C also shows a signature of hydrogen bonding in its chemical shift.

Appearance of resonances between 8-16 ppm in ^{1}H NMR arising from hydrogen bonded O-H and N-H protons, is a good indication that the polymer has acquired an ordered structure. Such resonances have relatively low temperature coefficients and are not affected significantly by solvent titrations. When a sample is dissolved in ^2H_2O, the protons locked in hydrogen bonds exchange slowly with ^{2}H. Some protons may not exchange for several days. This is also the case for protons which are not accessible to solvent, for example, those in the core of the molecule. In certain stretches of polymers involved in secondary structures, evidence of sequentially placed hydrogen bonded protons may be present. Therefore, sequential resonance assignments coupled with exchange effects substantiate the presence of a secondary structural element.

The constraints for a hydrogen bond are usually expressed in the form of a range of distance between the hydrogen atom and the acceptor atom (typically 1.8 to 2.1 Å). The hydrogen bonds are usually linear. This can be dictated by an additional distance constraint between the donor and acceptor atoms (typically 2.7 to 3.0 Å for N and O atoms). Algorithms used in structural prediction employ additional force fields for hydrogen bond optimization. Since the nature of hydrogen bonds is essentially electrostatic, inclusion of electrostatic and polarization terms greatly helps in achieving a good geometry for such bonds, even though several of these force fields are not equipped to bring in the directional nature of hydrogen bonds.

The hydrogen bond constraints are extremely valuable in structural simulation as these lock the chemically distant molecular fragments in an ordered state.

5.3 Nuclear Overhauser Effect (NOE)

The most powerful experimental constraint in any NMR based structural study is the ^{1}H-^{1}H NOE (η). Typical distances of up to 5 Å give rise to detectable NOE in NOESY. As has been discussed in Chapter 1, when intramolecular dipole-dipole interactions constitute the dominant relaxation pathways, a simple equation can be derived for the NOE (η_{ij}) between two spins i and j. In these cases, NOE depends on the distance (r_{ij}) between the spins and the correlation time (τ_c) of the motion of the inter-nuclear vector i-j. If the biopolymer is tumbling such that a uniform correlation time for interatomic vectors can be assumed, then the inter-proton distance r_{AX} can be estimated from equation 4.5, provided a reference distance (r_{ref}) is available in the molecule.

$$\eta_{AX}/\eta_{ref} = (r_{ref}/r_{AX})^6 \qquad 4.5$$

5.3.1 Limitations of NOE

Equation 4.5 has served as the backbone of 3D structure determination by NMR. There are several distances, which are relatively fixed and can be used as reference. Examples are: *geminal* protons in methylene groups (i.e. H2′/H2″ in deoxyribose), ortho protons in aromatic rings (e.g. H5-H6 protons in cytosine) and NOEs across β-sheets in proteins. However, this equation has been often misused while attempting to get accurate distances from NOEs data. The following limitations of equation 4.5 may be noted.

(i) Alternative relaxation mechanisms: It is assumed that the intra-molecular dipole-dipole interaction is the only mechanism for cross relaxation. In fact, the total relaxation rate of a spin j can be written as:

$$R_j\,(1/T_{1j}) = \Sigma_{i<j}\,(\rho_{ji}) + \rho^* \;;\; where\ \rho_{ji} = (\gamma_i\,\gamma_j\,\hbar)^2\ \tau_c(ji)\,/r_{ij}^{\,6} \qquad 4.6$$

Here ρ^* is the direct relaxation rate of the spin j by other pathways. One of the dominant relaxation mechanisms other than the dipolar interaction can be through paramagnetic metal ions. Such ions often evade purification procedures and may bind to certain sites in the macromolecule. The spins around this binding site will be dominated by paramagnetic relaxation. Contributions to ρ^* may also arise from quadrupolar relaxation and chemical shift anisotropy (CSA).

(ii) Use of a single correlation time: The motion and the effective correlation time τ_c for different pairs of spins may be different. Globular proteins tend to have isotropic tumbling motions. This may not be true for non-spherical molecules. In addition to the anisotropic tumbling, local motions in less rigid regions such as loops, unstructured regions and turns in a molecule can seriously violate the validity of this equation. Motions due to CH_3 group rotations, aromatic ring flips and other local conformational fluctuations can hamper accurate measurements of distances using equation 4.5.

(iii) Measurement of NOE volumes: NOE is usually measured in terms of volumes of NOESY cross peaks. Difficulties are encountered in the accurate measurement of such volumes. Most researchers use in-built software on the spectrometer, which uses the digitized data to estimate the volume. The accuracy of the measurements depends significantly on the quality of the spectrum, particularly the S/N ratio, the spectral editing procedures employed and artefacts such as base-line drifts and the t_1 noise. Besides, volumes of poorly resolved peaks can have significant errors. Such errors may not be reflected in the values estimated using software packages. Estimation of volumes is one of the important factors that may lead to errors in the corresponding distances.

(iv) Choice of a reference distance: Even though this may be a distance which is not subject to motional averaging, errors in the measurement of relative NOEs may skew the accuracy of estimated distances in favour of those which are close to the reference distance.

(v) Spin diffusion and *two-spin* approximation: Each spin can relax through interaction with several spins. When a two-spin approximation is used, this results in what is known as *spin diffusion* and introduces errors in distance measurement. The problem is particularly serious if NOESY spectra have been obtained with long mixing times. The solution of the Bloch equations for the system, taking into account all relaxation pathways, can be used to resolve this problem. A number of methods are available, which use full relaxation matrix. Examples are MARDIGRAS, IRMA, and MORASS. These should be preferred over the two spin approximation. An iterative procedure in repeated distance estimation and structure calculation may be used, which brings in greater accuracy in the constraints as one minimizes the target function during structure calculation.

(vi) Absence of NOE may not necessarily mean that the distance between the two spins is more than 5 Å. Several factors such as bound paramagnetic ion, spectral artefact and flexible regions can weaken NOE.

(vii) Existence of minor conformers: NOE depends inversely on the sixth power of the distance. Therefore, even a minor admixture of a fast exchanging conformer having short interatomic distances, can make a large contribution to the measured

NOE. This may distort the simulated structure, which reflects the time-averaged behaviour.

5.3.2 Introduction of NOE Constraints

Given the above mentioned problems with NOE based distances, the distance constraints should be introduced with great care, such that errors do not influence the simulated structure. Generally distance limits are introduced in the form of a potential function using a range for measured distances. The constraints may be refined through the use of relaxation matrix and made more accurate iteratively, as one approaches the final structure.

Out of the several different treatments for introducing NOE based distance constraints, the most common one uses lower and upper bounds (d_l^{obs} and d_u^{obs}) for distances derived from NOE data. The violations are treated with a flat-bottomed harmonic potential (Equations 4.7 to 4.9):

$$\text{For } d_{ij} < d_l^{obs}; \quad V_{NOE} = \sum K_{NOE}(d_{ij} - d_l^{obs})^2 \qquad 4.7$$

$$\text{For } d_{ij} > d_u^{obs}; \quad V_{NOE} = \sum K_{NOE}(d_{ij} - d_u^{obs})^2 \qquad 4.8$$

$$\text{For } d_l^{obs} \le d_{ij} \le d_u^{obs}, \quad V_{NOE} = 0 \qquad 4.9$$

An alternative approach involves use of an equilibrium distance derived from the experimental data. The structures can also be refined by making use of the NOE volumes and relaxation matrix techniques. The later approach has several advantages including the fact that they assure the best fit between structure and NOESY data. However, such an approach has its limitations, as it is computationally expensive, relies on accurate measurements of NOE volumes and requires a scaling factor between the experimental and calculated volumes.

5.4 Scalar Coupling Constants (J)

The next important parameter in conformational studies is the through bond scalar J-coupling. It helps in spin system identification and also in sequence specific resonance assignment. Scalar couplings are used to establish connectivity between chemically remote pairs of nuclear spins in multi-dimensional NMR, making use of one bond couplings (^1J) between hetero-nuclei. These large values of couplings provide efficient pathways for magnetization transfer, leading to different kinds of hetero-nuclear correlations (Chapter 3).

5.4.1 Torsion Angle Dependence of ^3J

One of the most important parameters in conformational studies in all classes of biological compounds is the *vicinal* or three bond coupling ^3J, between the spins P and S in a four atom molecular fragment P-Q-R-S. These couplings show a strong dependence on the torsional angle (θ) between the planes containing P-Q-R and Q-R-S. The relations, generally called Karplus type equations, have a functional dependence of the form:

$$^3J\,(P\text{-}Q\text{-}R\text{-}S) = A\cos^2\theta + B\cos\theta + C\,;\ \text{with}\ |A| > |B| > |C| \qquad 4.10$$

Table 4.2: Typical parameters for Karplus type relations for different types of ^3J.

Coupling type	A	B	C	Molecules of interest
H-C-C-H	10.5	-1.2	0	Proteins, nucleic acids, sugars, lipids
H-C-N-H	9.4	-1.1	0.4	Proteins
H-C-O-H	10.4	-1.5	0.2	Sugars
H-C-N-C	7.4	-4.4	-0.8	Proteins, nucleic acids
H-C-C-C	5.0	1.7	0.0	Lipids, proteins
H-C-C-N	-5.3	2.2	0.9	Proteins
H-C-C-F	54.0	-20.0	2.0	Labelled biological molecules
H-C-O-P	16.3	-4.6	0.0	Nucleic acids, phospholipids
C-C-O-P	9.5	-0.6	0.0	Nucleic acids, phospholipids
N-C-C-C	2.8	-0.4	-0.4	Proteins

Values of the coefficients A, B and C depend primarily on the four atoms involved in the fragment. The values are influenced by properties of the substituents, such as their electro-negativity and relative positions. The couplings also depend on bond lengths and bond angles within the fragment, which in turn are related to nature of hybridisation. Therefore, Karplus type relations for each situation are carefully calibrated using model compounds. Some typical values, along with the areas of their applications, are listed in Table 4.2. More specific values for different combinations of atoms are available in literature.

The functional form of ^3J coupling constants is shown in Figure 4.10. It can be seen that for each particular value of the observed coupling, there are generally four possible solutions for θ. The use of Karplus relations can be made in two ways. If it is known that the molecule has a rigid conformation, then the equation can be used to estimate value of the torsional angle θ. It is possible that out of the four possible solutions, some may be ruled out on steric considerations. Use of other geometrical constraints can also help in fixing a unique range for conformational angles, which are consistent with the experimental values of ^3J.

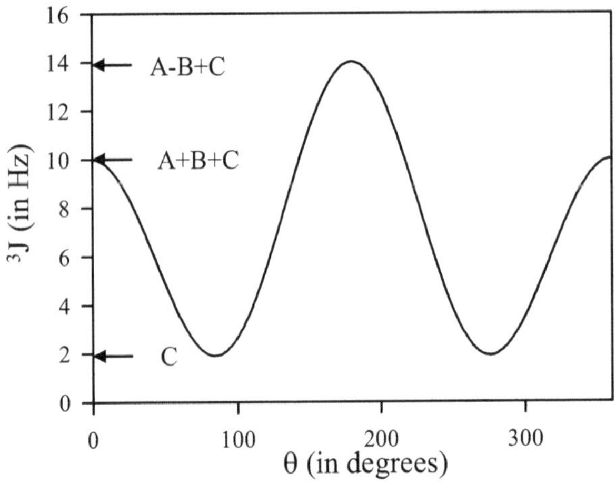

Figure 4.10: The values of three-bond coupling (^3J) as a function of torsion angle θ.

One of the approaches, which have been adopted in algorithms for structure determination, is to introduce a J-based potential function of the type:

$$V_J = \Sigma_J K_J [^3J^{obs} - {}^3J^{cal}(\theta)]^2 \qquad 4.11$$

Where the value of $^3J^{cal}$ is calculated from the value of θ at each stage of iteration and the difference from the observed value ($^3J^{obs}$) adds to other geometrical constraints in the structural function being minimized.

In several macromolecules, the side chains are more flexible than the backbone structure. In such a situation, the side chain may sample several conformations. Populations of different rotamers can be estimated from the observed ^3J values using the theoretically estimated J values for individual conformers.

5.4.2 Coupling Constants across Hydrogen Bonds

Hydrogen bonds are stabilised mainly by electrostatic interactions between the proton attached to an electronegative donor atom (N or O) and the lone-pair orbital on the acceptor atom. However, strong hydrogen bonds have a partial covalent bond character. This suggests that it should be possible to observe a small J-coupling across hydrogen bond. Though such a coupling was predicted early in NMR literature, it has been possible to observe and measure them only with the advent of isotope labelling of biomolecules.

Experimentally, such couplings were initially observed in ^{15}N labelled t-RNA molecules. $^2J(^{15}N-^{15}N)$ couplings of magnitudes as large as 7 Hz have been detected in Watson-Crick base pairs. Since then, J-couplings across hydrogen bonds have been detected in other biomolecules and their complexes. Such couplings provide a strong and direct source of information on the hydrogen bond interactions (Figure 4.11). They permit linking of hydrogen bond donor and acceptor atoms without ambiguity. The geometrical constraints can be placed by using parameters of hydrogen bonds between the donor and acceptor atoms as discussed earlier.

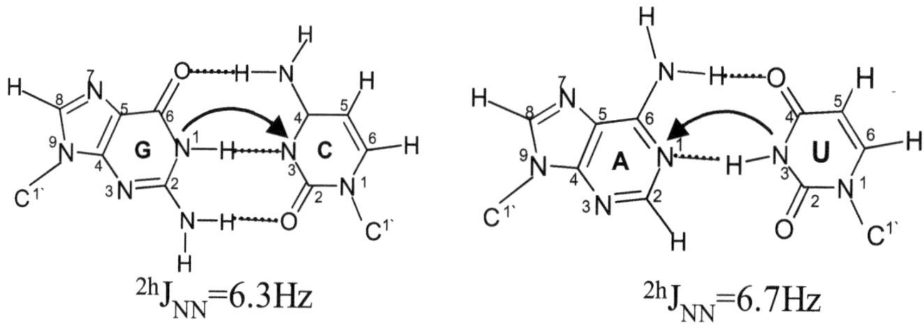

Figure 4.11: Trans hydrogen bond $^{2h}J_{NN}$ observed across G:C and A:U base-pairs.

5.5 Residual Dipolar Couplings (RDC)

The direct dipole-dipole interaction (D_{ij}), between two nuclear spins i and j is given by:

$$D_{ij} = K <(3\cos^2 \theta_{ij} -1)/2 >, \text{ where } K = -(\mu_0/4\pi^2)\hbar \gamma_i \gamma_j r_{ij}^{-3} \qquad 4.12$$

Here, θ_{ij} is the angle between the internuclear vector i-j and the magnetic field axis and r_{ij} is the internuclear distance, μ_0 is the magnetic permittivity of vacuum (Figure 4.12). The bracket < > in equation 4.12, indicates that the term is time-averaged over the molecular motions. In isotropic solutions, molecules move rapidly and isotropically. Under these conditions, the term $<(3\cos^2 \theta_{ij} -1)>$ is averaged to zero. This is the reason why one observes sharp lines in solution spectra. However, the direct information on dipolar couplings is lost. Indirectly, one can pick it up through NOE.

Dipole-dipole interactions are observed in solid-state and lead to splitting of dipolar coupled resonances into doublets (Chapter 8). The splitting depends on θ_{ij} and r_{ij}. When θ_{ij} is zero, the maximum value of D_{ij} (=K) is observed. At a magic angle of 54.7°, the value of $(3\cos^2 \theta_{ij} -1)$ and hence, the observed value of D_{ij} is zero. Thus, D_{ij} can be used to derive geometrical information, namely, r_{ij} and θ_{ij}. Both intra- and inter-molecular dipolar contributions are present in solids. Consequently, the lines are broad and the precision of measurements is low. In recent years, techniques have been developed for high-resolution NMR in solids where mechanical rotation of the sample around an axis at the magic angle together with

RF irradiation, is used to average dipolar interaction. These techniques are discussed in Chapter 8.

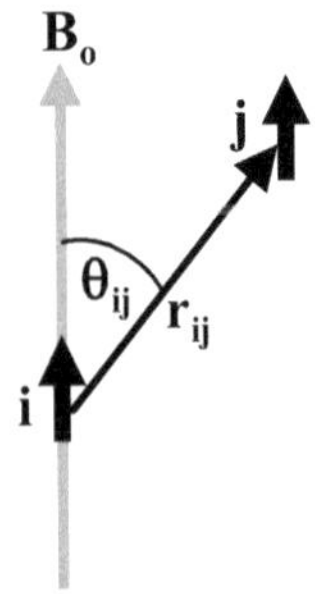

Figure 4.12: Direct dipole-dipole interaction (D_{ij}), between two nuclear spins i and j.

5.5.1 Partial Alignment

When there is a partial alignment, it is possible to retrieve dipolar interactions, at least partly. This is achieved by using conditions such that the molecules have a small net orientation with respect to the magnetic field and consequently, a partial order. The intermolecular interactions are averaged to zero, while a residual effect of the intramolecular interaction is present. The value of K is of the order of a few kHz. Therefore, dipolar interactions are potentially large interactions. Even a small net orientation of 10^{-3} is sufficient to observe detectable residual dipolar coupling (RDC) of a few Hz. The RDCs help in determination or the refinement of 3D structures of biomolecules.

Weak alignments lead to a finite value for the term $<(3\cos^2 \theta_{ij} -1)>$, which is denoted by the order parameter (S_{ij}) or the degree of orientation of the vector i-j. The order parameter is unity for perfectly ordered system and zero for a completely disordered one. The ^1H-^1H residual dipolar couplings are reflected as splittings in high resolution ^1H NMR spectra. The orientations of different internuclear vectors depend on the alignment of the molecule as a whole. To define the orientation of an unsymmetrical molecule, one needs to estimate five independent elements of the S matrix from the observed spectrum. The number of such elements is less if the molecule possesses symmetry. The remaining information on the residual dipolar couplings (RDC) can be used to determine relative internuclear distances.

This approach has been used extensively for determination of precise geometry of small organic molecules using nematic liquid crystals for alignment. However, the number of lines due to ^1H-^1H dipolar couplings increases sharply, as the number of ^1H spins increases. Spin systems containing more than six protons are difficult to analyse.

The situation is different if one uses heteronuclear NMR techniques using isotope labelled molecules. In such cases, the RDC between various ^1H, ^{13}C and ^{15}N spins can be measured. The alignment of the macromolecule can be described by an alignment tensor **A**$'$. In a molecular frame, where **A**$'$ is diagonal, its principle components reflect the probability for the x, y and z axes to be parallel to B_0. If the axes are chosen such that the magnitudes of $A_{zz} > A_{yy} > A_{xx}$, the dipolar coupling depends on the polar coordinates of the vector i-j, as described by the following equation:

$$D_{ij}(\theta,\phi) = K_a\,[(3\cos^2\theta - 1) + 3/2\,R\,\sin^2\theta\cos 2\phi] \qquad 4.13$$

In this equation, $K_a = 3/4KA_{zz}$ is the magnitude of the dipolar coupling tensor and $R = 2/3\,(A_{xx}-A_{yy})/A_{zz}$ is called the rhombicity. A given value of D_{ij} corresponds to a set of angles (θ, ϕ) defining the orientation of the vector i-j. The acceptable solutions for different sets of the values of (θ, ϕ) describe a cone. In other words, the observed value of D_{ij} restricts the orientation of the vector i-j on to the surface of a distorted cone. The inverted vector orientation i-j (Figure 4.13) also gives rise to the same value of coupling and cannot be distinguished.

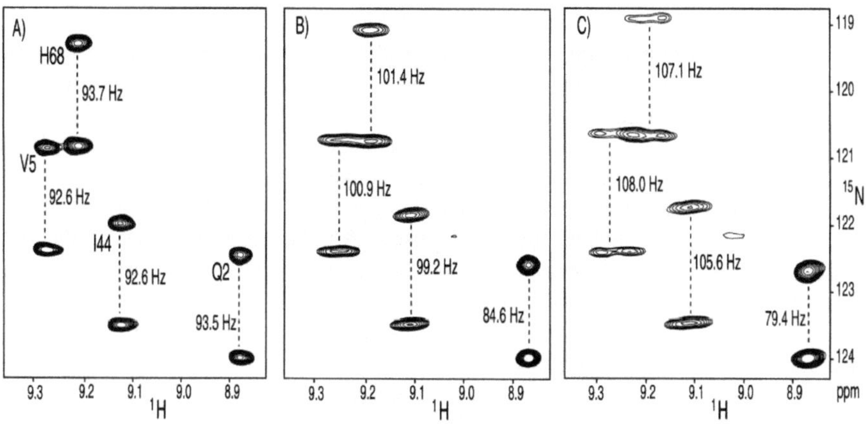

Figure 4.13: Selected region of [^{15}N-^{1}H]-HSQC spectra of ubiquitin, recorded in the absence of ^{1}H decoupling along the ^{15}N dimension, at different levels of molecular alignment. (A) Isotropic spectrum, with the marked splitting corresponding to $^{1}J_{NH}$. (B) Spectrum in 4.5% (w/v) bicelles. Marked splittings in panels B and C correspond to the sum of the $^{1}J_{NH}$ and RDC. The broadening in the ^{1}H dimension, observed in B and C relative to A is caused by ^{1}H-^{1}H dipolar couplings. A. Bax, Protein Science, 12, 1–16 (2003). (Reprinted with permission of Protein Science).

The RDC generally manifest as an additional field dependent contribution to the large one bond $^{1}J(^{15}N-^{1}H)$ splitting, which is around -94 Hz, or the $^{1}J\,(^{13}C-^{1}H)$ coupling which is around 130 Hz (Figure 4.13). These couplings are easily resolved in the heteronuclear correlation spectra of biological macromolecules. Hence, the magnitudes of RDC, allow determination of the internuclear vectors such as N-H and C-H with respect to the molecular alignment frame. This source of information leads to an improved precision and accuracy of 3D structure, and may be valuable in situations where the NOE data is inadequate for structural determination. In particular, RDC can provide long-range interactions.

5.5.2 Methods for Achieving Partial Alignment of Biological Molecules

The magnetic field can itself orient molecules because of the magnetic susceptibility of the later. For diamagnetic molecules, such an alignment is relatively small even at the highest magnetic fields currently available. However, for paramagnetic molecules, reasonable alignments can be achieved. Long rod shaped diamagnetic molecules such as nucleic acids and their complexes show sufficient alignment to enable the detection of RDC, since in such cases, the susceptibility of each residue adds up to a significant value.

To achieve molecular alignments, solvent molecules, which align in the magnetic field and restrict solute molecules to anisotropic motions, are usually employed. Work on organic molecules has been done using nematic liquid solvents. Bicelles consisting of mixture of long chain phospholipids and detergent have been used in early work on biological molecules. Due to high degree of bilayer ordering in the α-lamellar phase in such lipids, the anchored molecules show large dipolar couplings. The disadvantage with bicelles is their short life, sensitivity to pH and narrow range of the α-lamellar phase in terms of experimental conditions such as, composition, ionic strength, and temperature.

More user friendly liquid crystalline media include filamentous bacteriophage Pf1, which is fairly robust. It allows wide concentration range and is available commercially at reasonable cost. Other rod shaped viruses such as tobacco mosaic virus and bacteriophage have also been used. Another good medium is the one achieved by the anisotropic compression of polyacrylamide gel. The non-liquid crystalline gel is inert, and macromolecules can be readily recovered from the gel. In recent years, a number of other materials have been found useful to obtain orientation of biological molecules.

5.5.3 Use of RDC for 3D Structure Determination

RDCs place fairly strict restrictions on the orientation of inter-nuclear vectors. However, use of this information for structural studies is somewhat complicated. One of the difficulties is the inability to distinguish between a particular orientation direction and its inverse. Secondly, dipolar couplings tend to compete with each other. Therefore, the use of these couplings is generally made at the stage of structure refinement after a preliminary structure has been derived from NOE data, or for structural validation.

A method for incorporating dipolar couplings in programs such as X-PLOR and CYANA has been proposed. An energetic penalty function (V_{dip}) based on the RDC, between a pair of atoms i and j is introduced.

$$V_{dip} = \sum_{ij} K_{dip} (D_{ij}^{calc} - D_{ij}^{obs})^2 \qquad 4.14$$

When V_{dip} is included in simulated annealing protocol, the force constant K_{dip} is increased exponentially during the cooling stage of simulation.

5.5.4 Advantages of RDC

In recent years, extensive use of RDC has been made in structural investigations. Recall that the methods based on NOE, scalar couplings and chemical shifts generally have only short-range information about the structure. RDC provide both short and long-range information on the orientation of bond vectors. One such situation is in the determination of domain orientations in proteins. The inter-domain constraints are generally larger than 5 Å and hence cannot be derived from NOEs. Applications of RDC have been made in all classes of biological molecules, such as proteins, DNA, RNA and their complexes. Several applications have also addressed to unresolved structural discrepancies among previous structures derived from NMR and crystal structures and other biophysical properties and their relationship to structural genomics.

5.6 Use of Cross-Correlated Relaxation

There are other ways by which the information contained in the dipole-dipole interaction can be retrieved in solutions. One of the recent advances is the use of dipole-dipole cross relaxation. The method depends on relaxation through cross correlated dipolar coupling between the four nuclear spins that form two inter-nuclear vectors. A typical case is the interaction arising from nuclei in the bond vectors ^1H-^{13}C and ^1H-^{15}N. Such vectors form an integral part of ^{13}C and ^{15}N labelled biological molecules. The angle (θ), between the two vectors, can be measured using the effect of cross correlation on double and zero quantum coherences.

Even though the use of dipolar couplings to determine inter-tensor angles in solid-state is wide spread, the techniques for experiments to pick up such information in solutions are of recent origin. Several pulse-sequences have been developed to determine this important information, particularly in labelled proteins.

Consider the case of N-H and C-H vectors, represented by indices i and j, respectively. Double quantum transitions (DQ) involving ^{15}N and ^{13}C transitions can be excited and detected, yielding a four line spectrum. The four line pattern in the DQ coherence spectrum occurs as a result of the one bond scalar couplings between ^1H and ^{13}C and between ^1H and ^{15}N. These transitions can be labelled by the polarization states α or β of the two protons. The relaxation rates for the four transitions ($\Gamma_{\mu\nu}$) are different and are given by:

$$\Gamma_{\alpha\beta} = \Gamma^{a} - \Gamma_i^{CSA/DD} + \Gamma_j^{CSA/DD} - \Gamma^{e}_{i,j} \qquad 4.15$$

$$\Gamma_{\alpha\alpha} = \Gamma^{a} + \Gamma_i^{CSA/DD} + \Gamma_j^{CSA/DD} + \Gamma^{e}_{i,j} \qquad 4.16$$

$$\Gamma_{\beta\beta} = \Gamma^{a} - \Gamma_{i}^{CSA/DD} - \Gamma_{j}^{CSA/DD} + \Gamma^{c}_{i,j} \qquad 4.17$$

$$\Gamma_{\beta\alpha} = \Gamma^{a} + \Gamma_{i}^{CSA/DD} - \Gamma_{j}^{CSA/DD} - \Gamma^{c}_{i,j} \qquad 4.18$$

The first term Γ^a represents contribution from the auto-correlation relaxation and is same for all the four transitions. The next two terms are due to cross correlation relaxation arising from the interaction due to dipole and chemical shift anisotropy (CSA) for the N-H and the C-H vectors. The final term is due to dipole-dipole cross correlated relaxation. This is the term in which one is interested.

The intensities of these transitions [I(μν)] in a constant-time experiment, decay as exp($-\Gamma_{\mu\nu}$ T), where T is the time for which the signal has been allowed to decay and during which the DQ coherences evolve. $\Gamma_{\mu\nu}$ is the relaxation rate for the transition labelled as μν. The values of the cross correlation due to dipole-dipole and dipole-CSA cross correlation are given by:

$$\Gamma^{c}_{i,j} = 1/(4T) \ln [I(\alpha\beta).I(\beta\alpha)]/[I(\alpha\alpha).I(\beta\beta)] \qquad 4.19$$

$$\Gamma_{i}^{CSA/DD} = 1/(4T) \ln [I(\alpha\beta).I(\beta\beta)]/[I(\alpha\alpha).I(\beta\alpha)] \qquad 4.20$$

$$\Gamma_{j}^{CSA/DD} = 1/(4T) \ln [I(\beta\beta).I(\beta\alpha)]/[I(\alpha\alpha).I(\alpha\beta)] \qquad 4.21$$

For a slowly tumbling molecule such as a protein or nucleic acid, the cross correlated dipolar relaxation depends on the angle θ between the vectors C-H and N-H according to the equation:

$$\Gamma^{c}_{i,j} = (2/5)(\mu_0\hbar/4\pi)^2 [\gamma_H\gamma_N(r_{NH})^{-3}] [\gamma_H\gamma_C(r_{CH})^{-3}][3\cos^2\theta - 1]\, \tau_c \qquad 4.22$$

The most extensive use of the cross-correlation experiments has been to directly measure the angle Ψ in proteins. While ^3J (H^N-N-C^α-H^α) provides an estimate of the angle ϕ, the possibility of having a direct measure of angle Ψ is of great practical value. For a planar amide bond, the angle θ between the vector C^α-H^α for the amino acid residue k and the vector N-H^N for the next amino acid residue in sequence (k+1), is related to angle Ψ by the following equation:

$$\cos\theta = 0.163 + 0.819 \cos(\Psi - 119) \qquad 4.23$$

The dependence of the angle θ as a function of the torsion angle Ψ is shown in the Figure 4.14E. The experiments for measuring cross correlation relaxation $\Gamma^c_{i,j}$ allow a direct measurement of this angle using equations 4.23, since τ_c can be measured from relaxation measurements and other quantities in equation are known.

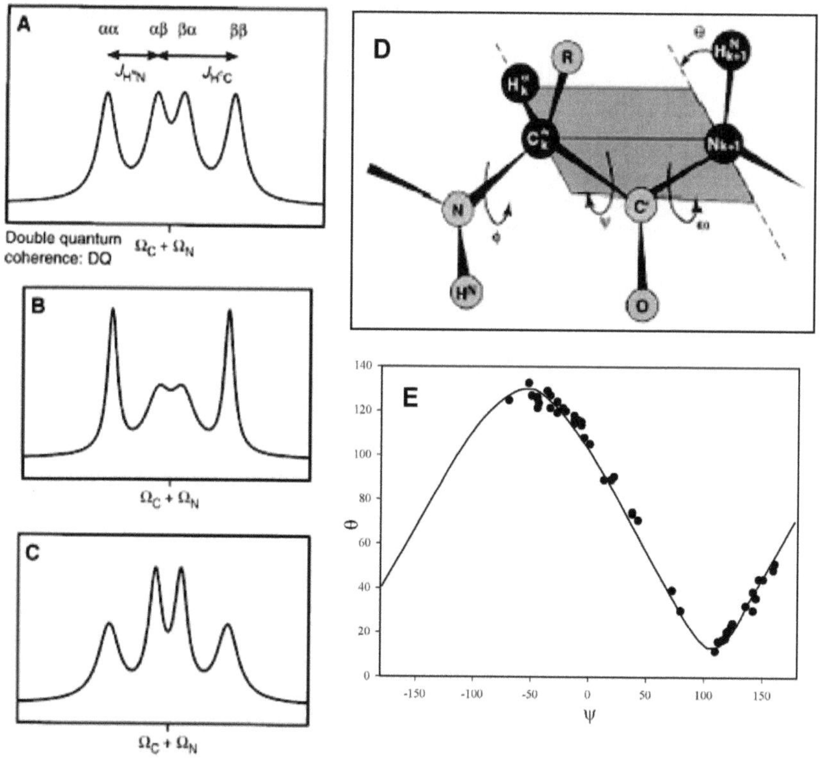

Figure 4.14: Qualitative representation of the measured cross relaxation effect. The signal in the double-quantum dimension is split due to $^{1}J_{HN}$ and $^{1}J_{CH}$. αα, αβ, βα, and ββ denote the four polarization states for the two protons H^{α} and H^{N}. The splitting of the signal without considering dipole-dipole cross-correlated relaxation is shown in (A). For (B) and (C), θ is assumed to be 90° and 0°, respectively. (D) Schematic representation of the peptide backbone. The measured angle θ together with the N-H^{N} and C^{α}-H vectors is highlighted. (E) Theoretical and experimental correlation between φ and ψ, assuming planar peptide bond. Values observed in a typical protein are indicated by black circles. The most populated regions are the α-helical and the β-sheet regions. B. Reif, M. Hennig, and C. Griesinger, Science, 276, 1230–1233(1997). (Reprinted with permission from AAAS).

5.7 Covalent Bond Distances

Input of covalent distances is an integral part of structural simulation algorithms. Covalent distances are generally well established. Once the primary sequence is known, introduction of such distances for proteins and nucleic acids is trivial. One of the important applications of covalent bonds is identification of the disulphide bridges in sulphur containing proteins. This can be established by an analysis of NOEs between β protons of cysteine residues participating in the S-S bridges. Once the Cys involved in such linkages are established, the S-S distances can be introduced as geometrical constraints.

Metal coordination constraints can be derived if the metal binding sites can be inferred from NMR. Otherwise, one can use information from sequence homology with related proteins to obtain metal binding sites. Such constraints have a major influence on the folding of the derived macromolecular structure and on its conformational flexibility.

The nature of linkages in sugar moieties in oligosaccharides is frequently established using NMR chemical shifts. This information can be fed into structure algorithms in the form of covalent distances.

5.8 Stereochemical Assignments

Introduction of the correct chirality for side chains is essential to obtain the correct structure. This is easy to introduce for most spins, along with the primary sequence of the biological structures. However, the methods for assignments do not provide specific resonance assignments for diastereotopic substituents attached to a prochiral center. Examples of such groups are the two protons in a methylene group (i.e. the 2′/2″ or the 5′/5″ protons in D-ribose or the two H^{α} protons in Gly), and the isopropyl groups of Val and Leu. The chemical shifts of the two sets of protons (δ_1 and δ_2) are distinctly different in most cases. So is the case for the coupling constants (J_1 and J_2) and the distances (b_1 and b_2) with respect to a third spin. This information is reflected in the COSY and NOESY. However, the use of this important information content is hampered in the absence of stereospecific assignments of the two sets of protons.

The most authentic way to establish stereo-specific assignments is through specific isotope labelling. This has been made possible in several cases through chemical and biosynthetic methods. Once such stereo-specific assignments are accomplished, the distances of such atoms/groups can be obtained specifically and introduced in structure calculations.

In some cases it is possible to assign the two groups of protons from the knowledge of the relative values of the coupling constants or the inter-proton distances. For example, one of the two ^3J couplings between H1′ and H2′ and H2″ in deoxyribose is invariably larger than the other. This is reflected in the COSY spectrum. A search using each of the two assignment possibilities may result in only one of them leading to an allowed conformation. Preliminary 3D structure for example, using X-ray crystallography, may also provide clues to such assignments. However, such approaches may not be possible in all cases.

In the absence of stereospecific assignments, the structural information is less accurate. Following strategies have been employed for structural calculations in such situations:

(i) The two substituents are replaced by a pseudo-atom (Q), located centrally with respect to the two atoms/groups 1 and 2. The upper distance bound (b_Q) to a third atom is increased by a pseudo-atom correction c_Q, $b_Q = \min(b_1+b_2)+c_Q$, where b_1,b_2 are the individual upper bounds of distances.

(ii) Use of a less restrictive of the larger of the two upper bounds max (b_1, b_2), simultaneously for both distances.

(iii) Use of a combination of the above two approaches by assigning a more restrictive upper limit to b_Q [= $\{(b_1^2+b_2^2)/2\}^{1/2}$ - c_Q]. This approach has been used in programs such as CYANA.
(iv) A distant restraint based on the sum average of the NOE derived distances [$<d> = (d_1^{-6} + d_2^{-6})^{-1/6}$] which is applied to both substituents. This approach has been used in XPLOR.

It is obvious that none of these approaches can make up for the information that is available once stereo-specific assignments can be achieved. The third approach, which uses two identical upper bounds, leads to fairly satisfactory structures.

5.9 Removing Undesirable Restraints

We note that there are several NOEs, whose information is already present in terms of bond lengths and bond angles given as input in primary structure. One example is the distance between *geminal* protons. It is desirable that such restraints are not used in structural algorithms.

6. PARAMAGNETIC MOLECULES AND REAGENTS

Like proton, electron also has spin ½. However, electron spins are paired during the formation of chemical bonds. Therefore, most molecules are diamagnetic. Paramagnetism arises from the presence of a free electron spin(s) in molecules. In several molecules, it arises from a metal ion having odd number of electrons. Paramagnetic metal ions are integral part of several biological molecules, such as heme proteins and metalloenzymes. A paramagnetic centre can also be introduced exogenously.

The magnetic moment (μ_S) due to unpaired electrons in a molecule is given by:

$$\mu_S = g\mu_B[S(S+1)]^{1/2} \qquad 4.24$$

Here, g the ratio of the angular and magnetic moment of electron, S is the electron spin quantum number (1/2 for single unpaired electron) and μ_B is electron Bohr magneton. Like the chemical shift, g is also a tensor quantity. Its value for free electron is 2.0023, but is significantly different in molecules.

In the presence of paramagnetic ions, both the chemical shifts and relaxation rates change significantly. The mechanism of interaction of electron spin with nuclear spin is similar to the nuclear spin-spin interactions. However, the electronic magnetic moment is several times larger than the nuclear moments. Therefore interactions between electron and nuclear spins, dominate over nuclear spin-spin interactions.

When a metal ion is bound to a particular site in the molecule, resonances in a large volume sphere around the metal centre are broadened beyond the level of detection, due to lower T_2 in paramagnetic molecules. At high magnetic fields, the separation between the electronic Zeeman levels is higher, and consequently the line-widths are larger. Therefore, high fields may not offer an advantage to study such systems. Even beyond the sphere of broadened resonances, chemical shifts and relaxation rates in paramagnetic molecules are dominated by the electron spins due to paramagnetic ions. Under favourable situations, such molecules show highly dispersed NMR spectra with chemical shifts extending much beyond the normal range observed in diamagnetic molecules. A number of methods have been designed to locate such broad resonances, and to exploit the advantages that such systems offer.

6.1 Contact Shifts

The chemical shifts in paramagnetic molecules are determined by two mechanisms: (i) a direct delocalization of the electron spin at various nuclear sites (called the contact shift δ^c), and
(ii) an effect arising from the anisotropy of the *g tensor* of the electron spin (leading to pseudo-contact shifts, δ^{pc}).

Both these shift contributions are large. The changes in the chemical shifts due to hyperfine interaction ($\Delta\delta_i^p = \delta^{pc}+\delta^c$), arise solely from the interaction between unpaired electron and the nuclear spins. These can be estimated from the difference from the corresponding shifts in a diamagnetic molecule (δ^d) with a similar chemical structure ($\Delta\delta_i^p = \delta^p - \delta^d$). A convenient way to estimate δ^d in a molecule is to replace the paramagnetic metal ion with a diamagnetic ion of similar size. However, it is generally difficult to separate contributions from δ^{pc} and δ^c, in the observed paramagnetic shift. Estimates of δ^c, can be made theoretically. In some cases, one of the two terms may be small. In such cases, the observed changes are due to only one of the two mechanisms.

The contact shift arises from the presence of finite electron spin-density at the site of the nucleus. The interaction of the nuclear and electron spins is given by:

$$H = A\mathbf{I}.\mathbf{S} \tag{4.25}$$

Where A is a constant, which is related to the spin density at the site of the nucleus and is called the Fermi contact hyperfine coupling constant. A general expression for the contact shift, taking into account of the anisotropy of the **g** and the magnetic susceptibility tensor $\boldsymbol{\chi}$ (which are related to each other), is given by:

$$\Delta\delta_i^c = \Delta\nu_i / \nu_0 = A[\chi_{xx}/g_{xx} + \chi_{yy}/g_{yy} + \chi_{zz}/g_{zz}] / B \tag{4.26}$$

where, B is a constant.

6.2 Pseudo-Contact Shifts

A more useful parameter for structural studies is the pseudo-contact shift. This interaction arises from dipole-dipole interaction between the electron and nuclear spins. It is similar in nature to the dipolar nuclear spin-spin interaction. For systems having axial symmetry, the change in chemical shift due to the pseudo-contact term is given by:

$$\Delta\delta_i^{pc} = \Delta v_i / v_0 = D\,(\chi_{zz} - \chi_{xx}) <(3\cos^2\theta_i - 1)/r_i^3> \qquad 4.27$$

In this equation D is a constant which depends on the properties of the ion, r_i is the distance between the nucleus i and the paramagnetic metal centre and θ_i is the angle between the vector $\mathbf{r_i}$ and the principle symmetry axis of the molecule. Thus, the ratios of pseudo contact terms provide information on molecular shapes, much in the same way as the nuclear dipole-dipole interactions. However, the interaction between nuclear and electron spin can manifest much beyond the 5 Å distance limit of in NOE measurements and can provide long-range information.

Equation for pseudo-contact shift can be generalised for a molecule having several paramagnetic centres. It is expressed in terms of the direction cosines l_{ij}, m_{ij} and n_{ij} of the position vector of the atom i with respect to the magnetic susceptibility vector tensor for the jth paramagnetic centre and the distance r_{ij}.

$$\Delta\delta_i^{pc} = \Sigma_j\ 1/(12\pi r_{ij}^3)\ [\Delta\chi^j_{ax}(3rr_{ij}^2 - 1) + 3/2[\Delta\chi^j_{rh}(l_{ij}^2 - m_{ij}^2)] \qquad 4.28$$

To use these equations in protocols for structure calculation, one needs to have a suitable diamagnetic reference compound such that the contribution to the experimental chemical shift arising from the pseudo-contact shift ($\Delta\delta_i^{pc}$) can be estimated. This problem may be critical for regions where $\Delta\delta_i^{pc}$ is small.

The $\Delta\chi$ tensor can in principle have five independent components. These have to be estimated as part of the structural protocol. Similar to the approach used for other NMR observables, a pseudo-potential term (V_{pc}) is introduced such that the difference between observed and calculated $\Delta\delta_i^{pc}$ is minimized as one approaches the final structure in the iterative process. Such constraints have been introduced both in DYANA (PSEUDODYANA) and energy minimization and molecular dynamics calculations.

6.3 Relaxation Rates

As has been discussed, nuclear spins in paramagnetic molecules relax mainly through the fluctuating fields arising from the interactions between electron and nuclear spins. The presence of electronic moment makes dramatic changes in both the T_1 and T_2 relaxation times of molecules. The contributions arise from three terms;

(i) the electron relaxation, corresponding to correlation time τ_s;
(ii) molecular tumbling (τ_r); and
(iii) chemical exchange (τ_M).

The total correlation time (τ_c) is therefore given by:

$$\tau_c^{-1} = \tau_s^{-1} + \tau_r^{-1} + \tau_M^{-1} \qquad 4.29$$

It is difficult to quantify different contributions to τ_c, though usually only one of the three terms dominates.

The relaxation contributions have important bearing on our ability to observe NMR spectrum. For example, the electron relaxation times τ_s of several metal ions are in the range 10^{-8} to 10^{-10} sec, which produce NMR line-widths of more than 100 Hz. These ions do not provide useful high-resolution NMR spectrum. From the view point of NMR, the useful ions are those which have τ_s of less than 10^{-11} sec. The list of such ions includes low spin Fe^{+3}, Co^{+2}, copper in tetrahedral sites and most of the trivalent lanthanides.

When the electron-nuclear spin dipolar term dominates the relaxation mechanism, both T_1 and T_2 depend inversely on the sixth power of the distance from the ion. This information can be used for obtaining relative distances between the paramagnetic centre and the nuclear spins.

6.4 Metallo Proteins

Paramagnetic ions exist naturally in a number of metalloproteins. High-resolution NMR studies on such molecules, is a flourishing area of research and has been discussed in Chapter 6.

6.5 Lanthanides Shift Reagents

Paramagnetic ions such as lanthanides (Ln^{3+}) can be used as exogenous probes to modify the properties of diamagnetic molecules. Because of similarity in the ionic radii, Ln^{3+} is an ideal probe for substitution of Ca^{2+}, in Ca^{2+}-binding proteins. The changes resulting from substitutions by Ln^{3+} in the chemical shifts and relaxation rates can be used to study conformations of large molecules.

A dynamic equilibrium may exist between the paramagnetic and the diamagnetic states of the molecules. In such situations, a small amount of bound ion may be sufficient to measure the large effects induced by paramagnetism. The use of lanthanides to replace diamagnetic metal ions in metalloproteins serves as a useful probe for obtaining additional constraints in structural studies.

The number of metal ions in the lanthanide series is large. Their magnetic properties vary significantly, such that different ions can be used for different purposes. The sign of $\Delta\chi$ changes as one goes to lanthanides with higher atomic numbers. Therefore ions with both positive and negative $\Delta\chi$ can be introduced. Certain paramagnetic lanthanides give relatively sharp lines. These can be used for measuring pseudo-contact shifts. Ions such as Gd^{+3} make large contributions to relaxation times and these ions can be used to enhance relaxation effects. Diamagnetic lanthanides such as Lu^{+3} can be used to estimate the properties of the diamagnetic metalloproteins.

6.6 Use of Spin Labels

One of the ways to obtain long-range constraints is to use a paramagnetic spin label. Nitroxide spin labels have been used in ESR spectroscopy for a number of years. The free electron in these labels is stabilized by the molecular frame in which nitroxide radical is embedded. A useful spin label reagent for NMR studies is 1-oxyl-2,2,5,5-tetramethyl-3-pyrrolidinyl (PROXYL). PROXYL can be covalently attached to a site in the macromolecule, where it may not influence the structure of the later. In proteins, Cys offers a convenient site for attachment.

The electron spin on PROXYL causes enhanced relaxation within a radius of about 15 Å. The changes in relaxation rates as measured from changes in the line-widths of the labelled molecule and the diamagnetic state have been used to estimate the distance of the nuclear site from PROXYL. Results can be analysed quantitatively to obtain ranges of distances. The method is particularly useful for information on intermolecular interactions and orientations of different domains in protein structures.

7. FROM NMR PARAMETERS TO STRUCTURES

A large number of NMR experiments on biological molecules are performed with the aim to determine their 3D structures. However, it is obvious that while NMR provides a wealth of information on the structure of biological molecules, the same cannot be translated directly into the 3D structure. The final step in structural studies is to unlock the indirect structural information in the form of NMR parameters, in order to obtain the 3D structures.

A number of software packages based on different approaches have been developed for this purpose. They can be divided into the following categories.
(i) Metric matrix distance geometry (programs such as DISGEO, DSPACE, EMBOSS, TINKER);
(ii) Variable target function (DIANA, DISMAN); Molecular dynamics in Cartesian space (AMBER, CHARMM, DISCOVER, GROMOS, SYBYL, XPLOR);
(iii) Torsion angle approaches (DYANA, CYANA, TANDY).
There is certain amount of overlap among these approaches. They all attempt to find a 3D structure of a given molecule which is consistent with the experimental NMR parameters, the correct stereochemical geometry, satisfactory covalent and hydrogen bonds, and is free from steric violations. Other experimental evidences are also incorporated as constraints in such programs.

7.1 Starting Structures

To reach a final conformation, an initial structure is often assumed as a starting point. This is altered by minimization of a suitable function, such that the final conformation is a *good fit* to the input data. In this endeavour, the initial structure has to be chosen carefully. An unsatisfactory starting structure may end up with erroneous results due to the minimization process leading to a false local minimum. There are several ways by which the initial structures can be generated.

One of the approaches to obtain a crude structure from NMR data is to use a distance geometry approach, discussed below. This provides a rough but correctly folded structure, which can then be subjected to more refined algorithms.

Another strategy is to use randomly generated values of the torsion angles, within the limits of allowed conformations. For example, one may start with a fully extended conformation. In such an approach, there is no operator bias. Several sets of starting conformations may be chosen. The family of acceptable solutions thus obtained is subjected to structure validation. Such an approach requires relatively large computational time. In molecular dynamics approach, Monte Carlo methods have been used to generate different random conformations.

A more systematic approach is to use grids of allowed conformational domains for the biomolecules and use these to generate a family of starting conformations. This approach ensures that no structures are missed.

If crystal structure is available then this may be chosen as a starting point. Such an approach has been used extensively for nucleic acids. For a new protein, the protein data bank (PDB) offers the possibility of obtaining a good starting point. Though the structures of almost 39,900 proteins are available in PDB, it turns out that the number of unique folds discovered so far is only about 1000. Thus it is likely that the new protein may belong to one of the known folding pattern. One of the ways to check this is to compare sequence homology with the listed proteins. If backbone dipolar couplings can be measured, then it is possible to search for structures in the PDB, which are compatible with the observed set of RDCs. Such an approach is in fact structural refinement, rather than structural calculation. It is highly useful in studies of proteomics, where fast through-put analysis is required.

7.2 Metric Matrix Distance Geometry

This was the first approach used for structural calculation of proteins using NMR data. The basic idea of such algorithms is to formulate the problem not in the Cartesian space but in a much higher dimension of interatomic distances. Note that information such as that obtained from NOE is in the form of interatomic distances. Also, the covalent and hydrogen bond distances are known with good accuracy. The dihedral angles from J couplings can also be transformed into distance information. The crucial step is to find a structure consistent with the observed distances and to embed such a structure into a Cartesian space. DISGEO was the first program, which was developed in 1981 for the structural calculations of NMR structures. It has been used for several years. More refined versions of this approach have appeared in later years.

Metric geometry or distance geometry (DG) algorithms allow distance information to obtain starting coordinates. For a molecule with N atoms, a $N \times N$ matrix containing lower [B(i,j)] and upper bounds [B(j,i)] is constructed from measured distances. The distances fed into the B matrix are based on bond lengths, bond angles, NOE constraints, information from J couplings, hydrogen bond distances etc. Thus we have N^2 elements of the matrix B. The computer time for solution is proportional to N^3. The metric geometry techniques allow these distance bounds to be *embedded* into a Cartesian coordinate system through certain mathematical manipulations.

The structure has an overall folding of the correct structure and is obtained without any operator bias, but it can be distorted. This is because DG methods can yield an accurate structure only if all the distances in the B-matrix are known accurately. This normally is never the case. The algorithms cannot distinguish between structures related by global inversion, translation and rotation. Thus, the method is not sensitive to chirality of the molecule. Torsional constraints cannot be introduced directly, but have to be translated in the form of geometrical constraints. However, DG does provide a reasonable starting structure.

7.3 Variable Target Functions: Torsion Angle Approaches

A somewhat different approach known as variable target function method was developed in 1985. The basic concept is to formulate the problem as one of minimization of certain objective or a target function (T). It was later superseded by torsion angle methods. One can start with a random structure and gradually fit to the restraints collected from NMR starting from intra residue constraints. The target size is increased in steps to include constraints between atoms further apart along the length of the polymer chain. Since differences in bond length and bond angles around their mean values are small, and in any case, cannot be measured by NMR, these can be taken as fixed. It may therefore be advisable to work in torsional space. The problem is reduced to a dimension covered by the torsional angles.

The local features of the conformational space are first established. The global fold is obtained towards the end of the calculation. The target size is increased incrementally once the local constraints are satisfied. Thus one uses a variable target function. This helps in preventing the conformation getting trapped in local minima. The calculations are started from a large number of random conformations to obtain a group of acceptable solutions.

Programs DIANA and DISMAN belong to this category. The initial methods in this group worked well with prediction of α-helical segments in proteins, which have sequential and short-range constraints. The prediction of β-sheets, which involve long-range constraints is less satisfactory. These methods have been superseded by the more efficient torsion angle dynamics (TAD) approach.

7.4 Molecular Mechanics and Molecular Dynamics Algorithms

More advance methods for solving the problem of structural calculations of biological molecules involve the use of classical mechanics of atoms in a 3D Cartesian space. The Newton's equations of motion are solved to obtain a trajectory for the molecular system. The degrees of freedom are the Cartesian coordinates of the atoms. The presence of kinetic energy terms in these algorithms allows the system to escape from local minima.

Calculation of interatomic interactions in a molecule is basically a quantum chemical problem. However, classical approaches make it possible to calculate energy, interatomic potentials, thermodynamics and dynamics of large biological molecules. Such computations are within the power of present day computers. The potential energy is written as a sum of several interatomic interaction terms, expressed in terms of interatomic distances (r), interatomic angles (φ) and dihedral angles (θ). To these, NMR constraints are added in the form of potential functions. The following function takes into account most of the inter-atomic interactions in a molecule:

$$V_{tot} = \Sigma_{bonds} \tfrac{1}{2} K_b(b-b_o)^2 + \Sigma_{angles} \tfrac{1}{2} K_\phi (\phi-\phi_0)^2 + \Sigma_{torsions} K_\theta [1+ \cos (n\theta-\delta)] + \Sigma_{nonbonded} [A/r^{12} - B/r^6] + \Sigma_{es} q_1q_2/(kDr) + \Sigma_{constraints} K_d (d-d_0)^2 \qquad 4.30$$

In the above equation, the first three terms cover bond stretching, bending and torsion in a molecule. The empirical constants, K_b, K_ϕ and K_θ are the respective force constants, and the terms represent deviation from the equilibrium values in each case. The fourth term takes care of van der Waals interactions using a Lennard-Jones (LJ) function for non-bonded interatomic interactions. The fifth term calculates the electrostatic interaction between atoms with partial charges (q_i) on various atoms restricted to monopole-monopole approximation. The charges are usually calculated from quantum chemical theories. The interactions are weighted by the dielectric constant of the medium (D), through a weighting factor k. To take into account the effect of water, D can be chosen differently for groups exposed to the medium and for those in the inner core of the macromolecule. In some potential functions, a term for hydrogen bonding is introduced. In the case of NMR, this is taken care by the geometrical constraints for the hydrogen bonded atoms.

The last term in equation 4.30, has geometrical constraints as obtained from NMR spectroscopy. This term consists of several components such as chemical shifts, NOEs, J couplings, RDCs, hydrogen bonds and cross-correlation relaxation. The values of K_d are chosen somewhat arbitrarily and can be changed during the course of structural calculations to provide variable emphasis to NMR constraints.

Once the potential function has been defined, it is possible to calculate energy and other dynamic properties of the molecules such as the *equations of motion* for each atom. Explicit equations for kinetic energy, acceleration etc can be written. The temperature of the system can be introduced. This can be done both in the Cartesian and the torsional space.

The parameters for different terms depend on the atom types and the type of molecules and are obtained empirically. A fair amount of ambiguity and variation exists in these values. Often these parameters are refined with experience. This implies that the terms have to be carefully calibrated for each molecular system. Potential functions for proteins may not work equally well for nucleic acids and polysaccharides. These functions are generally based on experimental results and quantum chemical calculations. However, there is a fair amount of variations in the parameters used in different softwares. The most commonly used computer programs are CHARMM, AMBER and GROMOS.

Two different approaches have been used to derive the conformation from an assumed starting point to the one that provides the best fit. Molecular mechanics (MM) approach is based on minimizing the potential energy function. In the molecular dynamics (MD) approach the total energy of the system is minimized. The directions of the atomic displacements are determined by the gradients of the potential energy. Molecular dynamics involves converting the potential energy into kinetic energy and solving Newton's equations of motion for atoms in the molecules. MD and MM approaches are similar. Both methods utilise information on the non-covalent interactions between atoms in a molecule. The basic difference between the two techniques is the approach to the displacement of the atomic coordinates. Both methods have been modified to introduce NMR geometrical constraints as outlined above.

Initially, these methods were developed for the calculation of conformational energies, spectroscopic and thermodynamic properties and dynamic behaviour of molecules. Later, by introducing terms containing NMR based geometrical constraints, these methods have been extended for structure determination by NMR. While using these programs for structural calculations, the objectives are different from what these algorithms were originally designed. Therefore suitable modifications have been made to make them work faster for structural refinement, while sacrificing the dynamic part. For example, simplified terms for Lennard-Jones and electrostatic interaction help to speed up the calculations.

The concept of combining MD with simulated annealing, through slowly lowering the temperature of the system, starting from a high value, is another important development for structural calculations. As stated earlier, the conformational energy landscape of even a medium size molecule is complex, and can have several minima. The temperature in MD allows a molecule to overcome barriers between various minima through the kinetic energy term. Thus, one can use several iterations of heating and cooling (called simulated annealing) to reach the global minima. Other parameters such as force constants can also be varied, during stages of heating and cooling.

Computer programs have been written to suit the requirements for structural calculations. For example, the original CHARMM program was modified to XPLOR, with the objective of structural calculation and refinement. XPLOR has been highly popular for NMR structure determination. Similarly, DYANA/CYANA incorporates the concept of simulated annealing.

7.5 Torsion Angle Dynamics (TAD)

Both MM and MD methods work in Cartesian space. As pointed out earlier, 3D structure determination is basically a problem of fixing the torsional angle as the bond lengths and bond angles do not vary significantly. Methods based on dynamics in the torsion space have been developed, which vary only the torsion angles using a variable target function. Examples of commercial software available for this purpose are DYANA/CYANA for proteins and TANDY for nucleic acids. These approaches are usually faster, which is an advantage for larger molecules. Another advantage is that the starting structures are random and the conformational space is better sampled.

In the MD approach, velocities of atoms are derived from the potential energy forces on the molecule. Initially, dynamics is performed at high temperatures to ensure extensive sampling of the conformational space. This is followed by *annealing*, which involves slow cooling of the system to yield lowest energy structures.

Both MD and TAD routinely use annealing. It involves coupling the system to a temperature bath, which is cooled down slowly (annealed) from its initial high temperature, thereby allowing the system to cross barriers between local minima of the target function. MD in the torsional space (TAD) provides an efficient way to calculate NMR structures. Strong potentials required in conventional Cartesian space to retain the covalent structure and, concomitantly, the high frequency motions caused, are absent in TAD. This results in a simpler potential energy function and in longer permissible time-steps for the numerical integration of the equations of motion.

To start, typically 100-200 random or extended structures of the macromolecule are generated at a high temperature (~10,000 K). TAD runs are then performed for 5-50 psec at temperatures of 10000 K to ensure efficient sampling of conformational space, followed by slow cooling to 0 K. Dynamics runs are performed at each cooling step. The kinetic energy in MD calculations helps in overcoming energy barriers and prevents molecules from being trapped in a local minimum. The final step typically consists of 1000 steps of conjugate gradient minimization to obtain the 3D structure.

7.6 Restraint Energy Minimization

As discussed, NMR structural calculations often involve simplified force fields. To determine the final structure(s), one uses a full potential energy function. The energy is minimized by varying the structural variants, to reach an acceptable structure. Such calculations refine the structures, starting from the acceptable solutions. Energy minimization can be performed using a box of water in which the molecule is placed. Thus, the effect of water can be taken into account explicitly. Further, suitable hydrogen bond potentials are introduced to obtain good hydrogen bond geometries.

It may be noted that the experimental constraints appear only through the last term. There is always a concern as to how much the final geometries are driven by the theoretical terms. It is therefore desirable to switch off, or underplay some of the theoretical terms at different stages of the minimization process. In the initial stages only the short-range geometrical constraints are included. The long-range constraints are usually introduced after the secondary structures have formed. This avoids trapping of the molecule into high energy local minima and knots.

It is important to have a wide conformational sampling. Usually a family of several structures satisfy NMR constraints. The objective is not to loose any one of the acceptable structures. This is achieved by using a large number of initial conformations, and enhancing sampling by using high temperatures during the dynamics. During annealing, temperature should be lowered slowly. The final steps in annealing are crucial for high quality structures. Typically a set of about 10-20 low energy conformations, which are consistent with the NMR constraints, are chosen as acceptable structures.

7.7 Validation of Final Structures

The last part of the exercise is to check the final structures for their validity. Has the structural calculation provided a conformation, which is satisfactory? Are the geometrical constraints satisfied? There are several ways by which such questions can be addressed.

The first indication for an acceptable derived structure for proteins is to obtain satisfactory Ramachandran (ϕ, ψ) maps. Information on acceptable range of values of torsion angles has been reported for all classes of biological molecules based both on quantum chemical methods and classical potential functions. A large number of points outside the allowed regions in the derived conformation make the structure a suspect. A convergence on this point indicates that there are no serious steric violations in the final structures. PROCHECK is often used to check the stereo-chemical quality of a protein structure or an ensemble of protein structures. The acceptable conformational spaces for nucleic acids and polysaccharides are also available in literature.

The second step is to check violations of the input constraints. If the number of violations is large and are of serious nature, then the structure should be rejected. Back calculation of NOESY spectrum, using relaxation matrix formalism, may be used to check quality of fit with the NOE data. Several research groups report root mean square (RMS) deviation between the observed and calculated NOE and the RMS deviations of the coordinates among the family of conformations to serve as indications of the quality of their structures.

There may be several NMR parameters, which have not been used as constraints during the minimization procedure. These properties provide good test for the validity of the converged structures. We have mentioned the use of chemical shifts, some of which can be calculated for the final structures and compared with experimental values.

Further, the parameters obtained from oriented systems can be used for validation if not introduced in the energy convergence. There are two such parameters which can be used for checking validity of a proposed structure. One is the change in the chemical shifts caused by magnetic field induced alignment. Secondly, if dipolar couplings have not been used in structural simulations then these can be employed for checking the correctness of the structure. A quality factor can be defined in terms of the root mean square deviation (RMSD) between the observed and calculated values of both the parameters:

$$Q_{shifts} = RMS\,(\Delta\delta^{cal} - \Delta\delta^{obs}) / RMS\,(\Delta\delta^{obs}) \qquad 4.31$$

$$Q_{RDC} = RMS\ (D^{cal} - D^{obs}) / RMS\ (D^{obs}) \qquad 4.32$$

Another approach is to use existing database and search for structural similarities. As pointed out earlier, it is likely that a similar fold of the polymer is already present in literature. This can be checked from databanks.

7.8 Presenting the Final Structures

The tools of molecular graphics provide means to project, visualise and analyse the 3D coordinates in molecular structures in a graphical form. Conformations derived from NMR data can be presented as ball-and-stick or space filling models. Using partial charges from quantum chemical methods, the electrostatic contours can be superimposed on such structures. When more than one acceptable solution is obtained, the family of such structures can be superimposed along with an average structure. Aspects such as location of hydrogen bonds, torsion angle distributions, restraint violations. can be visualised from graphical softwares. Some of the softwares available for this purpose are MIDAS, MOLMOL, GRASP and MOLSCRIPT.

Since NMR provides a family of acceptable structures, it is often desirable to obtain a mean or an average structure, so as to compare it with approaches such as X-ray crystallography. A crude structure for this purpose may be obtained by taking an average of the coordinates. Such a structure may be refined by restraint energy minimization. This provides a structure with good stereochemistry and closely represents the mean coordinates of the bundle of structures obtained from NMR.

The ability to visualise structures in 3D enables several studies to be made. These include interaction with other molecules. Docking procedures allows one to look for steric compatibility. The electrostatic interactions with ligands and cations can be studied using electrostatic surfaces. Such methods are important for probing ligand binding and drug design.

8. DATA BANKS FOR STRUCTURES AND NMR OF BIOMOLECULES

Once the 3D structure and NMR data have been obtained, there are international repositories where such data are deposited. These depositories serve as valuable sources of information, which can be downloaded by researchers and educators for various purposes.

8.1 Protein and Nucleic Acid Data Banks (PDB)

The protein data bank (PDB; www.pdb.org) is an important repository for 3D structures of biological macromolecules. Though the name suggests otherwise, it has structures of all classes of bio-molecules. PDB had structural coordinates for 46,973 macro-molecules, divided into 43,238 proteins, peptides and viruses, 1,799 nucleic acids, 1,903 protein-nucleic acid complexes, and 13 carbohydrates and other molecules. A majority of these structures are based on X-ray diffraction. NMR has contributed to 6,873 structures. Once a researcher has solved a structure, the researcher normally submits it to the PDB data bank so that it can be accessed by others (*deposit.pdb.org*). All structures follow a uniform protocol. Help desks such as *info@rcsb.org*, are useful sites for general enquiries and deposition questions for researchers.

There is another database called *Nucleic Acid Database* (NDB; http://ndbserver.rutgers.edu), which is an exclusive repository of 3D structural information about nucleic acids. NDB had structural coordinates for 3,557 structures.

8.2. Biological Magnetic Resonance Data Bank (BioMagResBank; BMRB)

BMRB *(www.bmrb.wisc.edu)* is an international repository for NMR data on biological molecules. BMRB and PDB have close links. BMRB works in collaboration with PDB to provide spectral data and atomic coordinates. *BMRB's mission is to collect, archive, and disseminate (worldwide in the public domain) the important quantitative data derived from NMR spectroscopic investigations of biological macromolecules.*

BMRB provides facilities to scientists in their quest to analyse NMR spectral data on biomolecules and to determine structure, dynamics and chemistry of biological systems. It is a publicly-accessible depository for NMR results on proteins, nucleic acids, polysaccharides, their ligands and cofactors. It is responsible for standardization of data-base on proteins and nucleic acids on behalf of the IUPAC-IUBMB-IUPAB inter-union task force. Its archives contain assigned chemical shifts, J-coupling constants, relaxation rates and chemical information.

A number of pulse sequences and software for biomolecular NMR data analysis and structural simulation can be down-loaded from its site. BMRB is an important site for anyone interested in biomolecular NMR.

8.3 Applications of Data Banks

PDB, NDB and BMRB data banks are used extensively by researchers and industries involved in drug-design and drug development. A researcher can dock new lead compounds to the target macro-molecule and see how well they may bind and look for suitable modifications. It is a major resource for several areas of biological research. Bioinformatics and structural genomics are two such areas (*www.rcsb.org/pdb/strucgen.html*) which have benefited from PDB. The data bank continues to be actively involved in developing information for such projects. It also provides resources for educational material and several aspects related to structural biology. Educators may find the site *www.rcsb.org/pdb/education.html*, useful for teaching and training programs. Computational biologists use the data to discover folding patterns and functional aspects of proteins starting from the 3D structures. Students may choose a structure create their own constraints and then starting from a random structure, try convergence to an identified structure using one of the algorithms discussed earlier in this Chapter. Alternatively, they can use available NMR data and try to see if they can arrive at the structure.

8.4 Where to look for Literature on NMR Structures

Journal of Biomolecular NMR published by Kluwer is devoted to the structures of biological molecules. In addition, structures of biological molecules are often reported in Journals such as Nature Structural & Molecular Biology, Biochemistry, Journal of Molecular Biology, Proteins and Protein Science etc. Methods in Enzymology and Quarterly Reviews of Biophysics often carry reviews on major advances in the field.

9. REFERENCES

9.1 Further Reading

C.E. Johnson and F.A. Bovey, *Calculation of nuclear magnetic resonance spectra of aromatic hydrocarbons*, J. Chem. Phys. **29**, 1012–1014 (1958).

W.S. Wishart, B.D. Sykes, and F.M. Richards, *Relationship between nuclear magnetic resonance chemical shift and protein secondary structure*, J. Mol. Biol. **222**, 311–333 (1991).

L. Banci, I. Bertini, G.G. Savelini, A. Romagnoli, P. Turano, M.A. Cremonini, C. Luchinet, and H.B. Gray, *Pseudocontact shifts as constraints for energy minimization and molecular dynamics calculations on solution structures of paramagnetic metalloproteins*, Protein-Structure Function and Genetics **29**, 68–76 (1997).

B. Reif, M. Hennig, and C. Griesinger, *Direct measurements of angles between bond vectors in high-resolution NMR*, Science **276**, 1230–1233 (1997).

A.J. Dingley, and S. Grzesiek, *Direct observation of hydrogen bonds in nucleic acid base pairs by internucleotide ^2J (NN) couplings*, J. Am. Chem. Soc. **120**, 8293–8297 (1998).

P. Gunther, *Structural calculation of biological macromolecules from NMR data*, Q. Rev. Biophysics **31**, 145–237 (1998).

J.L. Markley, A. Bax, Y. Arata, C.W. Hilbers, R. Kaptein, B.D. Sykes, P.E. Wright, and K. Wuthrich, *Recommendations for the presentation of NMR structures of proteins and nucleic acids*, Pure Appl. Chem. **70**, 117–142 (1998).

R.C. Stevens, *Design of high throughput methods of protein production for structural biology*, Structure **8**, R177–R185 (2000).

A. Bax, *Weak alignment offers new NMR opportunities to study protein structure and dynamics*, Protein Science **12**, 1–16 (2003).

R.S. Lipsitz, and N. Tjandra, *Residual dipolar couplings in NMR structure analysis*, Annu. Rev. Biophys. Biomol. Struct. **33**, 187–413 (2004).

9.2 Books

G. Govil, and R.V. Hosur, *Conformation of Biological Molecules*, Springer-Verlag (1982).

W. Saenger, *Principles of Nucleic Acid Structure*, Springer-Verlag (1984).

K. Wuthrich, *NMR of Proteins and Nucleic Acids*, Wiley, New York (1986).

M.P. Deutscher (ed.), *Guide to Protein Purification*, Methods Enzymol. **182**, 1–894, Elsevier Inc. (1990).

S.K. Sarkar (ed.), *NMR Spectroscopy and Its Application to Biomedical Research*, Elsevier (1996).

T.L. James (ed.), *Nuclear Magnetic Resonance of Biological Macromolecules*, Methods Enzymol. **394**, 1–631, Elsevier (2005).

CHAPTER 5

PROTEIN NMR: GENERAL PRINCIPLES AND RESONANCE ASSIGNMENTS

1. INTRODUCTION

With the availability of genomes of several organisms, including that of the human, the attention of biologists has moved towards understanding the significance of genetic variations within species and across various organisms. Proteomics is a new area of research, which focuses on the expression patterns of all proteins encoded by genes in a particular organism. The information contained in genomes helps in deducing sequences of proteins encoded by different genes. However, deciphering protein function requires a deeper understanding of the three-dimensional (3D) structure, dynamics and interaction between various proteins. Initiatives in this direction help to generate an atlas of protein-folds, structural modules and their relationship to different functions of proteins. In addition, structural genomics is expected to promote discovery of new structural modules, proteins with novel functions and in the design of new and modified proteins.

The techniques of genetic engineering have provided the ability to produce relatively large quantities of proteins. These can be labelled with isotopes that can be detected by NMR. Also mutations (substitution of one or more amino acid residues in the original *wild type* protein) can be achieved by altering the gene sequence. Both these innovative methods havc added new dimensions to studies of structure, dynamics and interaction of proteins.

Proteins are constructed from one or more polypeptide chains, synthesized by the condensation of L-(α)-amino acids. A typical single chain protein may contain 100–300 amino acid residues (M_r ~10-33 kDa). Several functional proteins lie in this range. Smaller molecules are called peptides. Larger proteins may be composed of several polypeptide chains. For example, haemoglobin contains four polypeptide chains. This provides more complexity in function and regulation. To date, the largest protein known is *titin* (M_r ~ 3000 kDa; 26,926 residues), which is found in skeletal and cardiac muscle. Proteins can covalently link to other biological molecules, such as lipids (forming lipoproteins) and sugars (in glycoproteins).

1.1 Functions of Proteins

Proteins form the physical and chemical basis of life. Proteins by themselves or in combination with other components, form building materials of living systems. Hence, they form the most important group of biological molecules. The proteins are responsible for the structure and functions of cells. Protein functions are multifarious, which is reflected in their diverse structural variations.

One of the important functions of proteins is to catalyze biochemical reactions. Such proteins are called enzymes. Enzymes have the ability to lower the activation energy of a biochemical reaction. They act with a high degree of efficiency and specificity and govern almost every chemical reaction in a living cell. Enzymes are classified and named according to the type of reactions that they catalyze. Some examples of the classes of enzyme catalyzed reactions (and the respective enzymes) are: oxidation-reduction reactions (*oxidases, reductases, hydrogenases, dehydrogenases)*; transfer of functional groups (*transferases)*, hydrolysis (*hydrolases*); transfer of groups within molecules generating isomers (*isomerases)* and formation of new chemical bonds (*ligases*). Note that the function is normally denoted by the name of the enzyme.

Certain enzymes require a metal ion or another small molecule (cofactor) for their function. Some proteins use a prosthetic group (such as the heme group in heme-proteins) for their function. The enzyme pocket where the biochemical reaction takes place is called the *active site*. Most proteins carry out their function by reversible binding to a ligand. Ligands generally bind to proteins at sites which are complementary in shape and physico-chemical properties.

The more complex functions of proteins include transport of materials in cells and across biological membranes, motion and locomotion of cells, regulation involving hormones, and metabolic control, immune and neuronal activity. Several of these functions require concerted and sequential interplay of a series of biochemical reactions and may involve several proteins.

Proteins also have a structural role. Several tissues such as joints, muscle, cartilage, nails and hair are built from proteins. Such proteins usually have large molecular weights and are multimeric.

1.2 Conformation and Dynamics of Proteins

Internal rotation around a large number of covalent bonds in proteins leads to their conformational variability. However, most proteins in their functional form have a relatively well defined 3D structure. The structure is determined by non-covalent hydrophobic and ionic interactions arising from the amino acid side chains. Under physiological conditions, proteins acquire a well defined conformation called the *native* structure which is thermodynamically stable. This structure may be destroyed under extreme conditions of temperature, pH or in the presence of reagents such as high concentration of urea (1 to 8 M) or guanidine hydrochloride, leading to a process called denaturation. Once the conditions are restored, proteins tend to fold back to their active form. In some cases, refolding may also result in alternative conformers.

What determines a protein structure? There are different levels at which the protein structure is deciphered:

(i) *Primary structure* that is dictated by the amino acid sequence.

(ii) *Secondary structure* refers to the peptide backbone conformation found in protein structures, such as α-helices, β-strands and turns.

(iii) *Tertiary structure* describes the packing and the fold of secondary structural elements and amino acid side-chains resulting into a compact 3D structure.
(iv) *Quaternary structure* is determined by the association of more than one polypeptide chain to form multimeric proteins, as is the case in haemoglobin, which contains four subunits.

Usually, the knowledge of protein primary sequence provides limited insight into its structure and function. An exception is the case where there is a strong sequence homology with another protein or a module having a known function. To understand the function of a protein, it is essential to determine its 3D structure. Even proteins with well defined 3D structure have considerable flexibility. The dynamic nature of proteins is crucial for their function. In fact, the activity of a protein is intimately linked both to its conformation and dynamic behaviour. In addition, one needs to unravel the conformation of the active site, amplitudes of molecular motions (dynamics) responsible for bringing functional groups into close proximity and disposition of solvent molecules. Proteins may change their structure upon binding to ligands and ions during their function. Protein-ligand binding is, therefore, an important property that needs to be explored.

Table 5.1: Some Milestones in Protein Structure Determination BMRB (http://www.bmrb.wisc.edu)

1930s	α-helix and β-sheet structures	Pauling and Corey
1955	Structure of collagen	Ramachandran
1955	First protein sequence	Sanger
1962	Protein structure of myoglobin	Perutz and Kendrew
1964	Structures of insulin, penicillin and vitamin B12	Hodgkin *et al.*
1964	Ramachandran map	Sasisekharan and Ramachandran
1980-	Protein-nucleic Acid complexes	Klug *et al.*
1982-	Photosynthetic centre	Michel, Deisenhofer and Huber
1985	Direct methods for the crystal structure determination	Hauptman and Karle
1986	α–amylase inhibitor (NMR structure)	Wüthrich *et al.*
1994	ATP Synthase	Walker
1998	Potassium channel	MacKinnon

One of the challenges in the field of protein science is to structurally characterise different proteins. There is a need for more and better high-resolution 3D structures of proteins and for detailed description of their dynamics. The information about the 3D structures thus derived, can have a profound impact on fundamental biology, medicine and biotechnology. Structures can be used to understand molecular basis of disease, develop diagnostics, design drugs and also in enzyme engineering.

1.3 History of Protein Structure Determination

Most of the 3D protein structures available in the Protein Data Bank (PDB) have been obtained using the technique of X-ray crystallography (~80%) and NMR (~16%). A few structures have been determined by theoretical modelling and other methods. NMR data provides the 3D structure of the molecule under conditions used in experimental measurements, which can be varied. Thus, the influence of pH, salt concentrations, temperature and other experimental conditions on the 3D structure can be studied.

Figure 5.1: Some eminent scientists involved in protein structure determination. (Courtesy: http://nobelprize.org).

A major contribution to protein structure determination was the establishment of α-helix and β-sheet structures by Linus Pauling and Paul Corey in 1930s. The first step in the determination of a protein structure is to determine the sequence of amino-acid residues. An ingenious method for this difficult task was developed by Frederick Sanger who reported the amino-acid sequence for the peptide insulin in 1955, for which he received the Nobel Prize in 1958. Max Perutz and John Kendrew reported the first protein structure (structure of the oxygen-carrying molecule myoglobin); the two shared the Nobel Prize in 1962. Two years later, Dorothy Hodgkin received the Nobel Prize for determining crystal structures of penicillin and vitamin B12. In recent years, there have been several Nobel prizes awarded for the discovery of the protein structures and their functions. Thus the study of the structure and function of proteins is an important field in modern biology, and is often referred to as structural biology.

2. ELEMENTS OF PROTEIN STRUCTURES

A polypeptide chain is composed of individual units known as L-α-amino acid residues. Each amino acid has a central tetrahedral carbon atom identified as C^{α} (sometimes referred as CA). This is covalently bonded to an amino (NH_2) group, a carboxyl (COOH) group, a hydrogen atom identified as H^{α} (or HA), and a variable side chain (R). A repertoire of twenty amino acids occurs in proteins. These differ from each other in the side chain (R). The various amino acid sidechains found in proteins are shown in Figure 5.2.

2.1 Nomenclature

In an inter-union joint venture, the International Unions of Pure and Applied Chemistry (IUPAC), Biochemistry and Molecular Biology (IUBMB) and that for Pure and Applied Biophysics (IUPAB) have recommended certain standard conventions for the representation of structures of proteins. Besides the C^{α} (CA), the amino nitrogen is denoted by N and its attached hydrogen by NH or H^{N}. The carbonyl carbon and oxygen are represented by CO (C′) and O respectively. In numbering position of the side chain atoms, the atom leading to the main chain has the highest priority. For cyclic amino acids such as Pro, the priority leads out from the main chain C^{α} atom rather than from the N atom (i.e. $C^{\alpha} > C^{\beta} > C^{\gamma} > C^{\delta} > N$).

Figure 5.2 depicts the abbreviation and numbering for different atoms in the twenty amino acid residues normally present in proteins. The hydrogens of an N-terminal amine are designated as H1, H2 (and H3 when the amino acid is protonated). The oxygen atoms of a C-terminal carboxyl are labeled O′ and O″ and the hydrogen of the carboxyl is identified as H. The hydrogens on methyl groups and protonated amines are designated as 1, 2 and 3 in the conventional way.

2.2 Backbone and Side-chain Torsion Angles

The conformation of the backbone of an amino acid in a polypeptide chain is defined by three torsion angles. For the ith residue, the backbone (ϕ_i, ψ_i and ω_i) and the side-chain (χ_i) torsion angles are represented as shown in Figure 5.3. Torsion angles ϕ_i, ψ_i and ω_i describe rotations around the N_i-C^{α}_i, C^{α}_i-C'_i and C'_i-N_{i+1} bonds respectively, in the protein backbone.

Besides the three backbone torsion angles, there can be several side-chain torsion angles depending on its length. These angles are represented by χ^j where j represents the rotatable bond. For example, j=1 corresponds to C^{α}-C^{β} bond; j = 2 denotes C^{β}-C^{γ} bond, and so on. Figure 5.4 shows Newman projection describing backbone torsion angles (ϕ, ψ, ω) and side-chain torsion angle (χ_i^1) in amino acid residues, with a methylene group at the C^{β} position. In residues with branched side chains, as in the case of Ile, two superscripts are used; the first indicates the position of the bond and the second indicates the branch. For example, the C^{α}-C^{β}-$C^{\gamma 1}$-$C^{\delta 1}$ torsion angle in Ile is denoted by $\chi^{2,1}$.

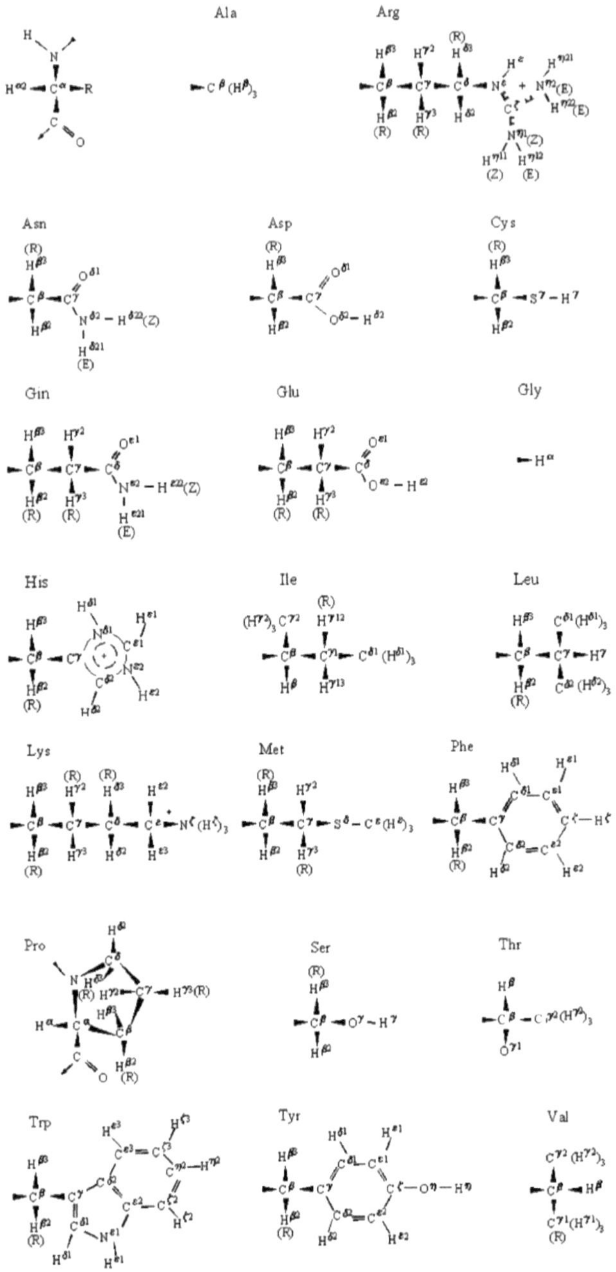

Figure 5.2: Abbreviation and numbering for atoms in amino acids.

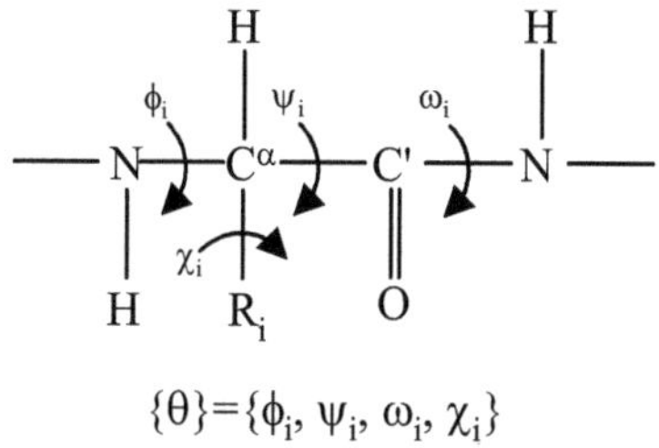

Figure 5.3: Backbone (ϕ, ψ, ω) and side-chain (χ^1) torsion angles in an amino acid residue. The π electron in the C=O bond are delocalised giving a partial double bond character to the C'-N bond and thus restricting the angle ω_i to a trans conformation.

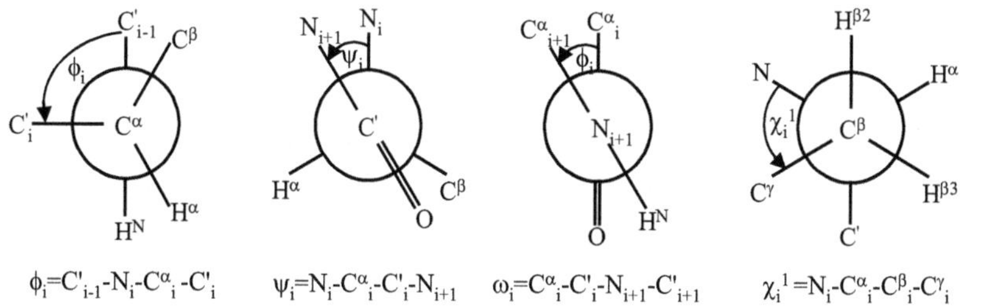

Figure 5.4: Newman projection diagram describing backbone (ϕ, ψ, ω) and the side-chain (χ^1) torsion angles in amino acid residues with a methylene C^β group.

2.3 Primary Structure

The amino acid residues in a protein are linked through peptide bonds in a specific sequence to form a polypeptide chain. Such a sequence of amino acids is called the *primary structure* of the protein. It was shown by Anfinsen in 1973 that the primary structure of a protein determines its *native* 3D structure. The 3D structure of the protein in turn, is responsible for its chemical and biological properties. A single point *mutation* (change in particular amino acid in the sequence) can sometimes change properties of a protein significantly.

A glance at the chemical structures of side chains in proteins brings out a number of interesting properties for these groups. Amino acids such as Gly, Ala, Val, Ile, Leu and Phe have non-polar (hydrophobic) side chains. Aromatic rings are integral parts of Phe, Trp and Tyr. Amino acids Asp and Glu are negatively charged at neutral pH, while Lys and Arg carry a positive charge. Such amino acids have groups which have acid-base properties. For example, histidine has a highly reactive five-member imidazole ring, which has a pK value close to physiological pH. It is often involved in the active sites of proteins and participates in electron and proton transfer reactions.

The amino acid Pro is unique and differs from rest of the amino acid side chains. It is bonded covalently to both the N and the C^α atoms of the peptide backbone and does not have the amide H^N. The cyclic five-member ring imposes constraints on the rotation around the N-C^α (angle ϕ). This has a significant influence on protein conformation. Amino acids Cys, Ser, Thr, Asp, Gln and Tyr are neutral but have a polar group, and therefore have hydrophilic properties. Two Cys residues in a polypeptide chain can be covalently linked forming S-S bond (disulphide bridges).

Thus, all the twenty amino-acids are endowed with certain specific properties. Further, certain amino acid residues in a peptide can be physiologically modified. These modifications play a major role in the regulatory function of a protein.

The primary structure of a protein is dictated by the gene which codes for it. Once the gene for a particular protein has been sequenced, it is possible to derive the primary structure from the code. The sequence can also be obtained by biochemical methods. Usually, NMR and X-ray crystallography methods are not used in determination of primary sequence. However, in some cases, these techniques have helped to detect an incorrect assignment of a residue in the amino acid sequence.

2.4 Conformational Freedom of Peptides

The peptide bond (C'-N) in a polypeptide chain has a partial double bond character. The amide group is therefore rigid. The six atoms that form the peptide group (atoms in the stretch $-C^\alpha_{i-1}-C'_{i-1}O_{i-1}-N_iH_i-C^\alpha_i-$) lie in a plane. Further, the conformation with a *trans* arrangement of C′=O and N-H groups (corresponding to $\omega_i = 180^o$) is favored. On the other hand, there is a fair degree of rotational freedom around the bonds $N_i-C^\alpha_i$, $C^\alpha_i-C'_i$. The dihedral angles ϕ and ψ dictate the conformation of the poly-peptide backbone and give rise to various secondary structures, as discussed later.

Ramachandran and Sasisekharan considered atoms in a dipeptide as hard spheres, with dimensions of the order of their corresponding van der Waals radii. They showed that large regions in the (ϕ, ψ) conformational space have high steric interaction because of overlap of such spherical shapes. Such conformations are therefore not allowed. For example, when (ϕ, ψ) are (0°, 0°), atoms in a dipeptide fold back and overlap. Therefore, such conformations have high steric repulsion. Thus, the ϕ_i and ψ_i angles show preferences only for certain conformational domains. Such preferred conformations are indeed observed in the structures of proteins. The allowed conformation regions for amino acids having a C^β carbon are shown in Figure 5.5 A. In particular, the regions with positive values of ϕ are largely in the disallowed regions. Only 18% of the total conformational spaces correspond to the allowed regions.

Such 2D (ϕ, ψ) maps are called Ramachandran maps (Figure 5.5 A). These maps provide an understanding of the conformational space available for proteins. Note that Gly does not have a C^β carbon. Because of this the (ϕ, ψ) map for Gly is symmetrical around the origin. Glycine has a large conformational freedom. On the other hand, the ring structure of Pro restricts ϕ in a range of -25 to -85°. In fact, C^β moeity linked to different types of side chains causes variations in the profile of the allowed regions of the conformational map. The energetic aspects of linked peptides have been studied extensively both by classical and quantum chemical methods. Such calculations form the basis of force fields used in structural algorithms for proteins.

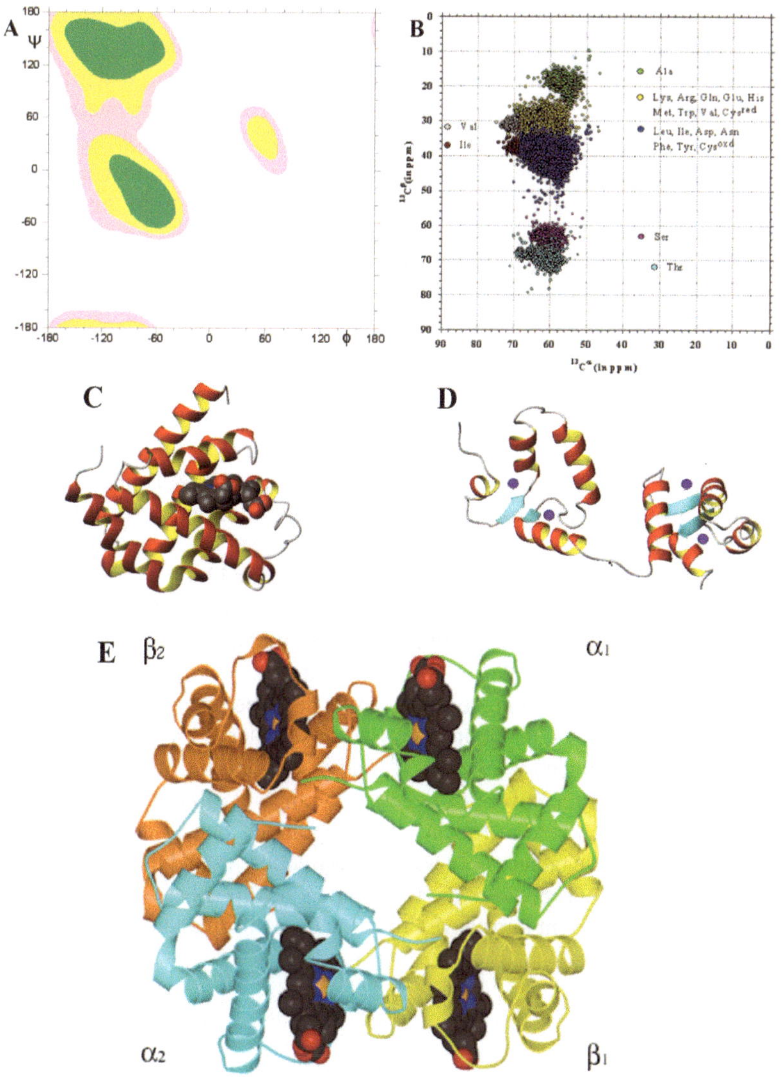

Figure 5.5: (A) Ramachandran map for a dipeptide containing a C^{β} atom. The shaded regions are the preferred ranges for β-sheets and α-helices. The backbone (ϕ, ψ) angles are partitioned into favoured (green), additionally allowed (yellow), generally allowed (pink) and disallowed conformations (white); (B) 2D plot of $^{13}C^{\alpha}$ and $^{13}C^{\beta}$ chemical shifts as deposited in BMRB. (C-E) Examples of higher level of structures in proteins: (C) Tertiary (myoglobin; PDB ID:2HHB); (D) Multi-domain (Calcium binding protein; Ca^{2+} are shown in blue spheres; PDB ID: 1JNX (NMR) and (E) Quaternary (haemoglobin; Four heme groups are shown in black; PDB ID= 1XXT (X-ray)) α-chains are shown in green and turquoise, and β-chains in yellow and orange.

Conformational angles of experimentally determined structures are usually represented in the form of such 2D maps. When the plot is constructed from database of known protein structures, the conformational map discriminates (ϕ, ψ) space into *allowed* and *disallowed* regions. The experimentally determined conformations are expected to lie in the allowed regions of (ϕ, ψ) space. Such conformational maps have proved to be indispensable tools in understanding the conformational space available for peptides and proteins and in the refinement and analysis of newly determined protein structures. This type of analysis is implemented in commonly used validation tools for protein structure such as PROCHECK (www.biochem.ucl.ac.uk/~roman/procheck).

2.5 Disulfide Bridges and Proline Rings

There are two amino acids which play distinct roles in stabilising protein conformations. Cys residues have sulfhydryl groups (S-H) in their side chain. Two Cys residues i and j can form a disulfide cross link (S_i-S_j) through an oxidation reaction. Disulphide bridges are important for providing conformational stability to proteins. Reduction of disulphide groups can lead to denaturation of protein structures. The internal rotation in this case is described by the torsion angles χ^1_i, χ^2_i, χ^3_i, χ^2_j and χ^1_j. The conformation around the S-S bond is usually *trans*.

The second residue is the proline. It is a five member ring formed by using the N and C^α atoms. Ring closure places additional conformational restrictions. Conformation of a five-membered ring can be described by a single pseudo-rotational angle (P), similar to the situation discussed in Chapter 7 for ribose and deoxyribose rings. The torsion angles χ^0_i, χ^1_i, χ^2_i, χ^3_i and χ^4_i describing ring pucker are therefore related. The angle ϕ for Pro is restricted in the range -25 to -85°. Pro does not possess an amide proton. Therefore it cannot form hydrogen bonds like other amino acids. Further, Pro can have either *cis* or *trans* conformation around the C′-N bond. The free energy difference (ΔG) between the two isomers of Pro is between -8.0 to 8.0 kJ/mol. Almost 6% of residues in turns have Pro in the *cis* conformation.

2.6 Secondary Structures

The detailed conformation of a protein is thus dictated by the backbone torsion angles ϕ_i and ψ_i angles and the orientation of the side chains. A number of non-covalent interactions, related to properties of amino acids, are responsible for the conformation of proteins. An important contribution to conformational stability comes from van der Waals and hydrophobic interactions. Recall that the twenty amino acid residues can be divided into hydrophobic and hydrophilic residues. Entropy factors associated with the interaction between these residues with water orient the residues such that the hydrophilic side-chains are exposed to the solvent. The hydrophobic residues generally lie in the interior of the protein structure. Reverse is the case when proteins are embedded in the hydrophobic environment of membranes. All amino acid residues (except Pro) have the capacity to form inter-residue

N-H---O=C hydrogen bonds. Residues which are exposed to the medium can also form hydrogen bonds with water. Charged amino acid residues can interact through electrostatic interactions. Such non-covalent interactions stabilise the conformations of proteins.

Protein structures are characterised by short ordered stretches. These are popularly known as secondary structural elements. Within the element of such secondary structures, the ϕ_i and ψ_i angles have approximately the same values. This gives a segment with ordered structure. Due to factors discussed above and to enable formation of optimum hydrogen bonds, certain sets of values for ϕ_i and ψ_i angles occur more frequently than others in protein structures. This leads to repeating conformations for several amino acid residues in such stretches. Three distinct secondary structural elements, called α–helices, β-sheets and β-turns are frequently observed in proteins. These structures are schematically shown in Figure 5.6.

2.6.1 α-helix

α-helix is a cylindrical right-handed helical structure. The residues participating in such a structure are characterised by values of ϕ and ψ torsion angles, which are around -57° and -47°, respectively. The stability of the structure is enhanced by hydrogen bonding between the backbone carbonyl oxygen of an amino acid i and the H^N of a residue situated four residues ahead (i.e. i+4), along the primary sequence. This leads to 5→1 [-N-H·····O=C-] hydrogen bonds (Figure 5.6 A), which lie parallel to the helical axis. The side-chains protrude towards the outside of the helix. The neighbouring residues in the helix are related to each other by a translation of 1.5 Å along the helix axis, and a rotation of 100° around the axes. This results in 3.6 residues per turn of the helix and a pitch of 5.4 Å. For L-amino acids, only the right handed α-helix is normally allowed. Short segments of left-handed α-helix have been observed in certain protein structures.

Since hydrogen bonds are in the same direction, the helix as a whole has a dipole moment. The amino end has a net positive and the C-terminal has a net negative charge. Therefore, negatively charged ions bind to the amino end. Certain amino acid residues such as Ala, Glu, Met and Leu promote formation of helical structure. On the other hand, Pro and Gly destabilise such structures. When α-helix occurs on the surface of a protein, it is usually made of hydrophobic residues on the side facing the interior. The hydrophilic residues reside on the exterior surface. Almost 25% of amino acid residues, in the data bank of all known proteins, have been found to exist in α-helical structure. A variant of the α-helical conformation is the so called 3^{10}-helix, where the residues are characterised by ϕ and ψ angles around -50° and -25°, respectively.

2.6.2 β-strands and β-sheet

β-strands are almost fully extended structures. The conformational angles ϕ and ψ in such secondary structures lie on the top left quadrant of Ramachandran map (Figure 5.5 A). The β-strands can interact with each other to form either antiparallel or

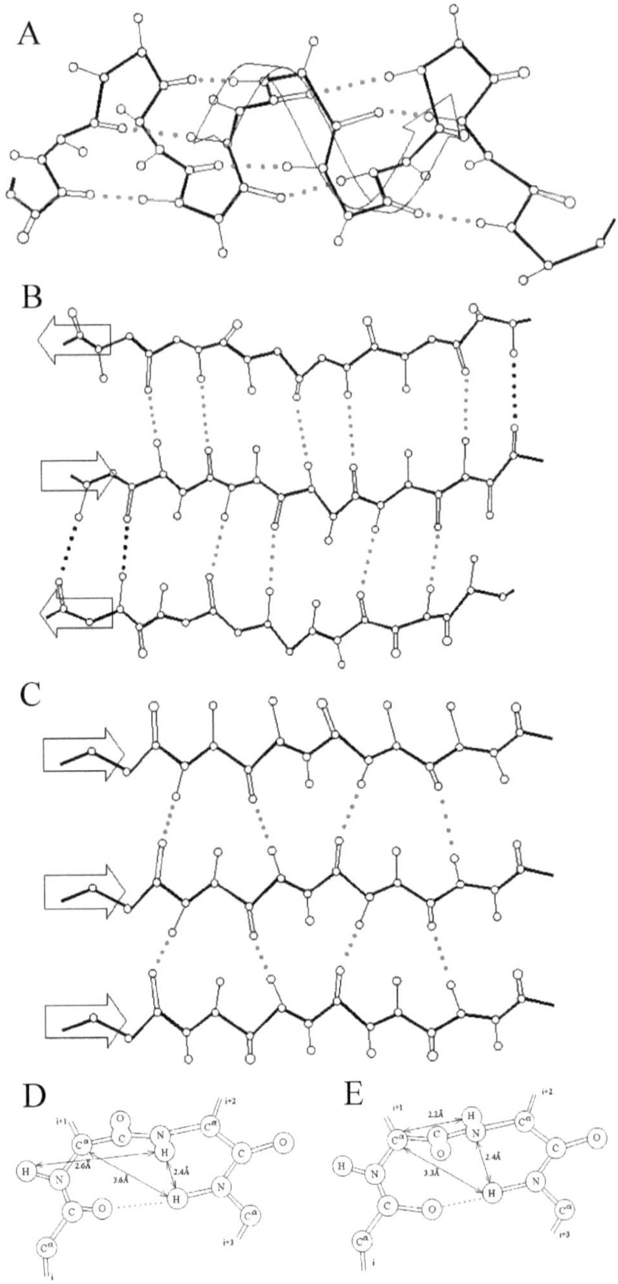

Figure 5.6: Secondary structural elements in proteins (A) α-helix, (B) anti-parallel β-sheet, (C) parallel β-sheet (D) Type I β-turn and (E) Type II β-turn.

parallel β-sheet structures (Figure 5.6 B and C, respectively). Such structures are stabilised by hydrogen bonds between the main chain amide (NH) and carbonyl (CO) groups belonging to different β-strands of the polypeptide chain. The axial distance between adjacent amino acid residues is around 3.5 Å. The side-chains project out on both sides of the β-sheets. The β-sheets are usually not flat but have a small right-handed twist. The β-strand structures occur widely in protein conformations.

2.6.3 β-turns

Another important secondary structure element is the β-turn (Figure 5.6 D and E). A polypeptide stretch can abruptly reverse its direction by taking a sharp turn. The β-turns are often found between two strands in a β-sheet. It allows the protein chain to turn by 180° in the space of only three amino acid residues. The protein chain may form one β strand, then turn back on itself, form yet another β strand adjacent to the first, and so on. This results in a β-sheet structure.

The β turns are hairpin like structures in which the carbonyl (CO) group of nth residue of a polypeptide is hydrogen bonded to the amide (NH) group of the $(n+3)^{rd}$ residue. β-turns also connect α and β structural regions in globular proteins.

Two types of turns, I and II (Figure 5.6 D and E), are observed in protein structures. In type I turn, the intervening residues at positions n+1 and n+2 adopt values of -60, -30, -90 and 0° for torsion angles ϕ_2, ψ_2, ϕ_3 and ψ_3, respectively. In the Type II turn the corresponding torsion angles are -60, 120, 80 and 0°, respectively.

Pro, with its capability of forming *cis* peptide bond and Gly, with its greater conformational freedom, are two amino acid residues, which are commonly involved in turn structures. A loop structure also allows reversal of polypeptide chain direction. However, loops involve larger number of amino acid residues. In a typical globular protein almost 33% of the amino acid residues are involved in forming turns or loops.

2.6.4 Collagen

Another type of secondary structure was discovered by Ramachandran and his group in the fibrous protein collagen. Collagen contains an unusual amino acid, hydroxy proline (Hyp). It forms a left-handed triple helix with (ϕ, ψ) angles of (-60°, 140°). A specific repeating sequence (Gly-X-Pro/Hyp) promotes such triple stranded helices. Collagen is the most abundant protein in animals and forms part of the connective tissues, such as tendons, cartilage, the organic matrix of bone.

2.6.5 Poly (L-Proline)

The structure of poly(L-Pro) is of interest since Pro can have both *cis* and *trans* conformations. It has a restricted freedom around the N-C^{α} bond. In solid-state, a regular structure called Poly (L-Pro) (I) is observed with *cis* peptide bonds. When dissolved in water, this isomerizes to a different structure (II), with *trans* peptide bonds. In concentrated aqueous solutions, the polymer chain has random sequences of *cis* and *trans* conformations. NMR has been effectively used to characterize such conformational transitions. Short segments of poly(L-Pro) (II) conformation occur in protein structures.

2.7 Fibrous Proteins

Proteins such as collagen in connective tissues are examples of fibrous proteins. Such proteins are also referred to as structural proteins in view of their roles. Proteins that make up hair, wool, nails etc. are also fibrous proteins but are built of α-helical segments (α-keratin). Silk fibroin and spider webs are largely composed of β-sheet structures (β-keratin). Fibrous proteins are insoluble in water and are not amenable to studies by solution NMR. However, they have been studied by solid-state techniques (Chapter 8).

2.8 Tertiary Structure: Globular Proteins

Most functional proteins such as enzymes, transport proteins and immunoglobulins have a folded or globular structure. The secondary structure elements such as α-helix, β-sheets, β-turn and loops make up the bulk of these structures. Quite often, disulphide bridges provide additional stability to the relatively compact structures of globular proteins. The tertiary structure describes the 3D organization of all atoms in a polypeptide chain. If a protein consists of only one polypeptide chain then the tertiary structure describes the complete 3D structure. Hence, this level of the structure is the determinant of the functional property of a protein. Tertiary interactions usually bring amino acid residues, which are far apart in the primary sequence of a polypeptide chain, close together in the 3D space. This leads to formation of loops, β-sheets and active sites.

The tertiary structure (Figure 5.5 C) is usually an assembly of the α-helical and β-sheet conformations linked by connecting segments. Proteins with more than 100 amino acid residues are known to fold into two or more globular units (Figure 5.5 D), called *domains*. A small polypeptide segment called the linker, joins such domains. The domain conformations may be preserved even when the protein is cleaved in the linker regions by specific reagents. Each domain may have a unique structure and function.

The complex 3D tertiary structures of large proteins can be classified into finite number of motifs or folds. For example, while it is estimated that human genome codes for ~25,000 proteins, the independent folding patterns may not be more than 1000. These complex motifs are built from simpler secondary structural units e.g. all α, all β, alternating α and β segments (α/β), segregated α and β segments (α+β) etc. A PDB identifier is used to classify different motifs present.

2.9 Higher Levels of Structural Organization

Some proteins consist of more than one polypeptide chain. Such proteins are called multimeric proteins. Their structures involve a higher level of organization. Quaternary structure describes the spatial arrangements of various subunits and the nature of contacts of individual subunits belonging to a multimer protein. Individual polypeptide chains in such proteins and their interactions provide regulation and control in their function. A typical example is haemoglobin (Figure 5.5 E), which is composed of four polypeptide chains. It carries and releases oxygen to different parts of the body in a highly controlled fashion. Myoglobin also carries oxygen in muscles, but it lacks the so called *allosteric* control to deliver oxygen present in haemoglobin.

In some cases, such as in structural proteins, protein-protein interactions help in stabilizing such structures. Large protein assemblies are used in biological systems for multi-step reactions. Such high M_r supra-molecular assemblies are currently beyond the scope of NMR investigations.

3. INTRODUCTION TO NMR OF PROTEINS

Protein structure determination by NMR spectroscopy involves several steps: sample preparation, data collection, data analysis, assignments, structure calculation and refinement. Except for the steps involving sample preparation and data collection, all other steps require extensive use of computational tools and are often carried out in an iterative fashion.

3.1 History of NMR of Proteins

Initial reports of low resolution ^{1}H NMR spectra of proteins appeared in late 50s and early 60s. These spectra did not provide much information and are of mainly academic interest. This was followed by the important work on paramagnetic heme proteins by Shulman and Wüthrich and on kinases by Cohn. However, the full glory of NMR was unravelled with the advent of multi-dimensional and multi-nuclear FT techniques. The use of 2D NMR made it possible to determine structures of large molecules, such as proteins in solution (in contrast to the structure in crystals by X-rays). Wüthrich and his colleagues started work on the use of NMR for protein structure determination with studies on NOE build-up, spin diffusion and sequential resonance assignments. A framework for NMR structure determination of proteins was formulated in 1986, with the simultaneous publication of crystal and solution structure of a globular protein, α-amylase inhibitor, *Tendamistat*. Since then, several molecules have been studied both by crystallography and NMR. In most cases the agreement between the two methods has been good.

NMR has become a powerful tool for unravelling protein 3D structures. In recent years, approximately one-fourth of the new structures have come from NMR studies. X-ray crystallography still has an advantage over NMR for studies on large proteins and the structures can be obtained with better precision. Then why use NMR? There are several reasons. NMR is unique for characterization of the dynamics of macromolecules in solutions under conditions close to those which are present *in-vivo*. It provides information on the stability of different regions of the protein. It helps in understanding the nature of unstructured states and on the conformational ensembles of inherently flexible stretches. NMR provides information on the hydration of biomolecules, which is critical for maintaining the functional state. It provides a structural and dynamic basis for understanding the way a protein sequence translates into a folded conformation. A large number of proteins undergo post translational modification for the purpose of signalling, modifying stability or activity. Such modifications can be monitored by NMR to achieve a functional understanding. Interaction and binding can be studied conveniently and thermodynamic parameters can be obtained. Finally, NMR is an ideal tool for studying biological activity of macromolecules.

In this Chapter, we shall introduce the rudiments of NMR of proteins. This is followed in Chapter 6, by the recent trends in the study of their structure, dynamics, interaction and function.

3.2 Current Status

Most NMR experiments on proteins are recorded in two or more spectral dimensions. The experiments fall in two classes. The first set contains experiments aimed at the identification and assignment of signals to the respective nuclear spins. The second set is used to obtain structural information. Homo-nuclear ^{1}H-^{1}H 2D NMR experiments can be used for proteins with M_r less than 10 kDa, both for assignments and structure determination. For larger proteins, 3D triple-resonance techniques are used. Such experiments are useful due to their high sensitivity and the ease of analysis. They work well for proteins with M_r up to ~ 30 kDa. Beyond this limit even 3D techniques do not suffice. Hence, 4D triple resonance experiments are used. However, the measurement time increases steeply with the number of dimensions.

The assignment of NMR signals to atoms in the macromolecule is termed as sequence specific resonance assignment (abbreviated as *ssr*). This forms an important step towards structural characterization. Until recently, the resonance assignment was considered as the most tedious and time-consuming step in protein NMR studies. With the advent of multi-dimensional triple-resonance strategies (using ^{1}H, ^{13}C, ^{15}N spins), it has become clear that the information content of protein spectra can allow automation of resonance assignments. This has resulted in development of NMR spectroscopy as a powerful tool in structural genomics, which involves high-throughput protein structure determination.

Structural information is obtained from NMR parameters as has been discussed in Chapter 4. Most of the information is derived from the NMR properties of ^{1}H. This information is useful only after the identity of the atoms that give rise to individual signals has been established. Thus, proteins with their complexity in structure have posed challenges both from computational and spectroscopic point of view. New experimental and computational NMR methods continue to be developed to aid in structural characterization of proteins and their interaction with other molecules.

3.3 Sample Preparation

Several aspects of sample preparation for solution NMR studies have been discussed in Chapter 4. For proteins, several special precautions are needed. Solubility, stability and aggregation are some of the limiting factors for solution studies. These arise partly because one needs relatively high concentrations due to the low sensitivity of NMR. The problems are more acute when one uses multi-dimensional spectroscopy. Presently, one generally uses 500 μM samples for recording complex 3D triple resonance spectra. As mentioned in Chapter 1, higher fields, super-conducting probes, low temperature coils and preamplifiers combined with better technology have led to significant enhancement in S/N ratio. Thus, one can overcome the problem of aggregation, which is prevalent with several proteins, through the use of lower concentrations. The total spectrometer time required to record 3D NMR spectra has been reduced several fold in recent years.

Special 3D and 4D NMR experiments on labelled proteins require several hours of acquisition. There can be chemical shift changes for the same spin in different spectra arising from changes in experimental conditions such as pH and temperature during the course of an experiment. These factors can contribute to incomplete or erroneous assignments. For proteins with low stability, it is imperative that all data be acquired in a short duration of time. Experiments should be performed under identical conditions of pH and temperature, preferably with the same sample. Protein samples may degrade with time, due to cleavage, unfolding or due to aggregation. Stability of protein samples is therefore, another important factor in NMR studies.

3.4 NMR Approach to Protein Structure

As discussed in Chapter 4, clues to the solution structures are embedded in the chemical shifts, coupling constants, NOEs, exchange rates of labile protons, RDCs, cross relaxation and paramagnetic shifts. The procedure for 3D structure determination of a peptide or protein involves the following steps:
(i) Sample preparation with suitable isotope labels.
(ii) Identifying and recording 2D/3D/4D spectra appropriate for the sample.
(iii) Grouping and identification of ^{1}H resonances belonging to different amino acid residues, which form characteristic networks of coupled spin systems.
(iv) Sequence specific resonance assignments by connecting adjacent networks of spin systems in the primary sequence; thus establishing chemical shifts.
(v) Use of chemical shifts thus obtained for secondary structure determination.
(vi) Estimation of ^{1}H-^{1}H distances using NOE.
(viii) Measurement of J-couplings using COSY or TOCSY; establish torsion angle ϕ.
(ix) Determination of ^{1}H/^{2}H exchange rates for labile protons; detection of hydrogen bonds and the solvent protected groups.
(x) Determination of long-range structural constraints using parameters such as RDCs, cross relaxation rates, contact shifts, long-range NOEs and pseudo-contact shifts (PCSs).
(xi) Setting covalent distances, primary sequence, disulphide bridges, metal coordination.
(xii) Stereo-specific assignments of pro-chiral groups.
(xiii) 3D structure calculation based on the NMR data thus acquired.
(xiv) Structural refinement, validation and presentation.

3.5 Classification of Chemical Shifts

With the different types of side chains in the twenty amino acid residues, protein NMR signals are generally well dispersed. Classification of chemical shifts of individual nuclear spins is an important step to achieve sequence specific resonance assignments. Information on chemical shifts of all amino acid residues can be obtained from BMRB (*http://www.bmrb.wisc.edu*).

As an illustration, Table 5.2 lists the chemical shifts of ^1H, ^{13}C and ^{15}N spins for two residues, Ala and Arg, as derived from BMRB. The table gives the observed values of an average chemical shifts with their standard deviations for the two amino acid residues. Similar information can be obtained for other amino acid residues.

Such data helps in classifying each kind of spin resonance into distinct spectral regions. For example, the amide proton of Ala [A(H^N)] in proteins on which data is available, resonates around 8.18 ppm with a standard deviation of 0.63. The hydrogen bonded amide protons resonate in the 7-9 ppm region. The corresponding range for H^α is 4-5 ppm. Some resonances may occur in overlapping regions, while others may have their own unique range. This data serves as starting points in assigning individual networks of coupled spin systems to various amino acid residues. The exact values of the chemical shifts of individual residues depend on the conformation of the protein. However, these are also influenced by factors such as temperature, pH, ionic strength, presence of paramagnetic ions etc.

Table 5.2: ^1H, ^{13}C and ^{15}N chemical shifts for Ala and Arg from BMRB.

Residue	Spin	δ (ppm)	Spin	δ (ppm)
Ala	H^N	8.18±0.63	C'	177.76±2.62
	H^α	4.26±0.45	C^α	53.19±2.11
	H^β	1.36±0.28	C^β	19.04±2.48
			N	123.21±4.54
Arg	H^N	8.23±0.62	C'	176.40±3.96
	H^α	4.29±0.48	C^α	56.84±2.52
	$H^{\beta2}$	1.80±0.30	C^β	30.71±2.46
	$H^{\beta3}$	1.78±0.29	C^γ	27.34±2.35
	$H^{\gamma2}$	1.58±0.28	C^δ	43.20±2.05
	$H^{\gamma3}$	1.56±0.30	C^ζ	159.57±2.17
	$H^{\delta2}$	3.11±0.29	N	120.74±3.97
	$H^{\delta3}$	3.11±0.27	N^ε	93.21±15.12
	H^ε	7.34±0.64	$N^{\eta1}$	74.12±7.53
	$H^{\eta11}$	6.85±0.48	$N^{\eta2}$	75.01±10.40
	$H^{\eta12}$	6.79±0.46		
	$H^{\eta21}$	6.78±0.45		
	$H^{\eta22}$	6.76±0.48		

3.6 $^{13}C^{\alpha}$ and $^{13}C^{\beta}$ Chemical Shift Statistics

The ranges of chemical shifts for both ^{13}C and ^{15}N are useful in the analysis of hetero-nuclear correlated spectra. Note the larger dispersions of ^{13}C and ^{15}N shifts (Table 5.2). For example, $^{13}C^{\alpha}$ of Ala resonates around 53.19 ppm with a standard deviation of 2.11 ppm. Figure 5.5 B shows a plot of $^{13}C^{\alpha}$ and $^{13}C^{\beta}$ chemical shifts based on the data from BMRB.

Depending on $^{13}C^{\alpha}$ and $^{13}C^{\beta}$ chemical shifts, amino acid residues can be classified into nine distinct categories:
(i) Ala which has $14 < {}^{13}C^{\beta} < 24$ ppm;
(ii) Arg, Cys^{red}, Gln, Glu, His, Lys, Met, Val, Trp have $^{13}C^{\beta}$ chemical shifts in the range 24-36 ppm;
(iii) Asp, Asn, Cys^{oxd}, Ile, Leu, Phe and Tyr which have $^{13}C^{\beta}$ shifts in the range 36-52 ppm;
(iv) Ser residues which have $^{13}C^{\beta}$ in the range 56-67 ppm;
(v) Thr residues which have $^{13}C^{\beta} > 67$ ppm;
(vi) Some Val residues have $^{13}C^{\alpha}$ shifts > 64 ppm and $24 < {}^{13}C^{\beta} < 36$ ppm;
(vii) Some Ile residues have $^{13}C^{\alpha} > 64$ ppm and $36 < {}^{13}C^{\beta} < 52$ ppm;
(viii) Gly are not shown in Figure 5.5 B as they do not have $^{13}C^{\beta}$.
(ix) Pro residues can be classified in a separate category. They lack amide proton and are not observed in most of the triple resonance spectra.

4. RESONANCE ASSIGNMENT STRATEGIES

As pointed out, the ^{1}H, ^{13}C and ^{15}N resonances in proteins fall into distinct spectral regions (Table 5.2). Such classification helps in identification of individual networks of coupled spin-systems due to various amino acid residues. The information helps in classifying each kind of ^{13}C resonances into distinct regions. This serves as searching points in the identification of ^{13}C spins belonging to various amino acid residues.

We shall first discuss assignments based on homo-nuclear 2D ^{1}H NMR alone, which can be carried out with small proteins. This involves three steps:
(i) identification of the amino acid type from their characteristic spin-system network;
(ii) connecting such networks to the corresponding amide protons and
(iii) sequence specific resonance assignments.

4.1 Identification of Networks of J-coupled Spin Systems

Protons belonging to each amino acid form a distinct network of J-coupled spin system. Each residue manifests its characteristic signature in the 2D J-correlation spectra, such as 2QF-COSY, 3QF-COSY, TOCSY and 2D-DQ. The networks of coupled spin systems formed by protons within individual amino acid residues are identified using such experiments. Methodologies for the identification of spin systems for different amino acid are described below.

4.1.1 Glycine

Gly has two H^{α} protons which form an AX (or AB) system. Due to the large *geminal* coupling (~ -16 Hz), it gives a strong cross peak in 2QF-COSY. Normally two H^{α}-H^{N} peaks are observed. The corresponding cross peak may not be observed if one of the two $^{3}J(H^{N}, H^{\alpha})$ is small. In such a case, relayed COSY or TOCSY through the *geminal* coupling can give a sufficiently intense cross-peak. 2D double quantum spectrum is useful in case of overlap of the two H^{α} resonances. Here one observes a remote peak at $\{\omega_1, \omega_2\} = \{2H^{\alpha}, H^{N}\}$, which helps to identify degeneracy in the shifts.

4.1.2 Amino acids Containing Methyl Group(s): Ala, Thr, Val, Ile and Leu

Ala contains a methyl group ($C^{\beta}H_3$) which is coupled to H^{α}. Such a spin system gives a strong cross peak in the 2QF COSY. The peak can be easily identified because of the large chemical shift difference between the spins involved (Figure 5.7).

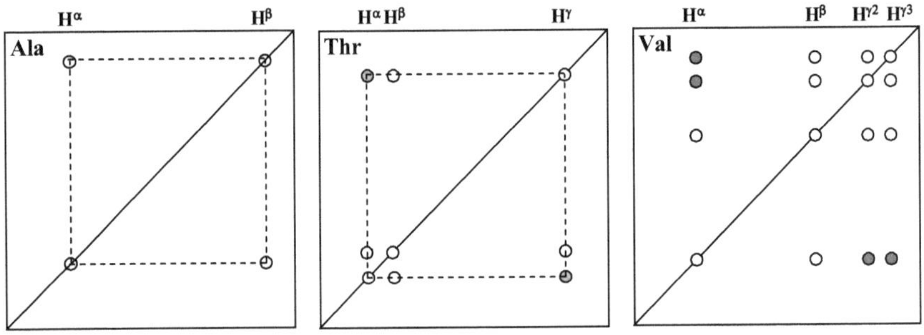

Figure 5.7: Schematic COSY (white) and TOCSY (grey) cross peaks in 2D NMR of Ala, Thr and Val.

Thr also contains a methyl group ($C^{\gamma}H_3$) which is coupled to H^{β}, resulting in an AX_3 spin system. This leads to a strong cross peak in 2QF COSY. The H^{β} in Thr is shifted downfield due to the presence of OH group attached to C^{β} and hence resonates close to its coupling partner H^{α}. Occasionally, there may be overlap between H^{α} and H^{β} resonances. The Thr spin system is identified in 2QF COSY from H^{α}-H^{β} and H^{β}-H^{γ} cross peaks. The H^{α}-H^{γ} cross peak in TOCSY and absence of H^{α}-H^{β} cross peak in 3QF COSY supplement these observations. The H^{α}-H^{γ} and H^{β}-H^{γ} correlations in TOCSY distinguish Thr from Ala.

Val forms an A_3B_3MX system. It is identified in 2QF COSY from the observation of one H^{α}-H^{β} and two H^{β}-H^{γ} cross peaks. The assignments are supplemented by the two H^{α}-H^{γ} cross peaks in TOCSY and the absence of H^{α}-H^{β} cross peak in 3QF COSY.

Leu forms an A_3B_3MPTX spin system. The H^{β} and H^{γ} protons appear in a crowded region, making it difficult to trace COSY connectivities from H^{α} to H^{δ}.

TOCSY helps in identifying the network of coupled spins from the observation of cross peaks from H^{α} to H^{γ}, $H^{\delta 1}$ and $H^{\delta 2}$.

Amino acid Ile also has two methyl groups and forms A_3B_3MPTX spin system. The two methyl groups (H^{γ} and H^{δ}) have different neighbours. Thus H^{γ} appears as a doublet while H^{δ} as a triplet. TOCSY helps in identifying the entire network of coupled spins from cross peaks ranging from H^{α} to H^{γ} and H^{δ}. Absence of H^{α}-H^{β} correlation in 3QF COSY spectrum forms an important criterion for identifying this spin system.

4.1.3 Asn, Asp, Cys and Ser

The NMR of the side chain of these amino-acid residues is characterised by three non-exchangeable protons (one H^{α} and two H^{β}) resulting in AMX systems. Their signatures in 2QF COSY are not distinguishable from one another. Two H^{α}-H^{β} cross-peaks are expected for such systems. In most cases only one is observed, because the second coupling is usually small. 3QF COSY helps in identifying the second correlation, since both the H^{α}-H^{β} peaks in 3QF COSY are the result of three active couplings. The 3QF COSY also helps in distinguishing *geminal* from the *vicinal* protons, due to their opposite phases. TOCSY can also be used where the large *geminal* coupling between the two H^{β} protons results in intense relay cross peaks between H^{α} and the two H^{β} protons, even when the values of $^3J(H^{\alpha}, H^{\beta})$ are small.

Out of the four AMX spin systems, Ser can be identified because of the large downfield shift of H^{β} protons arising from the OH group attached to C^{β}. Asn has two exchangeable side-chain NH protons ($H^{\delta 21}$ and $H^{\delta 22}$). These can be observed, provided their exchange rates with water are slow. Coupling between either of these two NH protons with H^{β} is small, although a COSY cross-peak is expected between the two NH protons. These protons are spatially close to the H^{β} protons and give rise to NOEs. This allows unambiguous identification of all resonances and helps in distinguishing Asn from other spin-systems. Once Ser and Asn spin systems are distinguished, rest of the AMX signatures can be identified as belonging to Asp and Cys residues.

4.1.4 His, Phe, Trp and Tyr

In addition to one H^{α} and two H^{β}, which form AMX system, Phe, Trp and Tyr residues possess several aromatic protons. The aromatic protons are not J-coupled to H^{β}. Therefore, such residues have two independent networks of coupled spin systems. The AMX systems of these residues manifest in both 2QF and 3QF COSY. The cross peaks appear in the same region as Asp, Asn and Cys residues. Since there is no observable coupling between H^{β} and protons in the aromatic rings, J-correlated spectra cannot be used to distinguish them. However, due to spatial proximity of H^{α} and H^{β} to the ring protons, these systems can be identified using NOESY. Similar strategy is used to distinguish H^{α}, H^{β} and the ring protons of His, though the corresponding distances between H^{α} and H^{β} to the ring protons are larger.

4.1.5 Long Side-chain Residues

Amino acid residues Glu, Gln, Met, Arg, Lys and Pro contain side-chains which extend beyond C^{β}. The methylene H^{β} of these residues resonates in the range of 1.5-2.5 ppm, while other methylene protons appear in the 1.5-4.0 ppm region. It is difficult to follow J-correlations in 2QF or 3QF COSY in such long side chains. However, TOCSY is a good option for the spin system identification in these cases.

Glu, Gln and Met contain one H^{α} and two methylene H^{β} and H^{γ} protons. Met has an additional methyl group (CH^{ε}_3), which is not coupled to H^{γ} protons. However, H^{ε} appears as a sharp line in the range 1.5-2.5 ppm, and shows strong NOEs to the rest of side-chain protons. As in the case of Asn, Gln also has two exchangeable side-chain NH protons ($H^{\varepsilon 21}$ and $H^{\varepsilon 22}$). There protons are observed provided their exchange rates with water are slow. The two protons are spatially close to the H^{γ} protons and give strong NOEs. This allows unambiguous identification of ^{1}H resonances of Gln residues and also distinguishing them from the rest of the long side-chain spin systems.

Arg, Lys and Pro contain the longest side-chains. Arg can be distinguished from Pro and Lys because it has side-chain NH (H^{ε}), which is lacking in the other two cases. Arg (H^{ε}) is J-coupled to H^{δ} protons. Hence, one can use this side-chain NH to assign remaining protons in Arg, using favourable conditions of pH. Lys with its side-chain methylene protons stretched up to C^{ε} is the longest residue and is easily distinguishable in the TOCSY spectra.

In favourable cases, Pro $H^{\delta 21}$ and $H^{\delta 22}$ have short contacts with H^{α} of its N-terminal neighbour, giving rise to detectable NOEs. Thus, one can use these NOEs to identify Pro spin system. Once $H^{\delta 21}$ and $H^{\delta 22}$ are identified, one can walk along the rest of the Pro side chain protons using TOCSY.

4.2 Linking Side Chains to the Respective Amide Protons

The next step is to assign the backbone amide (H^{N}) protons to their individual residues. This allows identification of the complete network of coupled spins with the individual amino acid residues. For this purpose, 2QF COSY and TOCSY spectra are useful. The spectral region where one observes H^{α}-H^{N} correlations is popularly known as the finger-print region of the 2QF COSY. This region is important for completing spin-system identification. TOCSY has been used in the case of H^{α} chemical shift degeneracy, wherein the respective relay peaks from H^{N} to H^{β}, H^{γ} etc. are observed and help to resolve such overlaps. Overlap in H^{N} chemical shifts can be handled by recording set of 2D spectra at different temperatures or pH.

4.3 Sequence Specific Resonance Assignments

The last step is to carry out sequence specific resonance (ssr) assignments. It is not possible to achieve ssr assignments with J-correlation spectroscopy alone, since there are no observable J-couplings between protons belonging to neighbouring amino acid residues. NOESY is used to connect neighbouring networks of coupled spin systems in the primary sequence by monitoring inter-residue NOE [$(H^N, H^\alpha, H^\beta)_{i-1} \rightarrow (H^N)_i$]. Other sequential connectivities such as $d_{\gamma N}$, $d_{\delta N}$ are also helpful. One differentiates between the self (spins belonging to the same residue) and sequential (spins between two adjacent residues) peaks by comparing the TOCSY (which provide only self peaks) and NOESY (which provide both self and sequential peaks). The finger print region is particularly useful (Figure 5.8) for this purpose. The sequential assignments are completed by walking from one residue to the next as shown in the inset of Figure 5.8.

4.4 Identification of Disulfide Bridges

In several cases, disulphide bridges are important for the protein to attain its *native* conformation. A downfield shift of individual $^{13}C^\beta$ resonances (in the range 38-48

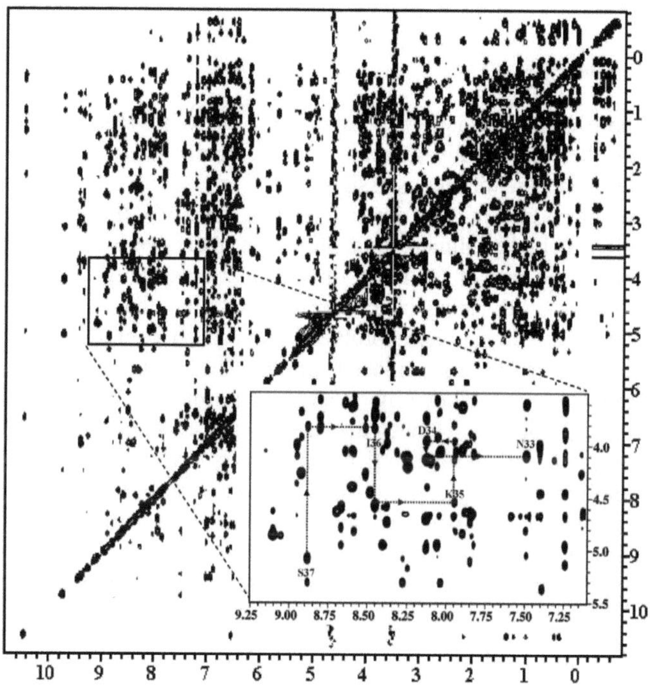

Figure 5.8: 2D NOESY spectrum of M-crystallin, an 85 amino-acid residue long protein from Methanosarcina acetivorans, recorded in a mixed solvent of 90% H_2O and 10% ^2H_2O. A part of the finger-print region is shown in the inset, which depicts the sequence specific resonance assignments for a stretch of amino-acid residues from N33 to S37.

ppm), indicates that Cys residues are in an oxidized state (i.e form S-S bonds). In the reduced state, the respective $^{13}C^{\beta}$ spins resonate in the range 24-36 ppm. The disulfide bridges can be identified with the help of inter-cysteine H^{β}-H^{β} NOEs. The H^{β} of the disulfide linked residues are typically 4-5 Å apart. The observed NOEs between such protons helps in establishing disulfide bridges.

4.5 Stereo-specific Resonance Assignments

Chemical shifts can be used for stereo-specific resonance assignments of protons on the same carbon but having different chemical shifts (pro-chiral groups). Examples are the pro-S and pro-R CH_3 groups in Leu and Val and the two Gly(H^{α}) protons. Use can also be made of intra-residue and sequential NOEs for stereo-specific assignments. Such information is valuable for structure determination and refinement. Experimental approaches for stereo-specific assignments use fractional ^{13}C labeling. Kainasho has developed methods for *Stereo-Array Isotope Labeling* (SAIL) of both amino acids and nucleotides. Such materials can be obtained at reasonable costs.

4.6 Higher-Dimensional NMR

Several problems arise with larger proteins (M_r > 10 kDa). Both the number of ^{1}H resonances and their line-widths increase with molecular size. This leads to spectral overlaps and prevents unambiguous assignments.

It was recognized during the development of 2D NMR that the FT techniques can be extended to higher dimensions. However, there were practical difficulties arising from the large size of the multi-dimensional data matrix $S(t_1, t_2, t_3)$ and the associated problems in devising useful experiments that can give good sensitivity and are free from artefacts. These problems have been overcome in a variety of hetero-nuclear 3D experiments proposed for studies of single ($^{13}C/^{15}N$) and double (^{13}C, ^{15}N) labelled proteins. They rely on large hetero-nuclear ^{1}J-couplings which are of the order of 90-150 Hz. Hence, the sensitivity of hetero-nuclear 3D and 4D experiments is high, and the spectral complexity is low. The methods have been extensively applied for larger proteins. These include both double- and triple-resonance experiments. Two of the double-resonance experiments (3D NOESY-HMQC and 3D TOCSY-HMQC) are described in Chapter 3. Certain 3D triple resonance experiments specific to proteins are described in the next section.

5. 3D TRIPLE-RESONANCE EXPERIMENTS FOR PROTEINS

3D *triple-resonance* experiments correlate three different kinds of spins, namely $^{1}H^{N}$, ^{13}C and ^{15}N using one and two-bond J-coupling interactions (Figure 5.9). In proteins, one can use these experiments to correlate $^{1}H^{N}$, $^{1}H^{\alpha}$, $^{13}C^{\alpha}$, $^{13}C^{\beta}$, $^{13}C'$ and ^{15}N spins. The experiments can also be used to correlate side chain spins with those in the backbone. This helps in the side chain assignments.

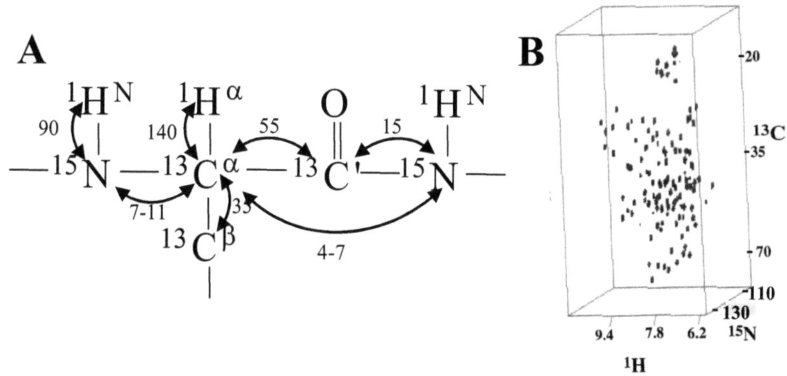

Figure 5.9: (A) One- and two-bond J-couplings (in Hz) which provide the basis for sequential resonance assignments in proteins by triple-resonance experiments. (B) A typical triple-resonance experiment [3D CBCA(CO)NH] where in one correlates ^{13}C, ^{15}N and $^{1}H^{N}$. Each peak is characterized by three chemical shifts (^{13}C, ^{15}N, $^{1}H^{N}$).

The nomenclature used to describe these pulse sequences is based on spins which are frequency labelled during various evolution and detection periods and the spins through which magnetization is transiently transferred to their coupled partners. The order in which the frequency labelling is done is made obvious from the name given to the pulse sequence. The most common *triple-resonance* experiments designed for connecting $^{1}H^{N}$, $^{1}H^{\alpha}$, $^{13}C^{\alpha}$, $^{13}C^{\beta}$, $^{13}C'$ and ^{15}N spins in proteins are: HNCA, HN(CO)CA, HNCO, HN(CA)CO, HN(CA)HA, HN(COCA)HA, CBCANH and CBCA(CO)NH/HNCACB. Though there are several other similar 3D techniques, it is not possible to discuss them all. A set of representative experiments are discussed below.

5.1 HNCA

HNCA represents a *triple-resonance* experiment in which coherence is transferred either from $^{1}H_i^{N}\rightarrow{}^{15}N_i\rightarrow{}^{13}C_i^{\alpha}(t_1)\rightarrow{}^{15}N_i(t_2;\ CT)\rightarrow{}^{1}H^{N}(t_3)$ or $^{1}H_i^{N}\rightarrow{}^{15}N_i\rightarrow{}^{13}C_{i-1}^{\alpha}(t_1)\rightarrow{}^{15}N_i(t_2;\ CT)\rightarrow{}^{1}H^{N}(t_3)$ (i.e. *out* from $^{1}H^{N}_{i}\rightarrow{}^{15}N_i\rightarrow{}^{13}C^{\alpha}_{i/i-1}$, shown as dotted arrows in Figure 5.10 A; and *back* from $^{13}C^{\alpha}_{i/i-1}\rightarrow{}^{15}N_i\rightarrow{}^{1}H^{N}_{i}$, shown as black arrows. CT stands for constant time evolution (Chapter 3). Figure 5.11 shows the pulse sequence for the 3D HNCA and 3D HNCO. WALTZ-16 is used to decouple ^{1}H during heteronuclear magnetization transfers and to decouple ^{15}N during acquisition.

HNCA provides information about $^{13}C^{\alpha}_{i}$ and $^{13}C^{\alpha}_{i-1}$ chemical shifts, from the observation of two ($^{13}C^{\alpha}$, $^{15}N_i$, $^{1}H^{N}_{i}$) peaks for a given pair of $^{15}N_i$ and $^{1}H^{N}_{i}$ chemical shifts. One peak belongs to ($^{13}C^{\alpha}_{i}$, $^{15}N_i$, $^{1}H^{N}_{i}$) and the other to ($^{13}C^{\alpha}_{i-1}$, $^{15}N_i$, $^{1}H^{N}_{i}$).

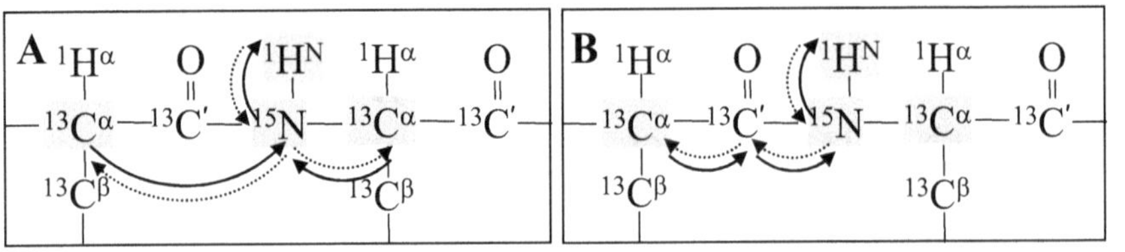

Figure 5.10: Coherence transfer pathways in (A) 3D HNCA and (B) 3D HN(CO)CA. In recording these spectra, the ^{1}H RF carrier is placed on the water resonance; the ^{15}N carrier is set around 120 ppm; the ^{13}C carrier is set in the middle of $^{13}C^{\alpha}$ (~ 45 ppm).

Using POF, one can calculate the transfer functions for both the self ($^{13}C^{\alpha}_{i}$, $^{15}N_{i}$, $^{1}H^{N}_{i}$) and the sequential ($^{13}C^{\alpha}_{i-1}$, $^{15}N_{i}$, $^{1}H^{N}_{i}$) peaks, which are dictated by the conformation dependent $^{1}J(^{15}N_{i}$-$^{13}C^{\alpha}_{i})$ and $^{2}J(^{15}N_{i}$-$^{13}C^{\alpha}_{i-1})$ values.

Figure 5.12 A shows an example of a ($^{13}C^{\alpha}$, $^{1}H^{N}$) plane of HNCA spectrum at a chemical shift of $^{15}N_{i}$. It may appear straight forward to identify ($^{13}C^{\alpha}_{i}$, $^{15}N_{i}$, $^{1}H^{N}_{i}$) and ($^{13}C^{\alpha}_{i-1}$, $^{15}N_{i}$, $^{1}H^{N}_{i}$) peaks, which have different intensities. The intense positive peak corresponds to the self peak ($^{13}C^{\alpha}_{i}$, $^{15}N_{i}$, $^{1}H^{N}_{i}$). This is because the involved couplings have different magnitude [$^{1}J(^{13}C^{\alpha}_{i}$-$^{15}N_{i}) = 11$ Hz and $^{2}J(^{13}C^{\alpha}_{i-1}$-$^{15}N_{i}) = 7$ Hz]. Due to overlap of self and sequential cross peaks ($^{13}C^{\alpha}_{i}/^{13}C^{\alpha}_{i-1}$, $^{15}N_{i}$, $^{1}H^{N}_{i}$), one may not observe distinct peaks. The distinction between the self and the sequential peaks in 3D HNCA spectrum can be made by comparison with the 3D HN(CO)CA spectrum, where in one observes only the sequential peak for a pair of $^{15}N_{i}$ and $^{1}H^{N}_{i}$ chemical shifts.

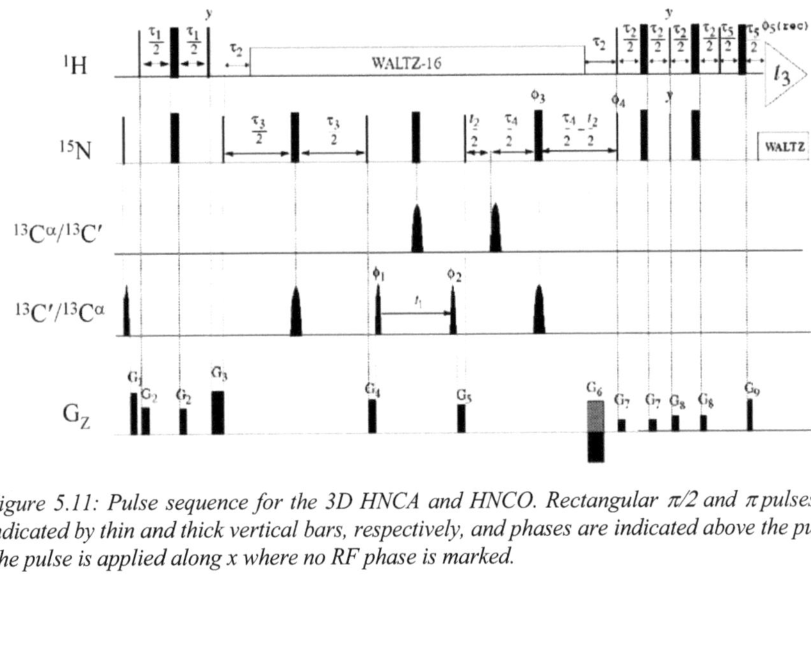

Figure 5.11: Pulse sequence for the 3D HNCA and HNCO. Rectangular π/2 and π pulses are indicated by thin and thick vertical bars, respectively, and phases are indicated above the pulses. The pulse is applied along x where no RF phase is marked.

5.2 HN(CO)CA

HN(CO)CA is essentially a sub-spectrum of HNCA. Coherence is transferred from ${}^1H^N_i \rightarrow {}^{15}N_i \rightarrow ({}^{13}C'_{i-1}) \rightarrow {}^{13}C^{\alpha}_{i-1}(t_1) \rightarrow ({}^{13}C'_{i-1}) \rightarrow {}^{15}N_i(t_2;$ CT$) \rightarrow {}^1H^N_i(t_3)$ (*out* from ${}^1H^N_i \rightarrow {}^{15}N_i \rightarrow$ (transiently on ${}^{13}C'_{i-1}) \rightarrow {}^{13}C^{\alpha}_{i-1}$; shown as dotted arrows in Figure 5.10 B and *back* from ${}^{13}C^{\alpha}_{i-1} \rightarrow$(transiently on ${}^{13}C'_{i-1}) \rightarrow {}^{15}N_i \rightarrow {}^1H^N_i$); shown as black arrows. The spectrum contains information only about the sequential ${}^{13}C^{\alpha}_{i-1}$ chemical shift for a given pair of ${}^{15}N_i$ and ${}^1H^N_i$ chemical shifts. The corresponding peak is observed at (${}^{13}C_{i-1}$, ${}^{15}N_i$, ${}^1H^N_i$). The transfer efficiency is dictated by values of ${}^1J({}^{15}N_i$-${}^{13}C'_{i-1})$ and ${}^2J({}^{15}C'_{i-1}$-${}^{13}C^{\alpha}_{i-1})$.

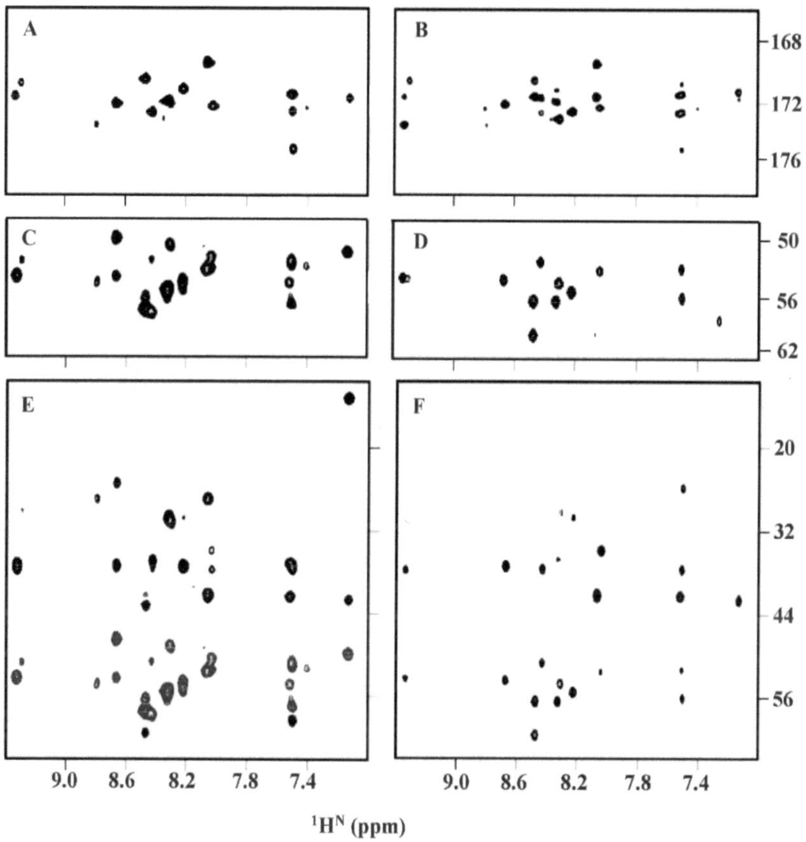

Figure 5.12: Representative (${}^{13}C$, ${}^1H^N$) planes of (A) HNCO; (B) HN(CA)CO; (C) HNCA; (D) HN(CO)CA; (E) CBCANH (-ve peaks are shown in grey); and (F) CBCA(CO)NH spectra at a chemical shift of ${}^{15}N$ of a 10 kDa protein. Positive contours are shown in black while negative contours are shown in red.

Figure 5.12 B shows an example of a ($^{13}C^{\alpha}$, $^{1}H^{N}$) plane of a 3D HN(CO)CA spectrum at a chemical shift of $^{15}N_i$. The sequential ($^{13}C^{\alpha}_{i-1}$, $^{15}N_i$, $^{1}H^{N}_{i}$) peak can be easily identified. This in turn helps in the distinction between the self and the sequential peaks in the HNCA spectrum for the same pair of $^{15}N_i$ and $^{1}H^{N}_{i}$ chemical shifts. For this reason, it is useful to record both HNCA and HN(CO)CA under identical conditions.

5.3 HNCO

This is the most sensitive *triple-resonance* experiment. Coherence is transferred along $^{1}H_i^{N} \rightarrow {}^{15}N_i \rightarrow {}^{13}C'_{i-1}(t_1) \rightarrow {}^{15}N_i(t_1;\ CT) \rightarrow {}^{1}H^{N}(t_3)$ (Figure 5.13 A). The pulse sequence is identical that for HNCA (Figure 5.11), with the difference that the $^{13}C^{\alpha}$ and $^{13}C'$ pulses are interchanged. HNCO spectrum has information only about the $^{13}C'_{i-1}$ chemical shifts, from the peak at ($^{13}C'_{i-1}$, $^{15}N_i$, $^{1}H^{N}_{i}$) for a given pair of $^{15}N_i$ and $^{1}H^{N}_{i}$ chemical shifts. The small value of $^{2}J(^{15}N_i$-$^{13}C'_i)$ is the reason for the non-observance of the self peak ($^{13}C'_i$, $^{15}N_i$, $^{1}H^{N}_{i}$). The coherence transfer efficiency depends on the value of $^{1}J(^{15}N_i$-$^{13}C'_i)$. Figure 5.12 C shows an example of a ($^{13}C'$, $^{1}H^{N}$) plane of a HNCO spectrum at a chemical shift of $^{15}N_i$.

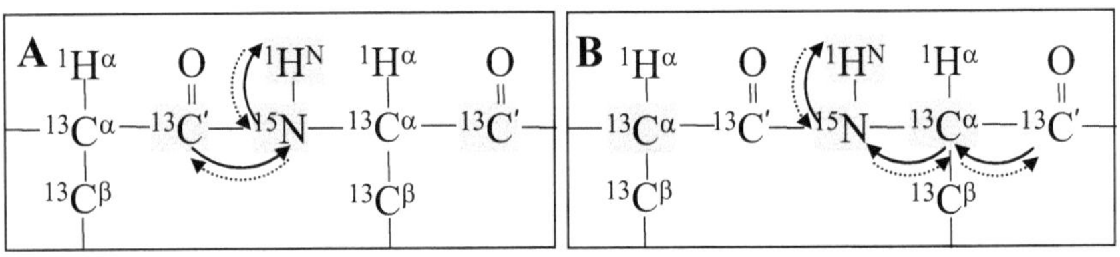

Figure 5.13: Coherence transfer pathways in (A) 3D HNCO and (B) 3D HN(CA)CO. In recording these spectra, the ^{1}H RF carrier is placed on the water resonance; the ^{15}N carrier is set around 120 ppm; the ^{13}C carrier is set in the middle of $^{13}C^{\alpha/\beta}$ (~ 45 ppm).

5.4 HN(CA)CO

This spectrum has information about $^{13}C'_i$ and $^{13}C'_{i-1}$ chemical shifts. For a given pair of $^{15}N_i$ and $^{1}H^{N}_{i}$ chemical shifts, two ($^{13}C'$, $^{15}N_i$, $^{1}H^{N}_{i}$) peaks are observed. Coherence is transferred from one of the following pathways: $^{1}H^{N}_{i} \rightarrow {}^{15}N_i \rightarrow$ (transiently on $^{13}C^{\alpha}_{i}) \rightarrow {}^{13}C'_i(t_1) \rightarrow$ (transiently on $^{13}C^{\alpha}_{i}) \rightarrow {}^{15}N_i(t_2;\ CT) \rightarrow {}^{1}H^{N}(t_3)$ (Figure 5.13 B); or $^{1}H^{N}_{i} \rightarrow {}^{15}N_i \rightarrow$ (transiently on $^{13}C^{\alpha}_{i-1}) \rightarrow {}^{13}C'_{i-1}(t_1) \rightarrow$ (transiently on $^{13}C^{\alpha}_{i-1}) \rightarrow {}^{15}N_i(t_2;\ CT) \rightarrow {}^{1}H^{N}(t_3)$. Hence, two ($^{13}C'$, $^{15}N_i$, $^{1}H^{N}_{i}$) peaks are observed. One belongs to ($^{13}C'_i$, $^{15}N_i$, $^{1}H^{N}_{i}$) and the other to ($^{13}C'_{i-1}$, $^{15}N_i$, $^{1}H^{N}_{i}$), which is relatively weaker (Figure 5.12 D). The sequential ($^{13}C'_{i-1}$, $^{15}N_i$, $^{1}H^{N}_{i}$) peak in HNCO helps to distinguish the self ($^{13}C'_i$, $^{15}N_i$, $^{1}H^{N}_{i}$) and the sequential ($^{13}C'_{i-1}$, $^{15}N_i$, $^{1}H^{N}_{i}$) peaks observed in HN(CA)CO for the same pair of $^{15}N_i$ and $^{1}H^{N}_{i}$ shifts.

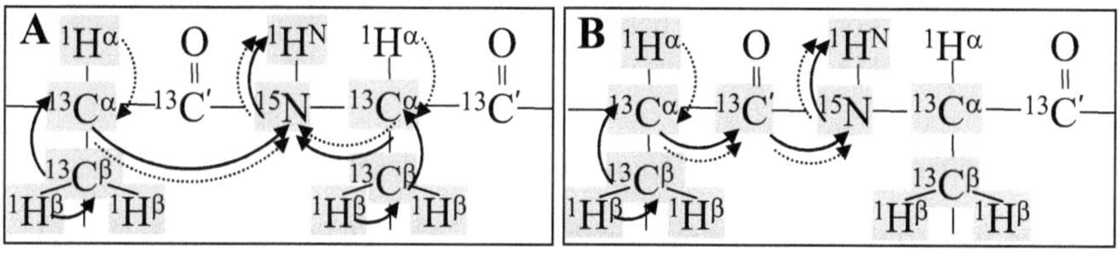

Figure 5.14: Coherence transfer pathways in (A) CBCANH and (B) CBCA(CO)NH. In these spectra, the ^{1}H RF carrier is placed on the water resonance; the ^{15}N carrier is set around 120 ppm; the ^{13}C carrier is set in the middle of $^{13}C^{\alpha/\beta}$ (~ 45 ppm).

5.5 CBCANH

Coherence is transferred either along $^1H_i^{\beta}\rightarrow{}^1C_i^{\beta}(t_1)\rightarrow{}^1C_i^{\alpha}\rightarrow{}^{15}N_{i/i+1}(t_2;$ CT$)\rightarrow{}^1H^N_{i/i+1}(t_3)$ or $^1H_i^{\alpha}\rightarrow{}^1C_i^{\alpha}(t_1)\rightarrow{}^1C_i^{\alpha}\rightarrow{}^{15}N_{i/i+1}(t_2;$ CT$)\rightarrow{}^1H^N_{i/i+1}(t_3)$ (Figure 5.14 A). The spectrum (Figure 5.12 E) shows four ($^{13}C^{\alpha/\beta}$, $^{15}N_i$, $^1H^N_i$) peaks (for a non-Gly dipeptide segment), for a given pair of $^{15}N_i$ and $^1H^N_i$ chemical shifts. One pair belongs to ($^{13}C^{\alpha}_i/^{13}C^{\beta}_i$, $^{15}N_i$, $^1H^N_i$) peaks and the other to ($^{13}C^{\alpha}_{i-1}/^{13}C^{\beta}_{i-1}$, $^{15}N_i$, $^1H^N_i$). It is easy to distinguish between ($^{13}C^{\alpha}_{i/i-1}$, $^{15}N_i$, $^1H^N_i$) and ($^{13}C^{\beta}_{i/i-1}$, $^{15}N_i$, $^1H^N_i$) peaks which are positive and negative in intensity, respectively. Due to overlap in cross peaks ($^{13}C^{\alpha}/^{13}C^{\beta}$, $^{15}N_i$, $^1H^N_i$) of self and sequential residues, one may not observe four distinct peaks. It is useful to record CBCA(CO)NH that provides information about only the sequential peaks ($^{13}C^{\alpha}_{i-1}/^{13}C^{\beta}_{i-1}$, $^{15}N_i$, $^1H^N_i$), and thus helps in resolving ambiguities. The HNCACB experiment which has been proposed later than CBCANH, is more sensitive and provides better information. In particular, when dealing with deuterated proteins, HNCACB is preferred.

5.6 CBCA(CO)NH

This is essentially a sub-spectrum of CBCANH. Coherence is transferred either along $^1H_i^{\beta}\rightarrow{}^1C_i^{\beta}(t_1)\rightarrow{}^1C_i^{\alpha}\rightarrow$(transiently on $^{13}C'_{i+1})\rightarrow{}^{15}N_{i+1}(t_2;CT)\rightarrow$ $^1H^N_{i/i+1}(t_3)$ or $^1H_i^{\alpha}\rightarrow{}^1C_i^{\alpha}(t_1)\rightarrow{}^1C_i^{\alpha}\rightarrow$(transiently on $^{13}C'_{i+1})\rightarrow{}^{15}N_{i+1}(t_2;$ CT$)\rightarrow{}^1H^N_{i/i+1}(t_3)$ (Figure 5.14 B). The spectrum (Figure 5.12 F) has information about $^{13}C^{\alpha}_{i-1}$ and $^{13}C^{\beta}_{i-1}$ chemical shifts, from the observation of two ($^{13}C^{\alpha/\beta}_{i-1}$, $^{15}N_i$, $^1H^N_i$) peaks (for a non-Gly dipeptide segment), for a pair of $^{15}N_i$ and $^1H^N_i$ chemical shifts. One peak belongs to ($^{13}C^{\alpha}_{i-1}$, $^{15}N_i$, $^1H^N_i$) and the other to ($^{13}C^{\beta}_{i-1}$, $^{15}N_i$, $^1H^N_i$). All peaks have positive intensity.

5.7 HN(CA)HA and HN(COCA)HA

HN(CA)HA spectrum has information about $^1H^{\alpha}_i$ and $^1H^{\alpha}_{i-1}$ chemical shifts for a given pair of $^{15}N_i$ and $^1H^N_i$ shifts. Two ($^1H^{\alpha}$, ^{15}N, $^1H^N$) peaks are observed for each pair of $^{15}N_i$ and $^1H^N_i$. One belongs to self ($^1H^{\alpha}_i$, $^{15}N_i$, $^1H^N_i$) and the other to sequential

($^1H^{\alpha}_{i-1}$, $^{15}N_i$, $^1H^N_i$). For distinction between the self and the sequential, it is useful to record HN(COCA)HA which shows only sequential ($^1H^{\alpha}_{i-1}$, $^{15}N_i$, $^1H^N_i$) peak.

5.8 3D Experiments used for Side-Chain Resonance Assignments in Proteins

Generally, four *triple-resonance* experiments, CBCANH, CBCA(CO)NH, HNCO and HN(CA)CO are sufficient for complete *ssr assignment* for all the $^1H^N$, $^{13}C^{\alpha}$, $^{13}C^{\beta}$, $^{13}C'$ and ^{15}N spins. HN(CA)HA and HN(COCA)HA enable assignment of $^1H^{\alpha}_i$ and $^1H^{\alpha}_{i-1}$ chemical shifts for a pair of $^{15}N_i$ and $^1H^N_i$ shifts. For complete side-chain resonance assignments one needs 3D experiments such as HCCH-COSY and HCCH-TOCSY. The HCCH-COSY represents a *double-resonance* experiment in which coherence is transferred between two vicinal protons which are directly bonded to ^{13}C. The transfer pathway can be from $^1H_i^{\beta}(t_1) \rightarrow {}^{13}C_i^{\beta}(t_2) \rightarrow$(transiently on $^{13}C_i^{\alpha}$)$\rightarrow {}^1H_i^{\alpha}(t_3)$ or $^1H_i^{\beta}(t_1) \rightarrow$ (transiently on $^{13}C_i^{\beta}$)$\rightarrow$ $^{13}C_i^{\alpha}(t_2) \rightarrow {}^1H_i^{\alpha}(t_3)$. The HCCH-TOCSY establishes through-bond correlations among all aliphatic ^1H and ^{13}C, where the TOCSY mixing is among the ^{13}C spins within each residue. The transfer pathway can be $^1H_i^{\alpha}(t_1) \rightarrow {}^{13}C_i^{\alpha} \rightarrow [{}^{13}C_i^{\alpha}, {}^{13}C_i^{\beta}, {}^{13}C_i^{\gamma}, ...(t_2;$ *due to* TOCSY mixing$] \rightarrow [{}^1H_i^{\alpha}, {}^1H_i^{\beta}, {}^1H_i^{\gamma}, ...(t_3)]$ and $^1H_i^{\beta}(t_1) \rightarrow {}^{13}C_i^{\beta} \rightarrow [{}^{13}C_i^{\alpha}, {}^{13}C_i^{\beta}, {}^{13}C_i^{\gamma}, ...$ (t_2; *due to* TOCSY mixing)$] \rightarrow [{}^1H_i^{\alpha}, {}^1H_i^{\beta}, {}^1H_i^{\gamma}, ...(t_3)]$.

All assignment strategies suffer from the fact that there can be chemical-shift variations for the same spin in different spectra because of changes in experimental conditions. In particular pH and temperature may be different for different experiments. These factors may contribute to incomplete or erroneous assignments. Further, for proteins with low stability, it is imperative that all data are acquired in a short time. Thus, it is desirable to restrict the number of experiments required as inputs to a minimum. All experiments should be performed under identical conditions of pH and temperature, with the same sample.

6. REDUCED DIMENSIONALITY: G-MATRIX FTNMR FOR PROTEINS

For large molecules, one needs several 3D/4D NMR experiments to accomplish resonance assignments. This requires long instrument time. In such a situation, one has to compromise either with the sensitivity or the resolution. While the sensitivity limitation dictates number of scans to obtain the desired S/N ratio, the sampling limitation dictates number of data points to obtain the required spectral resolution. One method of overcoming the sampling problem is the joint sampling of two or more indirect dimensions in a higher-dimensional experiment. This approach encodes the information content in a 3D-6D spectrum into lower dimensions (2D-3D) (Figure 5.15). The concept has led to the development of reduced dimensionality (RD) and later its generalization known as G-matrix Fourier Transform (GFT) NMR spectroscopy. Such experiments can be implemented without additional hardware.

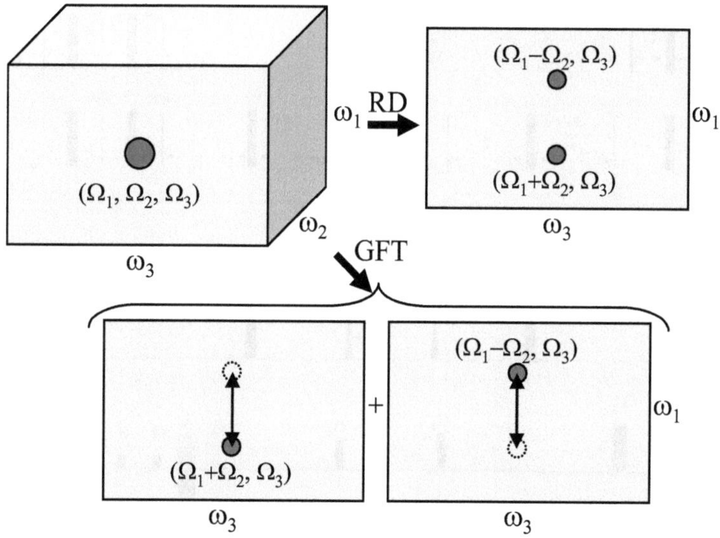

Figure 5.15: The concept of RD and GFT NMR.

6.1 RD NMR

The concept of reduced dimensionality was introduced in early 1990s. It lay dormant for a decade before being revived a few years ago. The objective is to obtain complete information content of a higher-dimension spectrum by recording a set of lower-dimension spectra. The basic idea is to simultaneously frequency label different spins such that the chemical shifts appear along a single spectral dimension. This results in less number of independent indirect dimensions; hence a reduction in the total instrument time.

For example, a 3D HNCO can be recorded as a RD 2D HNCO (Figure 5.16) by joint sampling of $^{13}C'$ and ^{15}N (the underlined nuclei in HNCO) chemical shifts. This is indicated by the underlined letters in the nomenclature of RD experiments. Notice the simple modifications that have been incorporated in the 3D HNCO pulse sequence to implement a RD 2D HNCO. These are the joint increment of $^{13}C'$ and ^{15}N chemical shifts along the indirect t_1 dimension and omission of phase-sensitive detection of the $^{13}C'$. As in the case of 3D HNCO, the ^{1}H RF carrier is placed at the position of the solvent line; the ^{15}N carrier is set around 120 ppm; the ^{13}C carrier is set in the middle of $^{13}C'$. Unlike in 3D HNCO, in RD 2D HNCO, quadrature detection in $^{13}C'$ is omitted, whereas that of t_1 (^{15}N) is achieved via gradient selection of coherences. WALTZ-16 is used to decouple ^{1}H during heteronuclear magnetization transfers and to decouple ^{15}N during acquisition.

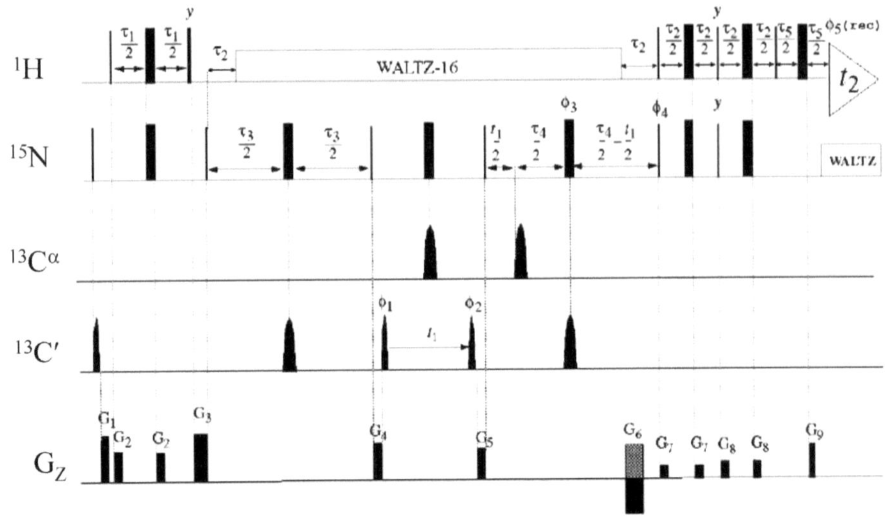

Figure 5.16: Pulse sequence for 2D RD HNCO. Rectangular $\pi/2$ and π pulses are indicated by thin and thick vertical bars, respectively. Phases are indicated above the pulses. Where no phase is marked, the pulse is applied along x. All π pulses applied on $^{13}C^{\alpha}$ are off-resonance hard pulses of duration ~40 μsec to minimize perturbation of $^{13}C'$ spins. The $\pi/2$ and π pulse lengths for the one lobe sinc pulse on-resonance at $^{13}C'$ are 71 and 64 μs, respectively. (The authors are thankful to H.S. Atreya and T. Szyperski for providing this figure)

Figure 5.17 shows the RD 2D HNCO spectrum for the protein Z-domain using a 1 mM u-[^{15}N, ^{13}C] doubly labelled sample in water. It comprises of in-phase doublets along the ω_1-axis for each ($^{15}N_i$, $^{1}H_i$) pair, with their chemical shifts at $\Omega(^{15}N_i)\pm\Omega(^{13}C'_{i-1})$ centred around the chemical shift of $\Omega(^{15}N_i)$ (Figure 5.17 B). The modulation of the $^{13}C'_{i-1}$ chemical-shift information over that of $^{15}N_i$, along the ω_1-axis, results in a spectrum that resembles HSQC. The centre of the doublet defines the δ of $^{15}N_i$ which is phase-sensitively detected. For a 3D HNCO experiment which required a measurement time of 3 hrs, the RD counterpart would take ~6 min using the same experimental conditions and parameters. The reduction in the measurement time results from the RD experiment requiring the simultaneous sampling of $^{15}N_i$ and $^{13}C'_i$ along one dimension instead of two samplings along the two indirect dimensions needed in 3D HNCO. The *central peaks* are the missing information in the RD HNCO experiment, compared to 3D HNCO. A [^{15}N, ^{1}H]-HSQC spectrum, which takes few minutes to record, serve to provide the information and helps to resolve ambiguities arising from two or more residues having degenerate ^{15}N and ^{1}H chemical shifts (Figure 5.17 C).

The RD principle can be extended to 4D and 5D NMR. For example, a 4D experiment can be acquired as a RD 3D which retains all the information about chemical-shift correlations in the original 4D.

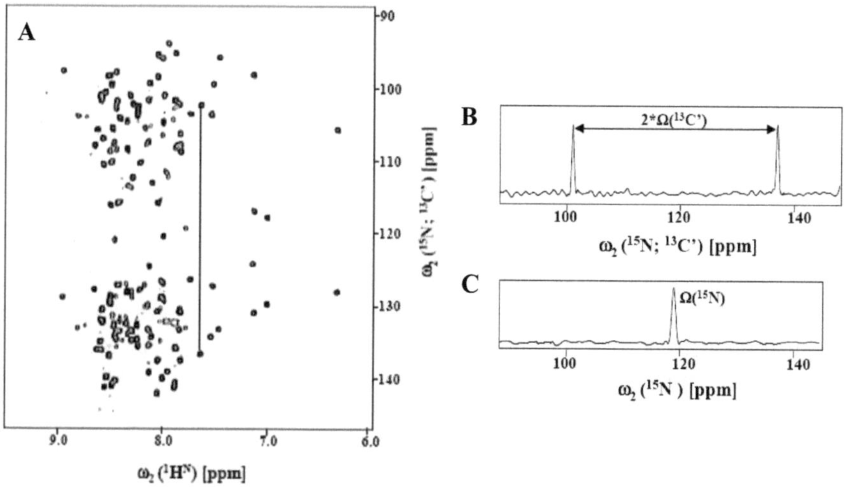

Figure 5.17: A 2D RD H̲N̲C̲O̲ spectrum (A) for the protein Z-domain. Cross sections along the ω_1 axis showing (B) a RD NMR peak pair ($\Omega(^{15}N)+\Omega(^{13}C')$) and ($\Omega(^{15}N)-\Omega(^{13}C')$) and (C) the corresponding central peak ($\Omega(^{15}N)$) seen in $[^{15}N, {}^{1}H]$ HSQC. (The authors are thankful to H.S. Atreya and T. Szyperski for providing this spectrum)

6.2 GFT NMR

One drawback of RD NMR is that the peak coordinates along the joint dimension represent linear combinations of chemical shifts of spins which have been simultaneously frequency labelled in the selected single spectral dimension. For example, when one reduces a ND to (N-2)D, then the number of peaks increases four fold. This increases spectral crowding. To resolve this problem, a conceptual addition to the idea of RD has been made. It involves post acquisition processing. The time-domain data is first multiplied by a *G-matrix* before the Fourier transformation (FT). Such processing is called GFT, primarily because of the order in which the *G-matrix* multiplication and FT are performed. GFT allows one to edit peaks into different sub-spectra such that the number in a given GFT sub-spectrum remains the same as in the parent ND FT experiment. In this way the problem of overcrowding is eliminated. The peaks encoded in a given type of linear combination of shifts for each residue falls into one sub-spectrum.

Consider the case where a 3D HNCO has been reduced to RD H̲N̲C̲O̲ by joint evolution of $^{15}N_i$ and $^{13}C'_{i-1}$. For a peak at ($^{13}C'_{i-1}$, $^{15}N_i$, $^{1}H_i$) in 3D HNCO spectrum, the magnetization prior to the acquisition of RD H̲N̲C̲O̲ spectrum is proportional to $\cos(\Omega_{13C'}t_1)\cos(\Omega_{15N}t_1)$. This can be expanded as

$$(\tfrac{1}{2})\{\cos(\Omega_{^{15}N}+\Omega_{^{13}C'})t_1+\cos(\Omega_{^{15}N}-\Omega_{^{13}C'})t_1\}$$

After FT two in-phase peaks centered along the projected dimension are observed at $\Omega(^{15}N_i) + \Omega(^{13}C'_{i-1})$ and $\Omega(^{15}N_i) - \Omega(^{13}C'_{i-1})$. In a GFT experiment the next scan is repeated with $\pi/2$ phase shift of the pulse at the beginning of the second (^{15}N) evolution period. This switches the cosine term for ^{15}N into a sine term resulting in $\cos(\Omega_{13C'}t_1)\sin(\Omega_{15N}t_1)$, which can be expanded as:

$$(½)\{\sin(\Omega_{^{15}N} + \Omega_{^{13}C'})t_1 + \sin(\Omega_{^{15}N} - \Omega_{^{13}C'})t_1\}$$

Hence, after the FT, two anti-phase peaks centered along the projected dimension are seen at $\Omega(^{15}N_i) + \Omega(^{13}C'_{i-1})$ and $\Omega(^{15}N_i) - \Omega(^{13}C'_{i-1})$ as in the case of the first scan. The dispersive peaks with opposite signs are converted to absorptive mode by changing the receiver phase by $\pi/2$. Addition of the two spectra or subtracting one from the other provides two simplified spectra. Each one of them has single peak corresponding to either $\Omega(^{15}N_i) + \Omega(^{13}C'_{i-1})$ or $\Omega(^{15}N_i) - \Omega(^{13}C'_{i-1})$ (Figure 5.15).

The GFT approach essentially includes an operation equivalent to the additions and subtractions of spectra. This is achieved by what is known as the *G-matrix* multiplication, which is performed in the time-domain. This is followed with Fourier transformation that results in two sub-spectra with peaks of the type ($^{15}N_i$ + $^{13}C'_{i-1}$, $^{1}H_i$) or ($^{15}N_i$ - $^{13}C'_{i-1}$, $^{1}H_i$), respectively. Such spectra are complemented by a conventional [^{15}N, ^{1}H]-HSQC spectrum, which yields ($^{15}N_i$, $^{1}H_i$) peaks.

Data acquisition time can be reduced by several orders of magnitude, by generalizing this concept and reducing the dimensionality of a ND experiment to N-j dimensions ($j \geq 1$). Such an experiment results in a spectrum comprising of peaks with all linear combination of the j+1 chemical shifts, $\Omega_0 \pm \Omega_1 \ldots \pm \Omega_j$, referred to as a *chemical-shift multiplet.* Hence, the number of peaks is increased by a factor of 2j compared to the parent ND experiment. Further, the chemical-shift multiplet is centred around Ω_0. The Ω_0 denotes the chemical shift of one of the j+1 jointly sampled spins and it turns out to be the phase-sensitively detected *centre shift.* This may result in spectral crowding. Such crowding is overcome by use of GFT NMR spectroscopy. In addition to joint incrimination of K+1 shifts in a single *GFT dimension,* the phases Φ_j of the pulses exciting spins of type j (j = 1...K+1) are systematically varied between 0 and $\pi/2$ to obtain both *cosine*- and *sine*-modulated data sets. This results in 2^{K+1} FIDs. The phase-sensitive detection of the K+1 shifts allows the editing of the 2^{K+1} FIDs into 2^K different sub spectra by multiplying the time domain data with the so-called *G-matrix*, followed by FT.

As discussed above, both 3D HNCO and RD HNCO have identical pulse sequences. Hence, losses due to relaxation when recording a GFT data set are about the same as in higher-dimensional experiments. However, each splitting of a peak reduces the S/N ratio by a factor of two. Some salient features of GFT NMR spectra are:

(i) The number of peaks in each sub-spectrum is the same as in the higher-dimensional spectrum. Thus the reduction in dimensionality does not increase spectral crowding.

(ii) A single dimension contains peaks with linear combination of different types of frequencies, resulting in a higher dispersion of signals. This means that the linear combination of N different frequencies spans a larger spectral range than that covered by the individual frequencies.
(iii) Chemical shifts are encoded multiple times in the linear combinations registered in GFT sub-spectra. For example, Figure 5.15 shows how the chemical shifts denoted as Ω_1 and Ω_2 are encoded in two GFT sub-spectra. This is equivalent to performing statistically independent multiple measurements for a given parameter, resulting in its over-determination and higher precision.
(iv) The underlying symmetry in the peak pattern of chemical-shift multiplets allows implementation of robust algorithms for spectral analysis. This feature makes spectra amenable for automation.

7. AUTOMATED NMR ASSIGNMENTS IN PROTEINS

Manual ssr assignment of ^{1}H, ^{13}C and ^{15}N spins is tedious and time-consuming. A number of automated assignments strategies have been proposed. This is particularly true for backbone resonance assignments (Table 5.3). The steps involved are often common and normally involve:

(i) Selecting an appropriate 3D spectral data set.
(ii) Picking peaks from the selected data set.
(iii) Classification of peaks to specific spin systems according to amino acids.
(iv) Carrying out *ssr* assignments.
(v) Mapping the sequentially assigned stretches onto the primary sequence.

The proposed approaches differ in the experimental inputs used and the algorithm for data analysis. The choice of particular software generally depends on the type and quality of the available spectral data. Most methods utilize information from a set of standard 3D triple-resonance experiments. These are coupled with algorithms such as simulated annealing, Bayesian statistics and artificial intelligence, characteristic ^{13}C$^{\alpha}$ and ^{13}C$^{\beta}$ chemical shifts, algorithms for threshold accepting, connectivity tracing and neural networks. Some strategies utilize prediction of chemical shifts for proteins which have high sequence homology to a previously assigned protein. Table 5.3 summarizes some of the popularly used programs for automated backbone protein resonance assignments.

The precision and accuracy of chemical shifts obtained from the spectral data play an important role in the success of an algorithm for automated assignments. This is primarily because the chemical shifts from different types of spectra and spin systems need to be grouped and matched. This requires introduction of limits for *chemical shift tolerances*, which are chosen based on respective spectral features. Large tolerances are required for shifts obtained with low precision, since errors can result in multiple and erroneous assignments. This is even more important for side-chain assignments which show higher spectral overlap. Experiments such as GFT

NMR, hold promise for providing chemical shifts with higher precision and accuracy.

Table 5.3: Programs for Automated Backbone Resonance Assignments.

S. No	Program	Source
1	AUTOASSIGN	http://www.nmr.cabm.rutgers.edu/NMRsoftware/nmr_software.html
2	CONTRAST	http://www.specres.com/contrast.asp
3	GARANT	http://www.biol.ethz.ch/IMB/groups/wuthrich_group/software/garan
4	IBIS	http://gwagner.med.Harvard.edu/ibis/
5	MAPPER	http://guentert.gsc.riken.go.jp/Software/Mapper.html
6	MARS	http://www.mpibpc.mpg.de/abteilungen/030/zweckstetter/_links/software.htm
7	PASA	http://www.lerner.ccf.org/moleccard/qin/pasa/
8	PASTA	http://www.org.chemie.tu-muenchen.de/people/jl/shell_pasta02/pasta_doc.html
9	RESCUE	http://www.infobiosud.cnrs.fr/
10	TATAPRO	http://www.tifr.res.in/~chary/TATAPRO.htm

7.1 Tracked Automated Assignments in Proteins (TATAPRO)

An approach for automated resonance assignments of backbone spins ($^{1}H^{N}$, $^{13}C^{\alpha}$, $^{13}C^{\beta}$, $^{13}C'/^{1}H^{\alpha}$ and ^{15}N) has been developed in our laboratory. The algorithm achieves assignments by utilizing the protein primary sequence and lists of peaks from a specific set of CBCANH, HNCO/HN(COCA)HA, CBCA(CO)NH and HN(CA)CO)/HN(CA)HA spectra. For a given pair of $^{1}H^{N}$ and ^{15}N chemical shifts belonging to a particular amino acid residue, such spectra yield $^{13}C^{\alpha}$, $^{13}C^{\beta}$ and $^{13}C'/^{1}H^{\alpha}$ chemical shifts of the same residue and the one preceding it. Peak lists from these spectra consisting of shift coordinates of the peaks, (ω_1, ω_2, ω_3) = ($^{13}C/^{1}H^{\alpha}$, ^{15}N, $^{1}H^{N}$), along with their intensities and phases are used as inputs. The information thus derived is used to create a *master-list* consisting of possible sets of $^{1}H^{N}{}_{i}$, $^{15}N_{i}$, $^{13}C^{\alpha}{}_{i}$, $^{13}C^{\beta}{}_{i}$, $^{13}C'_{i}/^{1}H^{\alpha}{}_{i}$, $^{13}C^{\alpha}{}_{i-1}$, $^{13}C^{\beta}{}_{i-1}$ and $^{13}C'_{i-1}/\,^{1}H^{\alpha}{}_{i-1}$ chemical shifts.

Based on the characteristic $^{13}C^{\alpha}{}_{i}$ and $^{13}C^{\beta}{}_{i}$ shifts, the 20 amino acid residues can be grouped into 9 distinct categories. Each is assigned a unique code. Such a code is used to tag individual sets of chemical shifts in the *master-list* and to translate the protein primary sequence into an array called *pps*-array. The program then uses the master-list to search for residues adjacent to the given amino acid and sequentially assigns maximum possible stretch of residues on either side. Each assigned residue is tracked into an array called assign-array, with the code mentioned earlier. The *assign*-array is then mapped onto the *pps*-array for ssr-assignments.

This method is based on a deterministic approach. Only four triple resonance experiments are required as inputs. In the event of degeneracy in all the $^{13}C^{\alpha}$, $^{13}C^{\beta}$ and $^{13}C'$ shifts, the approach explores all pathways to maximize the stretch of amino acid residues that can be assigned. Manual intervention is not required to check the grouping of peaks or residues which do not satisfy their characteristic $^{13}C^{\alpha}$ and $^{13}C^{\beta}$ chemical shift range. The program is robust to internal sequence homology and unusual chemical shifts.

Despite the demonstrated utility of triple resonance techniques, one encounters difficulties in resonance assignments. In principle, it should be possible to walk along the backbone of the polypeptide chain starting at C-terminal and ending at the N-terminus, by making use of various backbone spins that participate in magnetization transfer in triple resonance experiments. When the techniques are applied to larger proteins, rapid relaxation rates of the nuclear spins result in the broadening of several cross peaks thus hampering ssr assignments. Pro which lacks $^{1}H^{N}$ further aggravates the problem. It is therefore desirable, to have several good starting points in a polypeptide chain.

The spectral signatures of amino acid residues Ala, Gly, Ser and Thr are easily identifiable because of their characteristic $^{13}C^{\alpha}$ and $^{13}C^{\beta}$ shifts. Gly($^{13}C^{\alpha}$) resonates upfield around 50 ppm, in a region well separated from the $^{13}C^{\alpha}$ shifts of other residues where as Ala ($^{13}C^{\beta}$) and Ser($^{13}C^{\beta}$)/Thr($^{13}C^{\beta}$) resonate at less than 24 ppm and more than 58 ppm, respectively, in regions well separated from $^{13}C^{\beta}$ shifts of other residues. Such characteristic chemical shifts become significant in identification of these residues. In addition, if some more amino acid residues can be identified, then these may further aid in ssr assignments.

7.2 Selective Labeling/Unlabeling: Residue Specific Resonance Assignments

Residue specific amino acid labelling has been used by several researchers. In this procedure, a specific amino acid in the protein is labelled by feeding the host micro-organism with the desired ^{15}N or/and ^{13}C labelled amino acid. The remaining amino acids are supplied in unlabeled form. Specific labeling provides a direct assignment of the nuclei belonging to the particular residue, if it occurs only once in the sequence. If the labelled residue occurs more than once, then the assignment is residue specific. This provides additional starting points in ssr assignments.

This methodology is expensive if more than one amino acid residue is to be labelled. A less expensive methodology is based on *selective un-labeling* (Chapter 4) of residues in uniformly or fractionally labelled proteins. This simplifies multi-dimensional hetero-nuclear NMR spectra. A particular amino acid in the protein is prevented from getting labelled by feeding the host micro-organism with ^{15}N labelled NH_4Cl and ^{13}C labelled Glu as the sole source of nitrogen and carbon respectively, along with the desired amino acid to be assigned in an unlabeled form. This results in *un-labeling* of the desired residue. The technique can be used to simplify spectra of large proteins and enable stereo-specific resonance assignments of methyl groups in Val and Leu. As a result of unlabeling of specific residues, cross peaks due to such residues are not observed in NMR spectra. A comparison of a spectrum with that of a control having uniformly ^{15}N or/and ^{13}C labelled protein enables one to distinguish peaks (Figure 5.18) and the corresponding chemical shifts of spins belonging to the unlabeled amino acid residues. The approach can be used for residue specific assignments.

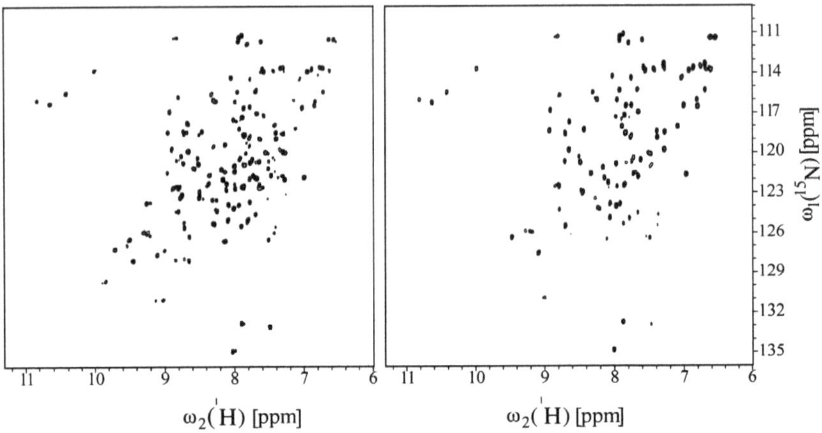

Figure 5.18: [^{15}N-^{1}H] HSQC of (u-^{15}N)-labeled and Lys unlabeled-^{15}N labeled Eh-CaBP. Cross peaks corresponding to the 16 Lys residues, which are absent in the Lys-unlabeled spectrum are readily distinguished.

7.3 Side-chain Assignments

Side-chain resonance assignments are more difficult due to two reasons. The number of spins in the side-chain far exceeds the number of backbone nuclei in a protein. This results in spectra which show a high degree of overlap and are not amenable for automation. Secondly, side-chains such as methylene, methyl and phenyl groups normally have different degrees of internal motion. This results in non-uniform signal strengths, and renders automated peak picking inefficient, which assumes uniform line shapes. Despite this, there have been attempts towards manual and automated side chain resonance assignments. The proposed methods are based on side-chain topologies of spin systems (Li and Sanctuary, 1997), ^{13}C chemical shift patterns of amino acid residues and semi-automated approaches.

7.4 Automated Chemical Shift Prediction Based on Sequence Homology

Though characterization of secondary structures can be achieved using different types of NMR data, chemical shifts have emerged as a simple and powerful mean for this purpose. Several automated methods have been proposed, that utilize conformational dependent chemical shifts to predict or identify the regular secondary structural elements.

It has been recognized that two proteins having a high degree of sequence homology share similar overall tertiary fold. Since chemical shifts depend on the 3D structure of the molecule, proteins with high sequence homology are expected to have similar shifts. This forms the basis for the program SHIFTY (*http://redpoll.pharmacy.ualberta.ca/shifty*).

The program proceeds in two steps. The query sequence is aligned against the primary sequences of proteins with known assignments using BMRB database. Search is performed using the dynamic programming method of Needleman and Wunsch. On completion of the alignment and scoring process, the highest scoring sequences are selected. Each pair-wise alignment is carried to the next step. The chemical shifts of the sequence, selected from the database are used to assign shifts to the query sequence. This is done in following ways:

(i) If the aligned residues match exactly, the chemical shifts are directly transferred from the database to the query protein without adjustments.

(ii) If the two alignments differ, then the database protein chemical shifts are subtracted from the corresponding residue's random coil shifts. These differences are added to the random coil shifts corresponding to the query residue.

(iii) No chemical shift predictions are made when a residue from the query protein lines up with a gap in the database protein.

The algorithm has been successfully tested for prediction of ^{1}H chemical shifts in 25 proteins. Robustness of the program depends on the degree of sequence alignment between the proteins. For proteins with high sequence identity, a good correlation is observed between the predicted and experimentally observed shifts. The quality of the chemical shift prediction falls rapidly for sequence identities below 35%.

8. TECHNIQUES FOR STUDYING LARGE PROTEINS

The power of the techniques discussed above, are limited to protein structures for M_r below 30 kDa. Study of very large macromolecules and assemblies by conventional techniques are hampered due to:

(i) extensive overlap and complexity of the spectra;

(ii) low solubility in such proteins and

(iii) low sensitivity and extensive rapid transverse spin relaxation resulting in line broadening.

8.1 Transverse Relaxation Optimized SpectroscopY (TROSY)

Methodologies based on TROSY (Chapter 3) have shown promise for studying larger proteins. As discussed, the success of manual or automated approach for ssr assignments in proteins is determined by two important parameters; good input of peaks in terms of resolution and sensitivity, and a reliable classification of individual spin systems. With the development of TROSY based approach for the implementation of various triple resonance pulse sequences and modification of pulse sequences for deuterated proteins, it is possible to acquire triple resonance spectra with high resolution and sensitivity for large proteins. For example, Figure 5.19 illustrates the [^{15}N, ^{1}H] HSQC and [^{15}N, ^{1}H] TROSY, recorded at 750 MHz under identical experimental conditions using of [u-^{15}N/^{2}H]-labeled gyrase-45 from *Staphylococcus aureus* in water at 25 °C, pH 8.6. The power of the technique is clearly brought out in these spectra.

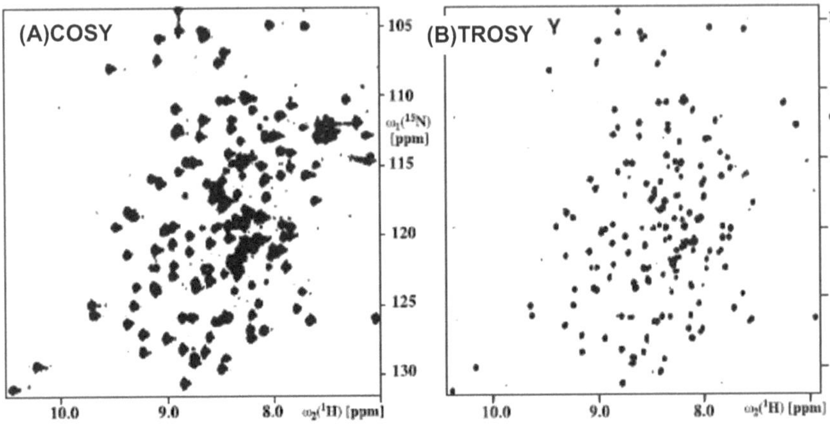

Figure 5.19: Comparison of (A) [^{15}N, ^{1}H] HSQC and (B) [^{15}N, ^{1}H] TROSY of [u-^{15}N/^{2}H]-labeled gyrase-45 from Staphylococcus aureus. R. Riek, K. Pervushin and K. Wüthrich. FEBS Letters, 504, 173-178, (2001). (Reprinted with permission from Elsevier).

Spectral overlap can be reduced by fractional deuteration, selective amino acid labeling and segmental labeling. The problem of sensitivity has been tackled by cryogenic probes and novel pulse techniques. In this section, we discuss methods for extending solution NMR studies to larger proteins.

8.2 ^{2}H labeling

Isotope substitution of some of the protons with ^{2}H greatly reduces the number of cross-relaxation pathways available to the un-labelled protons. This leads to longer T_2 relaxation and consequently narrower line-widths. Better spectral sensitivity and resolution in multi-dimensional spectral data are thus achieved. T_1 of deuterium (^{2}H) is 4.5 times more than that of ^{1}H and the average ^{1}H-^{1}H distance is longer in deuterated environment than in non-deuterated one. Therefore, longer NOESY mixing times can be used when working with deuterated proteins. This results in higher sensitivity of NOE cross peaks. As an illustrative example, Figure 5.20 shows two [^{15}N-^{1}H] HSQC spectra recorded with and without ^{2}H-labeling, under identical experimental conditions. The improvement in the quality of spectrum is obvious.

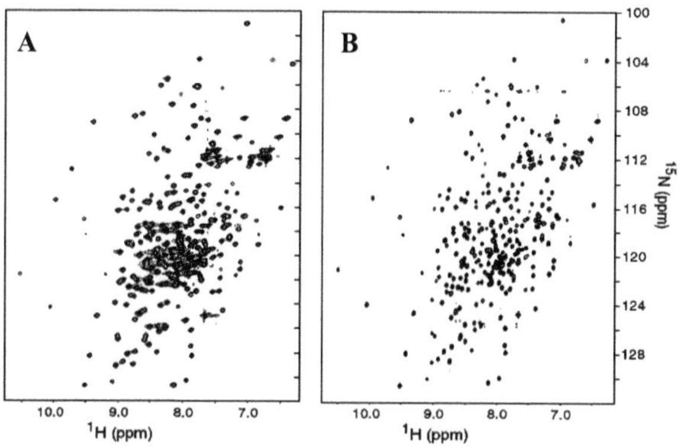

Figure 5.20: [^{15}N-^{1}H] HSQC spectra of recorded (A) without and (B) with ^{2}H-labeling, under identical experimental conditions. D.S. Garrett, Y. Seok, D.Liao, A. Peterkofsky, A.M. Gronenborn, G.M. Clore. Biochemistry 36, 2517–30. (1997). (Reprinted with permission from American Chemical Society).

By the combined use of ^{2}H-labeling and TROSY, good quality, high-resolution NMR spectra of proteins with M_r >100 kDa have been studied. These developments significantly extend the range of biomolecules amenable for study by solution NMR. Using a suite of 4D-TROSY triple resonance experiments, almost complete assignment of the backbone $^1H^N$, $^{13}C^{\alpha}$, $^{13}C'$ and ^{15}N chemical shifts and the side chain $^{13}C^{\beta}$ chemical shifts of a 723 residue long *malate synthase* G (82 kDa enzyme) has been accomplished. Clubbed with specific labeling, these methodologies have also been used in the complete assignment of methyl ^{1}H and ^{13}C chemical shifts of Ile ($^{13}C^{\delta1}$), Leu, and Val residues in the same protein. Such assignments throw light on the energetics and kinetics of substrate binding, domain reorientation upon ligand binding, and internal dynamics. The methods have also been used in the study of structure of membrane proteins and in protein-protein interactions.

9. PROTONLESS MULTI-DIMENSIONAL NMR

The enhancement in the sensitivity of ^{13}C has paved the way for its direct detection even in large biological macromolecules, as was initially proposed by Markley in 1988. This has recently led to the development of several so called *protonless* 2D and 3D NMR experiments on ^{13}C/^{15}N labelled proteins. In parallel, RF probes which are optimized for ^{13}C detection have been developed. Prototype probes with the inner coil optimized for ^{13}C sensitivity have initially been proposed. The advent of cryogenic techniques has then contributed to a dramatic increase in sensitivity of the ^{13}C channel in addition to the ^{1}H. These experiments have thus become alternate methods for studies on biomolecules. Direct detection of ^{13}C offers a valuable alternative to ^{1}H detection with the following advantages:

(i) Overcoming the problem of solvent (water) suppression.

(ii) ^{13}C spins in ^{2}H labelled proteins as well as quaternary carbons can be directly excited without requiring any polarization transfer steps.

(iii) Longer T_2 associated with ^{13}C allows large molecules to be investigated.
(iv) The amide H^N may be difficult to detect in segments undergoing chemical exchange. This leads to interruption in *ssr* assignment. Such protons are important for studying the protein function. In such situations the ^{13}C direct detection is favoured.
(v) There is unfavourable chemical-shift dispersion of ^1H resonances in unfolded protein systems. Such segments can be characterized.
(vi) In paramagnetic systems, the contributions to relaxation arising from the paramagnetic centre are reduced by more than 10 by opting for low-γ ^{13}C instead of ^1H.

As in the case of ^1H detected homo- and hetero-nuclear MD NMR, *protonless* NMR can be implemented in higher dimensions. They too enable one to observe homonuclear and heteronuclear correlations using one and two-bond J-couplings. For example, one can use these experiments to observe correlations among $^{13}C^{\alpha}$, $^{13}C^{\beta}$, $^{13}C'$ and ^{15}N spins as well as with the remaining ^{13}C and ^{15}N spins in the side chains. The nomenclature used to describe these pulse sequences is based on spins which are frequency labelled during various evolution and detection periods and the spins through which magnetization is transiently transferred to their coupled partners.

9.1 2D Protonless NMR

The 2D CON represents an experiment in which backbone $^{15}N_i$ and $^{13}C'_{i-1}$ spins are correlated. The excellent resolution provided by the ^{13}C detected 2D NMR spectra can be appreciated looking at the [^{15}N, $^{13}C'$] correlation experiment shown in Figure 5.21. This spectrum provides excellent resolution showing 161 cross peaks out of the expected 163. Unlike [^{15}N, ^1H]-HSQC this includes correlations from Pro residues and those of Asn and Gln side-chains.

Other 2D experiments include CACO, CBCACO, CANCO and CBCANCO. The CACO experiment provides all the $^{13}C^{\alpha}_i$–$^{13}C'_i$ correlations and the correlations of carbonyl/carboxylate spins and the attached carbons belonging to Asp/Asn ($^{13}C^{\beta}_i$–$^{13}C^{\gamma}_i$) and Glu/Gln[13] (C^{γ}_i–$^{13}C^{\delta}_i$). The CBCACO experiment provides $^{13}C^{\beta}_i$–$^{13}C'_i$ and $^{13}C^{\alpha}_i$–$^{13}C'_i$ correlations, while the CANCO experiment provides $^{13}C^{\alpha}_i$–$^{15}N_i$ $^{13}C'_{i-1}$ and $^{13}C^{\alpha}_{i-1}$–$^{15}N_i$ $^{13}C'_{i-1}$ correlations. One can even record a CBCANCO experiment which provides $^{13}C^{\alpha}_i$–$^{15}N_i$ $^{13}C'_{i-1}$, $^{13}C^{\alpha}_{i-1}$–$^{15}N_i$ $^{13}C'_{i-1}$, $^{13}C^{\beta}_i$–$^{15}N_i$ $^{13}C'_{i-1}$ and $^{13}C^{\beta}_{i-1}$–$^{15}N_i$ $^{13}C'_{i-1}$ correlations. The fact that the C^{β} chemical-shift dispersion is much larger compared to the C^{α} shifts, recording of both CBCACO and CBCANCO helps in overcoming the overlap problems. They also help in the identification of the peaks arising from Ala, Ser and Thr, because of their unique C^{β} shifts.

9.2 Problem of Large Homonuclear Couplings in Acquisition Dimension

A major problem associated with ^{13}C direct-detection is the large-homonuclear ^1J(C-C) couplings, and smaller-homonuclear scalar couplings, such as ^2J(C-C), ^3J(C-C) and ^4J(C-C) that evolve during the detection period. In experiments where $^{13}C'$ spins are detected, $^{13}C'$ resonances are split into two with a separation of 55 Hz. When

$^{13}C^{\alpha}$ are detected, the resonances are more complex with the splitting caused by not only the 55 Hz $^1J(C^{\alpha}\text{-}C')$ but also 35 Hz $^1J(C^{\alpha}\text{-}C^{\beta})$. Though such ^1J(C-C) are useful for coherence transfer, they reduce the S/N ratio by more than two, if they are retained along the detection period and spoil the resolution. The values of ^2J(C-C), ^3J(C-C) and ^4J(C-C) are within the resolution of the experiments.

Band-selective homodecoupling is one of the methods for removing the homonuclear couplings during the detection period. For example, while detecting C′ nuclei one can selectively decouple C^{α} (with an optimized RF power to avoid artefacts). Other schemes are the so called *virtual* decoupling schemes, namely, in-phase anti-phase (IPAP) and spin-state selective excitation (S^3E), which have been extensively used in both solution- and solid-state NMR.

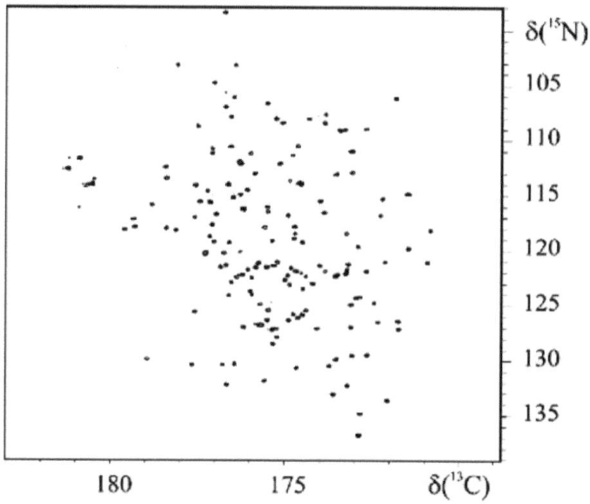

Figure 5.21: The 2D CON spectrum of u-^{13}C and ^{15}N labelled SOD sample (reduced monomer) A 700 MHz spectrometer equipped with a cryo-probe optimized for ^{13}C direct detection has been used to record this spectrum. W. Bermel, I. Bertini, I.C. Felli, M. Piccioli and R. Piaratelli. Prog. NMR Spectroscopy 48, 25-45. (2006). (Reprinted with permission from Elsevier).

In the IPAP method, illustrated for 1D experiment in Figure 5.22, two FIDs are acquired separately for each increment, one with the in-phase and the other with the anti-phase ^{13}C components. One then deconvolutes the splitting due to the $^1J(^{13}C\text{–}^{13}C)$ coupling seen along the acquisition dimension using a simple algorithm. In the S^3E method, the interconversion between in-phase and anti-phase is not complete and two different linear combinations of in-phase and anti-phase components are acquired and stored separately. These are manipulated to separate the two components and then shifted to the center of the original multiplet. The important difference between S^3E and IPAP methods are the delay time used for interconversion of in-phase components into anti-phase components, which in S^3E scheme is half of that of the IPAP.

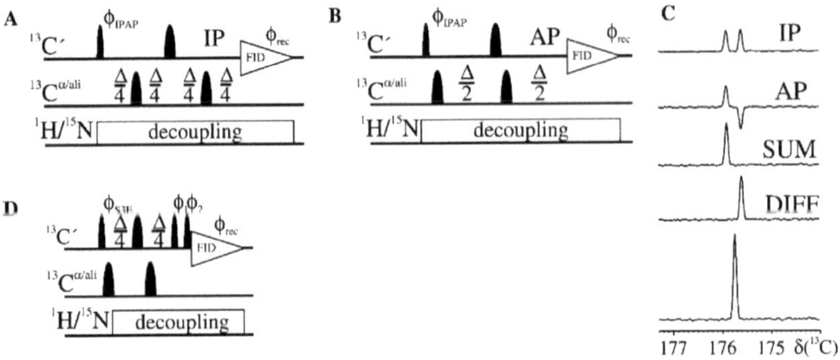

Figure 5.22: IPAP and S^3E approaches for $^{13}C'$- direct-detection to remove $^{13}C^{\alpha}$-$^{13}C'$ splitting in the direct acquisition dimension. Band-selective ^{13}C pulses denoted by narrow and wide ones, represent $\pi/2$ and π pulses, respectively. A and B show the two variants of the pulse scheme for IPAP to acquire $^{13}C'$(A for the in-phase and B for the anti-phase components, respectively). The results of the two experiments A and B on a sample of u-$^{13}C'/^{15}N$ doubly labelled Ala in ^2H2O are shown in panel C and indicated with IP and AP, respectively. D shows the scheme for the S^3E approach to acquire $^{13}C'$ W. Bermel, I. Bertini, I.C. Felli, M. Piccioli and R. Piaratelli. Prog. NMR Spectroscopy, 48, 25-45. (2006). (Reprinted with permission from Elsevier).

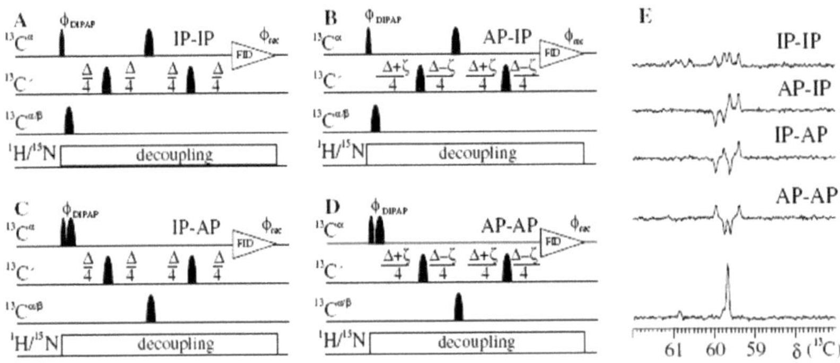

Figure 5.23: The DIPAP schemes for C^{α} direct-detection to remove the two $^1J(C^{\alpha}–C')$ and $^1J(C^{\alpha}–C^{\beta})$ splittings in the direct acquisition dimension. Band-selective ^{13}C pulses denoted by narrow and wide ones, represent $\pi/2$ and π pulses, respectively. The two lines labeled with C^{α} and $C^{\alpha/\beta}$ indicate band-selective pulses. Panels A–D report the four variants of the pulse scheme to acquire and store separately the four components indicated with IP–IP, AP–IP, IP–AP and AP-AP, respectively. The results of the four experiments performed on the set-up sample of ^{13}C, ^{15}N labeled Ala in ^2H_2O are shown in panel E.W. Bermel, I. Bertini, I.C. Felli, M. Piccioli and R. Piaratelli. Prog. NMR Spectroscopy, 48, 25-45. (2006). (Reprinted with permission from Elsevier).

The IPAP and S^3E approaches described above are suitable only for spins such as C', which are coupled to only one other homonuclear spin (C^α). When the detected spin has more than one homonuclear couplings, one has to rely on a scheme where IPAP or S^3E is applied repeatedly. For example, in detecting C^α, one has to decouple two couplings, $^1J(C^\alpha–C')$ and $^1J(C^\alpha–C^\beta)$. For this purpose, one can either combine two IPAP blocks which are specific to $^1J(C^\alpha–C')$ and $^1J(C^\alpha–C^\beta)$ or the two blocks can be concatenated into a shorter block in which the total duration is dictated by the smaller coupling, $^1J(C^\alpha–C^\beta)$. Such sequence is called double IPAP or DAPIP. A 1D version of the sequence is shown in Figure 5.23. Notice that the total length of the block is around 14 msec, which is rather long. For large proteins at higher magnetic fields, this approach is made convenient by ^2H-labelling, which drastically reduces C^α relaxation rates.

9.3 3D Protonless NMR

As in the case of ^1H detected 3D *double-* and *triple-resonance* experiments, the *protonless* ^{13}C-detected 3D NMR experiments enable one to observe homonuclear and heteronuclear correlations using one- and two-bond J-couplings (Figure 5.9 A).

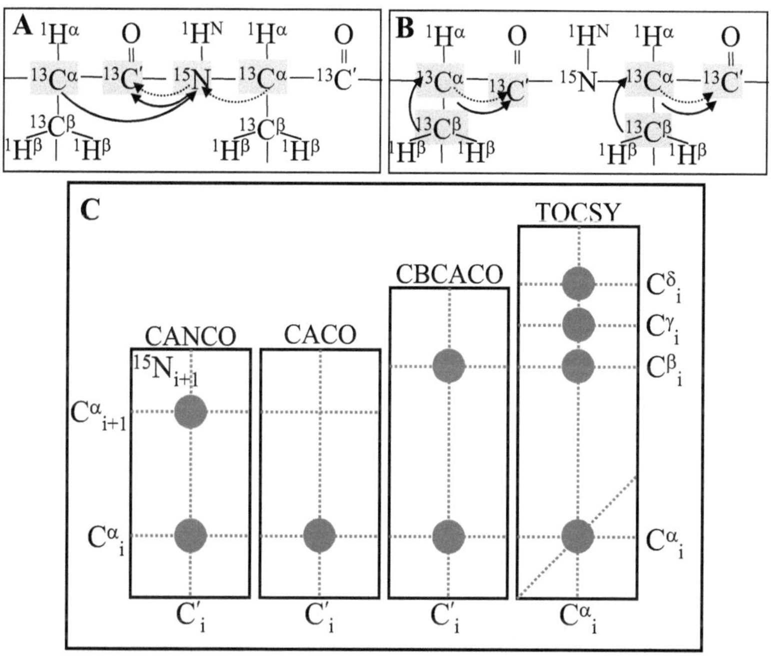

Figure 5.24: Coherence transfer pathways in (A) 3D CANCO and (B) 2D CBCACO. (C) Illustrative planes of 2D and 3D spectra showing respective cross peak signatures.

The first among such proposed pulse sequence is the 3D CANCO. It enables one to observe correlations among the backbone $^{13}C^{\alpha}$, ^{15}N and $^{13}C'$ spins within each amino acid residue in a protein. The coherence is transferred from $^{13}C^{\alpha}_{i+1}(t_1)\rightarrow{}^{15}N_{i+1}(t_2)\rightarrow{}^{13}C'_i(t_3)$ (solid arrows in Figure 5.24 A) and $^{13}C^{\alpha}_i(t_1)\rightarrow{}^{15}N_{i+1}(t_2)\rightarrow{}^{13}C'_i(t_3)$ (dashed arrows). Thus, the CANCO has information about $^{13}C^{\alpha}_i$ and $^{13}C'_{i-1}$ chemical shifts, from the observation of two ($^{13}C^{\alpha}$, $^{15}N_{i+1}$, $^{13}C'_i$) peaks for a given pair of $^{15}N_{i+1}$ and $^{13}C'_i$ shifts. One peak belongs to ($^{13}C^{\alpha}_i$, $^{15}N_{i+1}$, $^{13}C'_i$) and the other to ($^{13}C^{\alpha}_{i+1}$, $^{15}N_{i+1}$, $^{13}C'_i$). The transfer functions for both the peaks are dictated by the conformation dependent $^1J(^{13}C^{\alpha}_i - {}^{15}N_i)$ and $^2J(^{13}C^{\alpha}_i - {}^{15}N_{i+1})$ values. For example, Figure 5.24 C, shows a ($^{13}C^{\alpha}$, $^{13}C'$) plane of 3D CANCO spectrum at a chemical shift of $^{15}N_{i+1}$. The distinction between the two peaks can be made from a 2D CACO spectrum, where one observes only the ($^{13}C^{\alpha}_i$, $^{13}C'_i$) peak (CACO panel in Figure 5.24 C). Thus, the assignment of all $^{13}C^{\alpha}$, ^{15}N and $^{13}C'$ spins present in the protein can be carried out with the use of CANCO and CACO.

The CBCACO enables one to observe correlations among $^{13}C^{\alpha}$, $^{13}C^{\beta}$ and $^{13}C'$ spins within each amino acid residue in a protein. The coherence is transferred from $^{13}C^{\beta}_i(t_1)\rightarrow{}^{13}C^{\alpha}_i(t_2)\rightarrow{}^{13}C'_i(t_3)$ and $^{13}C^{\alpha}_i(t_1)\rightarrow{}^{13}C^{\alpha}_i(t_2)\rightarrow{}^{13}C'_i(t_3)$. With the prior knowledge of $^{13}C^{\alpha}_i$ and $^{13}C'_i$ spins one can assign the corresponding $^{13}C^{\beta}_i$. Further, CCCO or [^{13}C, ^{13}C] homonuclear 2D TOCSY can be used to assign the rest of the ^{13}C spins. Other useful 3D experiments for resonance assignments include CANCO-selective CBCACON and CBCANCO.

10. REFERENCES

10.1 Historical Foundation of Protein Structure and Function

A. Saunders, and A. Wishnia, *Nuclear magnetic resonance spectra of proteins*, Ann. N.Y. Aca. Sci. **70**, 870–874 (1958).

A. Kowalsky, *Nuclear magnetic resonance studies of proteins*, J. Biol. Chem. **237**, 1807–1809 (1962).

G.N. Ramachandran, C. Ramakrishnan, and V. Sasisekharan, *Stereochemistry of polypeptide chain configuration*, J. Mol. Biol. 7, 95–99 (1963).

C. Levinthal, *How to fold graciously*, Mossbauer Spectroscopy in Biological Systems, J.T.P. DeBrunner and E. Munck. (eds.) Proceedings of a meeting held at Alberton House, Monticello, Illinois, University of Illinois Press, pp. 22–24 (1969).

C.B. Anfinsen, *Principles that govern the folding of protein chains*, Science **181**, 223–230 (1973).

10.2 General Books on Proteins

C. Branden, and J. Tooze, *Introduction to Protein Structure*, Garland Publishing Inc., New York (1991).

M. Perutz, *Protein Structure*, W.H. Freeman and Company, New York (1992).

J.M. Berg, J.L. Tymoczko and L. Stryer, *Biochemistry, 5th International Edition*, W.H. Freeman and Co. (1995).

10.3 Books on Protein NMR

K. Wüthrich, *NMR of Proteins and Nucleic Acids*, Wiley (1986).

M.P. Deutscher (ed.), *Guide to Protein Purification*, Methods Enzymol. **182**, 1–894 (1990).

F.J.M. van de Ven, *Multidimensional NMR in Liquids – Basic Principles and Experimental Methods*, VCH (1995).

J. Cavanagh, W.J. Fairbrother, A.G. Palmer and N.J. Skelton, *Protein NMR Spectroscopy: Principles & Practice*, Academic Press Inc., San Diego (1996).

M.H. Levitt, *Spin Dynamics – Basics of Nuclear Magnetic Resonance*, Wiley (2001).

T.L. James (ed.), *Nuclear Magnetic Resonance of Biological Macromolecules*, Methods Enzymol. **394**, 1–631 (2005).

10.4 Study of Large Proteins

K. Wüthrich, G. Wider, G. Wagner and W. Braun, *Sequential resonance assignments as a basis for determination of spatial protein structures by high resolution nuclear magnetic resonance*, J. Mol. Biol. **155**, 311–319 (1982).

T.F. Havel and K. Wüthrich, *A distance geometry program for determining the structures of small proteins and other macromolecules from nuclear magnetic resonance measurements of intramolecular ^1H-^1H proximities in solution*, Bull. Math. Biol. **46**, 673–678 (1984).

G. Otting, and K. Wüthrich, *Studies of protein hydration in aqueous solution by direct NMR observation of individual protein-bound water molecules*, J. Am. Chem. Soc. **111**, 1871–1875 (1989).

M. Ikura, L.E. Kay, and A. Bax, *A novel approach for sequential assignment of ^1H, ^{13}C and ^{15}N spectra of larger proteins: heteronuclear triple-resonance three-dimensional NMR spectroscopy. Application to calmodulin*, Biochemistry **29**, 4659–4667 (1990).

D. Neri, M. Billeter, G. Wider, and K. Wüthrich, *NMR determination of residual structure in a urea-denatured protein, the 434 repressor*, Science **257**, 1559–1563 (1992).

J.R. Tolman, J.M. Flanagan, M.A. Kennedy, and J.H. Prestegard, *Nuclear magnetic dipole interactions in field-oriented proteins: information for structure determination in solution*, Proc. Natl. Acad. Sci. USA **92**, 9279–9283 (1995).

D.S. Garrett, Y. Seok, D. Liao, A. Peterkofsky, A.M. Gronenborn, G.M. Clore, *Solution structure of the 30 kDa N-terminal domain of enzyme Iof Escherichia coli phosphoenolpyruvate:sugar phosphotransferase system by multidimensional NMR*, Biochemistry **36**, 2517–2530 (1997).

N. Tjandra and A. Bax, *Direct measurement of distances and angles in biomolecules by NMR in dilute liquid crystalline medium*, Science **278**, 1111–1114 (1997).

G. Wider and K. Wüthrich, *NMR spectroscopy of large molecules and multimolecular assemblies in solution*, Curr. Opin. Struct. Biol. **9**, 594–601 (1999).

H.S. Atreya, S.C. Sahu, K.V.Chary and G. Govil. *A tracked approach for automated NMR assignment in proteins*, J. Biomol. NMR **17**, 125–36 (2000).

R. Riek, K. Pervushin and K. Wüthrich *TROSY and CRINEPT: NMR with large molecular and supramolecular structures in solution*, Trends Biochem Sci. **25**, 462–468 (2000).

R. Riek, K. Pervushin and K. Wüthrich, *TROSY and CRINEPT: NMR with large molecular and supramolecular structures in solution*, FEBS Letters **504**, 173–178 (2001).

J. Fiaux, E.B. Bertelesen, A.L. Horwich and K. Wüthrich, *NMR Analysis of a 900 K GroEL-GroES Complex*, Nature **418**, 207–211 (2002).

K. Wüthrich, (Nobel Lecture), *NMR Studies of Structure and Function of Biological Macromolecules*, Angewandte Chemie, Int. Ed. **42**, 3340–3363 (2003).

W. Bermel, I. Bertini, L. Duma, L. Emsley, I.C. Felli, R. Pierattelli and P. Vasos, *Complete Assignment of Heteronuclear Protein Resonances by Protonless NMR Spectroscopy*, Angew. Chem. Int. Ed. **44**, 3089–3092 (2005).

H.S. Atreya and T. Szyperski. *G-matrix Fourier transform NMR spectroscopy for complete protein resonance assignment*. Proc. Natl. Acad. Sci. USA **101**, 9642–9647 (2006).

H.S. Atreya and T. Szyperski, *Principles And Applications Of GFT Projection Nmr Spectroscopy*, Magn. Reson. Chem. **44**, Spec No: S51–S60 (2006).

I. Bertini, I.C. Felli, L. Gonnelli, R. Pierattelli, Z. Spyranti, and G.A. Spyrulias, *Mapping protein-protein interaction by 13C'-detected heteronuclear NMR spectroscopy*, J. Biomol. NMR **36**, 112–122 (2006).

W. Bermel, I. Bertini, I.C. Felli, m. Piccioli and R. Piaratelli, *^{13}C-detected protonless NMR spectroscopy of proteins in solution*, Prog. NMR Spect. **48**, 25–45 (2006).

CHAPTER 6

STRUCTURE, DYNAMICS AND FUNCTION OF PROTEINS

1. INTRODUCTION

After the completion of resonance assignments, one can use NMR signals to obtain information on several important aspects of biological chemistry. As discussed in the previous chapter, proteins play the most important functional role in living systems. Protein functions are closely linked to their relatively well folded conformational structure, their ability to adapt when binding to ligands and their dynamic behaviour. Significant changes take place in all these aspects, during the time-course that the proteins participate in biochemical reactions. Several of these important aspects have been studied by NMR. Such applications and their relationship with structure are discussed in this chapter.

2. CHARACTERIZATION OF SECONDARY STRUCTURAL ELEMENTS

The first step after the completion of sequence specific resonance assignments, towards the 3D structure determination, is to characterise secondary structural elements in a protein. While this can be derived using different types of NMR data, chemical shifts have emerged as a simple means towards this goal. Several automated methods have been proposed that utilize sequence specific NMR chemical shifts to predict or to identify the secondary structure elements in the protein. The development of these methodologies can be attributed to our understanding of NMR data vis-à-vis protein structures. In Chapter 4, we have already discussed the general approach to structural determination of biological molecules. In this chapter, the emphasis is on the methods specific to proteins.

2.1 Chemical Shift Index (CSI)

The $^{1}H^{\alpha}$ and $^{13}C^{\alpha}$ chemical shifts are the most sensitive NMR parameters, to changes in secondary structure. Hence, changes in chemical shifts have emerged as a means of characterizing secondary structural elements in a protein. The chemical shift index (CSI) was originally developed for the analysis of H^{α} chemical shifts. The approach was later extended to include shifts of $^{13}C^{\alpha}$, $^{13}C^{\beta}$ and $^{13}C'$. Wishart and others have made detailed analysis of the reported chemical shift data in proteins. A strong correlation has been observed between the changes in the chemical shifts and the nature of the secondary structure.

In this procedure, the differences between the chemical shifts of $^1H^\alpha$, $^{13}C^\alpha$, $^{13}C^\beta$ and $^{13}C'$ spins of each residue, in the ordered protein under study (δ_{obs}) and those for the same residue in a random coil state (δ_{rc}), are computed. The difference, $\Delta_s = \delta_{obs} - \delta_{rc}$, is called the *secondary shift*. Such chemical shift indices are sensitive indicators of the local conformation. The Δ_s values in a protein can be analysed using the program CSI: (*http://www-ccmr-nmr.bioc.cam.ac.uk/html/csi/*).

All amino acid residues experience an average $^1H^\alpha$ chemical shift change of ~0.39 ppm in the up-field direction in α-helices and a downfield shift of ~0.37 ppm for β-sheets or extended conformations. Similar changes are observed in the chemical shifts of carbonyl ^{13}C, $^{13}C^\alpha$ and $^{13}C^\beta$ spins. When a continuous stretch of three or more residues have their $^1H^\alpha$, $^{13}C^\alpha$, $^{13}C^\beta$ and $^{13}C'$ resonances shift in an expected direction, then these segments can be assigned to a particular secondary structure (i.e. the α-helical or β-sheet structure). Figure 6.1 shows an example of the observed CSI for a protein. A residue has been assigned to α-helical structure when two of the three CSIs agreed with those expected for an α-helix, and a β–strand segment when two of the three CSIs satisfied those expected for a β–strand. Eight α-helices and four β–strand segments have thus been identified in this protein.

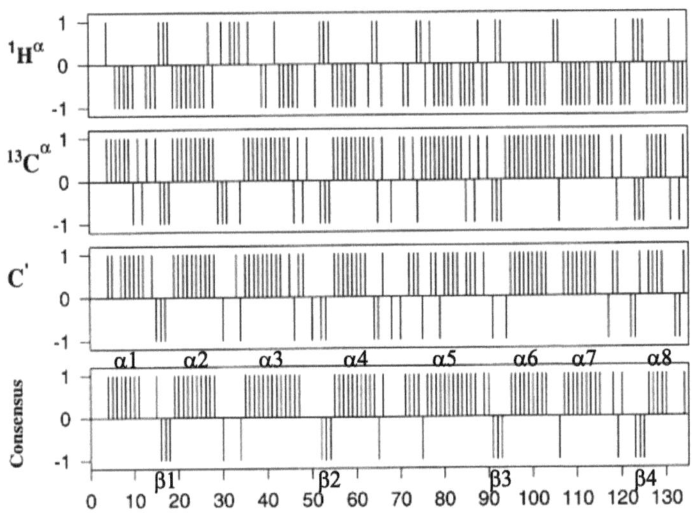

Figure 6.1: $^1H^\alpha$, $^{13}C^\alpha$ and $^{13}C'$ chemical shift indices for amino acid residues in a calcium binding protein. The + and – signs correspond to low- and up-field chemical shifts relative to random coil values (i.e. the sign of Δ_s). The signs are -, +, + and +, -, - respectively for α-helices and β-strands. S.C. Sahu, A. Bhattacharya, K.V.R. Chary, and G. Govil. FEBS Letters. 459, 51-56 (1999). (Reprinted with permission from Elsevier).

The expected accuracy of secondary structure prediction using CSI is around 92%. This suggests that the secondary structure in proteins can be obtained with confidence from the observed changes in the $^{1}H^{\alpha}$, $^{13}C^{\alpha}$, $^{13}C^{\beta}$ and $^{13}C'$ chemical shifts.

2.2 Secondary Structure using Chemical Shift and Sequence Homology

The program TALOS (http:/spin.niddk.nih.gov/NMRPipe/talos/), combines the information on the chemical shifts and the protein primary sequence for predicting backbone torsion angles ϕ and ψ. It is based on the principle that two proteins, with similar chemical shifts and local sequences, are likely to have similar local conformations. A moving window of three amino acid residues in the sequence of interest is searched against database of triplets of adjacent residues. The method uses database of H^{α}, $^{13}C^{\alpha}$, $^{13}C^{\beta}$, $^{13}C'$ and ^{15}N chemical shifts of 20 proteins for which X-ray structures with resolution better than 2.2 Å are available. Ten triplets, which have close similarity in secondary chemical shift and the amino acid sequence to those of the sequence of interest, are selected at a time. If the central residues in these ten triplets exhibit similar ϕ and ψ torsion angles, then their respective averages are used as angular restraints for the protein whose structure is under study. Tests carried out on proteins of known structures indicate that the RMSD between TALOS output and the X-ray derived backbone angles is about 15°. Almost 97% of the predictions made by TALOS are found to be correct.

2.3 Nuclear Overhauser Effect (NOE)

NOEs in proteins are used both for assignments and structure determination. In fact, NOE is the most important source of information for short and long range ordering in proteins. NOESY cross peaks between the H^{N} and the side chain protons (H^{α} and H^{β}), provide important information on secondary structures. The ^{1}H-^{1}H distances (d) in proteins are usually represented by the type of protons followed by their residue numbers. For example, $d_{\alpha N}(i, i+3)$ represents distance between H^{α} of the ith and H^{N} of $(i+3)^{rd}$ residues. Some useful sequential and medium-range ^{1}H-^{1}H distances are given in Table 6.1 along with their relationships to various secondary structural elements.

Depending on the position of the participating amino acid residues in the primary sequence, NOEs are usually divided into three categories: short-, medium- and long-range. The short-range NOEs are those between adjacent amino acids (Figure 6.2), such as $(H^{N}, H^{\alpha}$ and $H^{\beta})_{i-1} \rightarrow (H^{N})_i$. These are represented by d_{NN}, $d_{\alpha N}$, and $d_{\beta N}$, respectively. These NOE connect neighbouring networks of coupled spin systems in the primary sequence. Hence, such NOEs are also called sequential connectivities and are useful for sequence specific resonance assignments.

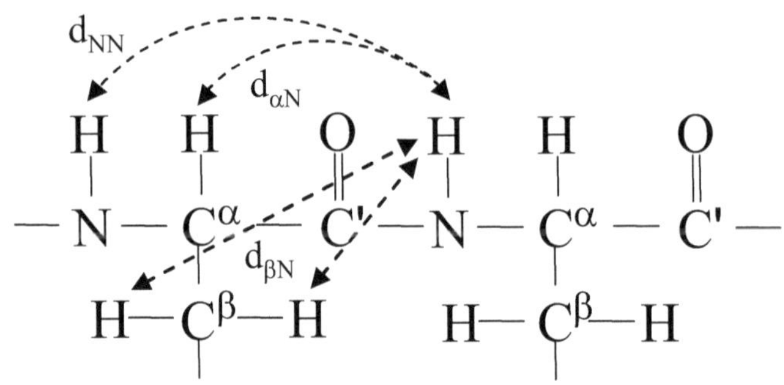

Figure 6.2: Sequential NOESY connectivities in proteins.

From the structural view-point, medium-range NOEs such as $d_{NN}(i, i+2)$, $d_{\alpha N}(i, i+3)$, $d_{\alpha\beta}(i, i+3)$ are useful for secondary structure characterization. The residues participating in α-helices are easily identified by a stretch of medium-range $d_{\alpha N}(i, i+3)$ and $d_{\alpha\beta}(i, i+3)$ NOEs. These stretches are also characterised by slow exchange rates for the backbone amide protons ($^1H^N$). In the 3^{10}-helix, the segments are characterised by a stretch of medium-range $d_{\alpha N}(i, i+4)$ NOEs and hence, can be distinguished from α-helical stretches.

Table 6.1: Sequential, short and medium-range ^1H-^1H distances in different secondary structures.

	α-Helix	3_{10}-Helix	β-strand	Type I β-turn	Type II β-turn
$d_{\alpha N}(i, i+1)$	3.5	3.4	2.2	3.4/3.2	2.2/3.2
$d_{\alpha N}(i, i+2)$	4.4	3.8		3.6	3.3
$d_{\alpha N}(i, i+3)$	3.4	3.3		3.1-4.2	3.8-4.7
$d_{\alpha N}(i, i+4)$	4.2				
$d_{NN}(i, i+1)$	2.8	2.6	4.2-4.3	2.6/2.4	4.5/2.4
$d_{NN}(i, i+2)$	4.2	4.1		3.8	4.3
$d_{\alpha\beta}(i, i+3)$	2.5-4.4	3.1-5.1			

The β-strand stretches show strong $d_{\alpha N}$ and weak d_{NN} NOEs. However, in order to distinguish between parallel and anti-parallel β-sheets, one needs to identify long-range interactions from inter-strand NOEs. As seen from the structures of β-sheets (Figure 5.6), both types of structures have the H^α and the H^N spins on neighbouring strands within the NOE observable distances. This is coupled with a good dispersion in chemical shifts. Long-range $d_{\alpha\alpha}(i, j)$ NOEs are therefore good indicators for the presence of anti-parallel β-sheet structure.

Type II β-turn shows a strong $d_{NN}(3, 4)$ connectivity (between the amide protons of the 3rd and 4th residues of the turn) and medium $d_{\alpha N}(2, 4)$ and $d_{\beta N}(2, 4)$

connectivities. Such turns can be further confirmed by the absence of $d_{NN}(2, 3)$ NOE. In the type I β-turn, one observes strong sequential d_{NN} cross peaks between the 2nd and 3rd, and between 3rd and 4th residues. Medium $d_{\alpha N}(i, i+1)$ connectivities are observed along this stretch. This is further confirmed by the absence of $d_{\alpha N}(i, i+2)$ NOEs.

Long-range NOEs between protons belonging to residues, which are far apart in a polypeptide chain, but are close in space, are crucial for describing the tertiary folds of proteins. Such NOEs provide the most important clues to the overall nature of protein-folding.

2.4 Three bond Scalar-couplings (^3J)

Several three bond couplings, both along the backbone and in the side chains of proteins, provide information on the local structure and dynamics of the backbone and the side chain conformations of proteins (Table 6.2). In fact, there are four coupling constants which depend on the rotation around the N-C^α bond (angle ϕ). It may be noted that the dihedral angle (θ) in Karplus relationship is related to ϕ by simple relationships. The most useful of the backbone couplings is $^3J(H^N\text{-}N\text{-}C^\alpha\text{-}H^\alpha)$, particularly since in ordered structures both H^α and the H^N are well resolved. The value for this coupling is less than 5 Hz for α-helix and more than 7 Hz for β-sheets. Values between these limits are observed for random coil conformations.

Table 6.2: Parameters for Karplus relationships for ^3J in proteins. [Adopted from Lindorff-Larsen et al. J. Biomol. NMR, (2005) 32, 273-280], except for the last row.

Coupling Constant	A	B	C
$H\text{-}N\text{-}C^\alpha\text{-}H^\alpha$	8.33	-1.69	0.44
$H^\alpha\text{-} C^\alpha\text{-}N\text{-}C'$	4.41	-2.14	0.77
$H\text{-}N\text{-} C^\alpha\text{-}C'$	5.5	-1.3	-0.16
$H\text{-}N\text{-} C^\alpha\text{-}C^\beta$	3.4	-0.75	-0.08
$N\text{-} C^\alpha\text{-}C^\beta\text{-}C^\gamma$	2.8	-0.4	-0.4
$H^\alpha\text{-} C^\alpha\text{-}C^\beta\text{-}H^\alpha$	9.4	-1.4	1.6

The side chain atoms are significantly more mobile. In most cases, the conformation around C^α-C^β is represented by a distribution of rotamers among the two *gauche* and one *trans* conformers. The two *gauche* conformers are usually more populated.

2.5 $^1H^N$-$^2H^N$ Exchange

The chemical shift and the temperature coefficient of the $^1H^N$ resonances involved in intramolecular hydrogen bonds such as N-H······O=C, are the most important properties used to detect the presence of such bonds. The effect of temperature on the chemical shift of $^1H^N$ involved in hydrogen bonding with the solvent is large. The coefficient is relatively small for the $^1H^N$ involved in intramolecular bonding.

The $^1H^N$-$^2H^N$ exchange rate provides an excellent means to examine the local conformational stability of proteins. For example, very high temperatures or addition of denaturing chemicals result in the *melting* or denaturation of ordered protein structures. This leads to up-field shift of the $^1H^N$ resonances following the collapse of hydrogen bonds. In small proteins, where spectral features are well resolved, $^1H^N$-$^2H^N$ exchange rates can be measured by monitoring the decay of $^1H^N$ resonance in ^2H_2O. A slow $^1H^N$-$^2H^N$ exchange is indicative of the involvement of the corresponding $^1H^N$ in intramolecular hydrogen bonding. For uniformly-^{15}N labelled protein, exchange rates of amide protons are measured indirectly by measuring the integral volumes of individual cross peaks in a series of 2D [^{15}N-^1H] HSQC spectra recorded immediately after the addition of ^2H_2O. HSQC is preferred over other types of 2D spectra for the purpose because this can be acquired in a few minutes. Therefore a series of spectra can be recorded within a limited time-frame.

The decay of the individual peak volumes, thus observed, is fitted to a single exponential decay. Protection factors (PF) for amide proton exchange are estimated from the equation: PF = k_{ch}/k_{ex}, where k_{ch} and k_{ex} are the calculated intrinsic and experimental exchange rates, respectively (Figure 6.3). The rates of exchange for individual protons can be related to equilibrium constants for opening and closing of individual hydrogen bonds. Thus, one can also calculate the free energy for the conformational transition which permits the $^1H^N$-$^2H^N$ exchange to occur.

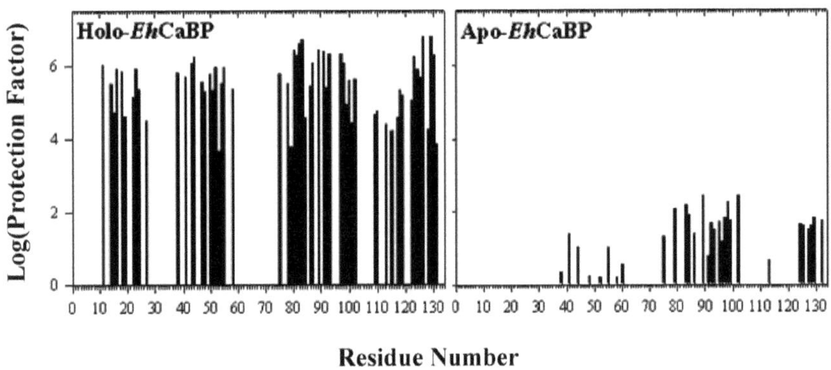

Figure 6.3: An example showing PFs for $^1H^N$ for holo and apo forms of a CaBP protein.

The $^1H^N$ – $^2H^N$ exchange can also be studied by what is known as the quenched exchange experiment. The proton exchange is allowed to occur for some time under certain experimental conditions. It is then stopped (quenched) suddenly by a change in solvent conditions. The degree of exchange is then monitored by recording a series of [^{15}N-^{1}H] HSQC. Protein folding intermediates, which may form on the millisecond time-scale, can be studied in this way.

3. OBTAINING THE FINAL STRUCTURES

While secondary structures can be easily deduced through chemical shifts, sequential and short-range NOEs, J-couplings and exchange rates, investigations of the long-range order in proteins require additional parameters. The availability of information about adequate number of long-range constraints is essential for characterizing loop regions, tertiary structures, domain orientations, quaternary structure and binding of proteins with other macromolecules. Long-range NOEs are therefore crucial in obtaining the final protein structures. RDCs have also emerged as potential source for structural clues. One can measure ^{15}N-^{1}H and ^{13}C-^{1}H bond-vector orientations from RDCs. Cross relaxation rates and use of paramagnetic probes are other sources of information for establishing long-range structure. However, the number of available long-range constraints is usually much smaller than the medium- and short-range information elements.

3.1 3D Structure Calculation

The final stage in 3D structure determination of proteins is to decide on a suitable algorithm for structure calculations using the observed NMR constraints. Computationally, this is the most intensive step in NMR based structural determination of a biomolecule. Structure calculation involves finding conformations of a protein, which are consistent with experimental data. Due to the complexity of the problem, it is neither feasible to do an exhaustive search of allowed conformations nor to find solutions by interactive model building.

The methods for structural calculations using NMR constraints have been discussed in Chapter 4. The process of protein assignments and structure determination depends heavily on ^{1}H-^{1}H NOE. The approach can be divided into two stages, which are employed iteratively:

(i) Assignments of signals in NOESY.
(ii) Structural simulation.

The first step involves the identification of ^{1}H spins in the protein which are close in space. This yields distance constraints, which fix ^{1}H-^{1}H distances. Since the 3D structure is usually not known *apriori*, it is not straight forward to assign protons involved in a particular cross peak. Hence, the assignment of NOESY peaks and structure calculation are done in an iterative manner. Several algorithms have been proposed to automate this process.

Chiral constraints are imposed either by defining improper torsion angles or with the use of a triple scalar product approach in algorithms used for structure determination. In the absence of stereo-specific assignments, a pseudo atom is defined which lie midway between stereo-related protons. This is a compromise from not using the NOE information at all and considerably weakens constraint information.

3.2 Automated NOESY Assignments

The input to most of the structure calculation programs includes the protein primary sequence, a list of assigned chemical shifts and a list of co-ordinates of cross peaks from the NOESY spectrum. In some approaches, an initial structure is calculated using preliminary NMR information or X-ray crystal structure (when available). This structure can then be used to aid assignments and validation of solution structure by NMR.

NOESY provides limited reliability on the chemical shifts and peak positions. Often NOE cross peaks cannot be attributed to a unique pair of spins resulting in an ambiguous NOE assignment, spanning multiple spin pairs. Obtaining a comprehensive set of distance constraints is thus by no means straightforward. An iterative process is used in such cases. A preliminary structure is calculated from a limited number of distance constraints. This is then used for better assignment of the cross peaks. Difficulties may also arise from spectral artefacts, poor signal to noise ratio and from the absence of expected signals. A complete and accurate assigned set of chemical shifts is usually a pre-requisite for minimizing erroneous NOESY assignments.

3.3 Structural Statistics and Quality

Out of the energy minimized structures, 10-20 structures with the lowest residual target function values are chosen for final analysis. Such structures are subjected to several tests for checking their quality. All pair-wise atom RMSDs are computed using standard softwares. For a good quality structure, the RMSD from experimental distance, dihedral angle and projection angle restraints should be minimal. Violations in distance restraints should normally be less than 0.5 Å. The number of *bad contacts* between pairs of atoms should be small. Most of the (ϕ, ψ) angles should be within the allowed regions of Ramachandran plot. A large number of angles in forbidden regions indicate poor structural quality. The quality of the structure is also evaluated by estimating the sum of energies contributed by Lennard Jones potential and the electrostatic interaction, which should be negative. Even a single bad contact in the structure can give rise to a large repulsive energy. The quality of these structures is finally analysed using softwares such as PROCHECK.

4. PROTEIN DYNAMICS

Since the early work of Anfinsen, which established the relationship between the primary sequence and the native state of proteins, the biochemist's view of protein structures has been often limited to the static 3D structures as determined by X-ray crystallography and NMR. These techniques provide snap-shots of the allowed conformations and overall architecture, reactive interfaces and active sites of a protein. At the same time, movement at a molecular level is essential for the function of biological molecules. Enzyme domains have to open up to bind substrates, convert these into product and release the products. Even an event such as binding of oxygen to haemoglobin requires the protein molecule to change conformation, at least transiently. Processes such as signal transduction requires a cascade of protein-protein interactions.

Thus, the dynamic nature of protein structures plays a key role in their functions. The amplitudes of motions during biochemical reactions may range from a fraction of an angstrom (Å) to the size of the proteins themselves. The time scale of motions can range from nanoseconds (nsec) to milliseconds (msec). Even faster motions in the picosecond (psec) time scales can be important. NMR offers different methods to study protein dynamics, which occur in different time-scales (Figure 6.4)

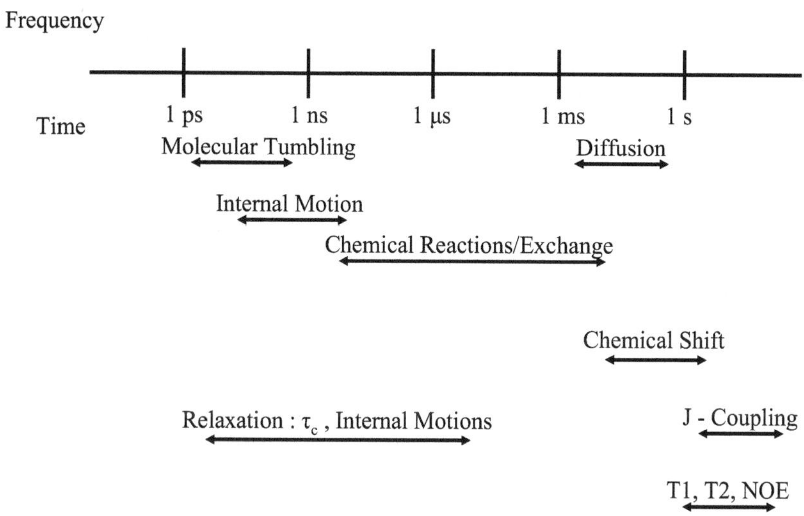

Figure 6.4: Time scales of molecular motions which can be followed using different techniques in NMR.

Study of structural dynamics is one of the most complex subjects, since there are thousands of degrees of freedom for a protein of average size. The water molecules in the solvation spheres also move with the motion of the protein. Local geometry continues to be maintained even when atoms move in a correlated fashion locally, while maintaining the overall folding pattern. However, when motions occur in an uncorrelated fashion, then not only the thermodynamics but also the chemistry and function are affected. These aspects cannot be explained by the static 3D structures obtained in the native state of a protein.

In crystal structures, certain aspects of protein mobility are embedded in the temperature factor (B-factors). However, B-factors may considerably underestimate protein flexibility in solutions. In some cases, the fluctuations for flexible side chains such as Lys as estimated from B-factors are found to be inaccurate by several Å. Some of these differences may be due to loss of flexibility in protein-structures in solid state due to crystal contacts.

In recent years, NMR and MD simulations have proved to be valuable tools for detailed analysis of molecular motions. The results have been used for exploring the relationship between dynamics and function.

4.1 NMR Parameters for Studying Dynamics

Relaxation parameters (R_1, R_2, $R_{1\rho}$ and NOE) depend both on the structure and motion of the molecules. These features have been introduced earlier as a function of the isotropic tumbling motions of the molecules in solutions, using a single correlation time τ_c. In such an approach, the molecule is modeled as a sphere undergoing Brownian motion.

The internal dynamics of molecules introduces an additional motional feature, which can be represented by a correlation time (τ_e) arising from the exchange between different conformational states sampled by the molecule. The second correlation time provides information on the fast internal motions, which may be in the nano- to pico-seconds time range, as compared to the slow tumbling motions of macromolecules.

It is usually assumed that the global tumbling and the internal motions are independent. In such a case, the spectral density functions can be expressed as a function of two terms, representing the two types of motions. This is the basis of the well known approaches developed by Woessner and by Lipari and Szabo, which correlate spectral densities and molecular motions.

4.2 Basic Theory

The most popular method for linking the internal motion in macromolecules and the NMR parameters is the approach of Lipari and Szabo. The formalism does not require a model for motion. It is therefore referred to as *model free* approach. The method allows NMR properties at each site to be expressed as independent motional parameters arising from the overall tumbling of the interaction vector and the local motion of the vector relative to the molecular frame. The formalism assumes that the two motions are separable.

Each motion gives rise to Lorentzian spectral density functions. The interaction vector can be due to dipole-dipole couplings, chemical shift anisotropy or quadrupole couplings. The relaxation parameters from the spectral densities are calculated in terms of order parameters S^2, which measure the spatial restriction of the internal motions, and the effective correlation time τ_e, which is a measure of internal motions. Thus the spectral density function $J(\omega)$ can be written as:

$$J(\omega) = \frac{2}{5}\left\{\frac{S^2\tau_c}{1+(\omega\tau_c)^2} - \frac{(1-S^2)\tau_M}{1+(\omega\tau_M)^2}\right\} \qquad 6.1$$

$$with\ \tau_M^{-1} = \tau_c^{-1} + \tau_e^{-1} \qquad 6.2$$

The order parameter S^2 indicates the contribution from the isotropic overall molecular motion. A physical picture of the motion can be obtained from the three parameters S^2, τ_M and τ_c. In deriving this expression, the two motions representing τ_c and τ_e are assumed to be independent. However, no assumptions are made about their time-scales. The formalism can be extended to include anisotropic motions as has been done in the original approach developed by Woessner.

NMR cannot detect a local motion which is slower than the global motion (i.e. $\tau_e > \tau_c$), as these parameters will be dictated by the later. Also, if τ_e and τ_c are comparable, then it is not possible to define the overall correlation time. If the local motions are very fast ($\tau_M << \tau_c$), then the second term in Equation 6.2 becomes negligible as compared to the first. Thus, the relaxation behavior can be interpreted in terms of model free approach for characterizing local motions which are ten to thousand times faster than the overall correlation time. For a value of τ_c of 10 nsec (typical values for correlation times of molecules with M_r around 20 kDa), τ_e of the order of 1 nsec to 10 psec can be handled by the formalism discussed above.

The Lipari-Szabo approach has been expanded to treat cases where one has two types of local motions, both of which are faster than the overall global motions: a slow motion and a fast motion characterized by correlation times τ_s and τ_f, respectively. Also, exchange broadening may be present in cases where the exchange rate between conformational states is slow on the NMR time-scale. In such cases, the relaxation properties are parameterized in terms of six variables S_s, S_f^2, R_{ex}, τ_c, τ_s and τ_f.

4.3 Protein Dynamics using NMR

The approach discussed above has become a standard tool for studying protein dynamics. The availability of labeled proteins and techniques of sequence specific resonance assignments have made it possible to measure relaxation behavior at individual sites in a protein molecule. A greater understanding of the field has been aided by the ability of site-directed mutagenesis, which allows study of changes in dynamic behavior following replacement of key amino acid residues.

The relaxation of the amide ^{15}N is dominated by the dipolar relaxation with its attached ^{1}H. At 500 MHz, about 80% contribution to the relaxation of the amide ^{15}N comes from dipolar interaction. The remaining 20% arises from CSA. The equation for amide ^{15}N relaxation contains spectral density functions at J(0), $J(\omega_N)$, $J(\omega_H)$, $J(\omega_H + \omega_N)$ and $J(\omega_H - \omega_N)$.

Specially designed pulse sequences allow measurement of NMR parameters R_1, R_2 (and $R_{1\rho}$) for ^{1}H and ^{15}N, and $^{1}H \rightarrow ^{15}N$ NOE at each amide site in a protein. This information is used to derive local motions. Spectral density mapping can be achieved using measurements of spectral parameters at several magnetic fields. Hence, several laboratories have invested in NMR spectrometers at more than one magnetic field.

Use can also be made of the dipolar mediated relaxation of $^{13}C^{\alpha}$ by the attached ^{1}H. However, in fully labeled proteins, the relaxation behavior is complicated by the additional interaction between directly bonded carbons. The $^{13}C^{\alpha}$-^{1}H spin systems therefore cannot be treated as ideal AX system in such cases. This problem is avoided if experiments are done on unlabeled proteins. However, the difficulty due to lower sensitivity, when using natural abundance, restricts studies to highly soluble proteins. Selective labeling has helped in overcoming this problem.

Partially deuterated proteins also provide information on dynamic aspects. An advantage of using ^{2}H is that in this case, the relaxation is dominated by quadrupolar interaction, which is relatively better understood. However, the smaller dispersion of the signals and efficient relaxation limits the resolution. A strategy which has proved useful is to use uniformly ^{13}C and fractionally ^{2}H labeled molecules. The relaxation properties of ^{2}H can be monitored by looking at the well-resolved ^{1}H-^{13}C spectra.

One example is the relaxation of methyl group. In case of an isolated CH_3 group, the ^{1}H-^{13}C *single-quantum* transition is exponential so long as the methyl group rotates rapidly around its three fold axis. Partial deuteration can be used to produce $^{13}CH_2D$ isotopomer and ^{2}H relaxation can be used to probe motions in the psec-nsec time-scales. A magnetization transfer scheme $^{1}H \rightarrow ^{13}C \rightarrow ^{2}H(T) \rightarrow ^{13}C(t_1) \rightarrow ^{1}H(t_2)$ has been employed where T is the time period during which ^{2}H relaxation occurs. ^{2}H relaxation has been used to study protein dynamics in a number of processes, ranging from molecular recognition, protein folding, stability and effect of mutations.

4.4 Experimental Results

The potential of using NMR as an experimental tool to study local dynamics of macromolecules has been realised in recent years. In general, these studies have established dynamic behaviour of proteins. That the loop regions are highly mobile and less structured as compared to secondary structural elements, is something that is expected. NMR provides details about such motions. Binding sites are generally more flexible than the rest of the domain structures. Segmental motions of protein sub-units, which get locked into specific structures on binding specific substrates and ligands, are an important component of binding specificity. Inter-domain movements are important from functional view-point and these can be studied. The site-specific dynamics, which is possible using NMR, has provided new insights into protein dynamics. These aspects are discussed in greater detail in Section 6.

5. PROTEIN FOLDING AND UNSTRUCTURED PROTEINS

An outstanding problem in structural biology is to understand how the primary sequence codes for the native 3D structure of the protein, and how this state is formed. If proteins were to reach their native conformation in a random fashion (or for that matter, through some limited transient states), then it will take years to reach an ordered structure. This was pointed out by Levinthal in 1968. On the other hand, most of the proteins fold to the native state, starting from unfolded state, in a matter of a few msec. The study of the route(s) by which a protein reaches its native structure is of great interest and is an active area of research. Understanding of protein folding is important not only from fundamental view-point, but also because defects in folding are responsible for a number of diseases.

There is another reason for studying unfolded proteins. Proteomics based studies indicate that an appreciable number of genomes code for proteins which have unstructured domains. Several of the unfolded and partly folded proteins play important roles in cellular processes and signaling events. In particular, proteins involved in transcriptional activation and cell cycle regulation contain proteins which are unstructured in solutions, but acquire structure on binding with their physiological targets.

5.1 Protein Folding Pathways

Protein folding involves multiple pathways, starting with the formation of secondary structural elements and concluding with long-range alignments of these elements. The partially folded protein (sometimes referred as *molten globule*) may collapse into a well folded structure due to hydrophobic interactions. In certain cases, the folding is assisted by special proteins called *chaperones*.

The folding process is often considered as going down a free energy funnel (Figure 6.5). At the top of the funnel, there is a high degree of conformational freedom and high entropy. The bottom of the funnel has the native structure, corresponding to the minimum Gibbs energy. The folding process is accompanied by decreasing number of available conformations, lower energies and entropy as one goes down the funnel. The intermediate states in the folding energy landscape are characterised by valleys of local minima. During folding, a protein can get trapped into these intermediate states. The process of transition between the intermediates and the native state is useful in getting insights into the problem of protein folding.

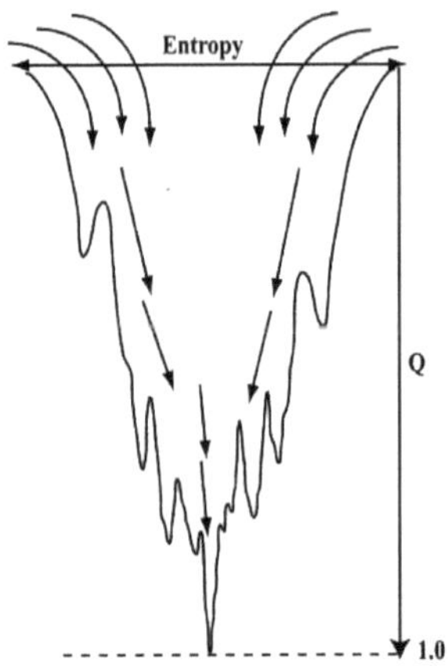

Figure 6.5: Folding funnel. A schematic view of protein folding pathways.

5.2 NMR Methodologies for Studying Unfolded Protein Structures

NMR is well suited to address the process of protein-folding, because it can probe structures in highly dynamic states. Site-specific probes for investigating formation and dissolution of structures are available. It has, therefore, been extensively used for studies in protein folding and for partly structured proteins. Study of proteins having fully or partially structured domains, are not amenable to X-ray crystallography. Another useful technique is CD, though this provides only a global picture of protein folding.

Studies of unfolded proteins are hampered by the relatively small dispersion in the ^1H spectrum. However, the availability of uniformly and specifically labeled proteins has provided means for indirect detection using hetero-nuclear NMR experiments. A number of new acquisition and assignment techniques have been introduced, primarily with the aim to study unstructured proteins.

One of the more frequently studied property, which is amenable to NMR, is the propensity of secondary structural elements in unfolded state. Such elements can be probed using chemical shifts or CD. With carefully calibrated shifts in the random coil state, secondary chemical shifts provide a sensitive indicator of α-helical propensities. The results can be supplemented by J-couplings, exchange rates and RDC. The β-sheet structures are relatively more difficult to establish, particularly in the upper part of the funnel, as these involve long-range constraints and interactions. Structural information using NOE is normally limited to short and medium-range order. However, spin labels and RDC provide more information on long-range structural information.

Characterizing intermediate states poses a major challenge to traditional biophysical and biochemical methods. In particular, NMR is a relatively slow

technique and certain experiments may take a few minutes to hours. Also, slow exchange between conformational states may broaden NMR signals. Several proteins fold within a few msec. However, there are several slow steps, such as proline isomerization and disulfide linkages, which can be conveniently probed on NMR time-scales. Certain problems can be studied through the use of quenched-flow $^1H/^2H$ exchange. In particular, innovative technique of a combined use of rapid mixing and NMR (stop-flow NMR) allows relative measurements and monitor different aspects of protein folding. One makes use of the equilibrium between the native and the intermediate states, through a two-process model describing the hydrogen exchange mechanism. According to this model, exchange occurs through two competing pathways:
(i) global unfolding followed by exchange from the unstructured protein, and
(ii) the local fluctuation.

Finally, dynamic information on unstructured states can be obtained using relaxation data. The model free approach using single overall correlation time and temporal de-convolution of internal motions and molecular tumbling is obviously not valid for un-structured proteins. This results in difficulties in interpretation of the data using model free approach. A distribution of correlation times is introduced for such studies.

5.3 Molten Globules

The *molten globule* is a term introduced to describe a stable, partially folded protein state, which is found in mildly denaturing conditions such as low pH or high temperature. Molten globule has native like secondary structures and fluctuating tertiary structure. Such a state may be formed prior to its collapse into the native conformation. It may however be mentioned that the definition of *molten globule* is loosely used to encompass highly ordered conformations to a state where only part of the structure is formed.

In such states, the NMR lines are broadened due to intermediate exchange between various conformational states. A number of NMR techniques have been developed to detect the properties of so called molten globules. There are several studies specifically dealing with such a state. For example, the dispersion in chemical shifts in apo-myoglobin show characteristics of a highly ordered structure similar to that observed in native holo-myoglobin. Several of the helices are well formed. It can be best described as a native structure with some fluctuations in the conformation (highly ordered molten globule). On the other hand, the A-state of α-lactalbumin is less structured and is often used as a model for molten globule.

5.4 Unstructured Functional Proteins

The number of studies on unfolded and partially folded proteins have been increasing rapidly. There are several examples where NMR methodologies have provided major insights in this important area of research. Such studies are difficult to conduct using X-ray crystallography.

5.5 *Protein Structures under High Pressure*

Pressure is an important environmental variable and is important for life under deep sea or high altitudes. The ensemble of conformational map sampled by a protein can be perturbed by pressure. This occurs because different low energy conformers have different effective volumes. Higher pressures shift the conformational equilibrium towards conformers with lower volume and thus perturb the ensemble of conformational states. This perturbation of conformational equilibrium can result in the population of states which have higher energies under normal conditions of pH, temperature and solvent conditions. These states may represent folding intermediates and unfolded states of a protein. Much of the information on this subject has come from the work of Akasaka and coworkers.

This technique has been used to study the conformational behavior of a number of proteins. For example, studies on apo-myoglobin have shown that the native protein is stable at low pressures. At high pressures (around 3000 bar), the protein gets unfolded. In general, structures that show the largest changes are those close to the buried water. This highlights the importance of water molecules as nucleation sites for volume fluctuations of proteins in their native conformation.

6. NMR STUDIES OF PROTEIN STRUCTURE, DYNAMICS AND FOLDING: SOME SPECIFIC EXAMPLES

Application of NMR to protein folding and intrinsically unstructured functional proteins is illustrated by choosing some specific examples of proteins of various sizes.

Early NMR literature in this area was focussed on the structure of cyclic peptides, which have important biological functions, such as oxytocin, vasopressin, gramicidin S and valinomycin. These peptides act as hormones, antibiotics, anti-toxins and ion carriers. The constraint imposed by the ring-cyclisation considerably reduces the conformational flexibility of such peptides, when compared with linear peptides of the same size. Such structures are not discussed here, as these are covered in several previous books. In 90s, the focus of NMR has been on medium size proteins. In recent years, much larger proteins have been investigated.

The focus here is on relatively larger peptides and globular proteins. It is impossible to give a complete review of this emerging area. We shall concentrate on some specific examples. The choice of the examples is based on systems which demonstrate a particular feature of the power of NMR spectroscopy. Comprehensive reviews in this area appear frequently in literature. The sequence of chosen examples, range from small to large proteins. The extent of available information using NMR decreases, in the case of larger proteins. These proteins range from a M_r of 6.4 kDa to 82 kDa, and contain between 58 and 723 amino acid residues (Figure 6.6).

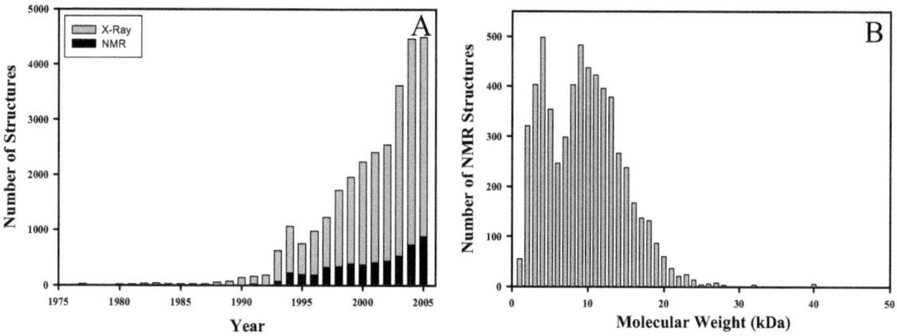

Figure 6.6: (A) Yearwise number of NMR and X-ray structures deposited in PDB. (B) Number of NMR structures deposited in PDB vs molecular weight.

6.1 Basic Pancreatic Trypsin Inhibitor (BPTI)

BPTI is a small peptide with 58 amino acid residues and has three disulphide bonds. It has served as a molecule of choice in early developments of 2D homo-nuclear NMR spectroscopy for sequence specific resonance assignments, structure and dynamics in proteins. In spite of its small size, it has a relatively rigid conformation due to the presence of disulfide bonds. The interior part of the molecule closely fits the crystal structure data, though disorder is observed near the periphery.

There are two important sets of information on this peptide derived from NMR, which are not possible from X-ray crystallography. Location of water molecules in the hydration sphere of the peptide has been established through the observation of NOE between water protons and hydrogen atoms of the polypeptide chain. The surface hydration is characterised by very short residence time of water on protein surfaces (typically 20 to 300 psec at 10 °C). The crystal structure of BPTI shows four water molecules in the interior. NMR results indicate a rapid exchange between these molecules and the bulk water, with a life time shorter than 1 msec. A rapid exchange of interior water of hydration seems to be a general property of proteins.

The second observation is the 180° ring flips of Tyr and Phe. These are again not reflected in the B-factors of the X-ray structures. The free energy of activation for such conformational transitions is in the range of 60-100 kJ/mole.

A number of mutants of BPTI have been examined. The mutants show dynamic behaviour which is significantly altered as compared to the wild-type protein. For example, a single point mutation A16V in the solvent exposed loop destabilizes the native conformation such that its melting point is lowered by 20° C. This corresponds to a lower free energy of 17 kJ/mole, although the native structure does not change with this mutation.

6.2 Human Ubiquitin

Ubiquitin is another small peptide (76 amino acid residue, M_r 8.5 kDa). It is an important protein that is tagged to another protein targeted for degradation by the cellular scavenging machinery. The NMR structure of this protein bears a close resemblance to the X-ray structure. However, from the view point of NMR spectroscopy, this molecule has been used as an important benchmark to test NMR protocols discussed in Chapter 3.

6.3 Ribonuclease

Ribonucleases (RNase) are the main enzymes for multi-facet activities in RNA metabolism, cleavage, transformation of precursors into matured RNA, production of regulatory RNAs, and degradation of certain types of RNA. A number of RNases have been studied by NMR. RNase A from bovine pancreas is a pyrimidine specific endoribonuclease. It consists of 124 amino acid residues. The residues His12 and His119 are critical for catalysis of RNA cleavage. The enzyme can be cleaved between residues 20 and 21 giving S peptide and S protein. It has served as the most important protein to examine structure-function relationship and to gain insights into intermediates in the protein folding process. The two histidines act as acid and base to attract a proton from the 2′ hydroxyl group of a ribose ring, and to donate it to the 5′ oxygen of the leaving phosphate group. NMR based titration studies have helped to assign pK values to these critical His residues.

The NMR solution conformation bears a remarkable similarity to the X-ray structure with some exceptions. The backbone torsion angles are well defined in the NMR derived structure, including those for the catalytically active His12 and His119. There are some differences in the side chain conformations particularly for His119, which is involved in the catalysis. There are also differences in the hydrogen bonds identified by the two techniques. The conformational dynamics has been observed at several points in the active site region both in the backbone and the side-chains.

6.4 Lysozyme

Hen lysozyme consists of a single polypeptide chain of 129 amino acid residues (M_r 14.3 kDa). It is one of the best characterised proteins and a good example of various motifs found in protein structures (Figure 6.7 F). It has four disulfide bridges and is made of two structural domains. The so called α-domain is located at the C-terminus and consists of four α-helices and a C-terminal 3_{10}-helix. The other domain (called the β-domain) consists of triple stranded β-sheet, a 3_{10}-helix and a long loop. A short antiparallel β-sheet links these two domains.

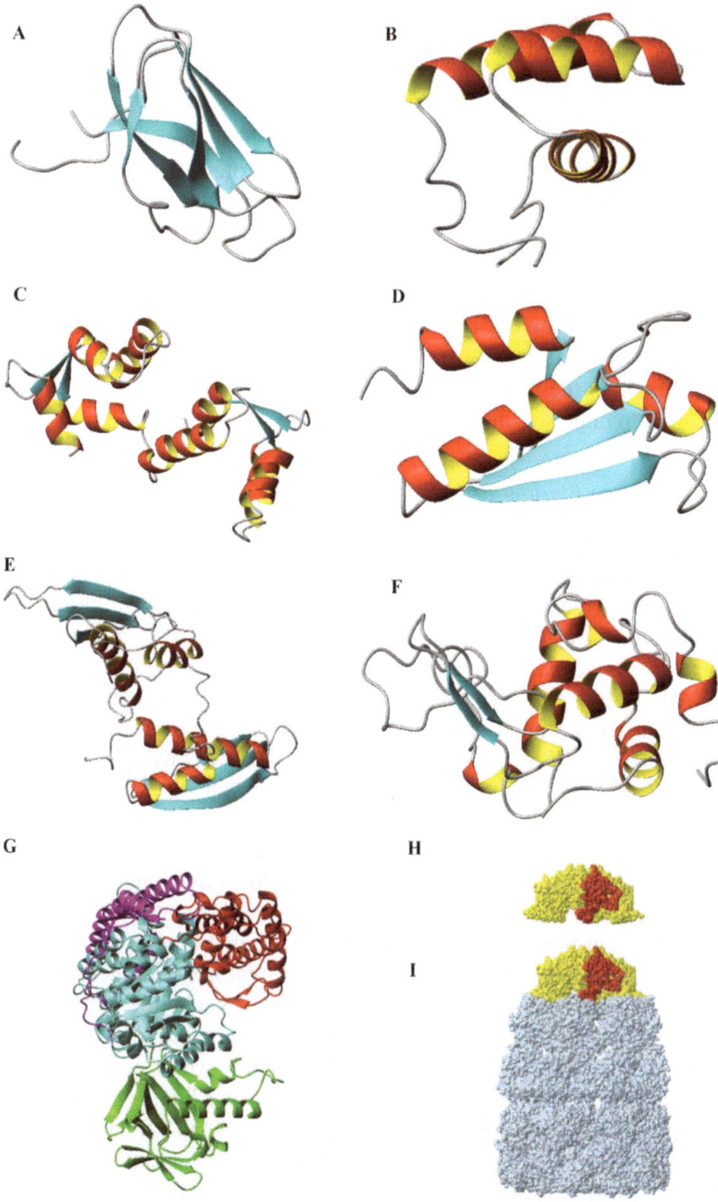

Figure 6.7: Some protein NMR structures from PDB. (A) α- Amylase (2AIT); (B) Homeodomain (2HOA);(C) RNAse (1JOX) (D) Glutaredoxin (1EGR); (E) dsRBD of PKR (1QU6); (F) Lysozyme (1E8L); (G) Malate Synthase G (1Y8B); (H) GroES (Yellow and red) and (I) GroES-GroEL(light blue) complex. The panels A-G were created using MOLMOL. Panels H and I adopted from J. Fiaux, E.B. Bertelesen, A.L. Horwich and K. Wüthrich, Nature 418, 207–211(2002). (Reprinted with the permission from Nature).

The structure, function dynamics and folding of lysozyme have been extensively studied by NMR. It has provided the means for exhaustive studies of rotamer populations in side chain conformations. The work on this protein has been instrumental in the development of several techniques for the analysis of unfolded proteins. Amide exchange and quenched-flow studies showed the presence of folding domains corresponding to those in the native protein.

Both the oxidized and reduced forms of lysozyme have been studied in highly unfolded states. The folding process involves parallel events with distinct kinetic profiles. Based on extensive NMR and other biophysical results, Dobson's group has proposed a schematic averaged energy surface for the folding of lysozyme. The study of intermediates provides considerable insight into the nature of incompletely folded species. In particular, the formation of α-domain shows that the emergence of different characters in the folded conformation, are decoupled. Variants of lysozyme have been found to undergo fibrillogenesis.

6.5 EF-hand Calcium Binding Proteins (CaBP)

Calcium binding proteins (CaBPs) play an important role in biological processes. These proteins are involved in storage and transport of calcium, signal transduction and pathogenicity of certain organisms. The most common Ca^{2+}-binding motif of these proteins consists of a contiguous 12-residue loop flanked by 2 helices, forming the so-called *EF-hand motif*. Hence, this sub-family of proteins are also called as EF-hand CaBPs. They are composed of around 140 amino acid residues ($M_r \sim 15$ kDa) and are mostly arranged in two domains, the N- and the C-terminal domains (Figure 6.7 C). The two domains are connected by a central linker region of about 6-10 amino acid residues. Each domain has two Ca^{+2} binding sites, called the EF hands.

Within the Ca^{2+}-binding loops, the residues that are involved in co-ordination with Ca^{2+} are at positions 1, 3, 5, 7, 9 and 12, and form a pentagonal bi-pyramidal geometry. The residues at position 1 and 12 are highly conserved Asp and Glu, respectively. Those at positions +3 and +5 can be either Asp or Asn. These residues coordinate directly to Ca^{2+}, through their side-chain carboxyl oxygens. The residue at position +7 co-ordinates with the Ca^{2+} through the backbone carbonyl oxygen. Residue at position +8 is invariably a hydrophobic residue that makes two hydrogen bonds to the equivalent residue of the paired EF-hand, a part of the short β-sheet connecting the two EF-loops. Residue at position 9 coordinates to Ca^{2+} indirectly through an intervening water molecule. Thus, in most of the EF-hand proteins Ca^{2+} is linked to seven oxygen atoms in pentagonal bipyramidal geometry. The flanking helices (E and F) of the Ca^{2+} binding loop are amphiphilic and possess 4 conserved hydrophobic residues at positions –1, -4, -5 and -8 in the first helix (E-helix), and positions +13, +16, +17 and +20 in the second helix (F-helix). These helices are mostly perpendicular to each other in Ca^{2+} bound form and are parallel to each other in Ca^{2+} free form.

A variant of this canonical EF-hand motif (the second most common one, called a pseudo EF-hand) is observed in the more recently determined structures. It belongs to the S100 protein family and consists of 14 residues.

NMR studies on CaBP reveal two helix-loop-helix motifs in each of the two domains (Figure 6.7 C), which provide the scaffolding for the EF hands. Thus, the proteins have eight α-helical segments in all. A short antiparallel β-sheet consisting of six residues helps to provide a stable conformation for each domain. On the other hand, there is a large degree of conformational mobility of the two domains relative to each other. This arises from the flexible central linker region connecting the two domains. In the crystal structure of calmodulin (CaM), this linker forms an α-helix. However, in solutions, NMR studies show large flexibility in the linker region. Experiments based on S^2, T_1, T_2 and exchange studies show that this region is highly disordered. Therefore the orientation of the two domains is ill defined in the NMR structures. Among the three proteins, EhCaBP shows the highest flexibility due to replacement of two amino acid residues in the central region (63 and 67) by Gly.

6.6 Maleate Synthase G (MSG)

In one of the most challenging problems, the group of Kay has shown the power of isotope labeling coupled with 4D TROSY-based triple resonance experiments on this 723 amino acid residue (M_r 82 kDa) enzyme (Figure 6.7 G). MSG catalyzes the condensation of glycoxylate with acetyl CoA to produce malate. MSG converts the two carbon unit acetate into malate (four carbons) during bioenergetic pathways. It is a monomeric protein and has four domains. These domains show significant changes in their relative orientation on ligand binding.

A combined X-ray and NMR analysis of this protein helped to establish domain orientations both in the apo- and the bound forms. Even though the NMR structure based on NOE and dihedral angle constraints is not as well defined as the X-ray structure, a number of details of biological importance have been obtained. Several domains are well defined and could be characterised by NMR. The location of the active site region is consistent with X-ray data. The kinetic and thermodynamic parameters for the ligand binding process have been determined. ^{15}N relaxation rates have provided information on how the protein dynamics changes upon ligand binding.

6.7 GroEL-GroES Complex

Molecular chaperones are involved in folding, transport and degradation of several proteins. In the *E. coli* chaperone system, a dome shaped homoheptamer GroES (M_r 72 kDa; Figure 6.7 H) interacts with a double ring shaped homotetradecamer GroEL (M_r 800 kDa; Figure 6.7 I) or with a single-ring GroEL variant SRI (M_r 400 kDa). The

GroES-GroEL complex (M_r 900 kDa) is among the highest molecular weight proteins studied using solution NMR techniques. The high symmetry of these systems considerably reduces the spectral complexity and analysis. Comparison of the [^{15}N-^{1}H] CRIPT-TROSY spectrum of free GroES with that bound to either SRI or GroEL shows that most of the amino acid residues in GroES have similar chemical shifts in the free form or when it is complexed with chaperone. The exceptions are residues 17-32, which show large changes in shifts. In free GroES the chemical shifts of these amino acid residues correspond to those of random coil values, reflecting a disordered conformation. The mobile loop becomes ordered on binding to GroEL. However, the rest of the amino acid residues do not show large conformational changes. These amino acid residues are therefore involved in making contact with GroEL.

6.8 Membrane Proteins Studied in Solutions

The techniques for studying large proteins have also opened the way to study membrane proteins in solution. The intact membrane bound proteins have solid-like properties and can be studied by high resolution solid-state NMR (Chapter 10). However, following the development of TROSY, it has become possible to solubilize and study such proteins in detergents. One such example is that of PaP, an enzyme that catalyzes the transfer of sn-1 palmitate chain from phospholipid to liposaccharides in certain organisms. Such enzymes can be solubilized in detergents to maintain their activity and studied by solution NMR techniques. Kay and co-workers have shown that the protein exists in two states, a dynamic state R that allows substrate entry and a rigid state T. The structural information, kinetic parameters and the equilibrium constants for the two states have been measured.

7. PARAMAGNETIC PROTEINS

As discussed in Chapter 1, the nuclear relaxation time T_2^* in the presence of paramagnetic centres in a molecule is dominated by paramagnetic relaxation due to coupling with the unpaired electrons. This results in considerable line broadening. In certain cases, detectable signals may not be observed at all. Even for proteins which yield signals, lines in a region close to the paramagnetic ions are too broad to be detected. Several cross peaks in multi-dimensional NMR are not observed. This causes difficulties in establishing connectivites between different spins.

The chemical shifts in paramagnetic proteins are dominated by contact and pseudo-contact shifts, which are considerably larger than the diamagnetic shifts. The ^{1}H signals often occur outside the normal 0-15 ppm range. Therefore, the characteristic chemical shifts used for spectral assignments of diamagnetic proteins, are of little value. The scalar coupling information is almost wiped out by broad resonances. Thus, sequence specific resonance assignment is a major problem. The contribution of dipole-dipole interaction to relaxation rates in paramagnetic systems is relatively small, which are mainly determined by paramagnetic relaxation. Strong RF fields are required to saturate signals due to short T_1 and T_2. Due to this limitation and large spectral widths, there is a need for associated changes in hardware, and suitably designed pulse sequences. Use of high fields is not helpful as signals are broadened further due to the effect of the magnetic field on T_2^* in such cases. Another severe limitation is the size of the protein. Normally, there is an intrinsic limit on the size and the magnetic field, which is appropriate for NMR studies. On the other hand, one advantage is that the signals are well dispersed.

Thus, there are several difficulties in the study of paramagnetic proteins. The approaches are distinctly different from those used in diamagnetic molecules. For these reasons, even though this area of research was the first one to attract the attention of NMR spectroscopists, the progress has been relatively slower. There is a renewed interest on NMR studies of metalloproteins because of their biological importance. Development of new techniques for isotope labeling, new mutants and multi-dimensional NMR, have been helpful in this endeavour. Through specially designed pulse sequences, it is possible to detect some of the broad signals.

7.1 Metal Ions of Interest

A large number of metalloproteins carry paramagnetic ions. The most common naturally occurring metal ions are Fe in its various oxidation states, Cu^{2+}, Ni^{2+}, Zn^{2+} and Ca^{2+}. While former three are paramagnetic, Zn^{2+} and Ca^{2+} diamagnetic. These are both diamagnetic. However, molecules containing $Zn^{2+}$and Ca^{2+}, have been conveniently studied both in their diamagnetic form and by substituting these ions by paramagnetic Co^{2+} and lanthanides, respectively. By using both the paramagnetic and diamagnetic analogues, additional information can be obtained. One of these is the ability to obtain additional structural information using paramagnetic effects on the relaxation times of various spins.

An interesting feature from the view point of chemical biology of metalloproteins is that several metal ions exist in more than one oxidation state. In each oxidation state there may be more than one spin state. These differences arise from the electronic configuration of such states. An interesting example is iron, which has oxidation states ranging from +2 to +4. The most common states are Fe^{2+} and Fe^{3+}. Each of these states can have more than one spin states, ranging from S= 0 to S= 5/2. Figure 6.8, shows the electronic configuration of *high* and *low* (H and L) spin states of Fe^{2+} and Fe^{3+}. This complexity provides a challenge to characterize each of these states using physical techniques. Each metal ion and its spin-state have unique NMR properties. Not all states give detectable NMR signals.

The metal ions which give satisfactory NMR spectra are tetrahedral Ni^{2+} (HS), penta- or hexa-coordinated Co^{2+}(HS), Fe^{3+}(LS) and Fe^{2+}(HS). Note that Fe^{2+}(LS) has S=0 and is therefore diamagnetic. All these states give relatively narrow signals.

On the other hand Cu^{2+} and Mn^{2+} give extremely broad signals. Hexa-coordinated Ni^{2+} and tetrahedral Co^{2+} are borderline cases. Most lanthanides give sharp signals. An exception is Gd, which has very short T_2^*. For this reason Gd based molecules have found applications as contrast reagents in MRI studies (Chapter 11).

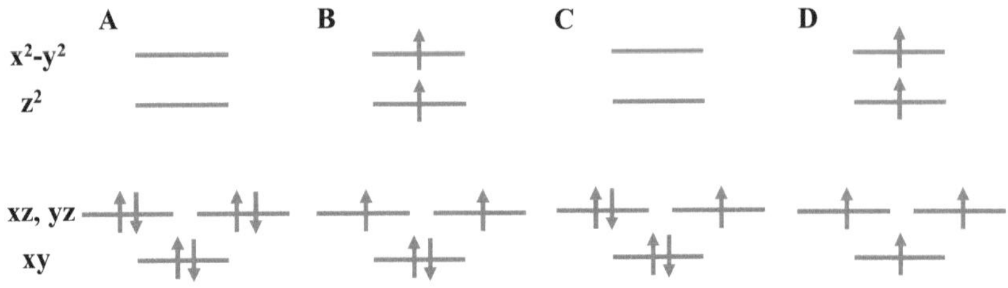

Figure 6.8: Electronic configurations of low (L) and high (H) spin states of Fe^{2+} and Fe^{3+}: (A) Fe^{2+} (L; S=0); (B) Fe^{2+} (H; S=2); (C) Fe^{3+}(L; S=1/2) and (D) Fe^{3+}(H; S=5/2).

The electronic properties of these metal ions depend not only on their electronic states but also on the nature of binding. Oxygen carrying protein haemoglobin is paramagnetic in the deoxy- form and diamagnetic in the oxygenated form, a property which forms the basis of fMRI.

7.2 Heme Iron Proteins

Hemoproteins use ferroheme (protoheme) as a prosthetic group. This is covalently linked to a polypeptide chain. Heme proteins are an important class of proteins involved in a number of biological functions and exist both in *prokaryotic* and *eukaryotic* cells. The three important functions of heme-proteins are transport of electrons (*cytochromes*), oxygen (*globins*) and catalysis of redox reactions (*peroxidases, cytochrome c oxidase, nitrite reductases*).

The heme group consists of a macrocyclic porphyrin ring with a general formula shown in Figure 6.9. The most common heme moiety present in natural proteins is proto-heme (called heme b), where R is a vinyl group. The four central nitrogen atoms in heme are at a distance of about 2.01 Å from the center. The group shows very little radial expansion on binding to a metal ion. In most cases, the metal ion is iron. For bulkier metal ions, the coordination is affected by the displacement of the ion, normal to the plane containing the four nitrogen atoms. Monodentate ligands can bind to the central metal ion in either one or both the axial positions, forming penta- and hexa-coordinated complexes.

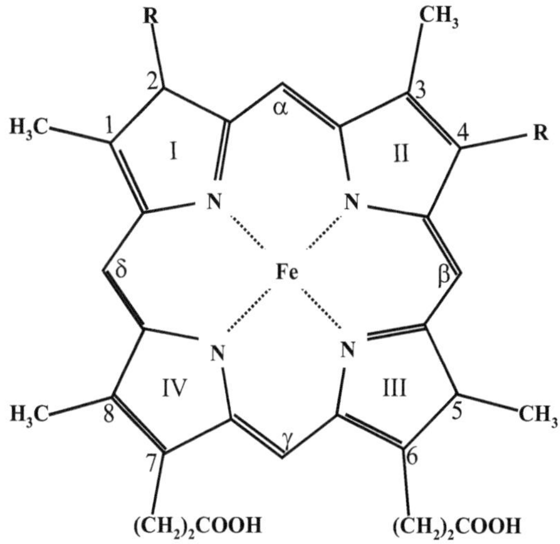

Figure 6.9: General formula for the heme group.

The wide difference in the chemistry and biology of iron containing heme proteins arises because of the manner in which the iron atom is linked axially to the heme group, its electronic configuration and the binding of the substrates. For example, in the case of myoglobin (Mb) and haemoglobins (Hb), two of the important proteins involved in oxygen storage and transport, iron is linked to the protein chain through a His residue. In some cases the binding is through a Cys. Hence, the minimum coordination number of iron is five. This is the situation in deoxy-Hb, where Fe^{2+} is in high spin state (S= 2). The sixth coordination number of iron can be satisfied by binding to oxygen (e.g. in oxy-Hb) or to other ligands (such as H_2O, OH, CO, CN). The geometry of heme group differs according to the state of iron. In high-spin Fe^{2+} (as in the case of deoxy-Hb), the central hole of the porphyrin ring is unable to hold the metal ion, which is displaced by about 0.75 Å out of the heme plane. The sixth ligand which approaches the metal ion from the other side, finds itself at a large distance from the ion. On the other hand, in the LS spin state, the metal ion is in the plane of the ring and is ideally suited to bind to the sixth ligand.

The π-electrons in the porphyrin rings give rise to large chemical shifts due to magnetic anisotropy arising from ring current effect. Therefore, even in diamagnetic compounds, the chemical shift dispersion of ^1H near these aromatic rings is large and the spectral features are often present beyond the usual 0-10 ppm range for ^1H. Secondly, a general property of the heme proteins is that in paramagnetic system, unpaired electrons are highly delocalized on the ligand both through σ and π orbitals. A larger π contribution occurs in the hexa-coordinated compounds due to co-planarity of the metal ion with the heme moiety. This results in down-field shifts of the H^{α} and the methyl resonances in such proteins.

7.2.1 High-Spin Iron (III)

High-spin Fe^{3+}, constitutes the resting state of several heme proteins, which exist in ^6A ground state. The magnetic anisotropy is small and the observed hyperfine shifts arise primarily from the contact term. However, the zero field Kramer's splitting gives rise to pseudocontact contribution, resulting in changes in chemical shifts and relaxation rates.

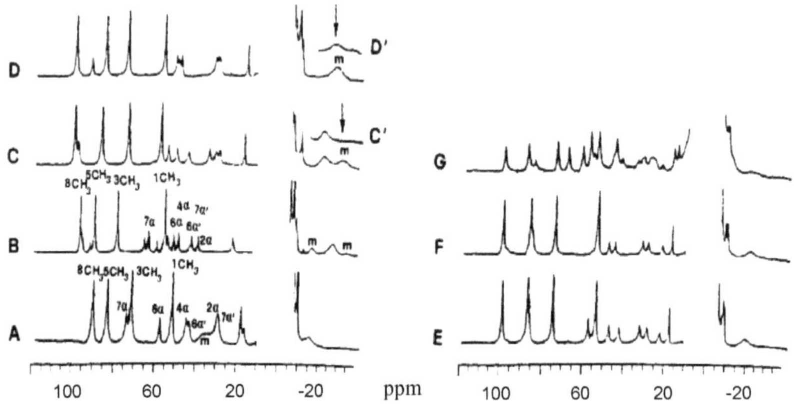

Figure 6.10: 1H NMR spectra of (A) sperm whale metMb; (B) Aplysia metMb; (C) HisiE71Phe metMb; (D) His[E7]Val metMb; (E) His[E71Gly metMb; (F) HisCE71-Gln; (G) elephant metMb. The insets C' and D' show the NMR spectra obtained on samples reconstituted with meso-$^2H_4$ hemin, which allows the meso signal assignment. K. Rajarathnam, G.N. La Mar, M.L. Chiu, M.L., S.G. Sligar, J.P. Singh, and K.M. Smith. J. Am. Chem. Soc., 113, 7886-7892 (1991). (Reproduced with permission from American Chemical Society.)

Figure 6.10 shows the ^1H NMR spectra of metmyoglobin (M_r 16 kDa) from a number of sources, recorded at 298 K in ^2H_2O solution (100 mM potassium phosphate buffer; pH 6.2). While there is an overall similarity in the patterns, large differences in the spectral features are observed for the meso protons. These are related to the differences in the structures of these molecules. An analysis of the chemical shifts in the heme pocket region suggests that the structure of metcyanomyoglobin is similar to MbCO.

Peroxidases (M_r ~ 40 kDa) are a class of proteins which oxidize biological substrates in the presence of H_2O_2. These proteins often contain non-covalently linked heme. Due to easy availability, cytochrome c oxidase (CcP) and horse radish peroxidase (HRP) have been extensively studied by NMR. It was found that pH has a substantial influence on the NMR spectrum. For HRP, the changes occur in three stages, corresponding to pK_a values around 4, 6 and 11. The alkaline transitions correspond to four different iso-enzymes. The low pH spectra reveal existence of multiple states. A variety of aromatic molecules are known to interact with HRP. ^1H spectroscopy has also been used extensively for studying interaction of substrates with peroxidases.

Lactoperoxidase (LPO, M_r 78 kDa) which is found in mammalian milk, saliva and tears, catalyses the oxidation of thiocyanate to a reactive species. It serves as an antimicrobial defense system. In this case, the heme group is covalently bound to the protein. Some of the other proteins in this group, studied by NMR, include Cyt P-450, chloroperoxidase, Cyt c′ and sulfmyoglobin.

7.2.2 Low Spin Iron (III)

This class includes haemoglobin, ferricytochromes, cytochrome P-450 and a number of other important proteins. The spectral features are more favorable and amenable to 2D and 3D homo- and hetero-nuclear NMR investigations. The assignments are also easier to achieve. Therefore, extensive NMR work has been done on all these proteins. Determination of reliable magnetic susceptibility tensor in recent years has allowed interpretation of pseudo-contact shifts. The contact shifts which dominate in such cases have been estimated for different sites.

Cytochromes are involved in electron transfer reactions. X-ray crystal structures of a number of such proteins are known. A representative example is Cyt-c (M_r 12 kDa). It has heme c and uses His and Met as axial ligands. The proteins occur both in oxidized and reduced forms. The reduced forms are diamagnetic (Fe^{2+}; S=0), give rise to sharper NMR signals. The sequence specific resonance assignments can be achieved through usual techniques. One strategy used in the assignment of signals of the paramagnetic oxidized forms is to use saturation transfer from well resolved signals of the reduced form to the paramagnetic state.

The oxidized form of Cyt-b_5 has been extensively studied using NMR. Cyt-b_5 is found both as bound to membranes in liver as well as in solution form in erythrocytes. The trypsin-solubilized protein from bovine liver consists of 84 amino acid residues. It forms a 1:1 complex with Cyt-c, myoglobin and haemoglobin. The binding has been studied using ^{1}H NMR.

Studies on the tetra-heme Cyt-c_3 (M_r 13 kDa) have provided interesting information on the redox mechanism, which is shown to be a four consecutive one-electron process. The intra-molecular electron exchange rate is fast on the NMR time-scale but the inter-molecular rate is slow. The five sets of spectra corresponding to the five oxidation states have been observed.

7.2.3 High Spin Iron (II)

High spin Fe^{2+} (S=2) is present in the penta-coordinated porphyrins. The NMR spectra in these cases are fairly broad, but meaningful information has been obtained in several cases. For example, the chemical shifts in sperm whale deoxy-Mb and deoxy-Hb have been analyzed. The temperature and pH dependences of chemical shifts have also been analyzed for both Mb and Hb.

7.2.4 Diamagnetic Iron Heme Proteins

Several heme proteins, including oxy-Hb and oxy-Mb, are diamagnetic. In these cases, complete sequence specific resonance assignments has been achieved. The information thus obtained can be used for analysis of the spectra for paramagnetic analogue. In fact, Mb was the first protein for which the crystal structure was determined. It has served as a paradigm for structure-function relationship. The details of Mb conformation are known to depend on packing forces and different geometries are obtained in different crystal forms. Mb has a single heme group which alternates between HS (S=2; paramagnetic) and LS (S=0; diamagnetic) states. The solution structure of sperm whale carbonmonooxy Mb (151 amino acid residues; M_r 18 kDa) has been determined using NMR distance constraints and chemical shifts. The ring current shifts arising from the heme pocket provide good dispersion of the chemical shifts and several of the backbone resonances could be assigned using homonuclear techniques alone.

The protein has also provides an excellent opportunity to use chemical shift restraints in structure refinement. A close similarity has been observed between the crystal and the solution conformations. NMR work suggests that the distal His is slightly displaced compared to the crystal structure and that the side-chain of Leu89 is disordered, but is in contact with the heme ring.

Extensive NMR studies have also been carried out on deoxy and oxy haemoglobin from horse and human. Haemoglobin consists of four polypeptide chains (α_1, α_2, β_1 and β_2), each containing a heme group. The four chains are held together by salt bridges and hydrophobic interactions. The oxygen binding capability of haemoglobin exhibits cooperativity and is also controlled by pH (Bohr's effect), concentration of inorganic phosphates and carbon dioxide. NMR spectroscopy has helped in detecting tertiary and quaternary changes in the structure of haemoglobin during oxygen binding, mechanism of interaction with ligands, Bohr's effect, and cooperativity in oxygen binding.

7.3 Iron-Sulfur Proteins

A number of iron-sulfur proteins participate in electron transfer reactions, though some also have catalytic role. These proteins contain polymetallic centers such as Fe_2S_2, Fe_3S_4, Fe_4S_4, besides the mono iron center. The iron ions have tetrahedral coordination and bridge sulfide ions and Cys residues (Figure 6.11). The Fe^{2+} and Fe^{3+} ions are in the HS states, i.e. S = 2 and 5/2, respectively. The cluster patterns of a number of iron sulfur proteins have been characterised.

NMR studies of iron-sulfur proteins have been very useful in characterizing cluster patterns. The spectra of such proteins normally show distinct peaks for α-CH and β-CH_2 protons. For example, in reduced rubredoxin (M_r 6 kDa) four signals in the 11-17 ppm range and four signals in the 150-260 ppm range are observed. The ratio of the intensities of the two groups (1:2) shows that the more shifted group of signals are due β-CH_2 protons. A quasi D_{2d} symmetry results in the observation of only four, instead of eight signals expected.

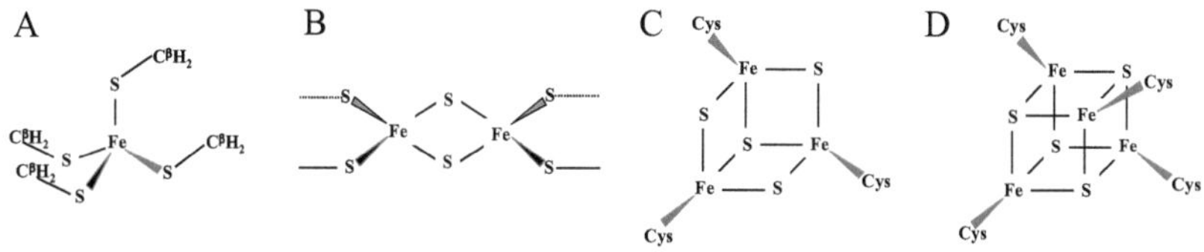

Figure 6.11: Iron-sulfur centers in proteins. Schematic representation of the structurally characterised iron-sulfur centers: (A) mono-iron center as present in rubredoxin, (B) Fe_2S_2 cluster, (C) Fe_3S_4 cluster, and (D) Fe_4S_4 cluster.

Proteins such as ferredoxins found in chloroplasts, contain Fe_2S_2 center. Each of the two irons is coordinated to two cysteines and is bridged by two sulfur atoms. In the oxidized state ferredoxin gives fairly broad NMR. However, sharp signals in the 15-45 ppm range are observed in the reduced form. NOESY experiments were carried out and partial assignments of the protons could be made.

7.4 Other Iron Proteins

In addition to the heme and iron-sulfur proteins discussed above, iron is also present in metalloenzymes bound to oxygen. These include single iron atom per unit and bi-nuclear oxo-bridged iron. Typical residues in such proteins are His, Tyr, Glu and Asp. Some of the important proteins in this family studied by NMR include catechol dioxygenase, isopenicillin N synthase and ribonucleotide reductase B2.

7.5 Other Metal Ions

Even though, iron proteins have been the focus of wide attention for NMR spectroscopists, proteins containing other metal ions have been studied. For example, Co^{2+}, though it is not present in natural proteins, has favorable relaxation properties. It has been extensively used as a paramagnetic substitute in diamagnetic Zn proteins, and for Cu proteins, which have unfavorable relaxation properties. In fact, the Co^{2+} substituted enzymes retain the activity of Zn-proteins. Typical examples of such studies include carbonic anhydrase, carboxypeptidase A, liver alcohol dehydrogenase, zinc finger peptide and azurin.

Ni^{2+} is another metal ion which has been used as a paramagnetic substitute to probe metal-enzymes. Finally, extensive use has been made of lanthanides using both the induced paramagnetic shifts and the effect on relaxation behavior.

8. FUNCTIONAL ASPECTS STUDIED BY NMR

In this section, we will discuss some applications of NMR, which illustrate the power of NMR spectroscopy in understanding protein function. Several of these applications are unique in the sense that other techniques may not be able to provide the information derived from NMR.

8.1 pH Titrations

It is known in the case of RNase A that His residues play a key role in its catalytic reaction. This is true for several proteins which participate in acid-base catalytic reactions. In addition, charged groups are involved in electrostatic interactions, hydrogen bonds and salt bridges. Changes in the ionization states of such residues may drastically affect the stability of the *native* structure. The intrinsic pK_a values may be drastically affected from their intrinsic values in the amino acid residues due to interactions in folded structures of proteins.

Histidines are particularly important in functional studies, as they have pK_a close to physiological pH. Several other amino acid residues having acid-base properties are also important. The pH titration of proteins by conventional means has therefore been used effectively to understand the stability and functional properties of proteins as a function of pH.

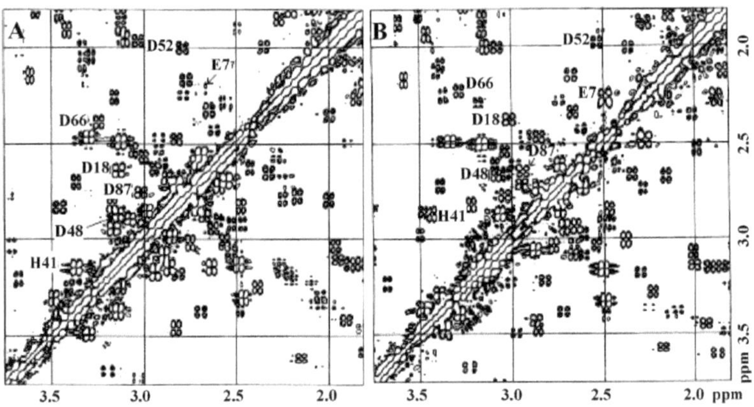

Figure 6.12: DQF COSY spectra of turkey egg white lysozyme at different pH (A)1.1 and (B) 5.9. Spectra were recorded on a 600 MHz spectrometer at a temperature of 35 °C. The changes in the positions of the $H^{\beta2}$–$H^{\beta3}$ cross peaks for His41, Asp18, Asp48, Asp52, Asp66, and Asp87 can be easily noticed. Also shown is the H^{β}–H^{γ} peak of Glu7. K. Bartik, C. Redfield, C.M. Dobson. Biophysical J. 66, 1180-1184 (1994). (Reprinted with permission from the Biophysical Society).

NMR is a powerful tool for such studies as it has the unique ability to provide pK_a values of individual residues. These can be determined by measuring chemical shift changes as a function of pH in a typical range of 3-9 (Figure 6.12). The results have been used to establish the contribution of electrostatic interactions in structural stability and the role of such amino acid residues during catalytic function and mechanistic description of substrate recognition.

8.2 Ligand Binding

Study of binding of small molecules (ligands; L) to enzymes (E) and other proteins is an important step in understanding enzyme structure and function, activation of proteins and a number of other biological processes. The changes in the NMR of protein on binding are usually limited to the binding region. On the other hand, almost all NMR parameters of the ligand change when it binds to a protein. The binding process can be studied both by looking at the spectrum of ligand and the protein (Figure 6.13). However, not all parameters provide useful information. For example, changes in coupling-constants may be too small to be of much use. The magnitudes of changes in the chemical shifts depend on the structure of the binding site.

The reaction between an enzyme and ligand can be represented by:

$$E + L \rightleftharpoons [EL] \qquad 6.3$$

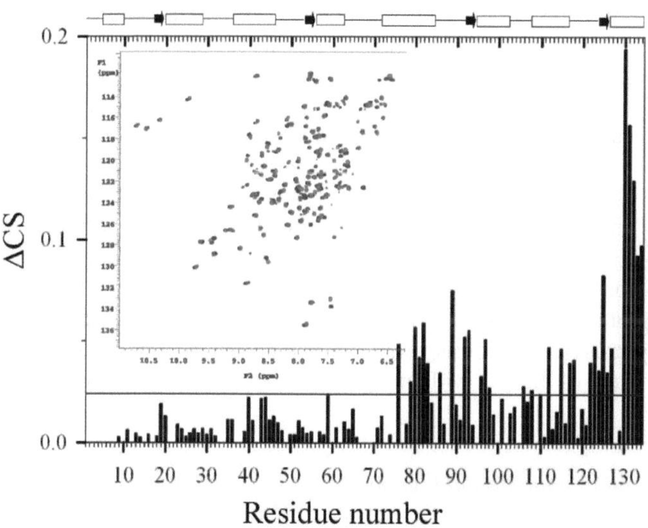

Figure 6.13: The chemical shift changes $\Delta CS = (sqrt((\delta H^N)^2+(\delta N/10)^2))$, where δH^N and δN are the changes in the respective chemical shifts upon complexation, for each amino acid residue as observed in the $[^{15}N\text{-}^{1}H]$ HSQC of a two-domain calcium binding protein upon binding to a peptide. The inset shows the overlay of HSQC spectra of the free form (black) and peptide bound form (grey) of the protein.

In majority of studies, the molar concentration of the ligand used is higher than that of the protein. The nature of the NMR spectrum of the E, L system depends on a number of factors. The appearance of the NMR spectrum depends on the exchange rate between the free and bound states of the molecules similar to the process discussed in Chapter 1. For weak binding (K_D typically > 10 μM), the exchange is fast on the NMR time scale and a weighted average of the two states is observed. For strong binding, separate signals from free and bound states are observed. Therefore, in case of fast exchange, the changes in the spectrum of protein are larger than those on the ligand spectrum. One usually follows the behavior of the protein-ligand complex by using NMR properties of the protein as modified by the binding. To be able to use the ligand spectrum, it is necessary that a sufficient amount of ligand is bound to the protein for the effect to be observed. Generally, while looking at a ligand spectrum, a small amount of protein is sufficient for obtaining binding information.

8.2.1 Diffusion Studies

The translational and rotational diffusion for a bound ligand is much slower than for free ligand. The diffusion rates can be measured using pulsed field gradient techniques (Chapter 3). The rotational diffusion is reflected in the relaxation behavior of the complex. For the ligand, the NOE can go from positive in the free-state, to a negative value when bound in the slowly tumbling complex.

8.2.2 Mapping of the Binding Site

Once the 3D structure of the protein is known, and the sequence-specific-resonance assignments have been made, the chemical groups involved in the binding with the ligand can be established. Only the resonances arising from the region around the binding site of the protein are expected to be affected. This can be followed using a simple 2D ^1H-^{15}N HSQC or HMQC experiment. In the case of strong binding, the peaks of the spins near the binding site will change position. The spectral features of the unbound protein will gradually disappear on the addition of the ligand and will be replaced by cross peaks corresponding to the complex. In case of weak binding, increasing amount of ligand will progressively move the position of such cross peaks.

8.2.3 Conformation of the Bound Ligand

The conformation of the ligand is expected to change significantly upon binding and can be conveniently established using NOE. Usually, small molecular weight compounds have a flexible conformation. Upon binding, the conformation is expected to become more ordered. In slow exchange limit, the NOE of the ligand generally become more pronounced and go from positive to negative values. Under fast exchange conditions, transferred NOE can be used to study the conformation of the bound ligand. In conjunction with docking experiments, both the binding and the conformation of the ligand can be optimized.

8.2.4 Changes in the Structure and Dynamics of the Protein

The ligand binding sites in the proteins are usually highly mobile. As we have discussed earlier, this dynamic feature allows ligands to enter and exit the binding sites. The protein-ligand complex however, leads to a more rigid conformation. These features can be picked using the site specific relaxation parameters of the protein as discussed earlier.

8.3 Enzyme Catalysis

Catalysis of a biological reaction is one of the most important functions of proteins. Classical enzymology has provided detailed insight into thermodynamics and biochemical mechanism of several enzymatic reactions. A great deal of our understanding comes from the interaction of enzyme (E), with the substrate (S), through the formation of an intermediate complex ES at a fast rate, which isomerizes to EP, resulting into the products (P). The last stage occurs at a slower rate.

$$E+S \rightleftharpoons [ES] \rightleftharpoons [EP] \longrightarrow E+P \qquad 6.4$$

This simple picture is the basis of the well known Michaelis-Menton equation in enzyme kinetics. It has provided a mechanistic view of the enzyme-substrate binding, and the lowering of the activation energy as a result of formation of [ES] complex. The model enables one to study changes in the kinetic parameters as a result of various experimental conditions. However, these methods do not provide a detailed mechanistic view of the molecular events during enzyme catalysis. The enzymatic reactions are inherently more complex than the one represented in the above model.

The ability of NMR to observe transient states in the micro- to milli-second time range provides important clues to molecular, dynamic and functional aspects of enzymes. Dynamic NMR techniques such as line shape analysis, rotating frame spin-lattice relaxation time and pulse sequences such as CPMG, help in studying enzyme dynamics and the conformational energy landscape of the catalytic processes. Binding of inhibitors, activators or slowly reacting substrates have been followed. The proton transfer or substrate exchange rates can be studied in favorable cases.

Such experiments have been performed on a number of enzymes. This enables us to learn about the conformational changes in enzymes during their function and to understand the role of such changes in enzyme activity. The power of NMR is illustrated by taking a few examples.

8.3.1 Dihydrofolate Reductase (DHFR)

DHFR (M_r ~ 18 kDa) catalyzes the reduction of 5,6-dihydrofolate to 5,6,7,8-tetrahydrofolate using NADPH as a coenzyme. It plays a vital role in the biosynthesis of several amino acids, thymidines and purines. Extensive studies on this enzyme have been made by enzyme kinetics, crystallography and NMR. Valuable information has been obtained using NMR on coenzyme-substrate-enzyme binding as well as on those with substrate analogues. Conformational changes in DHFR are induced by such a binding. It has been established that the Met20 loop and its dynamics play a crucial role in this catalysis. The active site loop interconverts among open, closed and occluded conformations. The kinetic scheme involves two-substrate, two-product reaction.

8.3.2 Triosephosphate Isomerase (TIM)

In the case of triosephosphate isomerase, NMR studies have shown that the ligand-binding is followed by loop opening/closing, and the time scale of this motion is in the same order as the turnover rates.

8.4 Multi-Domain Structures

NMR studies of proteins containing several domains (Figures 6.8 C and E) are more difficult, partly because these have much larger M_r and also because there may be a fair degree of molecular motions in the linker regions.

The first problem has been addressed by using gene cloning methods to obtain sufficient amount of material of individual domains. Since each domain in a protein has a unique conformation and function, such an approach allows high precision structures to be obtained for each domain. These domain structures are then coupled together by introducing the linker regions.

The second problem is more complex. There are only a few constraints to fix the inter-domain orientation. Residual dipolar couplings have proved useful in both structure calculation and refinement. Most importantly they provide information about long-range interactions. Thus, one finds RDCs as very useful structural parameters in defining domain orientation in proteins. The use of spin labels also provides useful information on protein conformation with more than one domain. However, both approaches have limited applicability when the linker regions are highly flexible.

9. REFERENCES

9.1 Historical Foundation of Protein Structure and Function

A. Saunders, and A. Wishnia, *Nuclear Magnetic Resonance Spectra of Proteins*, Ann. N.Y. Aca. Sci. **70**, 870–874 (1958).

A. Kowalsky, *Nuclear Magnetic Resonance Studies of Proteins*, J. Biol. Chem. **237**, 1807–1809 (1962).

G.N. Ramachandran, C. Ramakrishnan, and V. Sasisekharan, *Sterochemistry of polypeptide chain configuration*, J. Mol. Biol. 7, 95–99 (1963).
C.B. Anfinsen, *Principles that govern the folding of protein chains*, Science **181**, 223–230 (1973).

9.2 General Books on Proteins

C. Branden, and J. Tooze, *Introduction to Protein Structure*, Garland Publishing Inc., New York (1991).
M. Perutz, *Protein Structure* W.H. Freeman and Company, New York (1992).
J.M. Berg, J.L. Tymoczko and L. Stryer, *Biochemistry*, 5th International Edition, W.H. Freeman and Co. (1995).

9.3 Books on Protein NMR

K. Wüthrich, *NMR of Proteins and Nucleic Acids*, Wiley (1986).
M.P. Deutscher (ed.), *Guide to protein purification*, methods Enzymol. **182**, 1–894 (1990).
F.J.M. van de Ven, *Multidimensional NMR in Liquids – Basic Principles and Experimental Methods*, VCH (1995).
J. Cavanagh, W.J. Fairbrother, A.G. Palmer and N.J. Skelton, *Protein NMR Spectroscopy: Principles & Practice*, Academic Press Inc., San Diego (1996).
Malcolm H. Levitt , *Spin Dynamics – Basics of Nuclear Magnetic Resonance*, Wiley (2001).
T.L. James (ed.), *Nuclear magnetic resonance of biological macromolecules*, Methods Enzymol. **394**, 1–631 (2005).

9.4 Study of Large Proteins

K. Wüthrich, G. Wider, G. Wagner and W. Braun, *Sequential resonance assignments as a basis for determination of spatial protein structures by high resolution nuclear magnetic resonance*, J. Mol. Biol. **155**, 311–319 (1982).
T.F. Havel and K. Wüthrich, *A distance geometry program for determining the structures of small proteins and other macromolecules from nuclear magnetic resonance measurements of intramolecular 1H-1H proximities in solution*, Bull. Math. Biol. **46**, 673–678 (1984).
G. Otting, and K. Wüthrich, *Studies of protein hydration in aqueous solution by direct NMR observation of individual protein-bound water molecules*, J. Am. Chem. Soc. **111**, 1871–1875 (1989).
M. Ikura, L.E. Kay, and A. Bax, *A novel approach for sequential assignment of 1H, 13C and 15N spectra of larger proteins: heteronuclear triple-resonance three-dimensional NMR spectroscopy. Application to Calmodulin*, Biochemistry **29**, 4659–4667 (1990).
D. Neri, M. Billeter, G. Wider, and K. Wüthrich, *NMR determination of residual structure in a urea-denatured protein, the 434 repressor*, Science **257**, 1559–1563 (1992).
J.R. Tolman, J.M. Flanagan, M.A. Kennedy, and J.H. Prestegard, J.H. *Nuclear magnetic dipole interactions in field-oriented proteins: information for structure determination in solution*. Proc. Natl. Acad. Sci. USA. **92**, 9279–9283 (1995).
N. Tjandra and A. Bax, *Direct measurement of distances and angles in biomolecules by NMR in dilute liquid crystalline medium*, Science **278**, 1111–1114 (1997).
G. Wider and K. Wüthrich, NMR *spectroscopy of large molecules and multimolecular assemblies in solution*. Curr. Op. Struct. Biol. **9**, 594–601 (1999).
R. Riek, K. Pervushin and K. Wüthrich, *TROSY and CRINEPT: NMR with large molecular and supramolecular structures in solution*, Trends Biochem Sci. **25**, 462–468 (2000).
J. Fiaux, E.B. Bertelesen, A.L. Horwich and K. Wüthrich, *NMR Analysis of a 900 K GroEL-GroES Complex*, Nature **418**, 207–211 (2002).
K. Wüthrich, (Nobel Lecture), *NMR Studies of Structure and Function of Biological Macromolecules*, Angewandte Chemie, Int. Ed. **42**, 3340–3363 (2003).

9.5 Protein Structure, Dynamics and Folding

R.M. Brunne, E. Liepinsh, G. Otting, K. Wüthrich and W.F. van Gunsteren, *Hydration of proteins : A comparison of experimental residence times of water molecules solvating the bovine pancreatic trypsin inhibitor with theoretical model calculation*, J. Mol. Biol. **231**, 1040–1048 (1993).

K. Osapay, Y. Theriault, P.E. Wright and D.A. Case, *Solution structure of carbonmonooxy myoglobin determined from nuclear magnetic resonance distance and chemical shift constraints*, J. Mol. Biol. **244**, 183–197 (1994).

M.W.F. Fisher, A. Majumdar and E.R.P. Zuiderweg, *Protein NMR relaxation: Theory, applications and outlook, Prog. Nucl. Magn. Reson. Spectrosc.* **33**, 207–272 (1998).

A. Matagne and C.M. Dobson, *The folding process of hen lysozyme: a perspective from the new point*, Cell. Mol. Life Sci. **54**, 363–371 (1998).

N. Tjandra, *Establishing a degree of order: obtaining high-resolution NMR structures from molecular alignment*, Structure **7**, R205–R211(1999).

S. Doniach and P. Eastman, *Protein dynamics simulations from nanoseconds to microseconds*, Curr. Opin. Struct. Biol. **9**, 157–163 (1999).

S.C. Sahu, A. Bhattacharya, K.V.R. Chary, and G. Govil, *Secondary structure of a calcium binding protein (CaBP) from* Entamoeba histolytica, FEBS Letters. **459**, 51–56 (1999).

K. Akasaka and H. Li, *Low-lying excited states of proteins revealed from non-linear pressure shifts in 1H and ^{15}N NMR*, Biochemistry 40, 8665–8671 (2001).

H.S. Atreya, S.C. Sahu, A. Bhattacharya, K.V.R. Chary and G. Govil, *NMR Derived Solution Structure of an EF-hand Calcium-Binding Protein from* Entamoeba histoltica, Biochemistry **40**, 14392–14403 (2001).

H.J. Dyson and P.E. Wright, *Unfolded proteins and protein folding studied by NMR*, Chem. Rev. **104**, 3607–3622 (2004).

L.W. Kay, *NMR Studies of protein structure and dynamics*, J. Mag. Res. **173**, 193–207 (2005).

M. Oliveberg and P.G. Wolynes, *The experimental survey of protein-folding energy landscapes*, Q. Rev. Biophys. **38**, 245–288 (2005).

9.6 Paramagnetic Proteins

I. Bertini, and C. Luchinat, *NMR of Paramagnetic Molecules in Biological Systems*, Benjamin/Cummins, Menlo Park, CA, USA (1986).

K. Rajarathnam, G.N. La Mar, M.L. Chiu, M.L., S.G. Sligar, J.P. Singh, and K.M. Smith, *'H NMR hyperfine shift pattern as a probe for ligation state in high-spin ferric hemoproteins: Water binding in metmyoglobin mutant*, J. Am. Chem. Soc. **113**, 7886–7892 (1991).

I. Bertlni, P. Turano, and A.J. Vila, *Nuclear magnetic resonance of paramagnetic metalloproteins*, Chem. Rev. **93**, 2833–2932 (1993).

9.7 Protein Binding and Function

K. Bartik, C. Redfield, C.M. Dobson, *Measurement of the individual pKa values of acidic residues of hen and turkey lysozymes by two dimensional 1H NMR*, Biophysical J. **66**, 1180–1184 (1994).

G.C.K. Roberts, *Applications of NMR in drug discovery*, DDT **5**, 230–240 (2000).

M.J. Osborne et al., *Backbone dynamics in DHFR complexes: Role of loop flexibility in catalytic mechanism*, Biochemistry **40**, 9846–5998 (2001).

C.F. Josephine and V.J. Hilser, *Ligand induced changes in dynamics in the RT loop of the C-terminal SH3 domain of Sem-5 indicate co-operative conformational coupling*, Protein Science **12**, 982–996 (2003).

T.L. James (ed.), *Methods in Enzymology: Nuclear Magnetic Resonance of Biological Macromolecules Part C*, Elsevier Inc., (2005).

K. Lindorff-Larsen, R.B. Best, and M. Vendruscolo, *.Interperting dynamically averaged scalar couplings in proteins*, J. Biomol. NMR **32**, 273–280 (2005).

CHAPTER 7

STRUCTURE AND DYNAMICS OF NUCLEIC ACIDS

1. INTRODUCTION

Deoxyribonucleic acid (DNA) and ribonucleic acid (RNA) are hereditary molecules, which are responsible for the flow of genetic information in living systems. The genetic information of an organism is encoded in the form of a sequence of nucleotides in a DNA molecule. It is stored in the genomes or chromosomes of living systems. This information finally decides the nature of proteins needed for the function of a particular organism.

1.1 Chemical Constitution of Nucleic Acids

The backbone of a DNA molecule is composed of a linear chain of deoxyribonucleotide units (Figure 7.1). Likewise, a RNA molecule is built from ribonucleotide units. Each nucleotide in a DNA or RNA consists of a nitrogenous base, a five member sugar ring (furanose) and a phosphate group. The nitrogenous bases are usually derivatives of purines [adenine (A) or guanine (G); sometimes represented by the letter R] or pyrimidines [thymine (T), uracil (U) or cytosine (C); represented by Y]. The sugar ring is deoxyribose in DNA and ribose in RNA. A nucleoside is defined as a purine or pyrimidine base, bonded to the sugar through a glycosidic (C1′-N) bond. Nucleotides are the corresponding phosphate esters of nucleosides, obtained by condensation with either the 3′ or the 5′ hydroxyl group. The long RNA or DNA backbone consists of nucleosides linked by the 3′-5′ phosphate di-ester linkage. The variable part of a DNA or RNA molecule is its sequence defined by the four bases, namely, A, G, C and T for DNA, and A, G, C and U for RNA.

The proposal by Watson and Crick (Figure 7.2) in 1953 on the structure of DNA (commonly referred as B-DNA) changed the field of biology. It led to new disciplines such as molecular biology, biotechnology, bioinformatics and genomics. Using fibre diffraction and model building, they showed that DNA molecules occur as right-handed double-stranded (ds) helical structures. The chains of the two strands run in anti-parallel directions. The bases in the two strands are held together in the interior of the helix, by the so called *Watson-Crick* (WC) hydrogen bonded G:C and A:T base-pairs (Figure 7.3). In this sense, the two strands in DNA molecules are complementary. The stoichiometry of G and C and that of A and T in DNA, is close to 1:1. The genetic code of a living system is determined by the sequence of the bases, and is kept locked by the WC hydrogen bonds. The most important concept that we will repeatedly use in this Chapter is the base-complementarity in nucleic acids arising from the G:C and A:T base pairing.

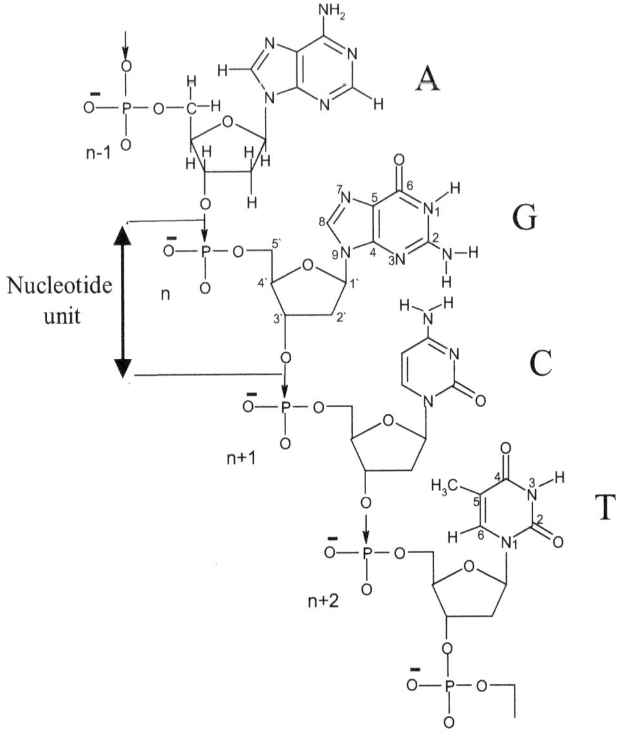

Figure 7.1: A linear chain of deoxyribonucleotide units showing the common bases.

James D. Watson F. H. C. Crick M. H. F. Wilkins Rosalind Franklin

Figure 7.2: Scientists who contributed to the discovery of the duplex model for DNA. Their contributions led to the emergence of modern biology. (Courtesy: http://nobelprize.org).

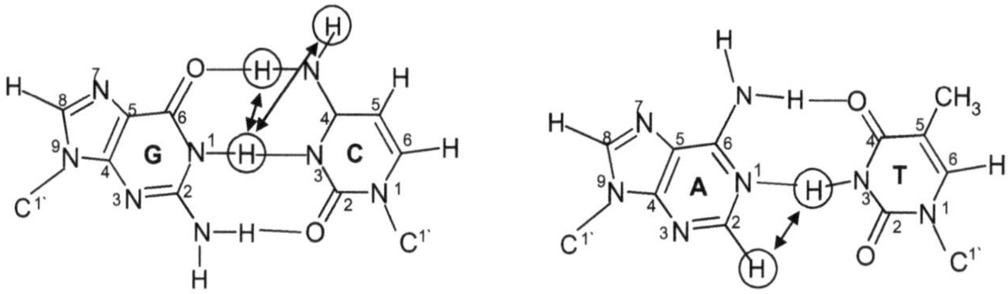

Figure 7.3: WC hydrogen bonded G:C and A:T base pairs. The short ^{1}H-^{1}H distances which can be picked up by using NOEs across the base-pairs are indicated by arrows.

The common bases for RNA are A, G, C and U. RNA molecules have more complex 3D structures. Frequently RNA contain bases other than those mentioned above. In fact these molecules have a high abundance of several unusual bases. The most common of these are β-pseudouridine (ψ), dihydrouridine (DhU), thio analogues, ionosine (I), aza-purines and 7-methyl-G. Thus the ratio of A:U and G:C deviates significantly from unity in RNA. The structures of RNA usually consist of a single strand, which is folded into a complex 3D structure. The main secondary structures in RNA are anti-parallel double-stranded helical stems with A:U and G:C base pairs. A double-stranded RNA (ds RNA) contains two complementary strands, similar to the DNA. The ds RNA forms the genetic material of some viruses. Such double stranded regions are interspersed with loops and bulges.

Under physiological concentrations, the phosphate group in nucleic acids is negatively charged. The short range phosphate-phosphate repulsions in a strand can destabilise the secondary structures. Mono- and di-valent cations such as Na^{+} and Mg^{+2}, stabilise the secondary structures of DNA and RNA.

1.2 Biological Role of Nucleic Acids

Nucleic acids have three main biological functions. Each strand of a ds-DNA acts as a template in the synthesis of its complementary strand using the principle of A:T and G:C complementarity. In this way, an exact replica of the ds-DNA is synthesized. This process is called *replication*. During replication, specific proteins such as *helicases* promote the separation of the two strands and help them to unwind, so that the bases are exposed to the medium for the synthesis of the complementary strands. *DNA polymerases* catalyse the replication reaction.

DNA also acts as a template for the synthesis of RNA. This process called *transcription*. RNA synthesis is catalysed by *RNA polymerase*. Transcription starts from certain sites called promoters. It is stopped by termination signals, which are usually a stretch of RNA having a hairpin loop with four or more U, followed by several uracil residues. Transcription results in the formation of several different RNA molecules. The most important RNA molecules are those involved in protein synthesis, namely, messenger RNA (mRNA), ribosomal RNA (rRNA) and transfer RNA (tRNA). In this process, nucleotide A on the DNA strand results in the insertion of U on the RNA, while other complementarities are the same as for DNA replication.

During both these processes, the DNA duplex unwinds and interacts with certain specific enzymes such as *polymerases* and *helicases*. The subsequent events facilitates the nucleotides bases to be exposed to the medium. Thus, biological expression of genetic information demands that DNA assumes various alternate shapes and structures.

The third function of nucleic acids is RNA controlled synthesis of proteins, a process called *translation*. Three types of RNA molecules are involved in this process. mRNA contains the code in the form of three bases (codon) in its sequence for each amino acid present in the corresponding protein sequence. tRNAs are relatively smaller molecules, which are unique for each amino acid and are specifically recognised by the codons through a loop called the anticodon loop. Each amino acid has a specific tRNA, which insert the right amino acid in a growing polypeptide chain depending on the codon on mRNA. Protein synthesis takes place on ribosomes, which are fairly large supramolecular structures, consisting of 30S and 50S substructures. The two sub-structures are held together through weak intermolecular interactions stabilised by Mg^{2+}. Ribosomes contain a large amount of rRNA along with certain associated proteins which are called ribonucleoproteins. Ribosomes act as physical machinery for protein synthesis. They also act as a support for holding the other two types of RNA (t-and m-RNA) involved in the process of translation. tRNA enters and exits the rRNA assembly, before and after the insertion of the appropriate amino acid residue in the growing polypeptide chain.

Recently, it has been shown that RNA can have a number of other functions. They are involved in translational regulation, gene silencing, RNA processing and modification, protein and RNA stability, protein translocation. They can also function as biological RNA catalysts (*ribozymes*) in certain biochemical reactions.

Both *polymerases* and *helicases* of DNA and RNA are the two essential enzymes which play a crucial role in the above biological processes. Thus, the key functions of biological molecules, namely, *replication, transcription* and *translation* are mediated by protein-nucleic acid interactions. Proteins also play an essential role in processing and repair of DNA and RNA. Therefore, the study of protein-nucleic acid interactions is important in understanding the fundamental biological processes underlying living systems.

For more details on these topics, standard text books on biochemistry may be consulted.

1.3 Historical Background of Nucleic Acid Structures

The functional roles of DNA and RNA are closely related to their structures. Hence, there is an active interest in their structure determination. Early work using fibre-diffraction led to several models for double-stranded DNA, namely A-DNA, B-DNA, and Z-DNA depending on the nucleotide base composition and experimental conditions. C-DNA and D-DNA are subclasses of B-form. These polymorphic forms can be obtained by growing fibres under different experimental conditions.

In recent years, improved synthetic methods have made it possible to choose nucleotide sequences at one's will. Thus one can introduce controlled perturbations in an oligonucleotide of interest. In solution studies by NMR, physico-chemical conditions such as concentration, pH, ionic strength and temperature and the presence of specific metal ions can also be controlled. Such perturbations have provided important insights into the structures of canonical as well as alternate forms of nucleic acids and the associated structure-function relationships.

As a result of these recent advances, it is now possible to look at the 3D structures of nucleic acids both by NMR and X-ray crystallography. Till date, the Protein Data Bank (PDB) holds NMR structures of 705 nucleic acids and 121 protein-nucleic acid complexes.

2. ELEMENTS OF STRUCTURE OF NUCLEIC ACIDS

The conformational variations in nucleic acids arise from: (i) the flexibility of the furanose ring; (ii) the rotation around the glycosidic bond; (iii) the possibility of non-WC hydrogen bonding; and (iv) rotational degrees of freedom generated by various single bonds present in each unit. The 3D structures of nucleic acids are stabilised by inter-base hydrogen bonds, stacking interactions between the base pairs and non-bonded and electrostatic interactions in the backbone.

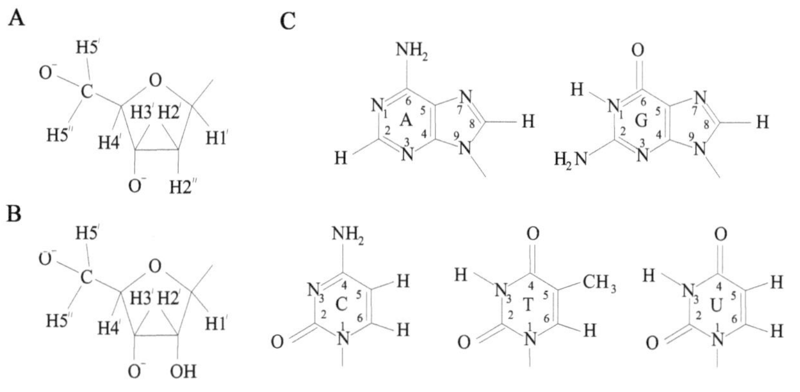

Figure 7.4: Numbering for atoms in (A) deoxyribose, (B) ribose and (C) the common bases.

2.1 Nomenclature

Following an inter-union joint venture, the International Unions of Pure and Applied Chemistry (IUPAC), Biochemistry and Molecular Biology (IUBMB) and Pure and Applied Biophysics (IUPAB), have recommended standard definitions for the presentation of NMR structures of nucleic acids. Figure 7.4 depicts the abbreviation and the recommended numbering for various atoms of ribose, deoxyribose and the common bases. The sequence of atoms in the backbone of an individual nucleic acid strand is as follows: $\{P\text{-}O5'\text{-}C5'\text{-}C4'\text{-}C3'\text{-}O3'\}_i\text{-}\{P\text{-}\ldots\ldots\}_{i+1}$. The strand follows the direction 5′ to 3′ . In the ribose and deoxyribose rings, the atom numbering is C1′-C2′-C3′-C4′-O4′-C1′. The bases are attached to the sugar through a C1′-N linkage (N9 for R and N1 for Y). The atoms in the base are numbered without the *prime* symbol e.g. C1, C2 etc.

The notation for the geminal hydrogen atoms (*two protons attached to the same carbon atom*) attached to C5′ in furanose is H5′ and H5″ for pro-S and pro-R protons, respectively. Likewise, the notation for the geminal hydrogen atoms attached to C2′ in deoxyribose is H2′ and H2″ for pro-S and pro-R protons, respectively. In the event of resonance ambiguity in NMR, the di-stereotopic substituents of pro-chiral centres, such as H2′/H2″ and H5′/H5″, are designated by slash. For example, when the chemical shifts H2′ and H2″ protons in a deoxyribose ring can not be assigned stereo-specifically, they are designated as H2′/H2″ = $\delta_1/\ \delta_2$.

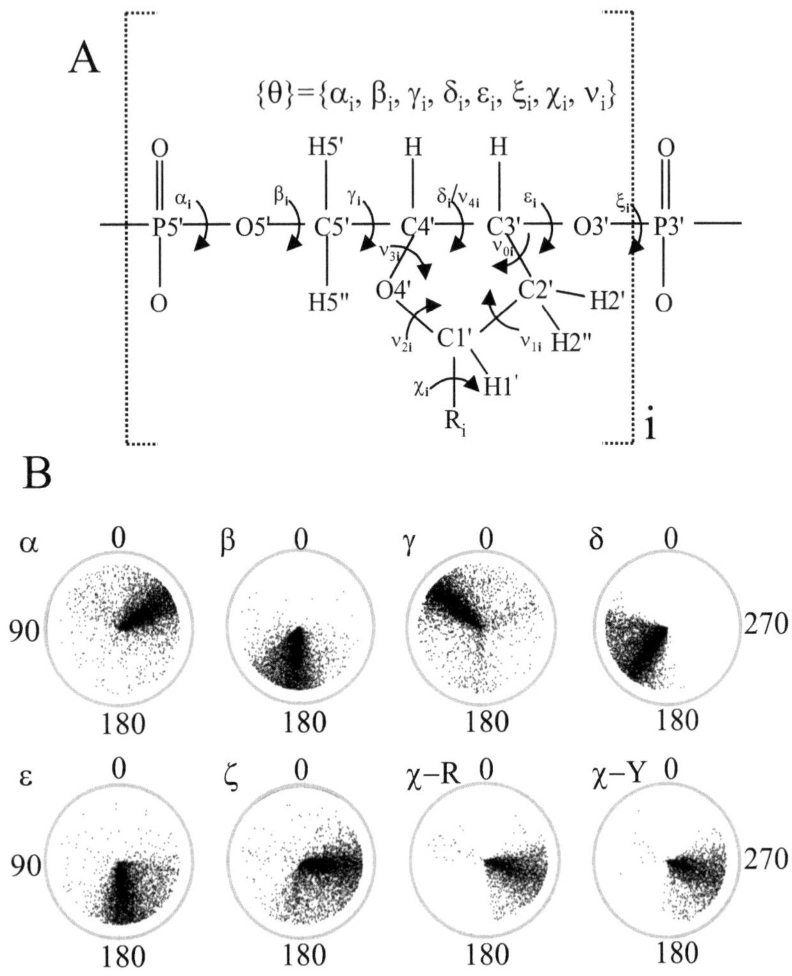

Figure 7.5: (A)The backbone ($\alpha, \beta, \gamma, \delta, \varepsilon, \zeta$), glycosidic (χ) torsion angles for purine (R) and pyrimidine (Y) bases and five sugar ring torsion angles.(B) Statistical distribution of various torsion angles in B-DNA derived from PDB files.

2.2 The Backbone Torsion Angles

There are six torsion angles in the sugar phosphate backbone of nucleic acids. For the ith residue, the backbone ($\alpha, \beta, \gamma, \delta, \varepsilon, \zeta$) torsion angles for purine (R) and pyrimidine (Y) bases are shown in Figure 7.5 A. Potential energy calculations show that even in absence of inter-base hydrogen bonding and stacking interactions, these angles show preferences for certain conformational domains, which are observed in the structures of nucleic acids. Figure 7.5 B indicates the preferred values of these torsional angles derived from the statistical distribution of all the B-DNA structures which have been deposited in the PDB.

2.3 Glycosidic Bond Rotation (χ)

Rotation around the glycosidic bond (C1′-N) dictates the orientation of the purine and pyrimidine bases with respect to the sugar ring. In most nucleic acid structures the bases are orientated in an *anti* conformation (χ with a preferred range of +90 $\rightarrow$ +270). The structures of both A-DNA and B-DNA have bases in the *anti* conformation. However, in certain cases, a *syn* conformation (χ with a range of -90 $\rightarrow$ +90) has been observed. For example, Z-DNA has alternating *anti* and *syn* conformation for the bases. WC hydrogen bonds are formed with the *anti* conformation in antiparallel-stranded (aps) duplexes. The *syn* conformation leads to other hydrogen bonding patterns.

2.4 Sugar Pucker: Pseudo-Rotation Angle (P)

The furanose ring is centrally located in the nucleotide fragment. It plays a key role in determining the conformation of nucleic acids. The ring conformation is defined by five torsion angles ν_o, ν_1, ν_2, ν_3, and ν_4, which are represented as shown in Figure 7.5. Conformations of cyclic systems are dictated by geometrical constraints arising from ring closure, which is not the case for non-cyclic molecules. There are five heavy atoms in furanose ring and one needs only two (i.e. N-3) torsion angles to define its geometry. Thus out of the torsion angles ν_o to ν_4, two are sufficient to uniquely define the geometry of the ring. For example, using the plane passing through C4′, O4′ and C1′, one may like to know the positions of C2′ (determined by the angle ν_o) and C3′ (the angle ν_1). However, since the bond lengths and bond angles in the furanose ring are fixed by the covalent bonds, even ν_o and ν_1 are not independent. For example, if $\nu_o = 0°$ (atom C2′ lies in the same plane as C4′, O4′ and C1′), then C3′ can lie about 0.6 A° above or below the plane describing the other four atoms. When C3′ is on the same side of the plane as C5′ and the base, the conformation is called C3′-*endo* and is denoted by ^3E. The other conformation is called C3′-*exo* (E_3).

The furanose ring is thus non-planar. Its structure can assume continuum of conformations depending on the values of ν_o to ν_4, which are interrelated. Altona and Sundaralingam have introduced a two-parameter description for the conformation of furanose ring in terms of a pseudo-rotational angle P (Figure 7.6). The five torsional angles, ν_o to ν_4 are related to P and the maximum angle of pucker ν_m by the following equations:

$$\nu_j = \nu_m \cos\,[\,P + (j\text{-}2)\ \delta\,],\ \text{with}\ j = 0\ \text{to}\ 4,\ \text{and} \qquad 7.1$$

$$\tan P = [(\nu_4 + \nu_1) - (\nu_3 + \nu_o)]/[\,2\nu_3\,(\sin 36^o + \sin 72^o)] \qquad 7.2$$

As P goes from 0 to 360°, the values of ν_0 to ν_4 varies from $+\nu_m$ to $-\nu_m$. For the five member ribose ring, the values of ν_m and δ are approximately 38° and 144°, respectively.

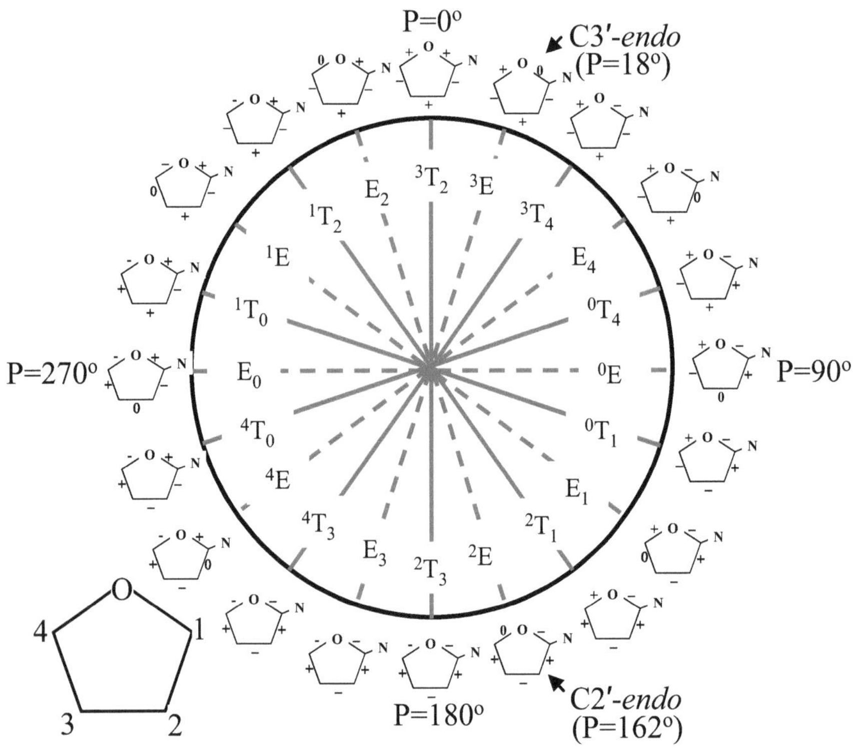

Figure 7.6: Relationship between the pseudo-rotation phase angle (P) and classical notations for a five member sugar ring. This is illustrated by the so called pseudo-rotation wheel.

Thus, the conformation of a furanose ring can be described by specifying a single parameter, namely, *the phase angle of pseudo-rotation* (P). Incidentally, ν_3 and δ, which characterize torsion angles around the same bond (C3′- C4′), are assigned different values for a given conformation, because of the way in which the two angles are defined (C2′-C3′-C4′-O4′ for sugar pucker and O3′-C3′-C4′-C5′ for backbone torsion angle δ). In fact, it is possible to obtain the value of δ from the value of P.

Potential energy calculations show that the conformation of the sugar ring is characterized by a low potential energy barrier. The energy differences between the various sugar puckered conformations are relatively low. The crystal structures of nucleic acids show a wide spread of sugar puckers. However, these are usually centered around C3′- and C2′-*endo* conformers ($P = 18^{\circ}$ or 162° respectively).

In several NMR reports, this conformational fluidity has been represented by a two state model consisting of a rapid equilibrium between the classical ^3E and ^2E conformation. This is obviously an over-simplification. An alternative approach uses another kind of two state model consisting of a range of conformation having P angles in the north part of the pseudo-rotational wheel (represented by letter N and centered around C3′ *endo)* and the south part (represented by S that includes C2′ *endo* conformation). The rapid conformational equilibrium is then defined as that between N and S conformers. However, it is difficult to assign the precise values of the NMR properties for these two wide ranges of conformers.

The standard models for A-DNA are built from sugars in the C3′-endo geometry, while those for B-DNA use C2′-endo geometry. The Z-form has alternating C3′-*endo* and C2′-*endo* geometries.

2.5 Inter-base Hydrogen Bonding

Nucleic acid bases have multiple sites for hydrogen bonding. The most common base pairing scheme is that proposed by Watson and Crick (WC) in their model for DNA structure. This scheme is the one used in models for A-, B- and Z-DNA.

With the stereo-chemically favourable backbone and side-chain torsion angles and sugar puckers, alternative hydrogen bonded structures can be stabilised by favourable stacking interaction and by hydrogen bonding schemes other than those involved in WC base-pairing. Figure 7.7 summarizes some of the possible hydrogen bonded schemes. Such alternative hydrogen bonding schemes determine the folding patterns of RNA and lead to the unusual structures of DNA.

The *aps* double helical structure is the most common conformation for DNA and RNA. In all the three model structures mentioned earlier, two antiparallel chains are held together by WC base pairs. Both the A- and B- forms of DNA are right-handed structures, while the Z-form is left-handed. The sugar backbone is on the outside of the helix. The base pairs are inside and are stacked on the top of each other.

B-DNA has a wide major-groove and a narrow minor-groove, which are linked by potential hydrogen bond donors and acceptor. The bases are stacked perpendicular to the helix axis at 3.4 Å distance. This is the most common form of DNA conformation in solutions. There are two variations of the B-DNA structure called B_I and B_{II}, which differ from each other in two backbone torsional angles but have similar overall structures. RNA and RNA-DNA hybrids generally adopt A-type structures with an axial rise of 2.8 Å.

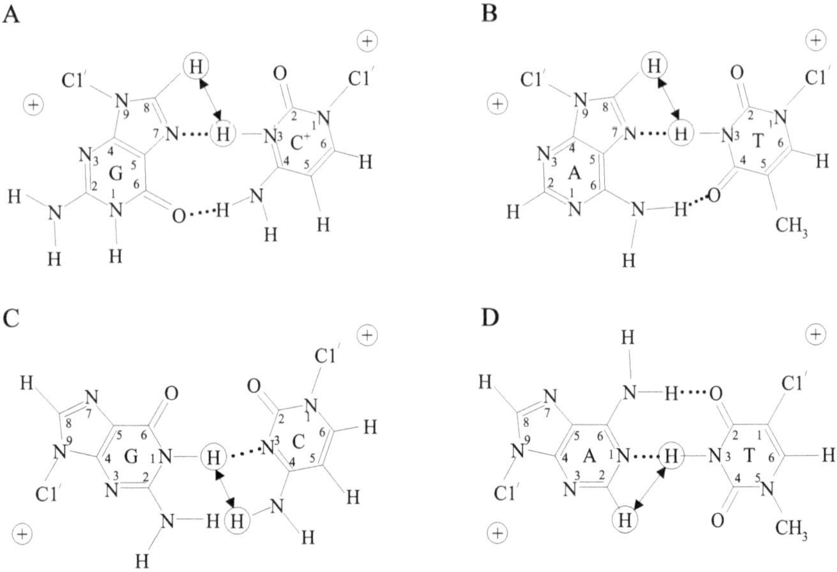

Figure 7.7: Some of the possible base-pairing schemes in nucleic acids, other than WC base pairs (A) Hoogsteen G.C^{+} base-pair; (B) Hoogsteen A.T base-pair; (C) Donohue G:C^{+} base-pair; (D) Donohue A:T base-pair. Several other base-pairs have been observed, particularly in RNA. The short ^{1}H-^{1}H distances picked up by NOEs across the base-pairs are indicated by arrows.

Nucleic acids can adopt a number of alternative conformations. Examples of non-duplex DNA structures are single-stranded hairpins, triplexes, tetraplexes and i-motif. DNA even has the capacity to form parallel-stranded (*ps*) structure. Such nucleic acid polymorphism is governed by factors such as sequence, concentration, temperature, pH, and other pertinent solvent conditions.

On the other hand, RNA, whose primary function is protein synthesis within a cell, is usually single-stranded. The polynucleotide chain of RNA molecules fold back on itself to generate double-stranded A- type helical regions interspersed with a single-stranded loop regions which adopt hairpin or bulge-like conformations. RNA structures are, therefore, more complex than those of DNA. The structural investigation of DNA and RNA has been a fruitful subject for NMR spectroscopists.

3. NMR SPECTROSCOPY OF NUCLEIC ACIDS

In this section, the general features of NMR of nucleic acids are discussed. It may be noted that nucleic acids contain an additional nuclear spin ^{31}P, which is present in 100% natural abundance.

3.1 Comparison with Protein NMR:

The protocols and procedures needed for structure determination of nucleic acids are similar to those for proteins. In practice however, there are considerable difficulties in solving 3D structures of nucleic acids as compared to those of proteins. On an average, the number of protons in each residue is approximately ten for both nucleic acids and proteins. However, the average M_r for nucleotide is three times greater than that of a peptide. This results in a three-fold lower ^{1}H density in nucleic acids. Consequently, for macromolecules of similar sizes fewer structural constraints are available in nucleic acids.

Nucleic acids are built from only four different nucleotides, as compared to twenty amino acids in proteins. This results in smaller chemical shift dispersion. Recall, that nucleotides have seven degrees of freedom for internal rotation as compared to two for proteins. The problem is further complicated by spectral overlaps due to the presence of large fraction of nucleotides in the A- or B- form DNA.

Seven of the protons in deoxy-nucleotides (H1′, H2′, H2″, H3′, H4′, H5′, and H5″) are contributed by the sugar rings. The information content on the sugar ring is therefore relatively rich. The 100% abundant hetero-nucleus ^{31}P shows J-coupling to H3′ at one end and to H5′/H5″ on the other. These factors lead to overlapping and strongly coupled spectra.

On the other hand, information on the backbone structure can be obtained from a limited number of nuclei, namely, H3′, H4′, H5′, and H5″ and ^{31}P. However, these protons and ^{31}P suffer from relatively low dispersions in chemical shifts. The role of ^{31}P in the phosphate group can be compared to the pivotal role played by ^{15}N in the amide group. Here also the situation is less favourable. For example, the chemical shift dispersion for ^{31}P in nucleic acids is less than 10 ppm, compared to 30 ppm observed for amide protons. The ^{31}P-^{1}H coupling constants are small (3-10 Hz) as compared to the values of ^{1}J (N-H) in proteins, which provides a convenient pathway for assignments and measurement of residual dipolar couplings. Further, ^{31}P has unfavourable chemical shift anisotropy resulting in shorter T_2 and broader lines.

The main secondary structures B-DNA and A-RNA are helical in nature and are predominantly rod-like. This results in fewer long-range interactions and NMR constraints, there by posing problems in understanding molecular structure and dynamics. The molecular tumbling rates are smaller and anisotropic, resulting in broader resonances.

The factors mentioned above lead to difficulties in resonance assignments. Apart from that, presently, very few experimental data is available for constraining the six backbone torsion angles. For these reasons, the complete structure determination of DNA and RNA from ^{1}H NMR has been limited to fragments containing less than 50 nucleotides. Structure determination of nucleic acids using ^{1}H is primarily based on estimation of short ^{1}H-^{1}H distances detected by NOEs, on dihedral angles that can be derived from the through bond couplings and on characterisation of base-pairing.

As in the case of proteins, the use of isotopic labelling of nucleic acids has opened new possibilities for studying their structure and dynamics. The measurement of residual dipolar couplings (RDC) in anisotropic media provides valuable source of additional distance and angle information. Another important observable is the J-coupling across the hydrogen bonds. These two parameters form an integral part of recent methodologies for the structural characterization of nucleic acids. In this chapter, the rudiments of NMR of nucleic acids are introduced, followed by recent trends in the study of structure and dynamics of nucleic acids.

The general approach adopted for the solution structure determination of nucleic acids involves:
(i) Classification of individual chemical shifts.
(ii) Identification of networks of coupled spin systems contained within individual nucleotide units using COSY, ECOSY, TOCSY etc., and
(iii) Sequential linking of the identified nucleotide units by NOESY spectrum in H_2O or ^2H_2O, where one observes inter-nucleotide NOE connectivities, such as imino (TH3/G1H)→imino (TH3/G1H) or $(H6/H8)_{i+1}$→ (H1′/H2′/H2″).

3.2 Classification of Chemical Shifts

As in the case of proteins, BMRB provides information about various non-exchangeable and exchangeable ^1H chemical shifts in DNA and RNA. The data-bank provides all the observed chemical shifts with their standard deviation for each of the observable resonance. The NMR spectral data of nucleic acid is available at the website:www.bmrb.wisc.edu/ref_info/statful.htm.

The histograms in Figures 7.8 A and 7.8 B depict percentage of chemical shifts for non-exchangeable (spanning 0-10 ppm) and exchangeable (spanning 6-16 ppm) protons in DNA and RNA, respectively. The histograms depict the percentage of nucleotide units having a particular chemical shift within a certain range, which is dictated by the observed maximum and minimum chemical shifts for each type of ^1H.

As is evident from Figure 7.8 A, $T(CH_3)$ resonates up-field around 2.28 ppm. This spectral region is well separated from the rest of the chemical shifts. Likewise, Figures 7.8 A and 7.8 B help in classifying each group of ^1H resonances into distinct spectral regions. Some of them overlap with others while others lie in distinct regions. Such spectral classification is an important step to achieve sequence specific resonance assignments as discussed later. Similar classifications of chemical shifts are possible for other spins, such as ^{13}C, ^{15}N and ^{13}P. This knowledge is useful in the analysis of hetero-nuclear correlated spectra.

The chemical shifts of aromatic protons are mainly influenced by stacking effects from neighbouring base pairs. Thus, they provide important information on the secondary and tertiary structures of nucleic acids, and are also crucial for sequential resonance assignments.

3.3 Identification of Networks of Coupled Spin-systems

Protons belonging to each sugar ring form a distinct network of coupled spin systems (Figure 7.9 A and B). Hence, one is able to group them as belonging to individual nucleotide units. Besides, T(H5)-T(CH_3) (box *a* in Figure 7.9 A) and C(H5)-C(H6) (box *b* in Figure 7.9 A) form yet another independent coupled spin systems. 2D ECOSY (Figure 7.9 A) and TOCSY (Figure 7.9 B) spectra in ^2H_2O prove useful for the purpose of identification of such spin systems. Some regions of interest, which aid in such identification are shown in boxes in Figure 7.9 A and B. The zoomed region in Figure 7.9 A shows a selected region of the ECOSY of a self-complementary dodeca-nucleotide. It depicts well resolved H2′/H2″-H1′ J-correlations. The zoomed region in Figure 7.9 B shows the selected region of the TOCSY of the same sample, which help in identifying twelve sets of resonances (H1′, H2′, H2″, H3′ and H4′) belonging to the twelve nucleotide units.

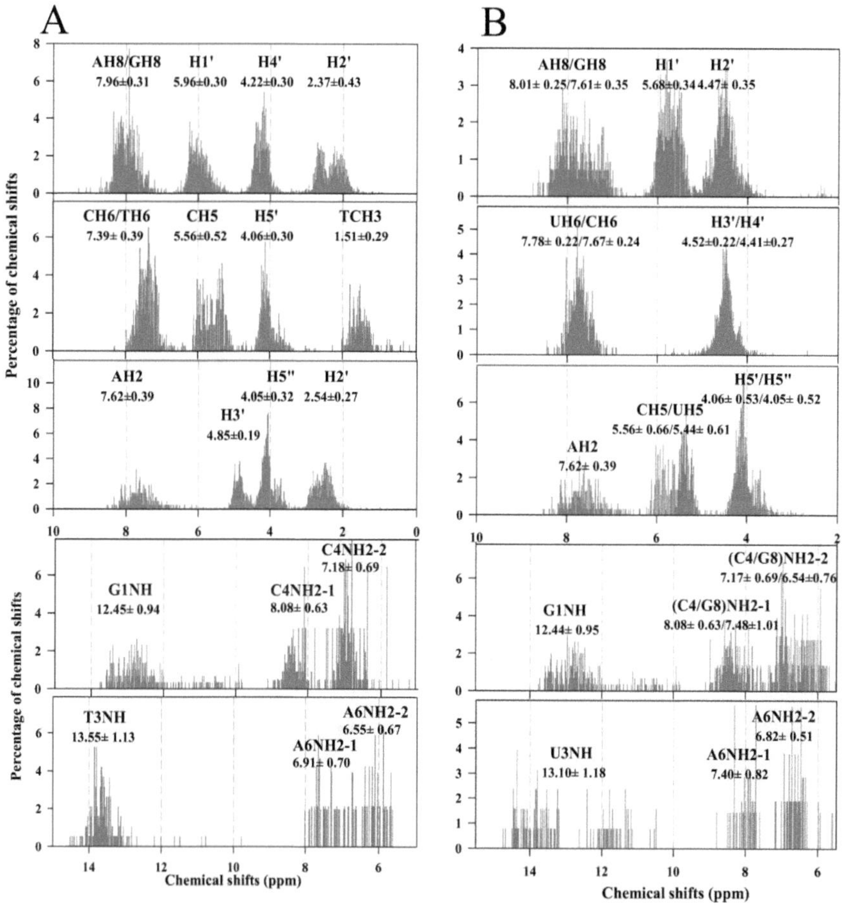

Figure 7.8: Distribution of various non-exchangeable and exchangeable ^1H chemical shifts in (A) DNA and (B) RNA, derived from BMRB. Number of bins has been set to 256 in generating these histograms.

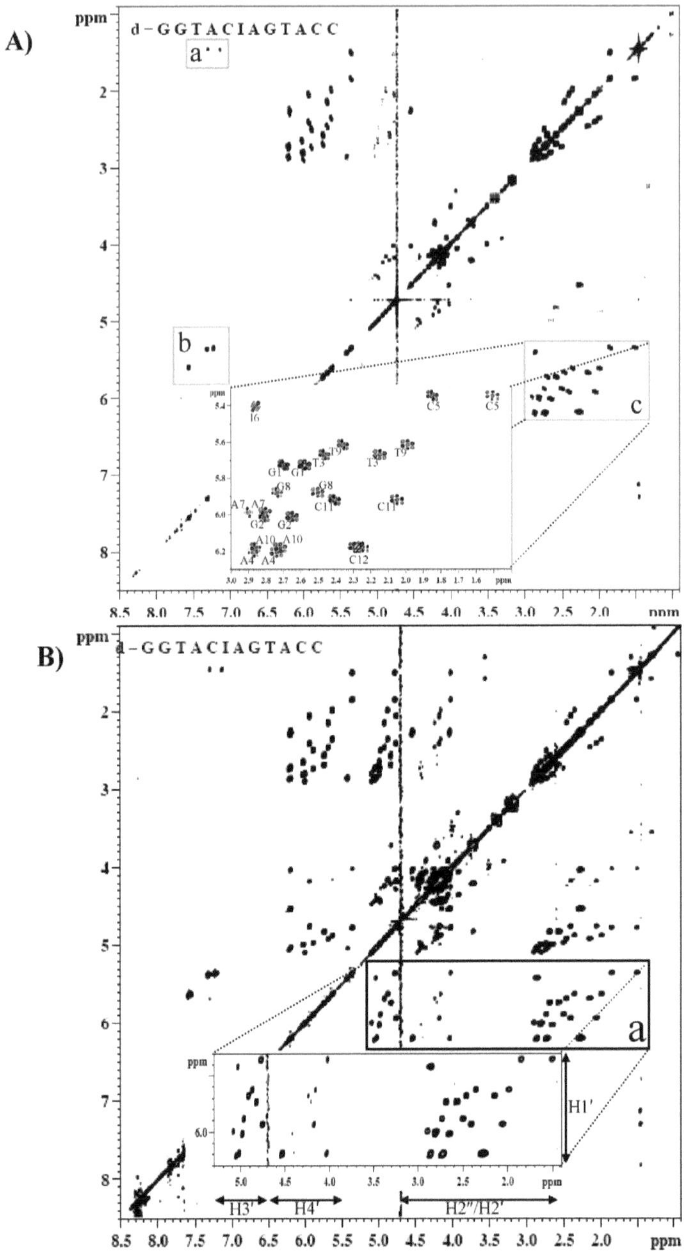

Figure 7.9: (A) ECOSY of d-GGATCIAGTACC (I stands for inosine, an analogue of G). The zoomed region depicts H2'/H2"-H1' J-correlations. Black and grey contours depict the +ve and –ve components of individual peaks, respectively.(B) TOCSY of d-GGATCIAGTACC. The zoomed region depicts TOCSY correlations from H1' to H2', H2", H3' and H4'.

3.4 Sequence Specific Resonance Assignment Strategies

Using 100% abundant ^{1}H and ^{31}P nuclei, one may use one of the following pathways to carry out sequence specific resonance assignments.

$$(H6/H8)_{i+1} \longrightarrow (H1'/H2'/H2''/H3')_{i}$$
$$\{T(N3H)/\,G(1NH)\}_{i} \longrightarrow \{T(N3H)/G(1NH)\}_{i\pm 1}$$
$$(^{31}P)_{i} \longrightarrow (H3')_{i-1} \longrightarrow (H4')_{i-1} \longrightarrow (^{31}P)_{i-1}$$

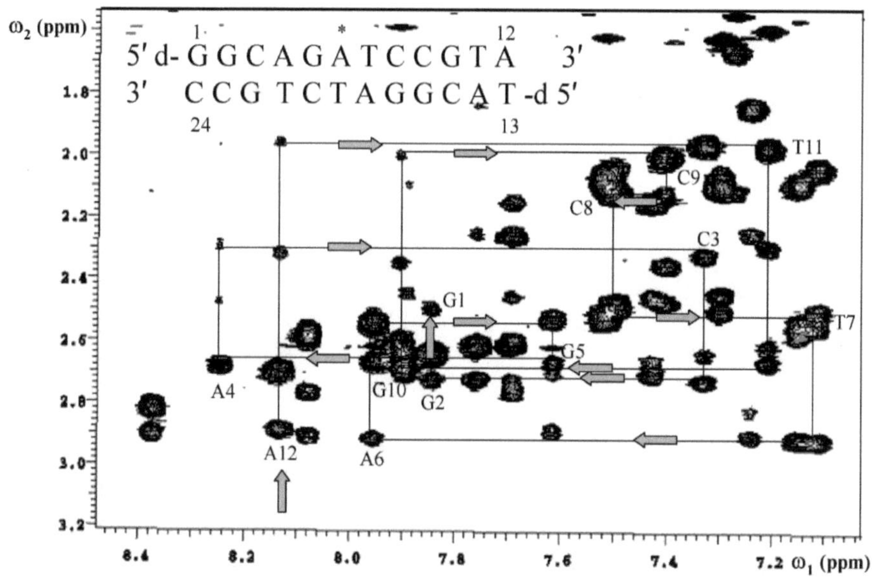

Figure 7.10: Selected region of NOESY of a non-self complementary DNA duplex, which depicts $(H6/H8)_{i+1} \rightarrow (H2'/H2'')_i$ NOEs for one of the two strands (top strand). Adenosine at position 6 (shown with an asterisk) is methylated.

As an illustrative example of sequence specific resonance assignment procedure, Figure 7.10 shows a selected region of the 2D NOESY of a non-self complementary DNA duplex, which depicts $(H6/H8)_{i+1} \rightarrow (H2'/H2'')_i$ NOEs for one of the two strands in a non-self complementary DNA duplex. Following the sequential pathway resonance assignments for the individual sugar ring H2′/H2″ and the base protons H6 and H8 are obtained. Once, H2′, H2″, H6 and H8 resonances have been assigned, remaining protons are assigned by the combined use of other regions of NOESY and TOCSY spectra. The stereo-specific assignment of H2′ and H2″ is usually achieved by comparing the intensities of the H1′-H2′ and H1′-H2″ NOE peaks. The latter is stronger than the former. On the other hand, the stereo-specific assignment of H5′ and H5″ is achieved based on their J-couplings to H4′ and NOEs to H3′ and H4′.

Whenever there is a major structural deviation from the standard form of DNA, some of the NOE cross peaks are either absent or overlap with other cross peaks. In these situations, the procedures based on ^{31}P-^{1}H correlations such as $(^{31}P)_i \longrightarrow (H3')_{i-1}$ correlations in ^{31}P-^{1}H hetero-nuclear correlated spectrum are found useful (Figure 7.11). For example, in an A-I mismatch base-paired self-complementary duplex d-GGTACIAGTACC, the above mentioned sequential (base → sugar) NOEs are not observed between I6 and A7. This is due to a bulge caused by the presence of a pair of purine-purine base-pairs in the middle of the oligomer. In such a situation, ^{31}P-^{1}H correlations (Figure 7.11) help in overcoming the problem of assignments.

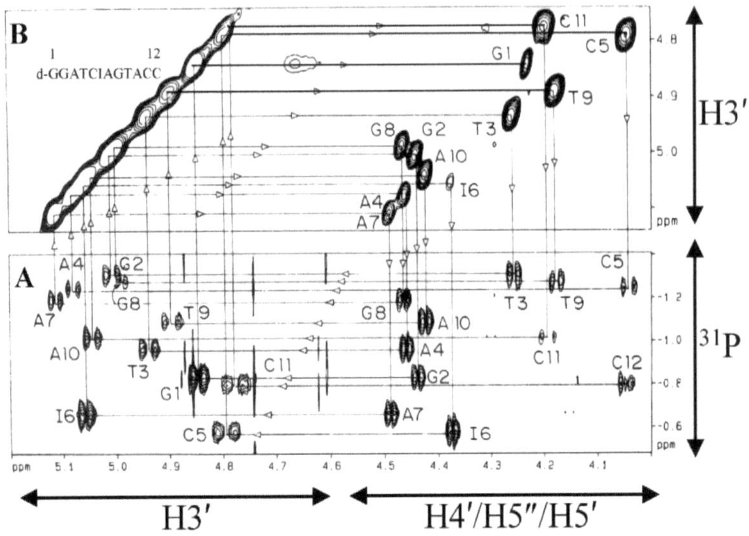

Figure 7.11: (A) ^{31}P-^{1}H hetero-nuclear correlated (HELCO) and (B) selected region of TOCSY spectra of a self-complementary dodeca-nucleotide d-GGTACIAGTACC.

3.5 Resonance Assignments using ^{13}C and ^{15}N Labelled Nucleic Acids

As discussed, spectra of nucleic acids suffer from severe spectral overlaps because of lower dispersion of individual chemical shifts. This makes it desirable to make use of additional nuclei such as ^{13}C and ^{15}N. Like the ^{31}P-^{1}H correlation experiments, one can use ^{13}C-^{1}H correlation in natural abundance for the assignment of ^{13}C spins. For example, through a concerted use of [^{1}H-^{1}H]-TOCSY, and [^{31}P-^{1}H] hetero TOCSY and [^{13}C-^{1}H]-HSQC, one can walk along a nucleotide chain and achieve complete sequence specific resonance assignments for all the sugar ^{1}H and the ^{13}C spins. [^{13}C-^{1}H]-HMBC can be used to monitor intra-base [^{13}C-^{1}H] long-range correlations between ^{1}H and ^{13}C spins.

Recent advances in labelling nucleic acids, have opened possibilities for using strategies based on $^{13}C/^{15}N$ hetero-nuclear multi-nuclear experiments. One of the approaches is similar to the one developed for protein NMR. For example, homo-nuclear (^1H-^1H) NOESY spectra with severe spectral overlaps can be resolved by recording [^{13}C/^{15}N] edited NOESY spectra such as 3D [^{13}C/^{15}N-^1H]-HSQC-NOESY or 3D [^{13}C/^{15}N-^1H]-HMQC-NOESY.

Several other techniques have been proposed for sequence specific resonance assignments in isotope labelled DNA and RNA. These techniques are based on through-bond coherence transfer. For instance, the 3D HCP experiment is a ^1H-^{13}C-^{31}P triple-resonance experiment specifically designed to assign the ribose H3′/C3′, H4′/C4′ on the 5′ side and the H4′/C4′ and H5′, H5″/C5′ resonances on the 3′ side of the intervening phosphorus in ^{13}C-labeled nucleic acids. In this experiment, magnetization is transferred *out* from $(H4')_i \rightarrow (C4')_i \rightarrow (P3')_{i+1}$, then *back* to $(C4')_{i+1}/(C5')_{i+1}$ and finally to $(H4')_{i+1}/(H5'H5'')_{i+1}$. Such experiments cause magnetization transfer following the $(C4')_i \rightarrow (P3')_{i+1} \rightarrow (C4')_{i+1}$ pathway. At the end of the sequential walk, one has the assignments for individual C3′, C4′, C5′ and C5″ and ^{31}P spins. Note that in the event of inefficient $(C4')_i \rightarrow (P3')_{i+1}$ coherence transfer (which may happen when the ε adopts g^- conformation) the $(C2')_i \rightarrow (P3')_{i+1}$ coherence transfer pathway helps one to walk along the backbone assigning respective spins. Once this is accomplished, assignment for remaining ^1H and ^{13}C can be obtained using H-C-C-H or CCH-TOCSY experiments involving a step of cross-polarization from $^{13}C \rightarrow {}^1H$.

With the availability of ^{13}C and ^{15}N doubly labelled RNA, a number of triple resonance experiments have been proposed to correlate ^1H, ^{13}C and ^{15}N spins of the sugar ring and the corresponding base. One such experiment is based on the correlation between H1′ and H6/H8, which uses the intervening ^{13}C and ^{15}N spins. The coherence transfer pathways used are:

$(H1')_i \rightarrow (C1')_i \rightarrow (N9)_i \rightarrow (C8)_i \rightarrow (H8)_i$ (for purine)

$(H1')_i \rightarrow (C1')_i \rightarrow (N1)_i \rightarrow (C6)_i \rightarrow (H6)_i$ (for pyrimidines)

In such experiments, semi-selective pulses are used to keep magnetization transfer to the unwanted pathways to a minimum. The most popular triple-resonance experiments are the 2D H_s (C_s N_b C_b) H_b or the 3D H_s C_s (N_b C_b) H_b. The suffixes ‘s’ and ‘b’ refer to sugar and base, respectively. For example, H_s represent a ^1H belonging to sugar, while N_b represents a ^{15}N spin belonging to base. These experiments have been used successfully in several (^{13}C/^{15}N)-double labelled RNA. The four magnetization transfer steps in these experiments make them less sensitive. This is particularly so in the non-helical regions, which are relatively more dynamic than the duplex regions. Such flexibility broadens resonances due to chemical exchange and hence transfer efficiencies are substantially reduced.

An alternate approach is to use a triple resonance 2D (^1H-^1H) $H_sC_sN_bH_b$ experiment to observe intra-nucleotide H1′-H8 connectivities in purines. This experiment utilizes ^2J(N9-H8), which is of similar magnitude as ^1J(N9-C8). It is possible to correlate H1′ with the corresponding H8 in three steps. The 2D $H_b(C_sN_b)H_b$ experiment has been successfully used to observe and assign H8 and H1′ resonances in duplex DNA. Other useful pulse sequences for observing such intra-nucleotide correlations are $H_sC_sN_b$, $H_bC_bN_b$, $H_sC_s(N_b)C_b$ and $H_bN_bC_b$. Further, [^{13}C-^1H]-HMBC or [^{15}N-^1H]-HSQC has been used to assign NMR spins in individual bases.

4. NMR PARAMETERS IN NUCLEIC ACIDS

In this section, we discuss the NMR properties of nucleic acids, which have been used to study structure, dynamics and interactions of various components of nucleic acids. The approaches listed are also applicable to complexes of nucleic acids with proteins, metal ions and other ligands.

4.1 Three Bond Coupling Constants (^3J)

Spins such as ^1H, ^{13}C and ^{31}P give rise to a number of three bond homo- [$^3J(^1H\text{-}^1H)$] and hetero-nuclear [$^3J(^1H\text{-}^{31}P)$ and $^3J(^1H\text{-}^{13}C)$] coupling constants in nucleic acids. The knowledge of the homo-nuclear $^1H\text{-}^1H$ coupling constants within the sugar ring can be used to fix the pseudo-rotation phase angle (P). Likewise, knowledge of $^3J(^1H\text{-}^{31}P)$ helps in fixing the backbone torsion angles, β (-P5′-O5′–C5′–C4′–) and ε (–C4′–C3′-P3′-O3′-).

There are six $^1H\text{–}^1H$ coupling constants in deoxyribose ring of DNA, namely ^3J(H1′–H2′), ^3J(H1′–H2″), ^2J(H2′-H2″), ^3J(H2′– H3′), ^3J(H2″–H3′) and ^3J(H3′–H4′). The two-bond *geminal* coupling ^2J(H2′–H2″), whose value ranges from -14 to -18 Hz, is independent of sugar conformation. However, the values of the other five *vicinal* couplings are dependent on the conformation of the ring. In the case of RNA, there are three $^1H\text{–}^1H$ coupling constants in the ribose ring, namely ^3J(H1′–H2′), ^3J(H2′– H3′) and ^3J(H3′–H4′), which show dependence on the conformation of ribose ring.

Quantification of the J values in furanose rings has been made possible by advances in 2D-correlated spectroscopy. A good example is the use of ECOSY, which has been extensively used to resolve the characteristic multiplet structures of H1′-H2′ and H1′-H2″ cross peaks and delineate useful ^3J(H1′-H2′) and ^3J(H1′-H2″) values. The features of the individual cross peaks in an ECOSY spectrum are influenced by the line-widths and magnitudes of J-couplings. Under a good digital resolution (<2 Hz/point), ECOSY cross peak patterns for the H1′–H2′/H2″ cross peaks appear as illustrated in Figure 7.9 A. The information about ^3J(H1′-H2′) and ^3J(H1′-H2″) is embedded along the ω_2 axis in H1′-H1′/H2″ cross peaks. It is also possible to obtain information about other *vicinal* couplings such as ^3J(H2′–H3′), ^3J(H2″–H3′) and ^3J(H3′–H4′). Difficulties arise because of the complex multiplicity of the H2′/H2″–H3′ and H3′–H4′ cross peaks, small dispersions in the H3′ and H4′ chemical shifts and relatively small values of coupling constants. Comparison of relative intensities of COSY cross peaks in low-resolution COSY/ECOSY spectra has been used for qualitative assessment of the magnitudes of the coupling constants.

An alternate approach for the quantification of coupling constants is simulation of selected COSY or ECOSY spectral regions with pre-defined chemical shifts (δ′s), coupling constants and line-widths as input. Simulated spectral regions are iteratively compared with the experimental counter parts to estimate J couplings.

This procedure is found useful in cases with broad resonance lines and overlap of individual cross peaks. Values of coupling constants can be estimated to within ±0.1 Hz, when spectra have narrow resonances and non-overlapping cross peaks.

There are five ^{1}H-^{1}H coupling constants outside the furanose ring, namely ^{3}J(H4′–H5′), ^{3}J(H4′–H5″), ^{2}J(H5′-H5″), cytosine ^{3}J(H5–H6) and thymine ^{4}J(TCH$_3$-H6). The cytosine ^{3}J(H5–H6) and thymine ^{4}J(CH$_3$-H6) coupling constants are not of structural interest since the intervening torsion angles are fixed. However, values of ^{3}J(H4′–H5′) and ^{3}J(H4′–H5″) show a strong dependence on the conformation of the intervening torsion angle, γ (-O5′–C5′–C4′–C3′-). Information on these couplings can be extracted from the characteristic multiplet patterns of the individual COSY cross-peaks between H4′ and H5′-H5″. Unfortunately, such cross peaks often suffer from extensive overlaps due to smaller dispersion in H4′, H5′ and H5″ chemical shifts.

4.2 ^{31}P–^{1}H Couplings

There are three ^{3}J(^{31}P–^{1}H) couplings which can be used to fix two backbone torsional angles, namely, β and ε. These are ^{3}J(^{31}P-H5′), ^{3}J(^{31}P-H5″) and ^{3}J(H3′-^{31}P). The three-bond couplings are related to their respective torsional angles by the relation:

$$^{3}J(HCOP) = 15.3\cos^{2}\varphi - 6.1\cos\varphi + 1.6 \qquad 7.3$$

The knowledge of ^{3}J(31P-H5′) and ^{3}J(31P-H5″) can be translated into information about the backbone torsion angle β. Similarly, ^{3}J(H3′-31P) can be used to estimate the value of the angle ε. Estimation of ^{3}J(31P-H5′) and ^{3}J(31P-H5″) couplings is generally complicated by relatively small dispersion in the H5′ and H5″ chemical shifts. However, there have been several successful attempts to delineate these couplings accurately, even in long DNA molecules.

4.3 Scalar Couplings across Hydrogen Bonds

Recently, J couplings have been observed between nuclear spins on the two sides of hydrogen bonds. These include ^{h1}J(HN), ^{h2}J(NN), ^{h3}J(NC) and ^{h4}J(NN) couplings. The values range from 0.15 to 11 Hz (Figure 7.12). Such couplings can be measured by COSY type experiments. The values of scalar couplings across hydrogen bonds are influenced by the relative geometry of the hydrogen bond partners. The measurement of the so called *trans* hydrogen bond couplings in labelled nucleic acids have become significant as these help to establish the nature of hydrogen bonded base pairs. In ds-nucleic acids, hydrogen bonds are generally seen across individual strands as inter-base hydrogen bonding. Once the spins involved in the J couplings have been assigned, the atoms involved in hydrogen bonding can be established. Knowledge of the hydrogen bonding pattern enables one to distinguish between various kinds of base-pairs (Section 2.5).

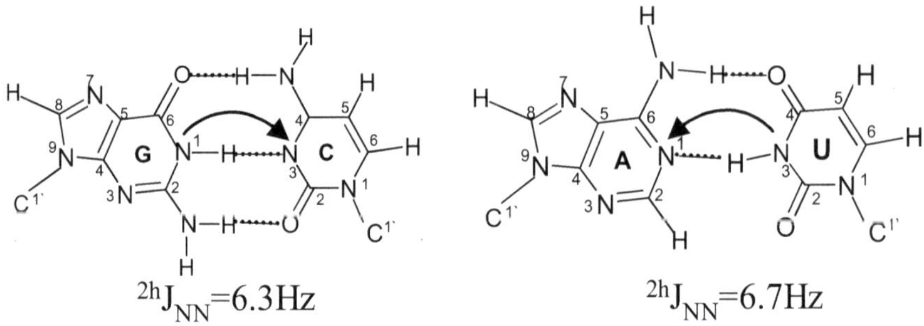

Figure 7.12: Two of the observed trans hydrogen bond coupling constants ($^{h2}J(NN)$). The values are 6.3 and 6.7 Hz, in G:C and A:U base-pairs, respectively .

4.4 Estimation of ^1H-^1H Distances using NOE

In nucleic acids, distances such as cytosine H5-H6 (2.48 Å) and thymine CH_3-H6 (3.0 Å) serve as useful yardsticks for the estimation of unknown ^1H-^1H distances using NOE. The H2′-H2″ (1.8 Å) is yet another useful distance in DNA. The choice of known distance is important in view of the segmental mobility and anisotropic motions associated with different spin systems. Thus, the thymidine H6-CH_3 distance may be used as a yardstick for NOEs involving CH_3 protons and the H2′-H2″ distance for all NOEs involving sugar protons. The cytosine H5-H6 distance serves as a yardstick for other ^1H-^1H distances. Definition of lower and upper bounds is a reasonable criterion placing experimental constraints for ^1H-^1H distances in structural simulations.

Alternatively, the σ_{ij}s obtained from the y-intercept of NOE/τ_m vs τ_m plots can be used in structural simulations. This method allows use of NOE data sets obtained at mixing times much longer than that permitted by the initial slope approximation. Such an approach helps in collecting more distance constraints and hence better description of the structure. Unfortunately, the NOE approach provides only short-range distances in nucleic acids and their complexes. Another limitation is the quenching of NOE interactions by dynamic processes at intermolecular interfaces in nucleic acids and their complexes, particularly in cases of weak binding.

4.5 Residual Dipolar Couplings (RDC)

Nucleic acids and their complexes often form elongated structures. Only a limited number of long-range restraints are available as compared to globular proteins that are compactly-folded. This results in a smaller number of NOE restraints for RNAs when compared to proteins of similar molecular weights. Residual dipolar couplings are particularly important in nucleic acids and their complexes. These measurements promise to significantly improve the accuracy and precision of the structures determined by NMR. In particular, in the case of RNA, which forms more extended structures and have several sub-domains, lack of inter-domain NOE restraints affects the global precision of structures. In such situations, RDC helps to overcome the problem of domain orientation, by providing orientation constraints for N-H, C-H and C-C bond vectors, and thereby improving the global structure.

Partial alignment can be achieved using nucleic acids dissolved in phospholipids, filamentous phages or purple membrane. Bacteriophage Pf1 has been recommended as an ideal liquid-crystalline medium for RDC measurements in nucleic acids. The negatively charged nucleic acids are aligned by steric interactions with negatively charged phage particles, without directly binding to the phage. This does not alter the overall tumbling time of the molecule. Hence, the line-widths are not affected. However, some of the commercial materials may contain *nuclease*s, which may cleave nucleic acids. Such contamination should be avoided. Also, samples may aggregate in certain solvents.

Unlike in the case of proteins, the ^1J (^1H-^{15}N) couplings provide only a limited number of RDC in nucleic acids. The ^1H-^{13}C and ^{13}C-^{31}P coupling constants are small and cannot be measured accurately. Thus, $^1J(^1H$-$^{13}C)$ couplings serve as a major source of information in nucleic acids. Generally, the distribution of measured couplings covers only a small part of all possible directions. It is therefore possible to estimate alignment tensor through a variational protocol.

4.6 Use of Paramagnetic Labels

The use of paramagnetic spin-labels has proved valuable for long-range global constraints for refinement of structures of nucleic acids and their complexes. Paramagnetic spin label such as proxyl can be covalently attached to nucleic acids. Distances from this position can be estimated by measuring the effect of electron-proton relaxation on T_2 as reflected in ^1H line-widths. Distances of about 15 Å from the site of the paramagnetic centres can be estimated by measuring the dipolar relaxation on the ^1H from the paramagnetic centre. The incorporation of certain lanthanides also provides structural information in the form of pseudo-contact shifts.

The concerted use of NOE, torsional restraints, paramagnetic labels and RDC derived restraints significantly improves the local as well as the global precision of the structural ensembles of nucleic acids.

5. STRUCTURE SIMULATIONS OF NUCLEIC ACIDS

The approach to studies on the structure and dynamics of nucleic acids is similar to that discussed for proteins. One attempts to get as much structural information as possible from NMR and then use these as constraints in a suitable structural simulation algorithm. However, the parameters in these algorithms are not as well developed as that for proteins. Difficulties in handling the electrostatic and stacking interactions in nucleic acids and their complexes make the situation more complex.

5.1 Conformation of Deoxyribose and Ribose Rings

Conformation of sugar rings of individual nucleotide units in several oligonucleotides is determined from the knowledge of the intra-sugar ^{1}H-^{1}H *vicinal* couplings and ^{1}H-^{1}H distances. Figure 7.13 shows the dependence of the five ^{1}H-^{1}H vicinal coupling constants in the furanose ring on the angle P, using the following Karplus relation.

$$^3J\ (H\text{-}C\text{-}C\text{-}H) = 10.2 \cos^2 \theta - 0.8 \cos \theta \qquad 7.4$$

A striking feature is the sensitivity of the values of ^{3}J(H1′-H2′), ^{3}J(H2″-H3′) and ^{3}J(H3′-H4′), which show significant variations as a function of angle P. Thus, one can compare experimentally derived coupling constants with theoretical values to determine the conformation of individual sugar rings. In particular, it may be noted that the models of A-, B- and Z-DNA using fibre diffraction use standard sugar conformers namely C3′-*endo* for A, C2′-*endo* for B and alternately C2′-*endo* and C3′-*endo* conformations for Z-DNA.

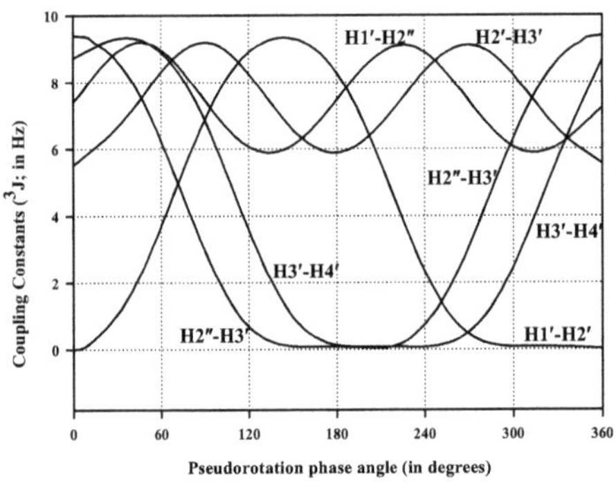

Figure 7.13: Dependence of the five ^{1}H-^{1}H vicinal coupling constants in deoxyribose ring on the pseudo-rotational angle P.

As discussed in Section 4.1, ECOSY helps in resolving conformation dependent characteristic multiplet patterns of H2′-H1′ and H2″-H1′ cross peaks. Though the estimated couplings may be error prone, even in difficult cases of poor resolution, one can determine as to which one of the two couplings is larger. This helps in fixing certain windows for the sugar puckers. For example, a ^{3}J(H1′-H2′) value larger than ^{3}J(H1′-H2″) qualitatively indicates that the sugar ring has a conformation in the S domain of the pseudo-rotational map, with P ranging from C1′ *exo* to C3′ *exo*. Further, as is clear from Figure 7.13, information about ^{3}J(H2″-H3′) and ^{3}J(H3′-H4′) is useful in further narrowing the domains of sugar puckers. However, it is not always straight forward to estimate J values from ECOSY spectrum, because of the low intensity of the corresponding cross-peaks.

The knowledge of a single coupling constant helps in fixing a range for the angle P. For example, a non-zero value for ^{3}J(H3′-H4′) rules out the possibility of classical C2′-*endo* geometry which has been used extensively in fibre-diffraction based model for B-DNA. Similarly, a non-zero value of ^{3}J(H1′-H2′) rules out the possibility of C3′-*endo* geometry. In most of the known B-DNA structures to date, the sugar puckers lie in the S domain of the pseudo-rotational wheel. The solution NMR data has clearly established such behaviour. Most of the nucleotide units assume a sugar pucker in the range of 90-150°. On the other hand, ribose rings in A-RNA stems generally prefer sugar puckers in the N domain.

In studies on solution conformations of mono- and di-nucleotides, the coupling constant data has often been interpreted in terms of a dynamic equilibrium between sugar puckers in the C2′-*endo* (or S domain) and C3′-*endo* (N domain) conformers. While a fast equilibrium on the NMR time scale between these two widely different conformers is unlikely in double-helical DNA, the proposition is nevertheless attractive since the two conformations form the basis for A and B forms of DNA. In such a situation, one can monitor the value of ^{3}J(H2′-H3′), which does not differ for the C3′-*endo* and C2′-*endo* conformers. When it is difficult to quantify H3′-H4′, one can monitor the relative intensities of the cross-peaks in low resolution COSY spectrum. One can then use the fact that, with the similar multiplet pattern, the intensities of the cross peaks are largely determined by the magnitude of the active coupling constant. Thus, dissimilar intensities of the H2″-H3′ and H3′-H4′ cross peaks would indicate that the sugar geometry is different from C3′-*endo* or C2′-*endo* and that there is no C3′-*endo*/C2′-*endo* equilibrium.

An alternative approach to obtain details of the sugar geometries is based on the ^{1}H-^{1}H distances in sugar ring. Such distances have been calculated theoretically as a function of P by making use of bond distances, bond angles, and torsion angles as input parameters. The plots (Figure 7.14 A) show that the H1′-H3′ and H2′-H4′ distances are greater than 3.7 Å for all values of P. Therefore, one does not expect NOESY cross peaks of appreciable intensities for such proton pairs. The H1′-H2″ and H2′-H3′ distances are short and vary in a narrow domain of 0.2 Å. Though these proton pairs give strong NOESY peaks, their intensities are relatively insensitive to the angle P. The H1′-H2′, H2″-H3′, and H3′-H4′ distances are also insensitive to sugar pucker.

On the other hand, H1′-H4′ and H2″-H4′ distances show strong dependence on sugar pucker. These distances vary between 2.8 and 4.2 Å. For a C3′-*endo* conformation one expects a strong H2″-H4′ and a weak H1′-H4′ NOE. For C2′-*endo* both H1′-H4′ and H2″-H4′ cross-peaks will be weak. In the case of a sugar pucker around O4′-*endo* conformation (P=90°), an intense H1′-H4′ peak is expected while H2″-H4′ may be absent. An important point is that there is only a narrow range of P angles (50-120°) where the H1′-H4′ peaks are likely to be significantly stronger than H1′-H2″ peaks. The fact that the sugar puckers can be established fairly well by using J values from COSY/ECOSY and ^{1}H-^{1}H distances from NOESY, highlights the power of 2D NMR techniques in obtaining the structure and dynamics of deoxyribose and ribose rings in nucleic acids. The cross-peaks discussed above fall in different spectral regions of the spectra, which help in avoiding resolution problems.

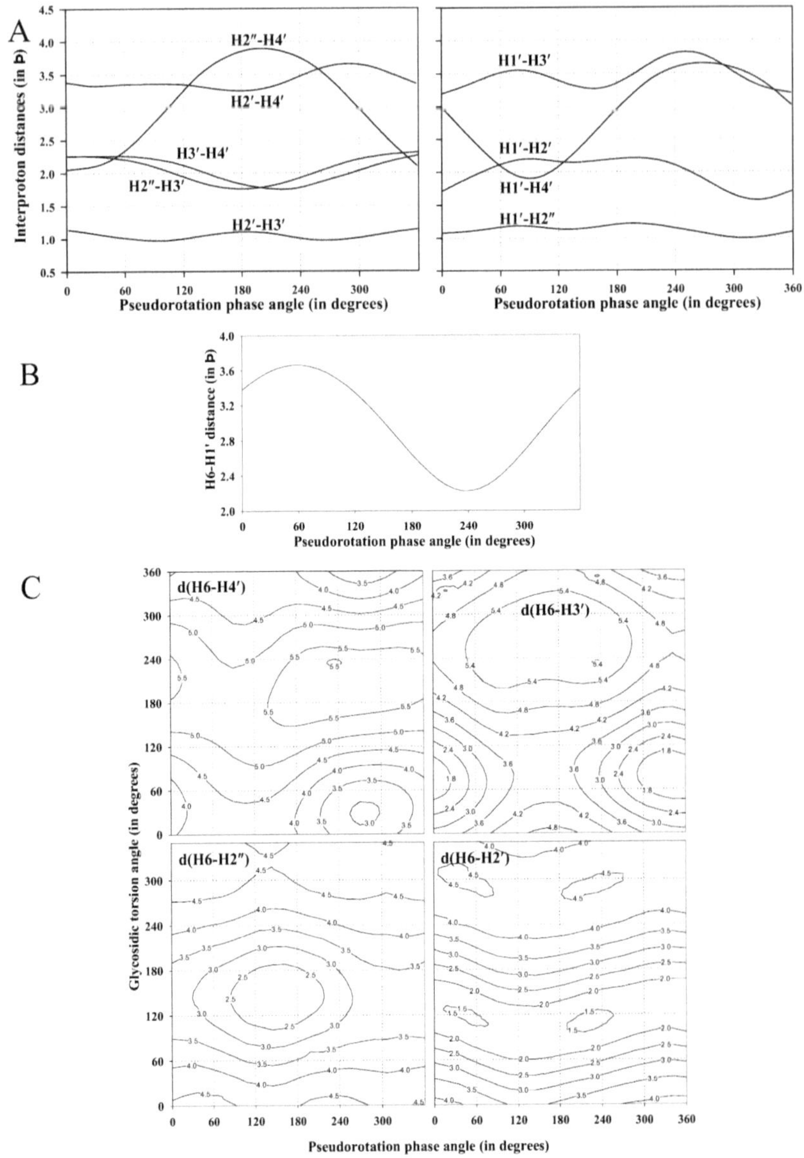

Figure 7.14: Theoretically calculated (A) intra-sugar ring ^1H-^1H and (B) intra-nucleotide H6-H1′ distances as a function of pseudo-rotation angle P. (C) Simulated iso-distance contours in the [P, χ] space.

5.2 Backbone Torsion Angles

The angle δ, which is also part of the sugar ring, can be fixed from the knowledge of pseudo-rotation angle P. The torsion angle β can be fixed by measuring values of the $^3J(^{31}P\text{-}H5')$ and $^3J(^{31}P\text{-}H5'')$, as described earlier. Similarly, ε can be estimated from the magnitude of $^3J(H3'\text{-}^{31}P)$. However, for a given value of $^3J(H3'\text{-}^{31}P)$ one obtains four values of ε. Further, the value of this J is nearly the same for the two preferred values of ε in DNA structures, namely 210° and 270°. The angle γ can be fixed from the knowledge of $^3J(H4'\text{-}H5')$ and $^3J(H4'\text{-}H5'')$. Unfortunately, H4′, H5′ and H5″ have chemical shifts in a narrow spectral range (3 to 4.5 ppm). The corresponding cross-peaks suffer from severe spectral overlap and it is difficult to estimate the J values in such cases.

There are no convenient spin-spin couplings which can fix the torsion angles α and ζ. This makes it difficult to establish sugar-phosphate conformation using coupling constant data. In such circumstances, inter-nucleotide distances are useful to discriminate between possible values. The overall geometry including α and ξ can be fixed by making use of the adjacent nucleotide $^1H\text{-}^1H$ distance constraints such as $(H8/H6)_i \rightarrow$ (H1′, H2′, H2″ and $H6/H8)_{i-1}$ coupled with structural simulation algorithms such as distance geometry, energy minimization or restrained molecular dynamics.

5.3 Glycosidic Torsion Angle (χ)

As seen from Figure 7.14 B and C, the intra-nucleotide $^1H\text{-}^1H$ distances between base protons (H6 for purines and H8 for pyrimidines) and the sugar protons (H1′, H2′, H2″, H3′ and H4′) depend on the sugar pucker as well as on the glycosidic torsion angle (χ). These distances, as measured from NOESY spectrum, can be used to fix angles χ for individual nucleotide units. The observations can be interpreted with the help of simulated iso-distance contours in the [P, χ] space. The χ values for nucleotides in typical B-DNA or A-RNA conformations are in the *anti* domain, ranging from 225° to 280°. However, *syn* conformations have been observed in several cases and have been used in the model of Z-DNA.

5.4 Hydrogen-bond Constraints for Base-pairs

Another useful parameter for structure calculation for nucleic acids is hydrogen-bond constraints. The 10-15 ppm ^1H region spanned by imino protons provides information not only on the type of base-pairs formed, but also on the sequential neighbours. This region is also used to study the melting behaviour of nucleic acids. The chemical shifts of imino and amino ^1H involved in hydrogen bonding depends on base stacking, hydrogen bonding, sequential neighbours and exposure to the solvent. For example, the imino proton of hydrogen-bonded G resonates around 13.5 ppm in a WC base pair. However, for a G:U base pair the resonance of G(1NH) lie in the range 10-12 ppm.

The amino protons in C involved in hydrogen bonding show narrow NOE cross-peaks where as the solvent exposed amino groups in G, A and C are relatively much broader. Sometime, it is difficult to observe such resonances. The base pair pattern can be deduced by looking at the NOE patterns. The short NOE distances depend on the hydrogen bond schemes as indicated in Figure 7.3 and 7.7. For example, in WC G:C base-pair, the G(1NH) shows a strong NOE contact with one of the amino ^{1}H of cytosine involved in hydrogen bonding. The A:U(T) base pairs are easily identified by a strong NOE cross peak between purine A(H2) and pyrimidine (3NH).

More recently, the coupling constants across hydrogen bonds (N-H...N), have emerged as a potential source of information to establish base pairing patterns in labelled nucleic acids. This allows one to fix the hydrogen bond donor and acceptors in nucleic acids through the HNN-COSY experiment.

Once the nature of base-pairs has been established, the constraints for hydrogen bonds can be fixed to a fair degree of accuracy. In fact, the pattern of hydrogen bonded base-pair is an important input for fixing the conformation of unusual (non WC base-paired) structures of nucleic acids. For example, for the usual WC base-pairing, the inter-atomic distances, G(O6)-C(H41), G(H1)-C(N3), G(H21)-C(O2), A(H61)-T(04) and A(N1)-T(H3), within an individual base-pair can be restrained in the range 1.7–2.0 Å. On the other hand, the heavy atoms in these base-pairs are restrained within a range of 2.8–3.2 Å. These constraints are generally relaxed during the final stages of the structure calculations. For strong NOEs generally observed between A(H2) and T(H3) belonging to A:T base pairs, and the analogous G(H1) and C(H4) in G:C base-pairs, the distances are restrained in the range 2.4–3.3 Å and 2.0–3.0 Å, respectively. Similarly, for non-WC hydrogen bonding schemes, which can be established from ^{1}H NMR, distance constraints are defined according to the base pair pattern.

5.5 3D Structure from NMR data

Computational procedures which have been developed over the years for protein structures rely on the implementation of the principles of distance geometry, restrained molecular dynamics calculations or interactive computer graphics. The NMR information is used as constraints to obtain a family of structures consistent with the experimental data. Similar methods have been used for nucleic acids. One of the difficulties with nucleic acids is that unlike in proteins, the NMR experimental constraints are usually those between neighbouring nucleotides. Thus, while one is able to obtain short-range structural details, long- range behaviour such as bends in the helical structures, inter-loop orientations, the helicoidal parameters such as twist, tilt, axial rise are difficult to obtain. Recently, several efficient and reliable protocols have been developed for nucleic acids and their complexes. Of the various approaches, simulated annealing protocol has been quite useful.

6. UNUSUAL DNA STRUCTURES

It is known that DNA is usually found in living organisms in the form of a B-type helical duplex structure. The usual conformation for RNA is A-type. Recent experiments have indicated that even in the canonical B-DNA, there are significant sequence dependent variations. Further, depending on the sequence and experimental conditions, DNA and RNA can adopt non-canonical structures. In fact, alternate structural forms of DNA and RNA are inherent properties of these molecules, which are utilized in their biological functions.

There have been several NMR studies on unusual forms of DNA structures. These include sequence dependent structure variations in duplex nucleic acids, single-stranded hair-pins, triplex, tetraplex and i-motif DNA. Complete description of these forms is beyond the scope of this book. In this section, NMR characterization of certain molecular systems has been presented. This provides insights into the polymorphism of DNA.

For RNA, conformational variations are even more frequent. Biologically active RNA molecules show unique 3D structures similar to proteins. Single and ds-RNA segments with loops and bulges as well as naturally occurring complex RNA structures have been discussed in Chapter 10.

6.1 Sequence Dependent Variations in B-DNA

If DNA had a uniform B-helix as is implied by WC model, then it would be difficult to understand why certain sequences are hot sites for recognition by drugs and certain enzymes. For example, restriction enzymes specifically cleave DNA at certain sites, by recognising specific segments, which may be only 4-6 nucleotides long in a long DNA molecule. Such segments have been studied in detail using NMR and interesting sequence dependent structural variations have been observed.

As an example, consider the structure of a 14-mer self-complementary duplex d-GGATTGCCAATCC. The GCCAAT segment is a recognition motif for several transcription factors. The torsion angles as determined from NMR coupled with distance geometry algorithm TANDY (Table 7.3) show interesting variations along the sequence. A significant B_I-B_{II} type deviation is observed in the central portion marking the beginning of the recognition motif. The base pairs are oriented as in A-DNA, but the sugar geometries and other torsion angles are similar to those in B-DNA. Similar sequence dependent variations have been observed in a number of other DNA sequences, which are specifically recognised by restriction enzymes.

6.2 Mismatch Base-pairs

Replication of DNA must occur with a high degree of precision for genetic information to be faithfully transmitted from one generation to the next. The WC base-pairing scheme provides a mechanism whereby this is accomplished. Incorporation of non-WC base pairs into duplex DNA may introduce errors in the genetic code. Such non-WC base-pairings are called mismatches. There have been several structural studies both by single crystal X-ray diffraction and NMR on R-R (A-A, A-G, A-I and G-G), R-Y (A-C and G-T) and Y-Y (C-C, T-T and C-T) mismatches in oligonucleotides. Most studies have used sequences designed to incorporate such non-WC pairs at specific sites. Note that, for clarity, WC base-pairs are represented as R:Y and mismatch base-pairs by R-Y.

Table 7.3: Torsion angles (in degrees) in self-complementary ds DNA GGATGGCCAATCC.

Residue	α	β	γ	*P*	χ	ε	ζ
G 1		208.5	30.8	21.0	227.3	170.4	275.2
G 2	315.5	204.2	32.7	152.6	251.0	166.4	261.1
A 3	313.4	219.7	25.5	180.2	249.1	158.6	256.7
T 4	308.5	190.7	74.6	139.9	224.3	161.4	261.7
T 5	271.7	208.7	71.9	119.1	233.3	160.2	277.2
G 6	314.0	215.1	36.9	156.3	248.4	293.5	173.9
G 7	234.8	140.4	73.6	168.1	240.6	185.1	262.7
C 8	291.8	216.1	55.0	129.4	225.1	195.8	242.3
C 9	278.5	164.0	97.0	126.5	225.6	163.7	246.4
A 10	334.8	193.6	39.4	164.0	249.2	165.8	244.4
A 11	342.0	181.7	28.3	161.1	234.7	177.2	218.5
T 12	326.5	160.7	70.7	177.4	231.4	183.6	268.1
C 13	316.8	186.6	31.7	122.5	240.8	169.7	274.4
C 14	301.7	210.4	36.0	148.8	260.2		

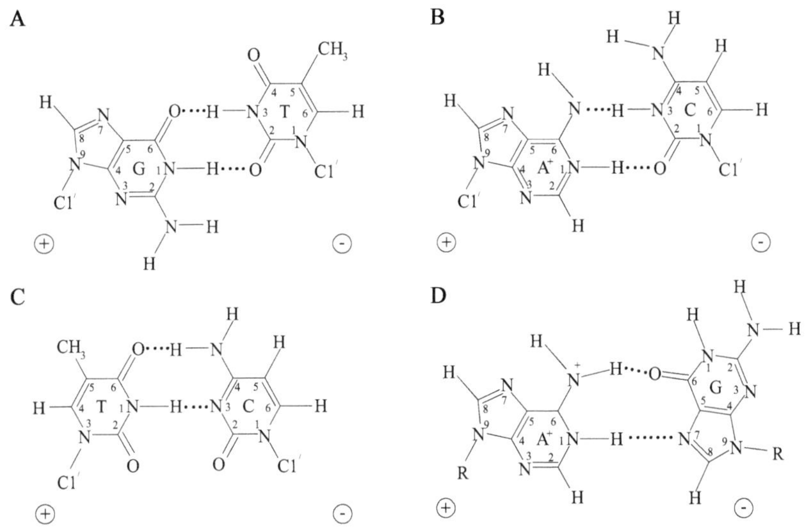

Figure 7.15: Some mismatch base-pairings established by NMR. (A) G-T, (B) A^{+}-C, (C) T-C and (D) A^{+}-G

One such mismatch base-pair, namely G-T is discussed below. The mismatch can be accommodated in a B-DNA duplex by forming a wobble-type structure such that the G ($4NH_2$) protrudes into the minor groove, while the T(4CO) group is displaced into the major groove, as compared to a normal WC base-pair. This requires an alteration in the overall conformation of the DNA duplex at the site of the mismatch. Such a structural change is reflected in up-field shifts of the imino ^{1}H resonances belonging to both G and T [with G(1NH) around 10.7 ppm and T(3NH) around 11.8 ppm)]. Further, the two protons show a strong NOE cross peak with one another. This provides evidence for their participation in hydrogen bonding and in the stabilisation of the G-T base pair (Figure 7.15), and rules out other possible structures. The observations are also indicative of the G base sliding into the minor groove and T into the major groove, simultaneously.

Other mismatch base-pairing can be characterised in a similar fashion from the assignment of imino and amino ^{1}H resonances belonging to individual bases, *trans* hydrogen bond couplings and their NOE correlations with other intra- and inter-strand base protons. Each mismatch base pair has one or two imino protons, which participate in hydrogen bonding. In addition, amino protons belonging to A, A^+, C, C^+ and G may be involved in hydrogen bonding. Thus, the number of observed imino and amino ^{1}H signals and their positions in a NMR spectrum depends on the pattern of base pairing.

The nature of conformational equilibrium when more than one structure is present, can also be investigated and information on thermodynamics and kinetics can be obtained.

6.3 Hairpin Nucleic Acids

Hairpins are single-stranded structures, which fold back to form a helical stem, leaving two or more nucleotides in the form of a loop. This class of DNA and RNA structures have important roles in biology. In DNA, these can act as regulatory sites in gene transcription and replication. Hairpins are integral part of natural RNA. Hence, such structures have attracted wide attention of structural biologists. There are a number of reports on hairpin DNA and RNA by spectroscopic and NMR methods. Chemical probing of symmetrical loops in cruciform structures suggest that the optimal loop size is around four nucleotides. However, some sequences such as d-GGTACIAGTACC form a loop with even two unpaired bases.

NMR studies reveal that the stems of hairpins are usually stabilised by adopting B-DNA or A-RNA conformations. The bases in the loop may undergo stacking and non-WC hydrogen bonding interactions. Studies of loop structures of several sequences indicate that there can be considerable structural differences within the loops of the same size. Interesting features have been noted on the conformation of DNA loops containing four pyrimidine nucleotides (Figure 7.16). The right-handed backbone continues through the 3′ top of the stem to the 5′ top of the stem, by taking a sharp turn, called the *turning phosphate*. The phosphate group where this sharp turn occurs has angles ζ and γ in g^+ and t conformations, respectively, while α prefers either g^+ or t conformation.

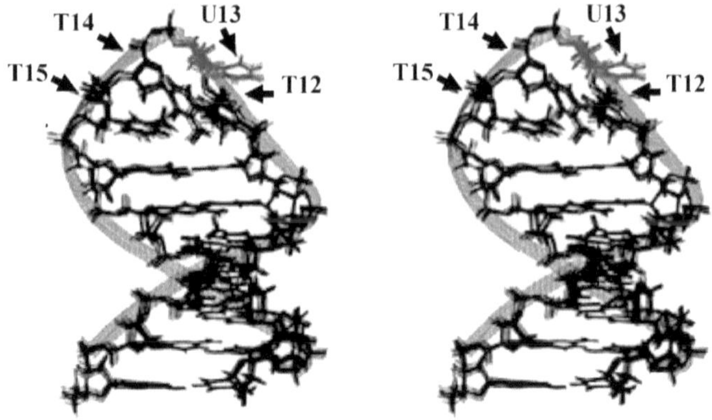

Figure 7.16: Stereo-view of the solution conformation of a hairpin DNA with a 4-base loop.

6.4 Parallel Stranded (ps) DNA Duplex

Nucleic acids also have the capacity to form *ps* structures. p*s*-DNA may play an important role in the regulation of replication and transcription, genetic recombination, chromosome folding, mutational processes and RNA splicing. There have been several reports of parallel stretches of DNA in different genomes. For example, a parallel complementary region between two Drosophila DNA sequences, fragments of the suffix and a 5′-non-coding sequence of the alcohol dehydrogenase gene has been reported. Parallel stranded stretches also form part of DNA and RNA triplexes, tetraplexes and H-DNA structures.

A number of DNA structures has been reported in which strands have the same polarity. Most of these structures require modifications of the chemical structures of DNA or a low pH environment. Parallel stranded structures have also been obtained by:

(i) designing hairpins with polarity reversal using a 3′/3′ or a 5′/5′ linkage
(ii) use of modified oligo-nucleotides with a bulky group substitution on the base, and
(iii) modification of backbone or glycosidic linkages.

Such structures often require conditions of low pH (such that the bases are protonated) and low temperature. In such cases ps-duplex formation is sequence dependent. At higher concentrations, higher orders of aggregated structures may be formed.

Parallel stranded duplexes exhibit spectroscopic, thermodynamic, and biochemical properties, which are significantly different from those of conventional B-DNA. The structures can be remarkably stable even under physiological conditions. Such duplexes are stabilised by reverse WC G:C and A:T base-pairs (Donohue base-pairs) or Hoogsteen A.T base-pairing, with the glycosidic bond in the *trans* orientation (Figure 7.7). Under mild acidic conditions, cytosine N3 is protonated (pKa ~ 4-5) and a Hoogsteen $G.C^+$ base-pair can provide additional stabilisation to ps-duplex. With appropriate sequences, the parallel-stranded structures can even form under physiological conditions and in the absence of chemical alterations of the bases or backbones.

5′ d-CCATAATTTACC 3′
5′ d-CCTATTAAATCC 3′

Figure 7.17: Alignment of the two strands in parallel orientation.

The structural feasibility of a ps-duplex with complementary bases has been demonstrated using sequence-constrained oligomers (CCATAATTTACC and CCTATTAAATCC). The sequences match perfectly with parallel polarity in a duplex but are extensively mismatched in anti-parallel orientation (Figure 7.17). Two pairs of $C:C^+$ residues at the ends of the duplex assert parallel polarity. Due to the presence of three hydrogen bonds in each $C:C^+$ base-pairs, such stretches constrain the intervening sequences also to be parallel. The constraints may be switched off by raising the pH. However, the duplex is found to be thermodynamically stable even under physiological conditions. The proposed 3D structure obtained from NMR studies, has backbone, glycosidic and pseudo-rotation angles similar to the one observed in an *aps* B-DNA. However, the helicoidal parameters and patterns of base stacking are significantly different.

6.5 Triple Stranded Nucleic Acids

The sequence-specific recognition of a duplex DNA by a third strand has implications in gene-regulation and site specific cleavage of genomic DNA. Such a complex formation can be achieved through the hydrogen bonding sites in the major and minor grooves of B-DNA.

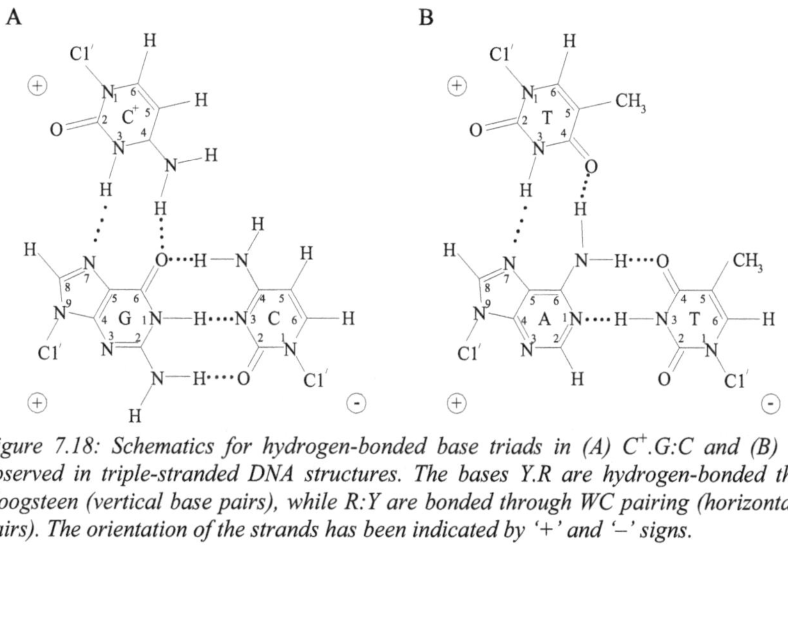

Figure 7.18: Schematics for hydrogen-bonded base triads in (A) C^+.G:C and (B) T.A:T, observed in triple-stranded DNA structures. The bases Y.R are hydrogen-bonded through Hoogsteen (vertical base pairs), while R:Y are bonded through WC pairing (horizontal base pairs). The orientation of the strands has been indicated by '+' and '–' signs.

Several years back, it was shown that poly(rA) and poly(rU) mixed in 1:2 molar ratio form triple stranded structures. The third Y strand binds in the major groove, parallel to the R strand of the WC A form RNA duplex, using Hoogsteen base-pairing [U.A:U] (WC base pairs are represented as R:Y and Hoogsteen by Y.R). In an attempt to form the DNA analogue of such an RNA triple helix, it was observed that a 1:1 molar mixture of $(GA)_n$ and $(TC)_n$ spontaneously transformed into a triplex, together with a free R strand at lower pH. The ease with which this triplex forms is attributed to the moderately high pK_a of C. Such protonated DNA structures were earlier identified as intra-molecular or inter-molecular H-DNA.

The base-base recognition in triplexes is achieved through hydrogen bonds between C^+ and G and between T and A through Hoogsteen pairing (Figure 7.18). Thus, C^+ recognizes C:G base-pair to form C^+.G:C triad, while T recognises A:T pair to form T.A:T triad. The Hoogsteen complementarity is identical to WC, with the exception of the polarity of the strands. Therefore, the sequences of purines in the homo-R strand can be arbitrary (any mix of A and G), whereas the sequences in the two homo-Y strands have to be in accordance with the WC and Hoogsteen complementarities. In contrast to WC pairing in which the two strands are antiparallel, the Hoogsteen pairing has strands in parallel orientation. Such a Y.R:Y triplex consists of an array of T.A:T and C^+.G:C base-triads. The distances between the two glycosidic bonds within each triad are almost equal.

There are several NMR studies on the structures of Y.R:Y triplexes. An interesting example of such a triplex is the one formed when a Y dodecamer (d-CTTCTCCTCTTC) and a homo-R hexamer (d-GAAGAG) is mixed in 1:1 molar ratio. The strands of oligomers combine to form a triple-stranded structure with six each of T.A:T and C^+.G:C base triads as shown in Figure 7.19. The two R strands are placed head to head, with their 3′ ends facing each other in the centre of the structure. One-half of each Y strand contains protonated and the other half contains non-protonated Cs. The two half segments containing protonated Cs are hydrogen bonded to each of the two R-hexamers through Hoogsteen T.A and C^+.G pairing. The segments containing non-protonated Cs are involved in WC (A:T and G:C) pairing. This leads to a palindromic triplex with C2-dyad symmetry with respect to the centre of the structure (Figure 7.19). This triplex structure has C2-dyad symmetry with respect to the center of the structure. In the top half of the triplex, yellow (R) and green (Y) strands are involved in Hoogsteen base-pairing, while yellow (R) and pink (Y) strands are involved in WC base-pairing. In the lower half, yellow (R) and green (Y) strands are involved in WC base-pairing, while yellow (R) and pink (Y) strands are involved in Hoogsteen base-pairing.

Information on the hydrogen bond schematics of such triplexes can be derived from imino and amino ^1H resonances, trans-hydrogen-bond couplings and NOE correlations with other intra- and inter-strand base protons. For example, two sets of cross peaks may be identified: one belonging to the amino protons of protonated Cs (8-10 ppm region), and the other to the non-protonated Cs (6-8.5 ppm). The hydrogen-bonded (HB) and the solvent-exposed (X) amino protons belonging to the 4-NH_2 group in Cs are separately identified. A unique feature of the NOESY spectrum of such a triplex is a network of intra- and inter-base cross peaks involving HB and X amino protons between themselves and with the H5 and H6 protons of Cs (Figure 7.18).

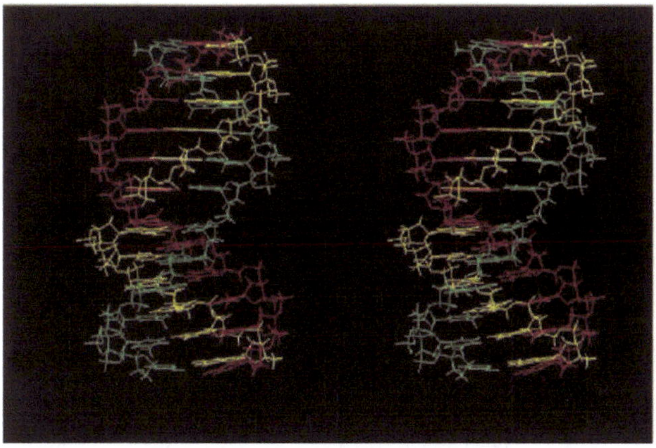

Figure 7.19: A stereo-view of a triple-stranded structure involving two strands each of d-GAAGAG and d-CTCTCCTCTTC derived from NMR. Both the purine strands are shown in yellow, while the two pyrimidine strands are shown in pink and green. S.R. Bhaumik, K.V.R. Chary, G. Govil, L. Keilang, and H.T. Miles. Nucleic Acids Res., 26, 2981-2988 (1988). (Reprinted with permission from Oxford University Press).

In a multi-stranded nucleic acid, each base pair has at least one imino proton (1NH/3NH), which is involved in the inter-base hydrogen bonding. Since the rise per residue along the helix axis ranges from 2.6 Å (A-DNA) to 3.4 Å (B-DNA), one observes NOEs between the imino protons of the sequential base pairs. Thus, one is able to walk along the helix axis from one end to the other end by monitoring inter-imino NOEs. With an unambiguous knowledge of the chemical shift of any one imino proton, the resonance assignment of the remaining protons can be completed. T(3NH)/C^+(3NH) resonances serve as starting points for sequential assignments of the rest of the imino proton resonances.

Though T.A:T and C^+.G:C triplexes have been extensively characterised, several alternative combinations have been proposed. There are instances, in which a homo-R strand binds in an anti-parallel orientation to the R strand of a R:Y duplex to form R.R:Y triplex. Early studies in this direction were focussed on R.R:Y base triads such as G.G:C (Figure 7.20) and I.I:C. More recently, combinations such as A.A:U/T, I.A:U/T, and I.G:C have been characterised. This followed identification of A.A:T and T.A:T base triples within the R.R:Y motif of DNA triplexes, while weaker interactions are detected for C.A:T, A.G:C, T.C:G combinations. Another feature of such studies been on the NMR characterisation of G.T:A, C^+.G –T and T.A^+- C triads.

A different class of triplexes are single stranded oligonucleotide sequences that form intramolecular triple-stranded structures with well defined strand orientations.

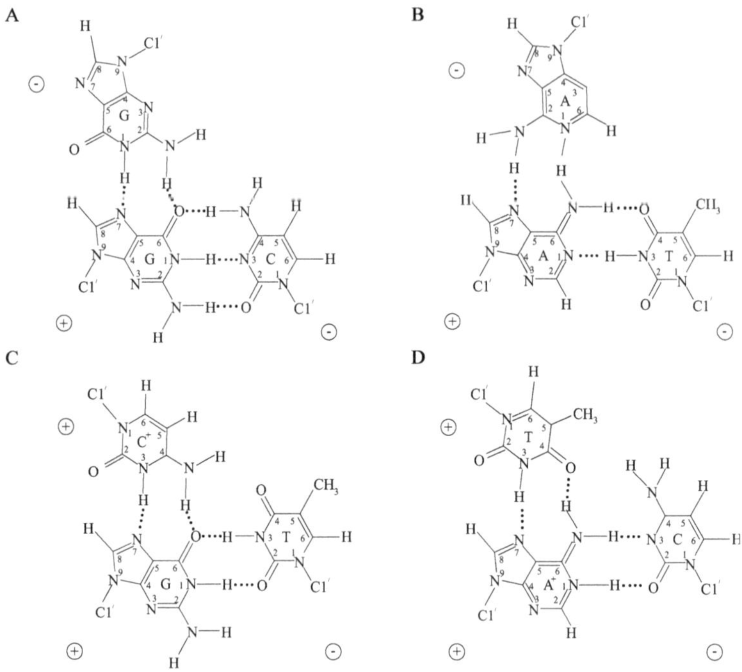

Figure 7.20: Schematics for hydrogen-bonded base pairing in A) G.G:C B) A.A:T C) C^+.G–T and D) T.A^+-C observed in triple-stranded DNA structures.

6.6 G-Quartet

Guanine has the unique property to form extensive *self* structures in solution. Hydrogen bonding between G from different nucleotides gives rise to G-G base pairs and G-quartets (Figure 7.21 A). The quartet structures are stabilised by specific cations which bind at the centre of the quartet structures. This behaviour was initially observed with G itself, 5′/3′-GMP, poly(I) and poly(G). It has been proposed that G-quartets occur in immunoglobulin switch regions. They are also found in the dimerization domain of HIV-1 RNA genome. The ends of chromosomes (called *telomeres*) have tandem repeats of G-rich regions.

There are four different variations in individual strand orientation in G-quadruplets, which have been characterised by NMR spectroscopy. Sequences such as d-TGGGGT adopt all parallel strand orientations. The other three conformations are adopted by dimeric or monomeric quadruplexes that contain loops joining two or more G-tracts, such as d-GGGGTTTTGGGG or d-GGGGTTTT-GGGGTTTT-GGGGTTTT-GGGG, respectively (Figure 7.21).

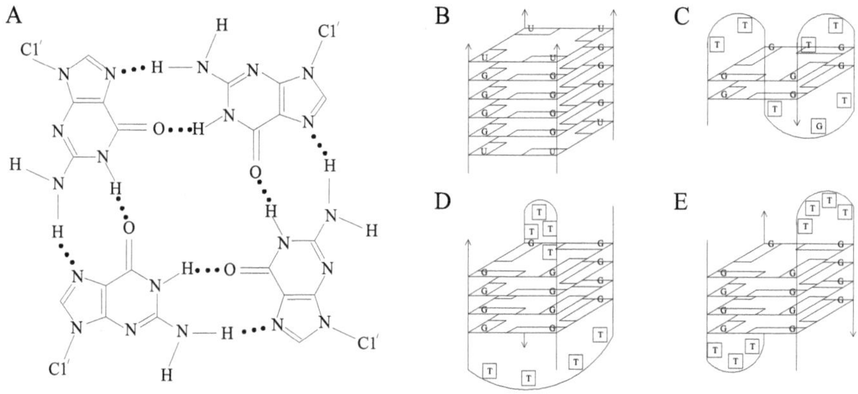

Figure 7.21: (A) Base-pairing in a G-quartet, and various schematic representations of G-quartet structures. (B) Structure formed by ribo(UGGGGU)$_4$. (C) Thrombin-binding DNA aptamer d(GGTTGGTGGTTGG). (D) Structure of d(GGGGTTTTGGGG) in Na^+ solution by NMR. (E) Structure of d(GGGGTTTTGGGG) in K^+ complex by X-ray crystallography.

As in the case of other DNA complexes, the imino and amino proton resonances provide vital information on the formation and structural details of G-quartets. For example, each G-quartet gives rise to four hydrogen-bonded imino proton resonances in the 10-13 ppm range which exchange with ^2H_2O very slowly. Each of these imino protons shows the expected NOE to H8 of G in its immediate neighbourhood (Figure 7.21 A).

6.7 i-motif Tetraplex

This structure (Figure 7.22 A) consists of two parallel duplexes aligned antiparallel to each other with twelve intercalating C-C^+ base pairs (Figure 7.22 B). The dyad symmetry along the tetrad axis limits the possible base positions, such that the centres of the C-C^+ base pairs are collinear, with successive base pairs offset by a rise and rotation around the tetrad axis. Note that the axes of 2-fold symmetry represented by arrows; one, along the helix axis through the centre of each C:C^+ base pair the second between the central C4:C^+4 base pairs. The sugar-phosphate backbones of the tetrad are stabilised by van der Waals contacts between the sugars. Individual H1′ protons are stacked above one another, separated on an average by 3.0 Å. Thus, H1′ protons form a means for sequence specific resonance assignments, due to strong H1′-H1′ NOEs, which form unique signature of this structure. This conformation also gives rise to unique H1′-H1′ NOEs between sugar protons belonging to different strands across narrow grooves and between H2′/H2″ and amino protons across the wide grooves.

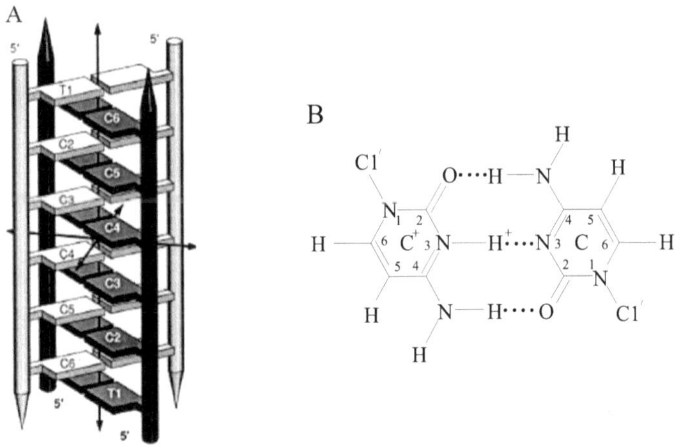

Figure 7.22: (A) Scheme of the two intercalated parallel stranded duplexes (d-TC_6 tetrad). (B) The base pairing scheme in a $C{:}C^+$ base-pair. K. Gehring, J.L. Leroy and M. Gueron. Nature. 363, 561-5 (1993). (Reprinted with permission from Nature Publishing Group).

6.8 Cruciform Structure

A cruciform structure comprises of three components: the loops, stems, and a four-way junction (Figure 7.23 A).

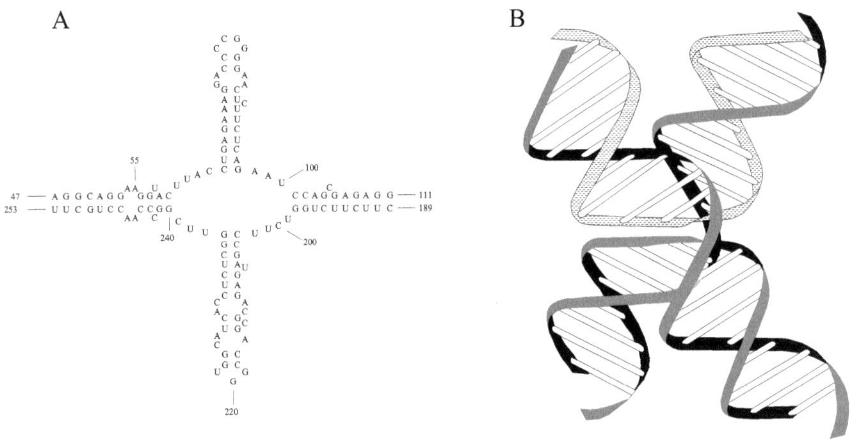

Figure 7.23: (A) A cruciform structure (B) A Holliday junction with 32 base pairs.

The stems consist of normal B-DNA duplex, which is cleavable by restriction enzymes. The cruciform stems are the only sections of the ds-DNA that are not subject to super-helical stress. Progress in understanding the structures of the loops and the junctions has come from the studies of such entities in isolation. It is difficult to study structures that require DNA super-coiling for their continued existence for two reasons. First, a cleavage within the DNA molecule releases the super-helicity and the structure is instantly destabilised. Second, super-coiled circles are relatively large molecules.

6.9 Holliday or Four-way Junction

Enzymes called *recombinase,* catalyse the exchange of DNA strands during the process called *recombination.* The Holliday junction provides clues as to how this takes place. It is a transient four arm intermediate DNA cross-over structure. It is formed during genetic recombination between DNA molecules with similar sequences and other cellular processes. Such junctions have been studied by NMR. Figure 7.23 B shows an illustrative example of a 32 base pair Holliday junction. A four arm DNA structure comprises the *junction* made of four base pairs at the centre. This junction folds into an X-shaped structure, in which the helical arms undergo pair-wise stacking to form two quasi-continuous coaxial helices. The stacked arms are rotated in order to generate the stacked structure.

Analysis of the structure of the four-way DNA junction, confirms that the right-handed, anti parallel stacked X-structure is likely to be the energetically favourable conformation. The two sides of the X-structure are not equivalent, as the four base pairs at the point of strand exchange are oriented in the same direction. NMR has been used to study Holliday junctions. For such studies, the sample preparation of DNA with Holliday junction is critical. One first mixes equimolar amounts of strand 1 with 4 and of strand 2 with 3, and record ^{1}H spectra for direct verification. This is followed by the titration of these two duplexes to form Holliday junction. Gel electrophoresis is used to estimate optimum ratios of the two pools to prepare the 1:1:1:1 admixtures.

7. POLYMORPHISM IN DNA

It is seen that DNA and RNA can acquire a number of conformations, other than the A- and B- duplexes. These conformations can coexist in equilibrium, or may be induced to go from one form to the other by changing experimental conditions such as concentration of nucleic acids, pH, temperature and salt and metal ion concentrations. NMR is an ideal tool to study such changes. In favourable conditions, the thermodynamic and kinetic parameters for such transitions can be measured using NMR. The melting profile during the transition at which an ordered DNA/RNA structure goes into a single-stranded random coil state can also be monitored from the imino- and amino- ^{1}H finger-print region.

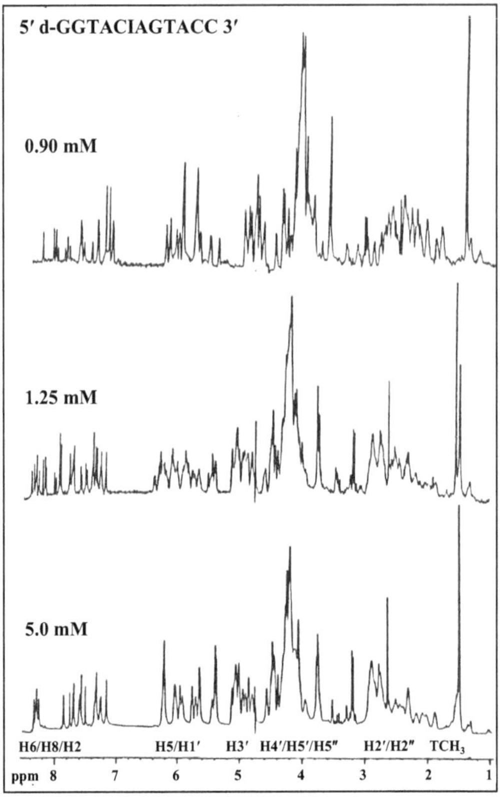

Figure 7.24: 1D ^{1}H spectra of d-GGTACIAGTACC as a function of DNA concentration.

7.1 Concentration-induced Conformational Transition

Figure 7.24 shows a set of 1D ^{1}H spectra of d-GGTACIAGTACC as a function of DNA concentration. At higher concentrations, one observes a set of resonances which have been assigned to a B-DNA duplex. The structure has a bulge in the centre due to bulky R-R base pairs. As the concentration is lowered, an additional set of resonances appear. This indicates the co-existence of two individual conformers.

The intensities of the resonances arising from the second conformer increase, with lower DNA concentrations. The second structure has been assigned to a hairpin, with two bases I and A in the centre, forming a loop. In such situations, ROESY spectrum provides a means for assigning resonances belonging to the second conformer from the knowledge of the duplex. Negative ROE cross peaks corresponding to the chemical exchange between the two conformers aid in identification of two distinct sets of cross-peaks. The observation of two separate sets of resonances and the associated chemical shift differences provide evidence for slow exchange on the NMR time scale between the duplex and the hairpin. Such data further aids in the estimation of thermodynamic and kinetic parameters for the two conformers.

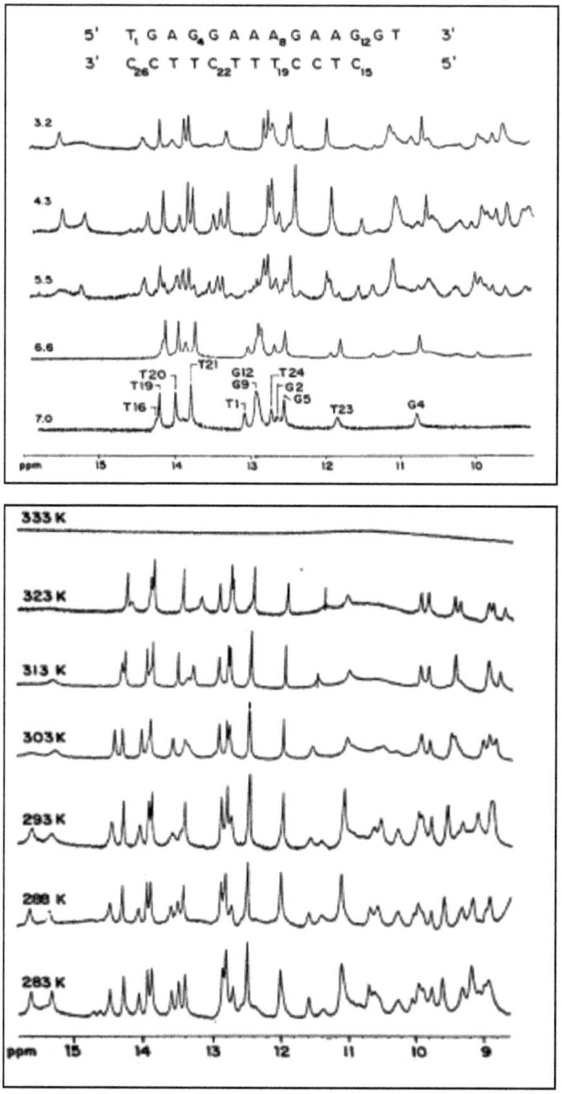

Figure 7.25: (A) pH dependence of imino and a part of amino ^{1}H region of the 1D NMR spectrum of a 1:1 molar mixture of d-TGAGGAAAGAAGGT and d-CTCCTTTCTTCC .(B) Temperature dependence of the imino proton resonances of the ^{1}H NMR spectrum a 1:1 molar mixture of d-TGAGGAAAGAAGGT and d-CTCCTTTCTTCC. S.R. Bhaumik K.V.R. Chary, G.Govil, K. Liu, and H.T. Miles. Nucleic Acids Res. 23, 4114-4121 (1995). (Reprinted with permission from Oxford University Press).

7.2 pH- Induced Polymorphism

As an illustrative example, the pH induced polymorphism in a 1:1 mixture of d-TGAGGAAAGAAGGT and d-CTCCTTTCTTCC oligomers is described. When mixed in a 1:1 molar ratio at neutral pH, the system adopts an anti-parallel duplex with three mismatched base pairs, namely A^+-C, G-T and T-C, as shown in in the following scheme:

```
5′ d-TGAGGAAAGAAGGT 3′
3′ d-   CCTTCTTTCCTC 5′
```

Using NMR parameters along with restrained molecular mechanics and dynamics simulation, the 12-nucleotide long antiparallel duplex DNA with three mismatched base-pairs has been structurally characterised. The results show that the overall conformation of the duplex is B-DNA with slight structural distortions near the mismatch sites. Figure 7.25 A shows the pH dependence of imino and amino ^{1}H region of the spectrum. There are six G, six A, six C and eight T (six in the Y and two in the R strand). With standard WC base-pairing, a maximum of twelve imino ^{1}H resonances are expected in the 12-15 ppm range. This is indeed observed in the spectrum at pH 7.0. The six Cs are expected to show twelve NH_2 resonances (six involved in hydrogen bonding and six exposed to water) in the region 6-8.5 ppm.

At pH below 5.5, Cs are protonated. The complexity in the imino and amino region increases with a multitude of lines between 8.5 to 16.0 ppm, as pH is lowered. Additional Hoogsteen base-paired imino (of G(1NH) and T(3NH)) and amino (of $C^+(4NH_2)$) ^{1}H resonances are observed in the region 14-16 and 8.5-10 ppm, respectively (Figure 7.25 A). Such a change in the 1D spectrum at pH below 5.5 and at moderate ionic strength, is due to the co-existence of a triplex as illustrated in the scheme below. The triplex is stabilised by five T.A:T, four C^+.G:C, two mismatched triads C^+.G-T and T.A^+-C and by a Hoogsteen C^+.G base-pair on one end, as shown in the following scheme:

```
5′ d- CTCCTTTCTTCC  3′
5′ d-TGAGGAAAGAAGGT 3′
3′ d-   CCTTCTTTCCTC 5′
```

7.3 Effect of Temperature

Temperature is another parameter which can induce conformation transitions. Ordered DNA structures go into single stranded random coils at higher temperature. This transition is reflected by changes in the chemical shifts and line widths of the hydrogen-bonded protons. The imino protons broaden at higher temperatures and finally disappear as the ordered structures melt into single strands existing as random coils. The melting curves of nucleic acids provide important information on the thermodynamics and stability of the ordered structures.

The multi-strand structures can melt in a single step or through several steps. For example, the temperature dependence of the imino ^{1}H resonances in the triplex discussed in previous section reveals that the triplex dissociates directly into single strands around 55 °C, without duplex intermediates (Figure 7.25 B).

7.4 Effect of Metal Ions

Metal ions play a key role in the stability of structures of nucleic acids. They bind to the negatively charged phosphate groups as well as to the grooves, and thus stabilise conformations of nucleic acids. Their concentrations have to be carefully chosen to provide good NMR spectra and to control the melting temperatures of ordered structures. Site-specific interactions can be followed using chemical shifts. In particular, the binding of Mg^{2+} has been of biological importance. For example, the quartet structures discussed in section 6.6 depend critically on metal ions.

8. DYNAMICS OF NUCLEIC ACIDS

As discussed in Chapters 4-6, the relaxation times and NOE are sensitive to the frequency and amplitude of biomolecular motions in psec to nsec time-scales. Measurement of NMR relaxation parameters, throws light on the dynamics of DNA and RNA. Such studies provide a physical description of the conformation space spanned by these molecules. The success of such studies has been primarily because of the ease with which one can produce large quantities of RNA/DNA by *in-vitro* transcription and also label them with ^{13}C or/and ^{15}N. However, the dynamics of nucleic acids is much more complex, as they possess seven degrees of rotational freedom per nucleotide. The motions in the backbone, the sugar pucker and the glycosidic bonds may show motions which are not correlated. The global tumbling cannot be handled by a single correlation time (τ_c) as molecules are rod-like.

^{13}C, ^{15}N and ^{31}P are the three NMR spins present in nucleic acids. All these have been used to study their dynamics. Motions of the sugar-phosphate backbone are difficult due to lack of NMR probes. The ^{31}P resonances show low dispersion and the relaxation is dominated by the chemical shift anisotropy. The number of proton coupled carbons per nucleotide residue is relatively more than the nitrogens or the single ^{31}P nucleus. The use of imino or amino group ^{15}N nuclei as probe suffers from the problems associated with the chemical exchange of the protons with water. Hence, ^{13}C is better suited for relaxation studies and dynamics of nucleic acids.

Though uniform ^{13}C enrichment is an attractive proposition and is fairly economical, it results in extensive overlaps mostly in the regions where the ribose-ring cross-peaks appear. Thus, one is left with C1′-H1′ and aromatic correlations (C2-H2, C6-H6 and C8-H8) for analysis of molecular dynamics. Uniform labelling also gives rise to ^{13}C-^{13}C couplings, which results in poor S/N ratio. This problem is overcome by using constant time HSQC correlation methods, which provide ^{13}C-homo-decoupling along the ^{13}C dimension or by random fractional ^{13}C labelling to an extent of 20-30%. Both these methods result in substantial loss in S/N ratio. With the advent of large magnetic fields and cryogenically cooled RF probes the sensitivity problem can be overcome to some extent.

Generally, it is advantageous to adopt selective ^{13}C enrichment of a specific site or certain number of sites, which do not show any ^{13}C-^{13}C spin coupling. The best sites for such site-specific labelling are C1′, C2, C6 and C8, primarily because of relatively large dispersion in their chemical shifts. This requires selectively labelled precursors, which may be generated by chemical synthesis.

The conformational freedom of furanose ring and that of individual bases can be examined by measuring ^{13}C relaxation parameters of one or more sites in each moiety. C1′ can be easily labelled with ^{13}C and its relaxation is dominated by dipole-dipole interaction with H1′, Therefore this isolated spin pair is particularly useful for dynamic studies. The ^{13}C-^{1}H pairs on the base pairs can be used for dynamic studies in this part of the nucleotides. The spin pairs in the purines are more suitable as the interpretation of the relaxation data is not complicated by the cross-relaxation between C5 and C6 in pyrimidines.

Examination of such local dynamics is correlated with the functional features of RNA and DNA under investigation. The model free formalism of Lipari and Szabo has been widely used for analysis. However, when the molecules exhibit motions on a wide range of time scales, such formalism becomes invalid and more refined methods have to be used. Molecular dynamics can complement NMR relaxation data in such cases.

9. REFERENCES

9.1 Further Reading

J. D. Watson and F. H. C. Crick, *Molecular Structure of nucleic acids: a structure for deoxyribose nucleic acid,* Nature (London) **171**, 737–738 (1953).

J. D. Watson and F. H. C. Crick, *Genetical implications of the structure of deoxyribonucleic acid,* Nature (London) **171**, 964–967 (1953).

9.2 Books

W. Saenger, *Principles of Nucleic Acid Structure*, Springer, New York, Berlin, Heidelberg, and Tokyo, 315 (1984).

T.M. Jovin, K. Rippe, N.B. Ramsing, R. Klement, W. Elhorst, and M. Vojtiskova, *Structure and Methods, Vol. 3: DNA and RNA* (R.H. Sarma and M.H. Sarma, eds.), Adenine Press, Schenectady, New York, 155 (1990).

T.L. James (ed.) *Methods in Enzymology. Nuclear Magnetic Resonance and Nucleic Acids*, Academic Press, San Diego (1995).

V.N. Soyfer, and V.N. Potaman, *Triple-Helical Nucleic Acids*, Springer - Verlag New York (1996).

V.A. Bloomfield, D. Crothers, I. Tinoco and J. Hearst, *Nucleic Acids: Structures, Properties and Functions*, University Science Books (2000).

David L. Nelson, David L. Nelson and Michael M. Cox, *Lehninger Principles of Biochemistry*, 3rd BK & CD edition, W.H. Freeman & Company (2000).

B. Lewin, *Genes*, 7th edition, Oxford University Press (2000).

Lubert Stryer, Jeremy M. Berg and John L. Tymoczko, *Biochemistry*, 5th Ed. W.H.Freeman & Co Ltd. (2002).

9.3 DNA Structure

G. Lipari and A. Szabo, *Nuclear magnetic resonance relaxation in nucleic acid fragments: models for internal motion*, Biochemistry **20**, 6250–6256 (1981).
J. Feigon, J.M. Wright, W.A. Denny, W. Leupin, and D.R. Kearns, *Use of two dimensional NMR in the study of a double-stranded DNA decamer*, J. Am. Chem. Soc. **104**, 5540–5541 (1982).
D.J. Patel. A. Pardi and K. Itakura, *DNA conformation, dynamics and interactions in solution*, Science **216**, 581–590 (1982).
D.R. Hare, D.E. Wemmer, S.H. Chou, G. Drobny and B.R. Reid, *Assignment of the non-exchangeable proton resonances of d(C-G-C-G-A-T-T-C-G-C-G) using two dimensional nuclear magnetic resonance methods*, J. Mol. Biol. **171**, 319–336 (1983).
D.J. Patel, S.A. Kozlowski, S. Ikura, K. Bhatt and D.R. Hare, *NMR studies of DNA conformation and dynamics in solution*, Cold Spring Harb. Symp. Quant. Biol. **47**, 197–206 (1983).
K.V.R. Chary and S. Mody, *Analysis of intrasugar interproton NOESY cross-peaks as an aid to determine sugar geometries in DNA fragments*, FEBS Lett. **233**, 319 (1988).
K.V.R. Chary, S. Mody, R.V. Hosur, G. Govil, C. Chen and H.T. Miles, *Quantification of DNA structure from NMR data: Conformation of d-ACATCGATGT*, Biochemistry **28**, 5240 (1989).
B.R. Reid, K. Banks, P. Flynn and W. Nerdal, *NMR distance measurements in DNA duplexes: sugars and bases have the same correlation times*, Biochemistry **28**, 10001–10007 (1989).
K.V.R. Chary, *DNA structure from NMR data*, Magnetic Resonance-Current Trends; Eds. C.L. Khetrapal and G. Govil, Narosa Publishing House, New Delhi, 71–104 (1991).
S.G. Kim, L.J. Lin and B.R. Reid, *Determination of nucleic acid backbone conformation by ^1H NMR*, Biochemistry **31**, 3564–3574 (1992).
K. Gehring, J.L. Leroy and M. Gueron. *A tetrameric DNA structure with protonated cytosine-cytosine base pair*, Nature **363**, 561–565 (1993).
R. Nibedita, R.A. Kumar, A. Majumdar, R.V. Hosur, G. Govil, K. Majumder and V.S. Chauhan, *Solution structure of GCAAT recognition motif by 2D NMR, spectral simulation, molecular modelling, and distance geometry calculations*, Biochemistry **32**, 9053–9064 (1993).
O.Y. Federof, B.R. Reid, and V.P. Chuprina, *Sequence dependence of DNA in solution*, J. Mol. Biol. **235**, 325–330. (1994).
S.R. Bhaumik and K.V.R. Chary, *Molecular dynamics and mechanics calculations on a DNA duplex with A^+-C, G-T and T-C mispairs*, J. Bio. Struct. Dynamics **3**, 199–206 (2002).

9.4 Parallel Stranded DNA

N. Pattabiraman, *Can the double helix be parallel?* Biopolymers **25**, 1603–1606 (1986).
J.H. van de Sande, N.B. Ramsing, M.W. Germann, W. Elhorst, B.W. Kalisch, V.E. Kitzing, R.T. Pon, R.C. Clegg and T.M. Jovin, *Parallel stranded DNA* Science **241**, 551–557 (1988).
C. Otto, K. Rippe, K. Thomas, N.B. Ramsing and T.M. Jovin, *The hydrogen-bonding structure in parallel-stranded duplex DNA is reverse Watson-Crick*, Biochemistry **30**, 3062–3069 (1991).
G. Raghunathan, H.T. Miles and V. Sasisekharan, *Parallel nucleic acid helices with Hoogsteen base pairing: Symmetry and structure* Biopolymers, **34**, 1573–1581 (1994).
V. Rani Parvathy, S.R. Bhaumik, K.V.R. Chary., G. Govil, K. Liu, B.H. Frank and H.T. Miles, *NMR structure of a parallel-stranded DNA duplex at atomic resolution*, Nucleic Acids Res. **30**, 1500–1511 (2002).

9.5 Triple-Helical Nucleic Acids

G. Felsenfield, D.R. Davies, and A. Rich, *Formation of a three-stranded polynucleotide molecule*. J. Am. Chem. Soc. **79**, 2023–2024 (1957).
J.S. Lee, D.A. Johnson, and A.R. Morgan, *Complexes formed by (pyrimidine)n. (purine)n DNAs on lowering the pH are three-stranded*, Nucleic Acids Res., **6**, 3073–3091 (1979).
N.B. Ramsing, K. Rippe, and T.M. Jovin, *Helix-coil transition of parallel-stranded DNA. Thermodynamics of hairpin and linear duplex oligonucleotides*, Biochemistry, **28**, 9528–9535 (1989).
S.M. Mirkin, and M.D. Frank-Kamenetskii, *H-DNA and related structures*, Ann. Rev. Biophys. Biomol. Struct. **23**, 541–576 (1994).

S.R. Bhaumik, K.V.R. Chary, G. Govil, K. Liu, and H.T. Miles, *NMR characterisation of a triple stranded complex formed by homo-purine and homo-pyrimidine DNA strands at 1:1 molar ratio and acidic pH*, Nucleic Acids Res. **23**, 4116–4121 (1995).
V.N. Soyfer, and V.N. Potaman, *Triple-Helical Nucleic Acids*, Springer - Verlag New York (1996).
S.R. Bhaumik, K.V.R. Chary, G. Govil, K. Liu, and H.T. Miles, *Homopurine and Homopyrimidine strands complementary in parallel orientation form an antiparallel duplex at neutral pH with A-C, G-T and T-C mismatched base pair*, Biopolymers **41**, 773–785 (1997).
S.R. Bhaumik, K.V.R. Chary, G. Govil, *Molecular mechanics calculations on a triple stranded DNA involving C^+.G-T and T.A^+-C mismatched base triples*, J. Biomol. Structure and Dynamics **16**, 527–534 (1998).
S.R. Bhaumik, K.V.R. Chary, G. Govil, L. Keliang, and H.T. Miles, *A novel palindromic triple stranded structure formed by homopyrimidine dodecamer d-CTTCTCCTCTTC and homopurine hexamer d-GAAGAG*, Nucleic Acids Res. **26**, 2981–2988 (1998).

9.6 Isotope labelling of Nucleic Acids

D.P. Zimmer, and D. M. Crothers, *NMR of enzymatically synthesized uniformly $^{13}C/^{15}N$-labeled DNA oligonucleotides*, Proc. Natl. Acad. Sci., USA **92**, 3091–3095 (1995).
D.E. Smith, J. Su, and F. M. Jucker, *Efficient enzymatic synthesis of ^{13}C, ^{15}N-labeled DNA for NMR studies*, J. Biomol. NMR **10**, 245–253 (1997).
J.E. Masse, P. Bortmann, T. Dieckmann, and J. Feigon, *Simple, efficient protocol for enzymatic synthesis of uniformly ^{13}C, ^{15}N-labeled DNA for heteronuclear NMR studies*, Nucleic Acids Res. **26**, 2618–2624 (1998).
G. Mer, and W. J. Chazin, *Enzymatic synthesis of region-specific isotope-labeled DNA oligomers for NMR analysis*, J. Am. Chem. Soc. **120**, 607–608 (1998).
R. Sunita, B.J Rao and K.V.R. Chary, *A novel approach for uniform ^{13}C and ^{15}N labeling of DNA for NMR studies*, Biochem. Biophys. Res. Com. **290**, 928–932 (2002).

CHAPTER 8

CARBOHYDRATES, LIPIDS AND LIPID ASSEMBLIES

1. INTRODUCTION

The structure and dynamics of proteins and nucleic acids have been discussed in the previous chapters. These two classes of biopolymers generally have a specific primary structure. The 3D structures, which are essentially dictated by the primary sequence, confer specific and unique functions on such molecules.

In contrast, carbohydrates and lipids are groups of biomolecules which have relatively modest specificity in terms of their primary sequences. These classes of compounds are built using a wider variety of chemical units with diverse structures. This brings in a greater variations in their functional properties, which are fine-tuned by nature to achieve specific objectives of living systems. In general, an assembly of several molecules are responsible for the biological functions of lipids and carbohydrates.

As in the case of proteins and nucleic-acids, the structure and dynamics of carbohydrates can be approached through solution state NMR techniques. However, lipids, which form self assembled multi-molecular structures and are an integral component of biological membranes, are often not amenable to solution studies. In such cases, solid-state NMR techniques are the best alternate. For this reason, some aspects of the NMR of lipids and molecular assemblies have been discussed later in this book, after an introduction to high resolution solid-state NMR.

In this Chapter, applications of solution-state NMR to carbohydrates, lipids, their conjugates with each other and with proteins, have been discussed. The molecular organization of lipids and carbohydrates in biological membranes, is also covered.

2. CARBOHYDRATES

Monosaccharides, polysaccharides and carbohydrates, are built from simple sugar molecules. The basic moiety of these units is glucose (Glc) and its epimers such as allose, altrose, mannose, idose, galactose and talose. These monomeric units have the molecular formula $C_6H_{12}O_6$ and hence the name carbohydrate. However, monosaccharides can have other chemical compositions including those which contain functional groups such as amino, carboxyl, acetyl, acyl, phosphate and sulphate. The monosaccharides can therefore be divided into several groups.

The oligosaccharides are often quite large in terms of their molecular weights. The monomeric building blocks are linked both in linear and branched fashion. However, unlike proteins and nucleic acids, the conformations of polysaccharides are relatively more flexible. The local conformation near the branch points often forms the recognition motif for such molecules.

Oligosaccharides also exist in nature, covalently linked with other classes of molecules. These include proteins, lipids, steroid conjugates and natural antibiotics. Most of the serum and membrane proteins have oligosaccharide chains attached to them. Conjugation of sugars with proteins and lipids lead to the formation of glycoproteins, proteoglycans and glycolipids. A large number of proteins in fact, exist in nature with attached oligosaccharide chains and are thus, glycosylated. Proteins in serum, cell matrix and cell membranes are present as glycoprotein and proteoglycans. Lipid conjugation exists in the form of glycolipids, glycophosphatidyl-inositol complex and bacterial polysaccharides linked to lipids. DNA-oligosaccharide interactions have also found to play a role in diseases such as cancer and infection. The linkage positions and sequences in oligosaccharides also have a wide range. Thus, there is a large diversity in the structural forms of these groups of molecules.

In many cases it is difficult to crystallise such molecules and NMR serves as a major source of structural information.

2.1 Biological Roles of Carbohydrates

One of the most important functions of carbohydrates is to act as a fuel in biological oxidation reactions. These molecules are the most important source for providing energy for biological processes. In fact, until recently, the functional role of these molecules was considered to be of that a general nature, such as structural, protective and for providing energy to the cells.

Recent discoveries on the role of sugar moieties in diseases and in cell recognition have created a new interest in the field. New functions for carbohydrates have been discovered. Characterisation of specific oligosaccharides as blood group antigens, which are recognised by the serum antibodies, has become a major area of biomedical research. Changes in the glycosylation patterns in HIV-1 infected cells seem to play an important role in immunological responses to such infection. Several glycoproteins have therapeutic value. Examples are anti-cancer products and tumour antigens, wound and ulcer healing molecules and those involved in cell transplantation.

Oligosaccharides carry out a number of other specific functions such as that of adhesion, binding of pathogens, antibodies and toxins, and clearance of circulating glycoproteins. In industries, polysaccharides have found use as stabilisers, thickeners, emulsifiers and sweeteners in food products. Cyclodextrin has been widely used in drug delivery systems. Several marine polysaccharides are being investigated for food and medical purposes. Bacterial polysaccharides have found applications as anti-viral agents.

For these reasons, studies on the sequence, 3D structure and dynamics of oligosaccharides and their interactions with other classes of biomolecules have received renewed interest. Very few oligosaccharides or glyco-conjugates crystallise easily or provide resolved electron density for the glycan moiety. This is mainly due to the flexibility and heterogeneity of sugar linkages. Therefore, X-ray diffraction data on carbohydrates has been comparatively more limited when compared to proteins and nucleic acids. In recent years, complexes of oligosaccharides with proteins such as carbohydrate antibodies, plant lectins and mammalian carbohydrate binding proteins such as *selectins, galectins* have received good attention through X-ray crystallographic methods. However, NMR has served as a potential tool for structural data on carbohydrates.

2.2 Building Blocks of Carbohydrates

Apart from the five-member ribose ring, which has been discussed in the context of nucleic acids, the most common sugars involved in biological systems are D-glucose (Glc), D-mannose (Man), D-fucose (Fuc) and D-galactose (Gal). In addition, several substituted sugars such as β-D-N-acetylglucosamine (GlcNAc) and α-D-N-acetylneuraminic acid (NANA) also exist frequently in living systems. Monomeric sugars exist as six member rings (hexoses), called hexa-pyranose or hexa-furanose (Figure 8.1).

Glc is one of the primary fuels for providing energy to living systems. It is stored in the form of glycogen, a branched chain polymer. Glc is involved in a large number of metabolic processes. It has two configurations, α and β, which are readily inter-convertible in these reactions. Several other sugars such as α-D-mannose and β-D-galactose, and their conjugates with proteins and lipids, also play important roles in biological systems. Modified sugars have been found frequently, both in bacterial and animal systems

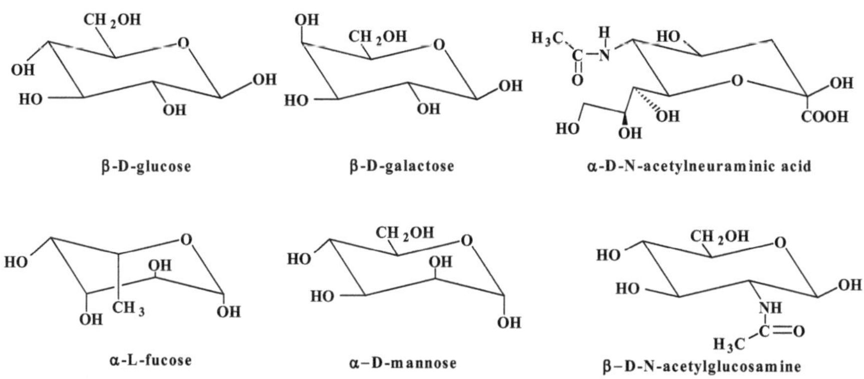

Figure 8.1: Structures of some of the common monosaccharides in the hexa-pyranose form, which are found in living systems.

2.3 Oligosaccharides

Oligosaccharides are condensation products of the six member sugar rings. A sugar unit such as Glc can be linked to another moiety by condensation through the hydroxyl groups in the two monomers. As seen from Figures 8.1 and 8.2, sugars have several hydroxyl groups. Therefore, there is more than one way to link these units. One of the hydroxyl groups which commonly participate in most of the condensation reactions is C1. The glycosidic C-O-C linkage that is formed during condensation may involve one of the other carbons. Accordingly, one may get 1→2, 1→3, 1→4, 1→5 or 1→6 linkages. The glycosidic bond so formed may be oriented either below (α) or above (β) the plane of the ring. In some carbohydrates, phosphate groups are involved in the inter-sugar linkage. With some exceptions, linkages such as 1→1 and 4→4 are absent.

In nature, the condensation reaction between monosaccharides is catalysed by a family of enzymes called *glycosyltranferase*. These are distinguished by the type of saccharides, the linkage position and the acceptor monosaccharide. The specificity of biosynthesis arises from the enzymes, which control interconversion of one sugar to the other during metabolic pathways and by *glycosyltransferases*, which control the condensation reaction. The chains are further processed by a series of other enzymes.

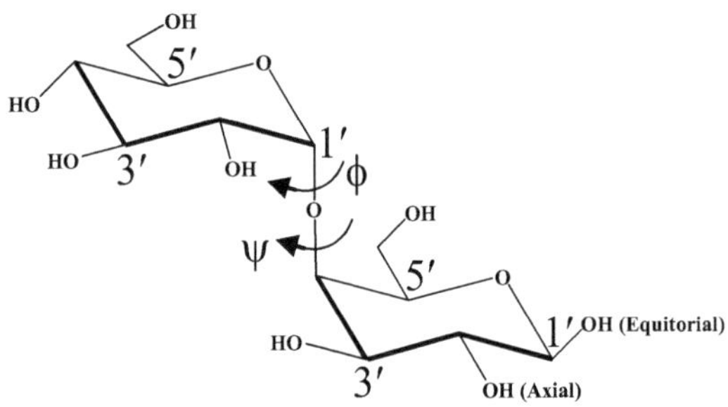

Figure 8.2: Numbering of atoms in a 1→4 linked disaccharide. Note that the orientation of hydrogen atoms, can take two possible directions with respect to the rest of the ring. These are referred as axial and equatorial.

Thus, in polysaccharides, one observes a wide variability in monomeric units and in their linkages. Starting from a few building blocks, oligosaccharides can provide a wider range of diversity and specificity.

2.4 General Features of NMR of Carbohydrates

The 1D NMR studies of carbohydrates were initiated in early 60s. However, for several reasons, applications of NMR to carbohydrates have not reached the same level of popularity and maturity as those of proteins and nucleic acids. One of the reasons is that carbohydrates are more difficult to synthesise, isolate, purify and characterize. They also have a wider variety of linkages. Further, usually one has access to only two NMR spins, namely ^{1}H and ^{13}C.

A typical NMR spectrum for a bacterial polysaccharide serotype 6B, having a sequence [-2-α-D-Galp(1→3)-α-D-Glcp(1→3)-α-L-Rha-p(1→4)-D-Ribotol(5PO_4-]$_n$

is shown in Figure 8.3. Several features may be noted from this spectrum. In fact, there are major differences between the multi-dimensional NMR approach when applied to polysaccharides as compared with proteins and nucleic-acids. It will be helpful to understand the differences between the spectra of polysaccharides and of other biological molecules.

(i) The C-H protons from the rings in carbohydrates lie in a narrow spectral range of 3-4.5 ppm (Figure 8.3). Therefore resonance assignments require use of protons in the methyl and aromatic regions as starting points, in COSY or TOCSY spectroscopy, wherever such peaks are available.

(ii) Several of these resonances are close to the water resonance. The chemical shift differences are relatively small. This results in most of the cross-peaks lying close to the diagonal in 2D spectra. This feature complicates interpretation of the data. Therefore, use of multidimensional NMR, high-fields, selective water suppression schemes and isotope-labelling are desirable for resolution and resonance assignments.

(iii) There are several protons in each sugar ring which form complex networks of J-coupled spin systems. While the involved J-couplings are moderately large in magnitude, the chemical shift differences for the ring protons are relatively small. This results in strongly coupled spin systems. The use of higher fields can minimise such problems by improving chemical-shift dispersions.

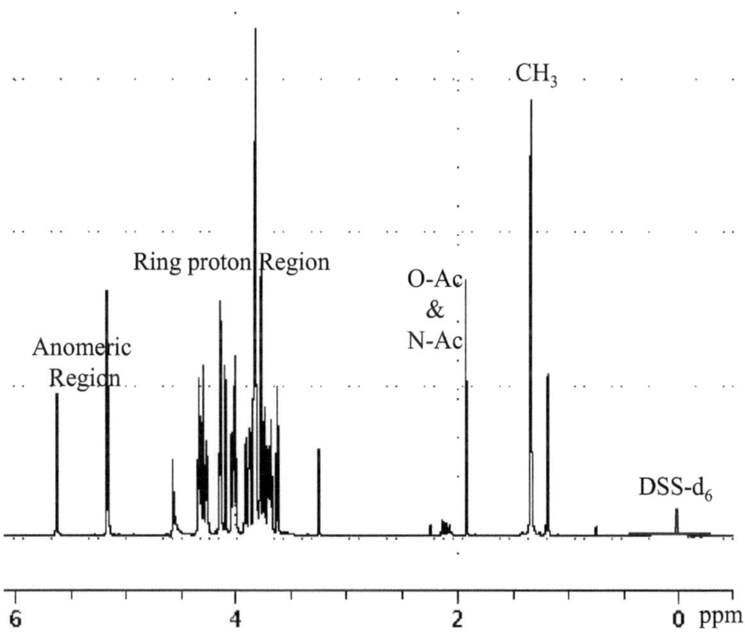

Figure 8.3: 800 MHz ^1H NMR spectrum of S.pneumoniae serotype 6B polysaccharide. (Thanks to S. Srivastava for recording this spectrum).

(iv)With some exceptions, there are no nitrogen and phosphorus atoms in polysaccharides. Therefore, the information has to be derived from the analysis of ^{1}H and ^{13}C spectra only. Further, unlike the gene cloning methods for proteins and the solid-state synthesis for nucleic-acids, incorporation of ^{13}C label in carbohydrates requires chemical synthesis. This is inherently more difficult.
(v) Use of deuteration is helpful in assignments. Use has been made of the relatively large isotope effects on ^{13}C chemical shifts when a ^{1}H is replaced by ^{2}H. The ease of $^{1}H/^{2}H$ chemical exchange for the hydroxyl protons has been utilised extensively.
(vi) Most of the oligosaccharides of interest have M_r around 1000 Da. Even longer polymers, have fair degree of conformational flexibility. Therefore such molecules fall in the intermediate motional range from the view-point of relaxation rates. This feature may result in either positive or negative NOE. The overall NOEs are therefore relatively small. It is more advantageous to use ROESY. In fact, the first application of ROESY (CAMELSPIN) was on a tetra-saccharide. Mixing times for such experiments are usually within the range of 100-300 msec.
(vi) The conformational flexibility around the glycosidic bonds in oligosaccharides is much higher. This creates serious problems in interpretation of the NMR data.
(vii) Oligosaccharides can have several types of linear and branched linkages. This causes problems in sequential assignments.
(viii) Oligosaccharides have several exchangeable hydroxyl groups. Such groups are usually observed using mixed water/acetone or water/methanol solvents and lower temperatures.
(ix) In view of above factors, the use of multidimensional NMR, high-fields selective water suppression schemes and isotope-labelling are desirable for resolution and resonance assignments.
(x) Complete assignments and clues to structural determination may be obtained using 3D COSY-ROESY and TOCSY-ROESY methods.

The assignments and determination of the 3D structure of a carbohydrate using NMR involves the following steps:
(i) Determination of the primary sequence, including branching.
(ii) Identification of the configuration of each residue: D, L, α or β.
(iii) Identification of the glycosidic linkage, e.g. $1\rightarrow2$, $1\rightarrow3$ etc.
(iv) Analysis of the conformation of individual sugar rings.
(v) Conformation of the glycosidic linkage. This is determined by two angles φ and ψ defining rotation around the C-O and O-C bond forming the glycosidic (C-O-C) linkage. In the case of $1\rightarrow6$ linkage, a third angle (ω) needs to be defined.

Interestingly, all these features have been handled by NMR. In fact, even before the work of Karplus on the stereochemical dependence of ^{3}J, Lemieux and his co-workers had established the dependence of ^{3}J couplings on the dihedral angles in carbohydrates. Before NMR became popular in proteins and nucleic-acids, extensive work had been done on carbohydrates, using 1D CW ^{1}H NMR.

2.5 Primary Structures of Carbohydrates

The primary structures of carbohydrates can be determined through a combined approach based on mass spectrometry, liquid chromatography (LC) and NMR. For example, using chemical degradation coupled with chromatographic methods, it is possible to analyse the composition of a carbohydrate. The concerted use of these techniques for analysing complex mixtures of biological compounds has been discussed in later Chapters. Such commercial instruments are ideally suited for exploring primary structures of polysaccharides, glycoproteins and glycolipids.

NMR has been used to analyse linkages and sequences of simpler polysaccharides. The location and type of inter-glycosidic linkage can be determined through long range ^{1}H-^{13}C correlation. The HMBC technique in particular has been used to correlates ^{1}H and ^{13}C on the two sides of the glycosidic linkage.

2.6 NMR Determination of the Conformation of Monosaccharides

There is extensive literature on the preferred conformations of pyranose rings, both in mono- and oligo-saccharides. In a classical sense, such rings can exist in six boat and two chair conformations. An approach similar to the pseudo-rotational treatment for ribose, allows one to treat the conformation on a more general basis, with lesser number of independent conformational variables. Such an approach has been reported for six member sugar rings. However, in most cases, one deals with the classical boat and chair forms. These classical structures have characteristic coupling constants.

Both the chemical shifts and coupling constants depend on the conformation of monosaccharides. The two parameters have been used in the conformational analysis of mono- and oligo-saccharides. The orientation of the C-H bond vectors with respect to the ring plays a major role in determining the nature of the ^{1}H NMR spectrum. When the C-H bond is approximately in the plane of the carbon atom, the ^{1}H is labelled as *axial* (a). The *equatorial* (e) protons points almost at right angles to such a plane. Equatorially oriented ^{1}H appears at lower fields compared to chemically similar but axially oriented ^{1}H. The α-anomeric protons resonate down-field of the β-anomeric protons and both resonate down-field of ring protons.

The ^{13}C chemical shifts like-wise, show characteristic behaviour, which is dependent on the conformation and configuration. These differences in chemical shifts arise from the chemical anisotropy of C-C bonds. Empirical relations for estimation of long-range shielding, which occur with changes in the conformation and configuration of pyranoses and their derivatives have been formulated, using β-D-Glc and β-D-xylose as references.

One of the crucial parameters for conformational studies of carbohydrates are the three bond ^{3}J (^{1}H-C-C-^{1}H) couplings. The most commonly used Karplus type relation for carbohydrates is given by:

$$^{3}J\,(H\text{-}C\text{-}C\text{-}H) = (7.8 - \cos\varphi + 5.6\cos 2\varphi)\,(1 - M\,\Delta\chi) \qquad 8.1$$

Here φ is the H-C-C-H dihedral angle, M is an empirical constant (usually taken as 0.1), and $\Delta\chi$ is the difference of the electronegativity of the substituent and hydrogen. The values of three bond proton-proton scalar couplings are in the range 2-3 Hz for *eq-eq*, 2-6 Hz for *ax-eq*, and 9-12 Hz for *ax-ax* proton pairs. More precise values can be obtained by taking into account the effects of electronegativity and geometrical arrangements of different chemical groups and the neighbouring residues. Details of such approaches are available in literature.

The use of three-bond ^1H-^1H J-couplings usually allows fixing of conformational structure of the rings. The inter-proton distances in the sugar rings also depend on the conformation and configuration of the rings. These can be estimated from ROESY spectra and are used to supplement conformational analysis.

The ^3J data and theoretical calculations show that only the chair conformations are preferred in solutions. It has been observed that in most cases, sugar rings are locked in the $^4C_1$ conformation. In some cases $^1C_4$ conformation or a mixture of the two chair conformers has been observed. In the later case, the populations of different conformers, has been estimated from the time-average coupling constants. NMR has thus established that the six membered sugar rings generally adopt a relatively well-defined conformation.

Two other groups of J-couplings have been applied to study conformations of polysaccharides. The ^2J(H-C-H) couplings also show conformation dependence. An important aspect of the conformation of pyranoses and furanoses is the orientation of the hydroxyl groups. This information can be obtained through the use of H-C-O-H couplings. The OH resonances can be observed by using suitable solvents (such as DMSO or mixed solvent consisting of water and acetone), which slow down the exchange rates of hydroxyl protons with water. These couplings provide information not only on the orientation of the hydroxyl groups, but also on the possibilities of their involvement in hydrogen bonding.

2.7 Assignments of NMR Spectrum of Carbohydrates

Strategies similar to those discussed for proteins and nucleic acids have been used for the assignments of NMR spectra of carbohydrates. Even though, most oligosaccharides of interest have fewer monomeric units, assignments are complicated by the fact that several ^1H peaks overlap in the homonuclear 2D spectra. The large variety of monomeric units and the presence of both linear and branched linkages complicate procedures for sequential assignments. The absence of ^{15}N and ^{31}P in most carbohydrates, limits the type of multi-dimensional methods that can be applied.

A concerted use of 2D spectroscopy such as 2QF-COSY, TOCSY and (^{13}C-^1H)-HMQC usually allows ^1H and ^{13}C assignments. 2QF-COSY is preferred because of better suppression of the residual water signal, and better resolution near the diagonal. Starting from a known resonance (say an anomeric ^1H), the entire cross-peak network due to intra-residue J network, can be identified through a TOCSY spectrum. Where 2D spectroscopy is not adequate, homonuclear and heteronuclear 3D methods such as TOCSY-ROESY and HMQC-NOESY have been used.

Selective deuteration, and use of its effect on changes in ^{1}H and ^{13}C relaxation rates and differential isotope shifts on ^{13}C chemical shifts have been helpful in assignment procedures.

Unlike the ^{1}H chemical shifts, ^{13}C chemical shifts in carbohydrates have good dispersion. There are several ^{1}H with large one-bond ^{1}H-^{13}C scalar couplings, which allow faster coherence transfer during multi-step, multi-dimensional spectroscopy. Use of ^{1}H-^{13}C correlation spectroscopy such as HMBC, allows correlation of ^{1}H and ^{13}C on the two sides of the glycosidic linkage, as well as within the sugar moiety. For establishing long range correlations, a 2D pulse sequence combining (^{13}C-^{1}H)-HMQC with TOCSY (2D HMQC-TOCSY) offers greater sensitivity and resolution.

2.8 Structure and Dynamics of Polysaccharides

Both quantum chemical and classical potential functions have shown that there exists a broad region of allowed conformations in disaccharides. The allowed range is spread around the *trans, trans* orientation in the (ϕ, ψ) conformational maps (Figure 8.4). Thus, there is a high degree of conformational freedom around the glycosidic linkage leading to greater flexibility in polysaccharides.

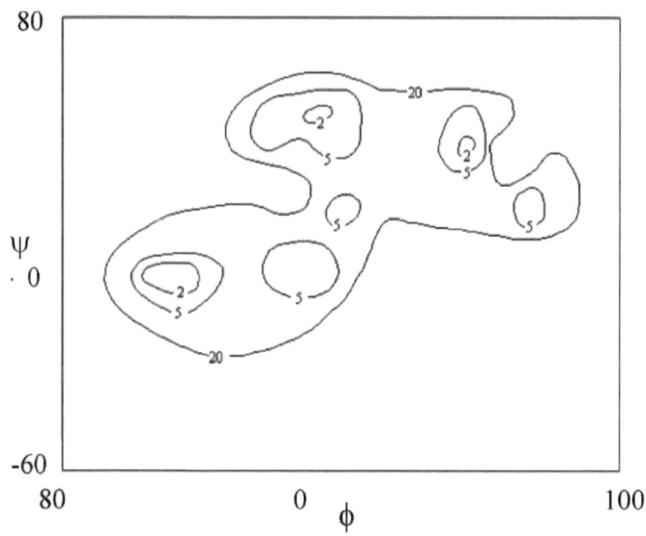

Figure 8.4: A (ϕ, ψ) conformational energy map of a disaccharide obtained by the CNDO method. (Thanks to A. Saran for providing this figure).

One of the families of inter-residue constraints, which can be measured by NMR, is the inter-glycosidic C-O-C-H couplings. These $^{3}J(^{13}C-^{1}H)$ couplings are typically in the range of 4-6 Hz and these values provide information on the nature of glycosidic linkage and their conformation. However, a definitive Karplus type relation for such couplings is lacking. In addition, some of the one and two bond ^{13}C-^{13}C couplings are sensitive to the nature and conformation around glycosidic linkages. Only a limited number of inter-residue distances can be estimated using ROESY or NOESY. Since the conformation of individual monomeric units can be well established using the J-coupling networks, the intra-residue ^{1}H-^{1}H distances can be used as standards for obtaining the (ϕ, ψ) angles from inter-residue distance constraints. Typically, a total of only 2-5 geometrical constraints are available for each glycosidic linkage. The problem is further complicated since these may be time-averages over several possible conformations.

The number of available NMR constraints linking two neighbouring sugar units, are thus limited. Considering the broad minimum in the (ϕ, ψ) space, the precise fixing of the conformation is much more difficult in the case of carbohydrates. Therefore, generation of the family of allowed conformations from the time-averaged NMR properties, coupled with the use of force fields and energy minimization, may lead to ambiguous results. Advances in NMR techniques, particularly for the measurements of weaker NOE, and heteronuclear couplings coupled with better force fields for carbohydrates, has improved the situation considerably. However, in general, oligosachharide backbone conformations are quite flexible and the analysis of experimental and theoretical results need not lead to a closely related family of conformations.

The principles for studying dynamic motions in carbohydrates are similar to those discussed for proteins and nucleic-acids. However, the overall tumbling motions and internal rotations around the glycosidic bonds are super-imposed. The motions relating to internal rotations are not very well characterised by the NMR constraints. One has to depend heavily on theoretical considerations. Therefore, attempts to determine conformation and dynamics of polysaccharides are limited.

2.9 Glycoconjugates

The importance glyco-conjugates (glycan moieties linked to proteins and lipids) in specific recognition during cellular interaction has been recognised recently. The term *glycobiology* was introduced in 1988, to focus attention to the oligosaccharide recognition motifs found on glycoproteins, mammalian glycolipids and bacterial polysachharides. Almost 70% of proteins have Asn-X-Ser sites, which participate in N-glycosylation of proteins. Glycan can modify properties of proteins to which they attach. They occur in living cells as hormones, immunoglobulins and are integral part of membranes and bacterial cell walls.

The linkages and primary structures of glyco-conjugates can be established using a combined approach based on MS-LC-NMR, as in the case of carbohydrates. As expected, the interpretation of NMR spectrum of such molecules in terms of structure and dynamics is even more difficult than those of oligosaccharides. The conformation of the linkage between an N-glycan and its Asp side-chain has been very well characterised using NMR and a statistical analysis of X-ray diffraction data. The Asp linkage is relatively rigid and planar. The first glycan residue has a tendency to extend away from the peptide backbone. N-glycosylation in small peptides generally induces more compact conformations near the site of glycosylation. For larger proteins, glycosylation has a considerable stabilising effect on the backbone structure, and alters proton mobility. On the other hand, the average structure of the glycan moiety in glycoproteins does not alter much and it is much more mobile than the structure of the linked proteins.

The cell surface glycoproteins play a crucial role in cell-cell recognition. A majority of glycoproteins are found on the cell surface. Strong inter-cellular interactions are a result of collective interactions on the two complementary cell surfaces and may be manifested only in *in-vivo* situation. Another area of recent interest is the interaction of an oligosaccharide with a large receptor, such as an antibody. Transfer NOE has been used for such studies. In view of the importance of glycoconjugates in biological processes and in anti-viral therapy, this area is receiving good attention.

While a large amount of work on carbohydrates and glycoconjugates has been done using solution NMR techniques, the more important problems in glycobiology involving complex glycoproteins and membrane bound cell markers, can be tackled much more effectively, through the use of solid-state NMR.

2.10 Applications in Development of Bacterial Polysaccharides as Vaccines

A relatively important area of pharmaceutical research is the application of NMR techniques in the formulation and validation of pneumococcal polysaccharide vaccines. Several polysaccharide based vaccines have been developed for prevention of bacterial diseases arising from infections due to *streptococcus pneumoniae, neisseria meningitidis, haemophilus influenzae and salmonella typhi.* In some vaccines, the polysaccharide is covalently attached to a carrier protein. The assay for the chemical identity of such polyvalent vaccines is a challenging task.

The bacterial polysaccharide vaccines contain a number of carbohydrate units and show structural diversity in terms of sugar composition, glycosidic linkages and presence of labile side-chains such as O-acetate and pyruvate. In some cases nitrogen and phosphate groups are also present.

NMR has proved to be a useful and powerful tool for the characterization of such polyvalent polysaccharide vaccines. Arising from the differences in their repeating unit structure and their dynamic properties, such molecules give well resolved and unique spectra (Figure 8.3). NMR techniques help in determination of composition, chemical identity, structural features and other properties of such vaccines. It serves as an alternative and more informative tool as compared to the usual colorimetric methods, which have limited specificity. The NMR based techniques are extremely selective for discrimination between serotype-specific polysaccharides, even when they differ by just a single glycosidic linkage. Subtle structural changes and the common impurities of the samples (for example, those resulting from purification procedures) can be checked readily from their characteristic ^{1}H NMR.

2.11 Recognition Motifs in Polysaccharides

One of the major biological functions of recognition in biological systems is the role played by oligosaccharides. Recognition motifs are commonly found in glycoproteins, but these are also present in the form of mammalian glycolipids and bacterial polysaccharides.

The characterization of blood group antigens recognised by natural occurring serum antibodies (immunoglobulins) has emerged as an important area in biological recognition by oligosaccharides chains. For example, the blood group A individuals have the sequence Fucα1-2[GalNAcα1-3]Galβ1- on glycoproteins and glycolipids in cells and tissues. They have anti-B positive sera. Such individuals cannot accept B positive red blood cells, which have the sequence Fucα1-2[Galα1-3]Galβ1-, and *vice versa*. Blood group O individuals have Fucα1-2Galβ1- motif and as the precursor H-antigen, and both anti-A and anti-B serum antibodies. Such individuals are therefore universal blood donors but can only accept group O blood.

Studies have been carried out using NMR to determine the conformational structure of blood group polysaccharides and their specificity during protein-binding. Chemical shift assignments have been achieved using 3D COSY/TOCSY-ROESY methods. The ROE/NOE data has been coupled to molecular modelling and dynamics simulation to obtain conformational preferences of such molecules.

The analogues of sialic acid (NeuAc) present a greater challenge in conformational analysis since the glycerol side-chain is flexible. The interest in these systems arises from their role in tumour antigens and its high affinity in cell adhesion molecules. These NeuAc residues are important in specificity of certain serum antibodies. NMR has been used extensively to characterise such moieties. The (ϕ, ψ) glycosidic angles as obtained from NMR data show a wide range.

2.12 Proteoglycans

Proteoglycans are conjugates of oligosaccharides with proteins where the former is composed of long repeats of sulphated uronic acids and amino-sugars. One of the interests in proteoglycans arises from their use as a blood anti-coagulant due to their affinity for binding to proteins involved in coagulation such as thrombin III. Detailed NMR studies on oligosaccharides obtained after enzyme cleavage of such molecules have provided valuable information on the recognition determinants, conformational properties and biological activity of this class of molecules.

3. LIPIDS AND MODEL MEMBRANES

The boundaries of living cells are formed by semi-permeable barriers, which separate the inside from the outside of the cells. Biological membranes act as barriers for molecules and allow their passage from inside to outside and vice-versa in a highly selective manner. A selective permeability is brought in by proteins, which serve as pumps, channels, receptors, energy transducers, and enzymes. Membranes also have a functional role in information transfer and storage of energy.

They mediate cell-cell recognition. In addition to the external membrane (plasma membrane), eukaryotic cells also contain internal organelles such as mitochondria, chloroplast and lysosome. Each of these cell organelles are also surrounded by membranes.

3.1 Elements of Biological Membranes

Biological membranes, which constitute boundaries in cells, are essentially built of two groups of molecules: lipids and proteins. The two classes of molecules are present in comparable amounts by weight. However, the exact composition of the two classes of molecules varies significantly from one membrane system to the other. This distribution strongly depends on the functional role of different membrane structures. For example, mitochondrial and chloroplast membranes, which are involved in energy transduction, have almost 75% of their weight in the form of protein component. Myelin, which acts as insulator around nerve fibres, has a high content of lipids (almost 80%). The M_r of lipids is between 500-800 Da, while proteins have M_r in the range of several kDa. Thus, there are hundreds of lipid molecules for each individual protein molecule.

Membranes can be viewed as non-covalently linked molecular assemblies. These are essentially 2D structures consisting of oriented protein molecules, embedded in organized lipid assemblies. The cell membranes also contain some sugars, which are present mainly in the form of conjugates namely, glycoproteins and glycolipids.

3.2 Lipids

Lipids are the essential components of cell membranes. Lipids constitute a wide range of relatively small molecules, with diverse chemical groupings. A common distinguishable property of lipids when compared to other biological molecules is that they are insoluble in water and highly soluble in organic solvents such as chloroform. Lipids play crucial roles in cellular function. They act as cofactors for enzymes, electron carriers, hormones, anti-oxidants, light-absorbing pigments and intracellular messengers. Lipids act as highly concentrated energy sources for the cells. However, their main function is to provide a fluid matrix for larger proteins and glyco-protein molecules, which are embedded in the membrane.

A large fraction of cellular lipids are phospholipids (Figure 8.5). These in turn, are built from two fatty acid chains, condensed to a glycerol moiety at C-1 and C-2 positions. The length of the alkyl chains in fatty acids in lipids can vary between 14 to 24 carbon atoms. However the most common chains found in natural lipids have 16 and 18 carbon atoms. Alkyl chains containing one or two double bonds may also be present. A phosphorylated alcohol, such as choline, serine, inositol, glycerol or ethanolamine is attached to the C-3 carbon of glycerol.

The most common lipids used in experimental studies of biological membranes are 1, 2 dipalmitoyl-sn-glycero-3-phosphocholine (DPPC) and 1, 2 dimyrositoyl-sn-glycero-3-phosphocholine (DMPC). Palmitate has 16 carbon saturated alkyl chain, while myristoyl has 14 carbons. A model lipid for studying effect of un-saturation in the acyl chain is POPC (1 palmitoyl-2-oleoyl-sn-glycero-3-phosphocholine).

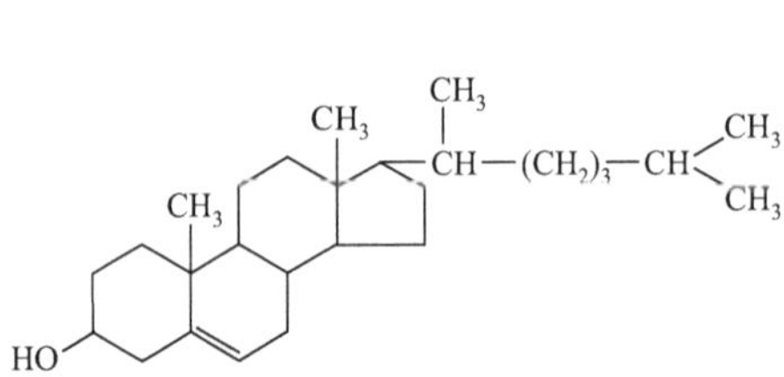

Cholesterol

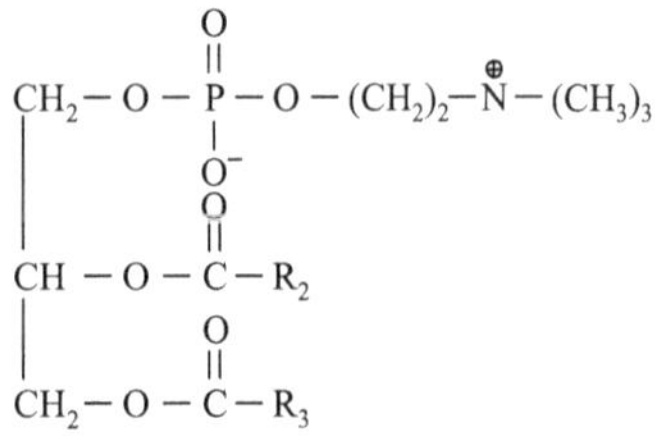

Phosphatidylcholine

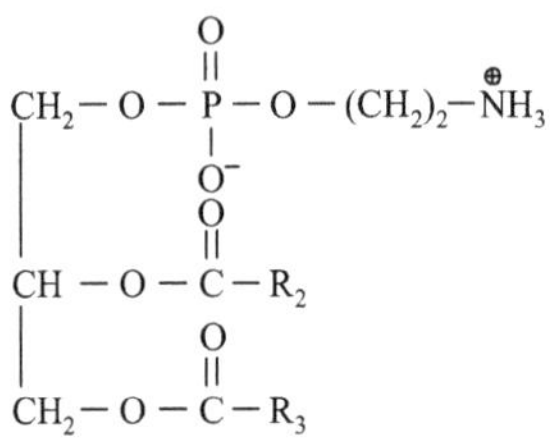

Phosphatidyl ethanol amine

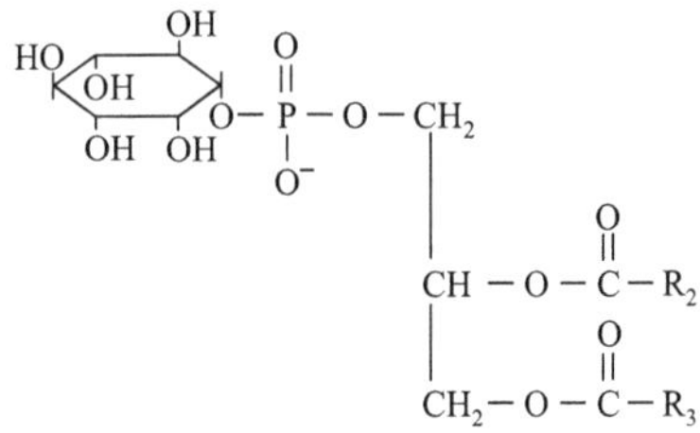

Phosphatidylinositol

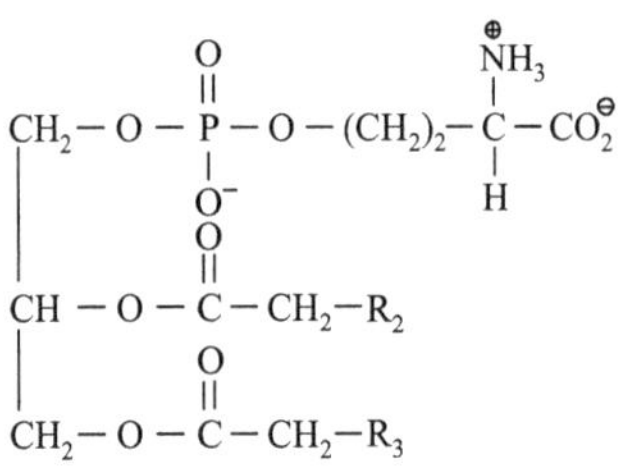

Phosphatidylserine

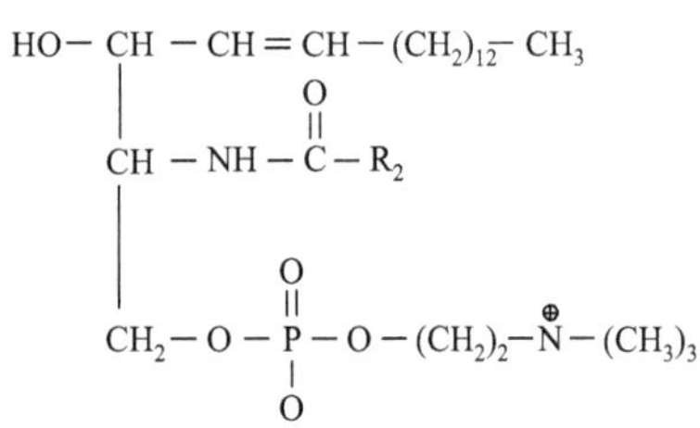

Sphingomyelin

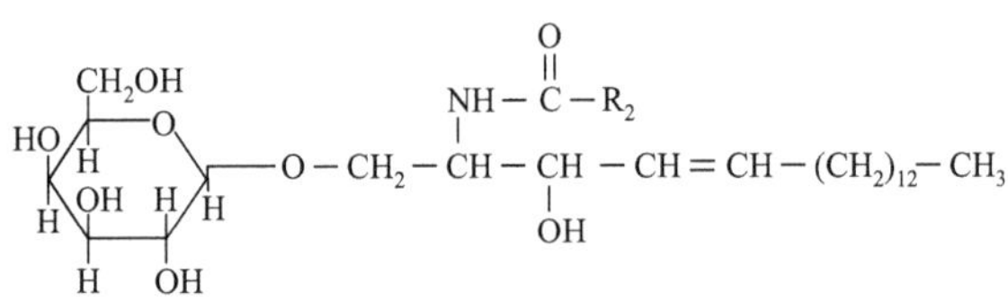

Glycolipids

Figure 8.5: Some of common lipids found in biological membranes. The lipids molecules have well defined hydrophilic and hydrophobic regions. They are often represented by a "head" representing the water soluble polar group and two "tails" representing the hydrophobic alkyl groups. They spontaneously form a 2D-bilayer, when dissolved in water.

A different group of phospholipids are composed of a sphingosine backbone instead of glycerol. Membrane lipids also include glycolipids, which are sugars containing lipids derived from sphingosine. The third type of lipid, which is involved in membranes of animals, is cholesterol, a molecule which modulates membrane fluidity. It has a steroid nucleus, which has four linked hydrocarbon rings.

As seen from Figure 8.5, a characteristic feature of lipids is that they usually have two well defined regions: one with polar chemical groups which are hydrophilic (water soluble), and the other containing hydrocarbon chains, which are hydrophobic (water insoluble). The hydrophobic regions endow membranes with ability to solubilize non-polar molecules in their matrix, and act as regions for the location and assembly of biological molecules that are not soluble in aqueous phase.

3.3 Conformation of Phospholipids

The conformation of individual phospholipid molecules is dictated by internal rotation around a large number of single bonds (Figure 8.6). The dihedral angles involved in defining the conformation are (θ_1, θ_3) for the glycerol moiety, α_1-α_6 for the polar head group, and β_1-β_n and γ_1-γ_n for the two acyl chains. In spite of a large degree of conformational flexibility, these molecules show preferred conformation around most of these bonds. It is difficult to crystallize phospholipids and a large amount of evidence for the conformation of lipids has been achieved through solution NMR and theoretical calculations.

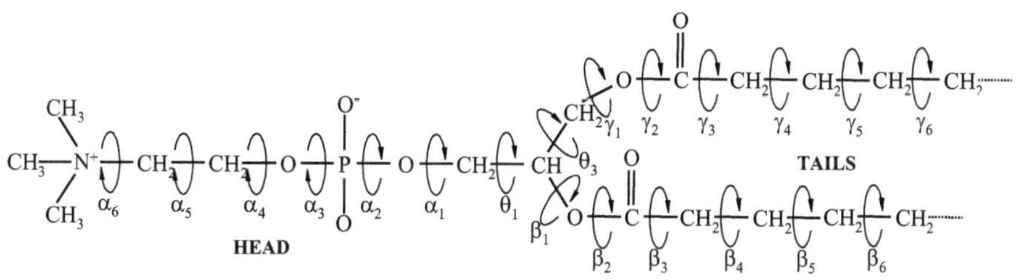

Figure 8.6: Conformational angles in phospholipids.

In chloroform solutions, the spectra of phospholipids are well resolved (Figure 8.7). The conformational populations around each bond can be estimated from the three-bond couplings and NOE data. Evidence indicates that more than one conformation exists around most of the bonds. A general feature of the preferred conformations phospholipids is that they lead to parallel hydrocarbon chains. The preferred conformation around the C-C bond is *trans*, thus leading to extended chains. The polar head group is well separated from the hydrocarbon chains as implicated in the head and tail model (Figure 8.6). The preferred structures can thus, readily lead to the formation of bilayers.

3.4 Lipid Vesicles and Bilayer Structures

An important property of lipids is that when these are immersed in water, the hydrophobic interactions between the fatty acid chains, and the hydrophilic interactions between water and the polar head group, drive them spontaneously to form a bilayer structure (Figure 8.8). Such a structure enables polar head groups of lipids to contact water on the two sides through hydrophilic interactions. The hydrocarbon chains are packed close to each other. Lipids on the two sides of the bilayer are placed tail to tail, to avoid any contact with water. Such an arrangement is stabilised by van der Waals interactions between hydrocarbon chains, and the hydrophilic interactions between water and the polar head group in the head group region. Formation of bilayers is a spontaneous self-assembling process and leads to closed structures, such as vesicles and liposomes. The weak molecular interactions also prevent any hole or gaps, and allow self-annealing.

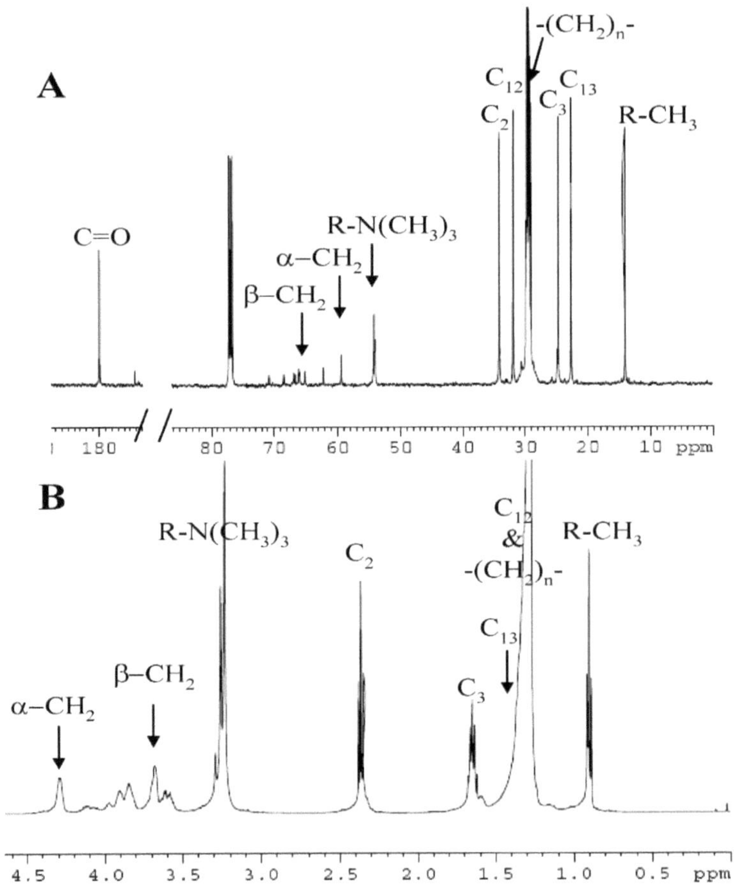

Figure 8.7: (A) ^{13}C and (B) ^{1}H spectrum of DPPC in $CDCl_3$. (Thanks to S. Srivastava for recording this spectra).

Lipid assemblies provide a fluid environment for other functional localization of other functional molecules such as proteins, and help in maintaining an ionic balance in the cells. Sonication helps in obtaining spherical vesicles of uniform size (typically 500 Å diameter), which are used in studies of model membranes. Typical width of lipid bilayers is 60 Å. The lipid bilayers are impermeable to most ions, polar molecules and water. Thus, intermolecular lipid-lipid and lipid-water interactions help in the formation of compartments and permeability barriers in cells. They separate the aqueous phases on the two sides of the bilayer.

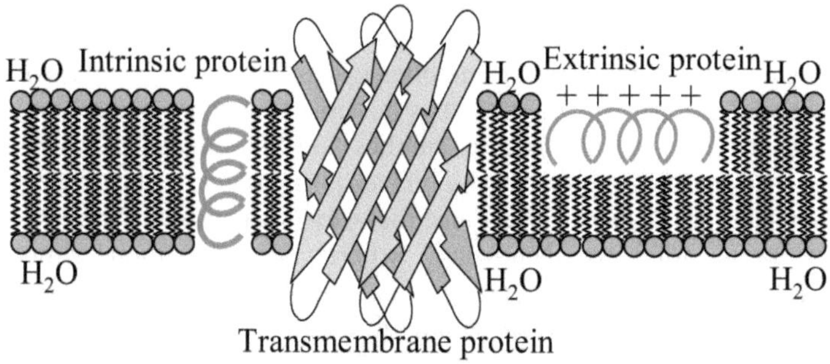

Figure 8.8: A schematic representation of biological membrane. The lipids form a 2D bilayer structure, in which the proteins are embedded. The fluid mosaic model, shown in the Figure, postulates three types of proteins. Those which are weakly bound on the surface (extrinsic or peripheral proteins), those which are embedded deep inside the bilayer (intrinsic proteins), and those which span both side of the lipid bilayers (trans-membrane proteins).

3.5 Membrane Architecture

Membrane proteins are embedded into the phospholipid bilayer structures. Proteins are responsible for most of the functions of biological membranes. Therefore the type of proteins and their relative concentration vary significantly depending on the type and function of the membranes. One distinguishes between integral proteins, which are embedded throughout the bilayer structures and peripheral proteins, which are loosely bound to the surface. A single integral protein may have several trans-membrane segments spanning the bilayer structures. The protein chains may also extend themselves on both sides of the membranes. Such proteins are called trans-membrane proteins. The most common structural motif of membrane proteins is α-helix. Such molecules are composed of hydrophobic amino acids. However, β-sheet structures can also be embedded into lipid bilayers, by the formation of hydrophobic surfaces exposed to the lipids, and the hydrophilic surfaces lying in the interior of the protein structures.

The proteins and lipids on the two sides of the membranes can be quite different. Such an asymmetry controls the biological functions on the two sides, which are different.

4. NMR STUDIES OF STRUCTURE AND DYNAMICS OF LIPID BILAYERS

The techniques of NMR have proved valuable in studies of role of lipids in controlling the structure and dynamics of biological membranes. This is particularly important, since neither the membrane proteins nor the lipid assemblies can be crystallised. Therefore, the diffraction methods are of limited use. Secondly, as stated earlier, the membranes have a very dynamic structure. NMR can probe such dynamic aspects over a wide range of time scales (typically 10^{-3} to 10^{-9} seconds). The molecular motions in lipids in biological membranes lead to relatively sharp lines and therefore several experiments can be done on static samples. MAS can be used for further narrowing of resonances.

One important feature of lipid assemblies is their bilayer behaviour, which leads to closed overall structures. However, it is possible to prepare planar lipid bilayers on glass plates. This way the lipid assemblies and the embedded proteins can be oriented at a desired angle between the bilayer normal and the magnetic field (B_0) axis. Such studies are of great value, since the overall behaviour of the NMR of such assemblies show solid-like behaviour, where the orientation terms in the Hamiltonian are important. In general, it is preferable to introduce well-aligned membranes with the bilayer normal parallel to the B_0. With such an orientation, the axis of rotational diffusion of the lipids, coincide with the direction of the B_0 and results in relatively narrow lines.

4.1 Molecular Motions in Model Membranes

One of the important properties of biological membranes is their highly dynamic nature, at the ambient temperature of living systems. As has been discussed in Chapter 1, the hydrocarbon chains can undergo *trans-gauche* isomerization around C-C bonds, at a relatively fast rate. A *gauche* conformer interferes with the intermolecular packing of the hydrocarbon chains. The segmental motions of alkyl chains arising from the $t \rightarrow g$ isomerisation provide a loose packing of lipid molecules, imparts fluidity and leads into the disorder of in the membrane structure. Unsaturated alkyl chains can also add to the disorder in the packing of hydrocarbon chains on a larger scale. Both, lipids and proteins can also diffuse laterally (lateral diffraction), at a relatively fast rate. However, a flip-flop motion involving transfer of molecules from one side to the other of the bilayer is relatively slow. This allows preservation of asymmetry in the chemical composition of the two sides of the membranes. Thus, membranes at body temperatures of living systems exhibit a 2D anisotropic motion.

Magnetic resonance techniques (NMR and ESR) have been used extensively to study structure and fluidity in model membranes. An important aspect of MR methods is the availability of wide range of techniques. These can be used to study different aspects of organisation biological membranes and dynamic processes occurring at different time-scales. For example, techniques based on electron spin resonance (ESR) using spin labels, provide information on the phase transitions and fluidity of membranes. ^{1}H NMR provides information on the conformation and intermolecular interactions. ^{2}H NMR has been used to study order parameters in different regions of bilayers. ^{13}C relaxation rates are useful in studies of segmental motions and the structures of proteins embedded in lipid bilayers. ^{31}P CSA provides vital information on lipid polymorphism. We shall discuss some of the experiments which have been used to study the structural and dynamic behaviour of lipids in biological membranes.

4.2 Phase Transitions

Model bilayer structures show a typical liquid crystalline (L_α) to gel (L_β) phase transition, as the temperature is lowered. Below a particular transition temperature (T_m) the molecular motions are frozen and membranes loose their capacity to transport molecules and ions across the cell membranes. In the temperature range covering the liquid crystalline behaviour, membrane-lipids show an anisotropic motion, with fast lateral and segmental motion. There is a narrow range of temperature at which lipid molecules exhibit liquid crystalline behaviour.

ESR has been widely used to study phase transitions in model membranes. A nitroxide spin label called TEMPO has been used for this purpose. Signals arising from TEMPO in aqueous phase (P) occur at slightly different positions compared to those in the hydrophobic phase (H). The relative extent of partitioning (f) of TEMPO between the hydrophobic and hydrophilic phases can be measured from the ratio of signal intensities, f = H/ (H+P). The partition coefficient varies as a function of fluidity of membranes.

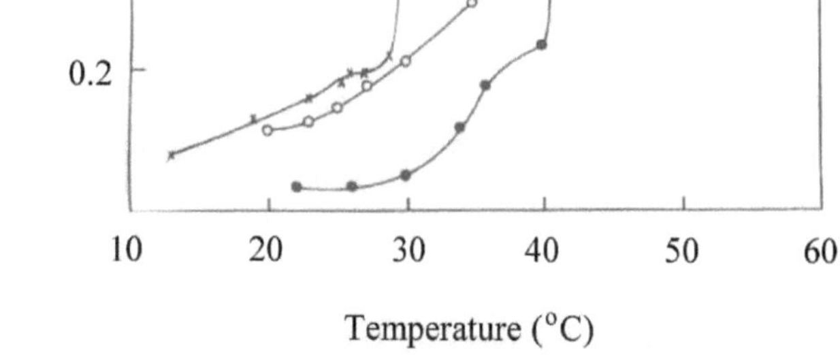

Figure 8.9: Partition coefficient of TEMPO, f= H/(H+P) as a function of temperature. (a) corresponds to pure sonicated DPPC vesicles (100 mM in water); (c) and (b) correspond to DPPC vesicles, incorporated with vitamin E acetate and α-tocopherol (20 mM), respectively. (Thanks to S. Srivastava for providing this figure).

Figure 8.9 shows the change in the partition coefficient of lipid vesicles prepared using DPPC. The partition coefficient *f* shows a gradual increase with temperature. At ~ 35° C, there is a sudden increase in *f*, which is followed by another sharp point of inflexion at 41° C. The two points correspond to a transition from the gel to liquid crystalline phase of the lipid bilayers. The pre-transition at 35° C, corresponds to pre-melting, where the polar head groups start showing fluidity. The hydrocarbon chains also become fluid above 41° C. Above around 55° C, the lipid vesicles are no longer able to maintain their order in the form of 2D vesicles, and melt completely. The behaviour of the partition coefficient curve arises from the fact that the spin-label goes into the hydrophobic phase only in the liquid crystalline phase.

Figure 8.9 also shows the phase transition curves in the presence of α-tocopherol (vitamin E), a well known anti-oxidant and the corresponding acetate derivative. The characteristic features of the phase transition curve are maintained in the presence of vitamin E acetate, except that the T_m is lower. However, in the presence of α-tocopherol, the transition curve is broad and devoid of sigmoidal behaviour. This is because in the case of vitamin E acetate, only van der Waal's interactions contribute to inter-molecular interactions. The replacement of the acetate group by a hydroxyl group, leads to formation of a strong intermolecular hydrogen bond between the vitamin and the polar head group of the lipids. This leads to a loss of the cooperative melting behaviour.

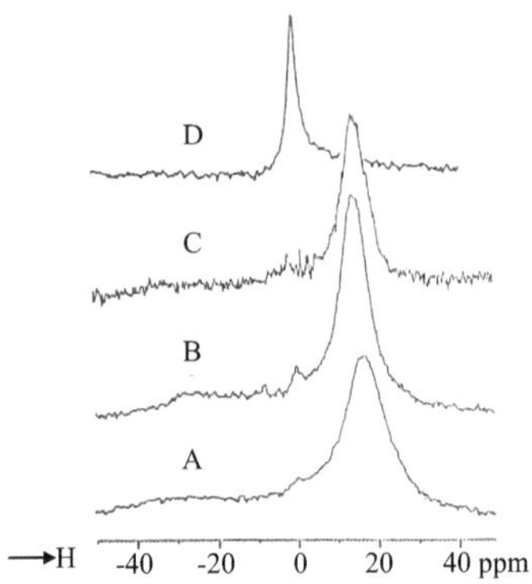

Figure 8.10: ^{31}P CSA patterns at 202.457 MHz, with broad band ^{1}H decoupling for: (A) unsonicated DPPC, at 323 K; 100 mM DPPC dispersed in H_2O); (B) DPPC + 10 mM α-tocopherol; (C) DPPC + 10 mM epinephrine (D) DPPC + 1 mM tryptophan. In sonicated vesicles only a sharp peak at 0 ppm is observed. (Thanks to S. Srivastava for providing this figure).

Lipid mobility is crucial for the survival of living systems. Living system carefully choose lipids such that the biological membranes have a fluid or liquid crystalline behaviour at their body temperature. This can be achieved by changes in the head group and the length and unsaturation in the alkyl chains. For example, DMPC (14 C alkyl chain) has a T_m of 29° C, while the T_m for DPPC is 37⁰ C. Other molecules such as cholesterol and vitamins, peptides and proteins, as well as exogenous molecules such as drugs and vitamins can also modulate the phase transition behaviour.

4.3 Lipid Polymorphism in Model Membranes

When dispersed in excess water, lipids, normally assume closed spherical structures. However, several other structures such as hexagonal and cubic phases have been shown to co-exist. Such departures from normal bilayer organization are caused by local heterogeneities as a result of the presence of other molecules and a distribution of different types of lipids.

Lipid polymorphism has been studied using a number of techniques, including electron microscopy and ^{31}P NMR. Each phospholipid molecule contains only one phosphate group. Therefore, the ^{31}P spectrum consists of a simple single line resonance from each lipid (Figure 8.10). The line-shape is governed by CSA, organization of phospholipid molecules and molecular motions. For example, the fast axial rotation of the phospholipids around their long axis in the liquid crystalline (L_α) phase, averages the ^{31}P chemical shift to a symmetrical tensor, where the tensor singular axis ($\sigma_{\|}$) coincides with the rotational axis. The resonance signal at 30 ppm is indicative of DPPC molecules with their long axis oriented parallel to the B_0 direction, while for perpendicular alignment the ^{31}P signal appears at -15 ppm. In a randomly oriented sample such as DPPC dispersion in gel phase, the overall rotation rate is slow. One observes a broad shoulder and a relatively sharp peak. The total span of the ^{31}P signal arising from its CSA is around 46 ppm. As the temperature is increased above T_m, the anisotropic chemical-shift tensor undergoes partial averaging.

An important aspect of lipid organization and fluidity from biomedical view-point is the effect of drugs, vitamins and other molecules on such properties. For example, similar bilayer organization is observed when the neurotransmitter epinephrine or α tocopherol is incorporated at a lipid to drug molar ratio of 10:1. However, a dramatic change in the line-shape is observed in presence of a much smaller amount of tryptophan. The shift of the signal to 0 ppm is indicative of fast lateral and overall tumbling motion, and of a change in molecular organization of lipids as a result of DPPC-Trp interactions. This leads to formation of hexagonal (H_{II}) phase. Lipids in hexagonal phase experience additional motional averaging due to lateral diffusion along the cylindrical axis. These observations have been confirmed by electron microscopy.

4.4 ^{13}C NMR Studies of Segmental Motions in Alkyl Chains

As discussed earlier, the ^{13}C spin-lattice relaxation times are dominated by intra-molecular ^{1}H-^{13}C dipolar interactions between directly bonded atoms. The ^{13}C relaxation times (T_1) have been used to study segmental motions along the alkyl chains in lipids organised in lipid bilayers. The NT_1 values (where, N is the number of ^{1}H attached to the carbon atom) can be measured at different depths in the lipid bilayers as a function of distance from the head groups. Such data have been used to study motional freedom. For example, when vitamin E acetate is embedded in DPPC lipid bilayers, the values of NT_1 show a significant decrease, as a result of restricted motion. The terminal methyl groups show significantly faster motions (Figure 8.11).

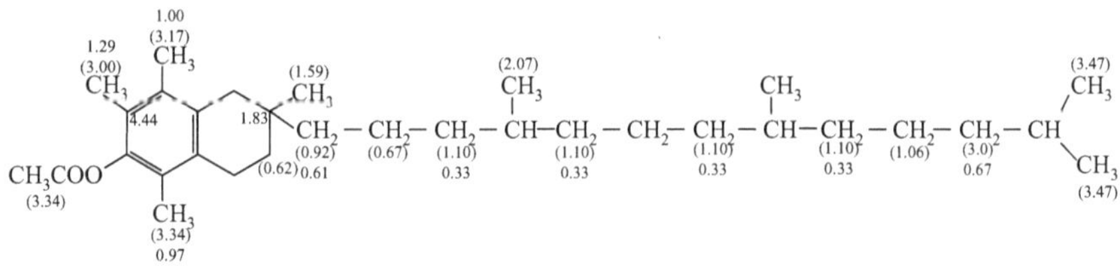

Figure 8.11: NT_1 values of vitamin E acetate embedded in lipid bilayers as compared to the values in $CDCl_3$ solutions (given in parenthesis). (Thanks to S. Srivastava for providing these values).

4.5 Molecular Order and ^2H NMR Studies of Membrane Organization

We have discussed some examples, which show how small molecules can perturb the mobility and ordering of lipids in biological membranes. The ^2H quadrupolar coupling constant is a sensitive measure of internal molecular motions. Since the quadrupolar coupling depends on the orientation of the C-D vector with respect to axis of the magnetic field B_0, a reorientation of the vector reduces the width and changes the shape and pattern of the ^2H resonance.

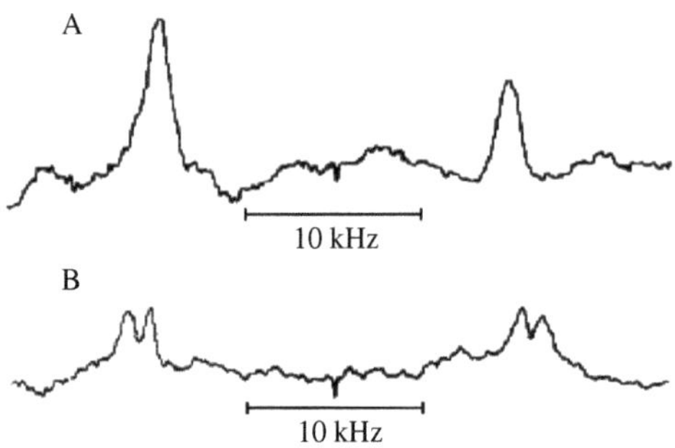

Figure 8.12: ^2H NMR spectra of bilayers of L–α–dipalmitoyl lecithin selectively deuterated in both the fatty acyl chains. (A) (2-10 deuterated DPL; 57 °C; 20,000 free induction decays. (B) C-3 deuterated DPL: 65 °C. A. Seelig and J. Seelig, Biochemistry, 13, 4839–4845 (1974). (Reprinted with permission from American Chemical Society).

Deuterium (^2H) NMR is an ideal tool for studies of mobility and ordering in lipid bilayers. For this purpose, deuterated lipids are used. The shape and width of the ^2H resonance provide information on the properties and segmental motions at different positions of the hydrocarbon chains. For example, in the classic work of

Seelig and Seelig, quadrupole splittings were measured using hydrated DPPC in the liquid crystalline phase. The lipid was deuterated at nine different acyl positions. The order parameter has been measured as a function of the position in the acyl chain. The order parameter is constant for the initial segments, but the splitting increases and the order decreases towards the methyl terminus of the alkyl chains. The presence and the number of *gauche* isomers have been determined by these experiments. The thickness of the hydrocarbon region of the bilayer has been estimated as 35 Å. Even at 20° C below the T_m, the order parameter and the line shape suggest considerable segmental motions. Around T_m, there are abrupt changes in the ^2H NMR pattern. Heating above the transition temperature causes an overall decrease in chain order (Figure 8.12).

4.6 Modulation of Motion and Order of Lipids by Larger Molecules

Some examples of how small molecules such as drugs and hormones affect the dynamic motions and order in lipid bilayers have been discussed above. A classic example of the effect of larger molecules such as proteins on the ^2H NMR spectrum for tri-deuterated (at C-14) DMPC lipid bilayers is shown in Figure 8.13. Both the line widths and the splittings change, dramatically. Molecules such as cholesterol decrease the mobility of the terminal alkyl chain, while some of the proteins increase the fluidity.

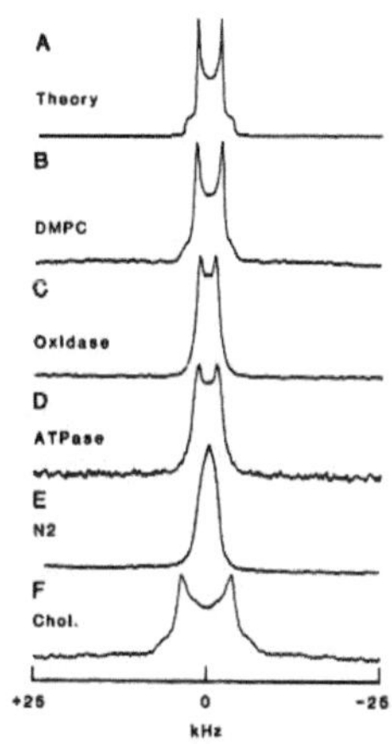

Figure 8.13: ^2H NMR spectra of [D_3-] DMPC in presence of (a) excess water; (b) cytrochome c oxidase (c) beef brain myelin proteolipid apoprotein (d) cholesterol. R.L. Smith, and E. Oldfield. Science, 225, 280-287 (1984). (Reprinted with permission from AAAS).

In view of the importance of cholesterol in modulation of membrane fluidity in animals and its biomedical consequences, this subject has received wide attention. Both ^2H and ^{13}C NMR, using specifically labelled cholesterol have been used for this purpose. The cholesterol orients parallel to the lipid bilayers and the steroid group is buried in the hydrocarbon chains of the phospholipids. The small head consisting of a single O-H group is located such that it is in the vicinity of the polar heads of phospholipids. It increases the overall ordering of the phospholipids in the liquid crystalline phase. On the other hand, it has a disordering effect on the hydrocarbon chains, in the gel phase. The rate of rotation and internal motions in cholesterol indicate that its hydrophobic tail undergoes rapid internal motions.

Figure 8.13 also illustrates use of NMR for monitoring changes in lipid mobility and order as a result of lipid-protein interactions. In fact, lipids orient around the transmembrane proteins. In this process, several lipids bind through hydrophobic interactions to the protein molecule.

4.7 Measurement of Lateral Diffusion using NMR

The translational diffusion of water and lipids in macroscopically aligned lipid bilayers in the liquid crystalline L_α phase has been determined using pulsed magnetic field gradients. The technique has potentials for biophysical applications in the field of membrane biology.

5. REFERENCES

5.1 Further Reading

R.U. Lemieux, R.K. Kullnig, H.J. Bernstein, and W.G. Schneider, *NMR Studies of Carbohydrates*, J. Am. Chem. Soc. **80**, 6098–6105 (1958).

A. Saran and G. Govil, *Conformational Structure of Polysaccharides: EHT Calculations on Maltose*, Indian J. Chem. **10**, 547–549 (1972).

S.J. Singer, and G.L. Nicolson, *The fluid mosaic model of the structure of cell membranes*, Science, **175**, 720–731 (1972).

A. Seelig and J. Seelig, *The dynamic structure of fatty acyl chains in a phospholipid bilayer measured by deuterium magnetic resonance,* Biochemistry **13**, 4839–4845 (1974).

R. L. Smith, and E. Oldfield, *Dynamic structure of membranes by deuterium NMR*, Science **225**, 280–287 (1984).

S. Srivastava, R.S., Phadke, and G. Govil, *Effect of incorporation of drugs, vitamins and peptides on the structure and dynamics of lipid assemblies*, Mol. Cell. Biochem. **91**, 99–109 (1989).

F.M. Marassi, and S.J. Opella, *NMR structural studies of membrane proteins*, Curr. Opin. Struc. Biol. **8**, 640–648 (1998).

M.R. Wormald, and R.A. Dwek, *Glycoproteins: Glycan Presentation and Protein-fold Stability*, Structure **7**, R155–R160 (1999).

F.M. Marassi, and S.J. Opella, *A solid-state NMR index of helical membrane protein structure and topology*, J. Mag. Res. **144**, 150–155 (2000).

B. Bechinger, and C. Sizun, *Alignment and structural analysis of membrane polypeptides by ^{15}N and ^{31}P solid-state NMR spectroscopy*, Concepts Magn. Reson. **18A**, 130–145 (2003).

L.E. Lerner, *Carbohydrate Structure and Dynamics from NMR, in NMR Spectroscopy and Its Application to Biomedical Research*, S.K. Sarkar (ed.) pp. 313–342 (1996).

5.2 Books and Reviews

S. Srivastava, R.S. Phadke, G. Govil and C.N.R. Rao, *Fluidity, permeability and antioxidant behaviour of model membranes incorporated with α-tocopheroland vitamine E-acetate* Biochim. Biophys. Acta **734**, 353–362 (1983).

S. Srivastava, R.S. Phadke and G. Govil, *Effect of incorporation of drugs, vitamins and peptides on the structure and dynamics of lipid assemblies*, Mol. Cell. Biochem. **91**, 99 (1989).

H.J. Allen, and and E.C. Kisalilus (eds), *Glycoconjugates: Composition, Structure and Function*, Marcel-Dekker (1992).

E.F. Hounssell, *^{1}H NMR in the structural and conformational analysis of oligosaccharides and glycoconjugates,* Prog. Nucl. Magn. Reson. Spectrosc. **27**, 445–474 (1995).

R. Lipowsky, and E. Sackman, *The Structure and Dynamics of Membranes*, Elsevier (1995).

L. Styer, *Biochemistry*, 4th Edition, W.H. Freeman and Co. (1995).

V.S. R. Rao, P.K. Qasba, P.V. Balaji and R. Chandrasekharan, *Conformation of Carbohydrates*, Taylor and Francis (1998).

C. Abeygunawardana, T.C. Williams, J.S. Summer and J.P. Hennessey, *Development and validation of an NMR-based identity assay for bacterial polysaccharides*, Anal. Biochem **279**, 226–240 (2000).

A.L. Lehninger, *Principles of Biochemistry*, W.H. Freeman and Co. (2000).

CHAPTER 9

HIGH-RESOLUTION SOLID-STATE NMR

1. INTRODUCTION

The structure, dynamics, interaction and function of biological molecules, when present in aqueous solutions, have been discussed in previous chapters. Even though, most biochemical reactions take place in cytoplasm and are thus in a solution phase, there are several situations when studies in solid state become important. One example pointed out already, pertains to molecules embedded in biological membranes. As discussed, lipids form self assembled multi-molecular structures and are integral components of biological membranes. Molecules embedded in membranes are generally not amenable to solution studies. A number of other macromolecular assemblies have restricted translational motions. As the size of the assembly increases, the molecular motions get more restricted. The immobilisation of large molecules and molecular assemblies results in the broadening of resonance lines and the behaviour of molecules approaches that in solid state. Several biological structures such as the bones are solids, while muscles and hard tissues are closer to solid rather than to solution state.

The normal NMR spectra in solid-state have fairly broad resonances and do not provide much information. In recent years, techniques have been developed which allow high-resolution spectra to be obtained in solids and solid-like biological materials. In this context, the term *solid-state NMR,* need not refer to materials in the usual sense of solid-state. For NMR investigations, this term usually covers situations under which molecules do not tumble isotropically, and the molecular motions are restricted to time scales of less than one msec. Such materials include both solids and semi-solids. They give very broad spectra due to line-broadening caused by dipole-dipole interactions and chemical-shift anisotropy. The spectra are normally devoid of much information.

In this Chapter, high-resolution solid-state NMR techniques are discussed. This is followed by the application of solid-state NMR for structure determination of proteins. Applications to other biological materials have been discussed in subsequent chapters.

2. NMR HAMILTONIANS IN SOLID-STATE

In solutions, where the molecules have fast isotropic motions, the NMR Hamiltonian is considered as composed of only two terms; one representing the Zeeman term (modified by the isotropic chemical shift and represented by $\mathcal{H}_\delta$), and the other representing the J-coupling ($\mathcal{H}_J$). Other interactions such as through-space dipole-dipole interactions, chemical-shift anisotropy (CSA) and the quadrupolar interaction (for nuclear spins with I > ½), are ignored. Due to isotropic and fast molecular motions in liquids, the time averages of such interactions are zero. Thus, for rapidly tumbling molecules in isotropic situations, such as in non-viscous solutions, one ends up with relatively simple spectra and fairly narrow lines. The orientation information contained in these interactions is no longer present. Further, rapid tumbling and translational diffusion also ensure that the molecules have identical time-averaged environment.

The information on chemical shift anisotropy and dipolar couplings can be partially retrieved using solvents with anisotropic motion, or by using very high magnetic fields, which partially orient the solute molecules, as has been discussed in Chapter 4.

2.1 NMR Hamiltonian in Solid-State

In solid-state, it is necessary to consider all the terms involving interaction of nuclear spins. Equation 9.1 represents the complete NMR Hamiltonian, which consists of the following terms:

$$\mathcal{H} = \mathcal{H}_Z + \mathcal{H}_J + \mathcal{H}_\delta + \mathcal{H}_{het(D)} + \mathcal{H}_{hom(D)} + \mathcal{H}_Q + \mathcal{H}_{e,n} \qquad 9.1$$

The terms which are common to both solid and solution states are those representing $\mathcal{H}_Z$ and $\mathcal{H}_J$. These two terms are independent of the orientation of the molecule with respect to the direction of the magnetic field. In most cases, one can neglect J anisotropy since it is small. However, interactions such as the CSA [$\mathcal{H}_\delta$], homo- and hetero-nuclear dipolar couplings [$\mathcal{H}_{hom(D)}$ and $\mathcal{H}_{het(D)}$], quadrupolar interactions [$\mathcal{H}_Q$] and electron-nuclear interactions $\mathcal{H}_{e,n}$ contain terms which depend on the orientation of the molecule relative to the direction of the magnetic field.

The last term in Equation 9.1, arises because of interaction between the electronic and nuclear spins. As has been pointed out, this term is dominant in the presence of a free electron spin, for example, in paramagnetic molecules. The interaction between electron and nuclear spins is similar to inter-nuclear interactions, except that the electronic magnetic moment is almost 2000 times that of nuclear moments. When present, electron-nuclear interactions may dominate over the inter-nuclear interactions. NMR applications of paramagnetic ions in metallo-enzymes have been discussed in Chapter 6. The spectral changes arising from this term will not be discussed in the present Chapter.

Figure 9.1 shows approximate relative magnitudes of various intramolecular components of nuclear spin interactions.

In isotropic solvents all terms in equation 9.1 except for $\mathcal{H}_Z$, have time dependence. The time-average of such terms is zero, except for $\mathcal{H}_\delta$ for which it is one third of the trace of the chemical shift tensor. This is the isotropic chemical shift observed in solutions, along with the scalar coupling term $\mathcal{H}_J$. All other interactions are averaged out in solutions but are retained in the solid-state, in semi-solids such as liquid crystals, biological membranes and systems having anisotropic motions.

Further, in solids, a molecule may be present in more than one conformation and may have different chemical environments.

Because of the large number of hetero- and homo-nuclear interactions, solid-state NMR spectrum shows broad features. The lines widths may exceed 200 ppm for heavier nuclei. The line width for ^{13}C and ^{15}N in crystalline solids are typically in the range of 0.5 and 10 ppm, depending on molecular order and motion. The signals of ^{31}P show an asymmetric line shape due to CSA. For ^{1}H, the dipolar interactions are much larger and the line-widths exceed the normal range for ^{1}H chemical shifts of 10 ppm. Thus, the advantage of the higher sensitivity in ^{1}H spectra is lost when dealing with solid materials. The broad lines in normal solid-state NMR spectra have limited information. In the past, the so called wide-line NMR spectra were used for studies of phase transitions utilizing changes in the line-shapes and their associated second moments.

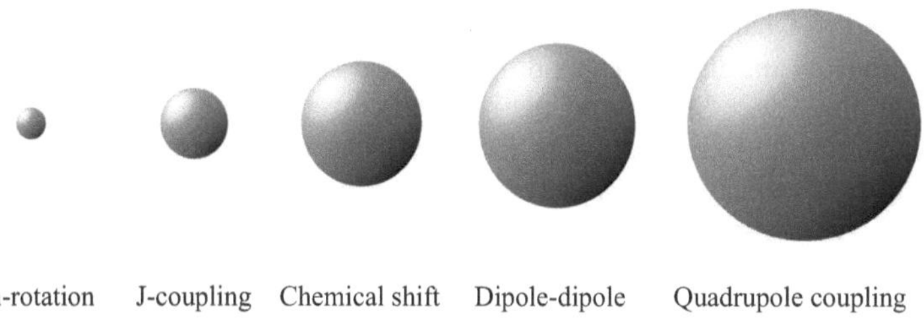

Figure 9.1: Relative magnitudes of various intramolecular components of the internal spin interactions.

In recent years, techniques have been developed which enable one to obtain spectra with a resolution somewhat similar to those obtained in solution.

The first demonstration of a technique called magic-angle spinning (MAS; discussed later in detail) was made by Andrew and his co-workers and Lowe independently in 1959 to record spectra of solids with lower line-widths, which enabled measurement of isotropic chemical shifts. In this technique, the solid sample (in polycrystalline, powder or single-crystalline state) is rotated at a fast speed around an axis that makes an angle of $\cos^{-1}(1/\sqrt{3}) \sim 54.74^{\circ}$ with the external magnetic field (B_0). This angle is called the magic angle. Since then, the field has advanced rapidly.

The advent of techniques for dipolar decoupling (DD), cross polarization (CP) and CP-MAS has facilitated recording of good quality and high resolution ^{2}H, ^{13}C, ^{15}N and ^{31}P spectra of biopolymers in solid-state. These techniques are now routinely available on commercial instruments. These have been applied to study molecular geometry, molecular dynamics, phase transitions, ordering and orientation of molecules in various biological materials. In recent years, solid-state NMR has evolved as a powerful alternative to X-ray crystallography for studies on solid materials. It has shown an edge over the later, in the studies of membrane bound proteins, fibril proteins, amorphous materials, supra-molecular assemblies and tissues, which cannot be crystallised easily.

A large amount of information in solid-state has come from the use of less abundant nuclei, such as ^{13}C and ^{15}N. In these cases, the homonuclear interactions are weak due to the low gyromagnetic ratio (γ) of the involved nuclear spins and their low natural abundance in unlabelled samples. The main contributions to line-width in such cases arise primarily from their interactions with protons and the CSA.

On the other hand, the use of ^1H NMR spectroscopy in solid-state is relatively more difficult. The ^1H-^1H interactions are stronger due to the large γ of proton. The situation is further complicated by the large number of inter- and intra-molecular interactions in a given molecule. Among the quadrupolar nuclei, the most useful probe is ^2H, due to its low quadrupole moment (Q).

2.2 Chemical Shift Anisotropy ($\mathcal{H}_{CSA}$)

As discussed in Chapter 1, the chemical shielding (σ, the absolute value of nuclear shielding tensor) and the corresponding chemical shifts (δ, measured with respect to a reference line) arise because of the induced currents of the electronic clouds in a static magnetic field (B_o), and due to the effect of magnetic susceptibility of the neighbouring chemical groups. Generally, the electronic cloud around atoms is not spherically symmetric. It is usually elongated along the direction of the chemical bonds or along the nonbonding p-orbitals. Consequently, the induced screening has a directional dependence. Therefore, σ is dictated by the orientation of the electronic cloud and consequently by the orientation of the molecule in the external magnetic field. In fact, as stated in Chapter 1, the screening constant σ is a second-rank tensor, which can be described by a 3×3 matrix of real numbers, with nine components.

$$\sigma = \begin{pmatrix} \sigma_{xx} & \sigma_{xy} & \sigma_{xz} \\ \sigma_{yx} & \sigma_{yy} & \sigma_{yx} \\ \sigma_{zx} & \sigma_{zy} & \sigma_{zz} \end{pmatrix} \qquad 9.2$$

In solutions one measures the trace of σ, which is given by

$$\sigma = \frac{1}{3}(\sigma_{xx} + \sigma_{yy} + \sigma_{zz}) = \sigma_{iso}. \qquad 9.3$$

The trace is invariant under coordinate rotation. Conventionally, the axes are chosen such that σ_{zz} is the largest followed by σ_{yy} and then σ_{xx}. The tensor σ is represented by an ellipsoid. It can be transformed into any other coordinate system by applying rotations using three Euler angles. Thus, it can be related directly to the molecular system and the magnetic-field axis. The three principal values of the chemical-shielding tensor, and the orientation of the PAS in terms of the three Euler angles, provide all the information on the CSA of a nuclear spin.

In practice, what is measured is the chemical shift δ (which is the frequency difference in ppm scale with respect to a reference compound) rather than the chemical shielding which is measured with respect to the bare nucleus. It follows that chemical shift in solids is also a tensor. It is possible to find a coordinate system, called the principal axis system (PAS), in which, only the diagonal terms are present. The values of the components of this tensor in the PAS are given by δ_{11}, δ_{22} and δ_{33}, with the convention $\delta_{11} \geq \delta_{22} \geq \delta_{33}$.

In general, $\mathcal{H}_\delta$ can be divided into isotropic and anisotropic terms. For a chemical group with axial symmetry, such that $\delta_{xx} = \delta_{yy}$, the CS tensor has a cylindrical symmetry and the Hamiltonian $\mathcal{H}_\delta$, can be written as:

$$\mathcal{H}_\delta = \gamma B_0 I_z \left[\delta_{iso} + \tfrac{1}{2}\delta_{CSA}\,(3\cos^2\theta - 1)\right] \qquad 9.4$$

Here, $\delta_{CSA} = \delta_{zz} - \delta_{iso}$ *and* $\delta_{iso} = (1/3)(\delta_{xx} + \delta_{yy} + \delta_{zz})$ *9.5*

The angle θ is the orientation of the z-axis in the PAS tensor, with respect to B_0 and δ_{CSA} is the magnitude of the CSA. Isotropic molecular motions in solutions result in averaging of the term $<3\cos^2\theta - 1>$ to zero. Thereby, only δ_{iso} is observed.

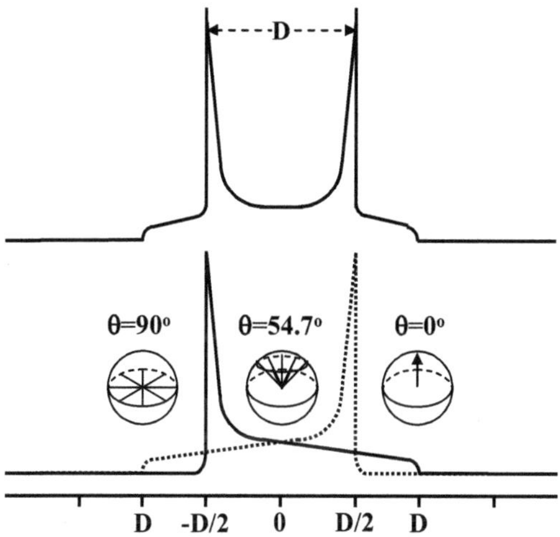

Figure 9.2: Pake pattern due to dipolar interaction between two unlike nuclear spins. The nuclear spin I (or X) exhibits two sets of signals corresponding to α and β spin states of the coupled partner.

In the case of a single crystal, there is a fixed orientation of the principal axis of the crystal with respect to B_0. The position of the resonance line in the spectrum therefore depends upon the magnitude of CSA and the angle between the PAS of the shielding tensor and the magnetic field axis. This angle can be changed by the experimenter, using a goniometer. By observing changes in the spectral pattern as a function of this angle, the CSA and the direction of the principal axis can be measured.

However, most molecules of interest may not yield single crystals. Therefore, one has to work on powder or amorphous materials. A powder sample can be thought of composed of several micro-crystallites, each one with different orientation with respect to B_0. This result in a broad NMR spectrum, which is often called the *powder pattern* (Figure 9.2).

The values of CSA can be large and as high as 1000 ppm for nuclei with high atomic numbers. The values of the components of screening tensor (σ), can be estimated for each functional group of interest either from experiments on model compounds or by using quantum chemical methods. The range of CSA for ^{1}H is around 10 ppm and for ^{13}C it can be up to 200 ppm. Like the isotropic chemical shift the value, when expressed in frequency units, depends on the magnetic field.

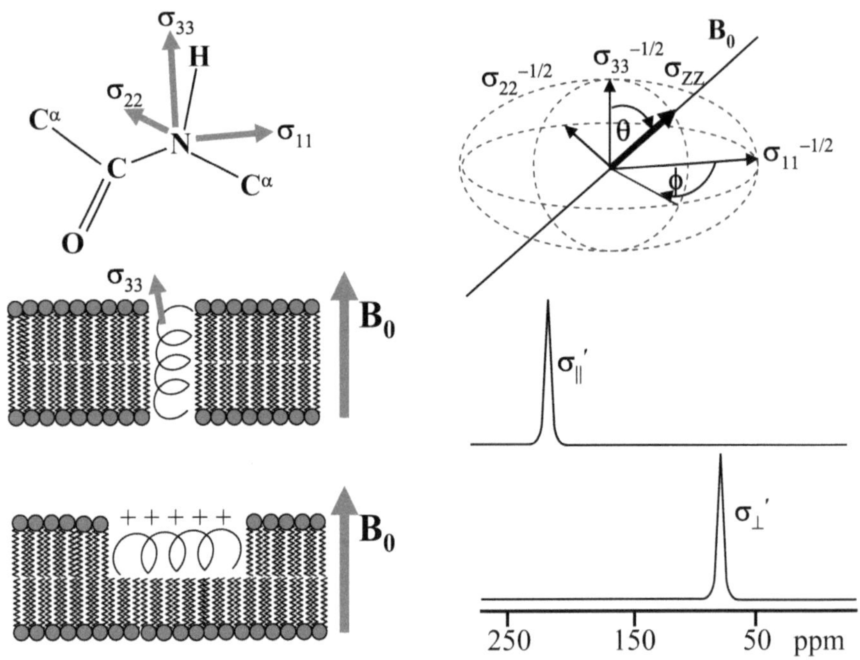

Figure 9.3: The ^{15}N chemical-shift tensor is represented as an ellipsoid. For the peptide bond, σ_{xx}, σ_{zz} are aligned in the plane of the paper, while σ_{yy} is in the perpendicular plane. For a protein embedded in oriented lipid bilayers, the orientation of the α-helical segments with respect to bilayer normal, can be obtained directly from the ^{1}H decoupled ^{15}N spectrum.

In protein NMR, one is often interested in the orientation of the amide group and consequently, in the ^{15}N CSA of the N-H bond and the ^{13}C CSA of the C=O bond. The value of ^{13}C CSA for the C=O group is ~120 ppm. The group has an axial symmetry. The direction of σ_{33} is perpendicular to the plane of the amide group, and is oriented along the p_z orbitals of the C=O group.

Figure 9.3 shows the properties of the ^{15}N chemical-shift tensor belonging to the amide group. The ranges of the components of σ in the PAS are $\sigma_{xx} \sim 65$ ppm, $\sigma_{yy} \sim$ 85 ppm and $\sigma_{zz} \sim 230$ ppm. In proteins containing α-helical segments, the NH vector and the σ_{33}-axis, cover an angle of about 17-18°. Both are oriented within a few degrees of the helical axis. This property enables one to fix the approximate alignment of the helix axis of proteins embedded in oriented membranes, simply by recording the ^{15}N chemical-shift interaction.

In cases of lipids and nucleic-acids, the CSA of ^{31}P is of interest, as it helps in fixing the alignment of each phosphate group.

2.3 Heteronuclear Dipolar Interactions [$\mathcal{H}_{het(D)}$]

It has been shown in Chapter 1, that the presence of a spin I results in a small magnetic field on a neighbouring spin X. This result in a positive or negative contribution to B_0, depending on the state of spin I as defined by the value of m_I (½ or -½ for spin ½ systems). Such a heteronuclear dipolar interaction between two spins I and X can be expressed by the Hamiltonian,

$$\mathcal{H}_{IX}(D) = -D_{IX}(3\cos^2\theta - 1)I_zX_z \qquad 9.6$$

The dipolar coupling constant between the two spins (D_{IX}) is given by:

$$D_{IX} = (\mu_0/4\pi)\,\hbar\,\gamma_I\,\gamma_X\,r_{IX}^{-3} \qquad 9.7$$

The angle θ defines the angle between the direction of the vector I-X joining the positions of the two spins and the direction of the magnetic field B_0, μ_0 is the permeability of free space, γ_I and γ_X are the gyromagnetic ratios of the two nuclei and r_{IX} is the internuclear distance.

One is often interested in the heteronuclear interaction [$H_{het(D)}$] between the abundant spin ^1H (denoted by I in this Chapter) and one of the rare spins, such as ^{13}C and ^{15}N, which is denoted by X. Note that the interaction is inversely proportional to the third power of internuclear distance, and directly proportional to the gyromagnetic ratios of the two nuclei. Therefore, the dominant contribution to this term comes from protons directly bonded to the heteronucleus (X). For directly bonded ^1H and ^{13}C (in a C-H bond) the dipolar coupling is about 23 kHz, while the interaction between directly bonded ^1H and ^{15}N (for the N-H bond) is around 12 kHz. This may be compared with the one bond dipolar coupling for ^{13}C-^{13}C which is 3 kHz, while the directly bonded ^{13}C-^{15}N dipolar coupling is 0.9 kHz. These values are much lower than the *geminal* dipolar coupling (across two bonds) for ^1H-^1H interaction, which is around 33 kHz. The typical distance between backbone carbonyl group in β-sheets is around 4.8 Å. For a pair of ^{13}C spins at this distance, the dipolar coupling is 69 Hz. Similarly, for ^{15}N-^{13}C pair at a distance of 4.1 Å, which corresponds to the distance between the hydrogen bonded carbonyl carbon and the amide nitrogen in β-sheets, the coupling is 44 Hz. Thus, the dipolar couplings of interest are much smaller.

In solutions, the orientation term in the dipolar coupling ($3\cos^2\theta$-1) is averaged to zero due to isotropic tumbling of molecules, as is also the case of CSA. For a crystal oriented at a fixed angle to the B_0, the spectrum of the spin X results in a doublet. This is often referred to as Pake doublet. The separation between the two lines of the Pake doublet depends on the value of the angle θ (Figure 9.2). For $\theta = 90^\circ$ ($\cos\theta = 1$), the splitting is equal to D, while for $\theta = 0^\circ$ the separation between the two components has the value of 2D. At $\theta = 54.74^\circ$, $\cos^2\theta = 1/3$, and therefore the splitting is zero. This angle plays a very important role in solid-state NMR and is called the *magic angle.*

In powder or amorphous materials different crystallites, constituting the sample, lie in different orientations. The powder spectrum is governed by the splitting for a particular orientation, and the number of molecules which are present in each of such orientations. The largest numbers of crystallites are present with $\theta = 90^\circ$. Therefore, this orientation contributes to the maximum signal. The spacing between the points of maximum intensity is equal to D, while the total span of the spectrum is 2D.

The dipolar pattern of a powder spectrum is shown in Figure 9.2. In powder solids, one may get line-widths of the order of 50 kHz in ^{13}C NMR, resulting from the one bond ^1H-^{13}C dipolar interactions.

2.4 Homonuclear Dipolar Interactions [$\mathcal{H}_{homo(D)}$]

The dipolar coupling interaction between two like-spins, for example, two ^1H or two ^{15}N nuclei, is similar to Equation 9.8, with one modification. In this case, spins can exchange magnetization through a *flip-flop* mechanism, where one spin flips from the state α to β, while the other spin simultaneously goes from β to α. The energy in such a transition is conserved for *like* spins. In the case of *unlike* spins, such a term does not conserve energy. For example, on a spectrometer operating at 500 MHz for ^1H, the ^{13}C resonance frequency is 125 MHz. The energy quanta corresponding to the two frequencies are thus, widely different. This flip-flop term is therefore not important for heteronuclear dipolar interaction or when the chemical shift difference between the two interacting spins in a homonuclear system is large.

The Hamiltonian for the dipolar interaction in a homonuclear spin system is given by:

$$\mathcal{H}_{12} = -D_{12}/2\ (3\cos^2\theta - 1)[2I_{1z}I_{2z} - \tfrac{1}{2}(I_1^+I_2^- + I_1^-I_2^+)] \qquad 9.8$$

Where, D_{12} is the dipolar coupling between the two like-spins 1 and 2 and I^+ and I^- are the raising and lowering operators, respectively. These two operators are given by:

$$I^+ = I_x + iI_y;\ I^- = I_x - iI_y;\ and\ D_{12} = (\mu_0/4\pi)\,\hbar\,\gamma_I^2\, r_{12}^{-3} \qquad 9.9$$

The term $I_1^+I_2^- + I_1^-I_2^+$ provides an exchange of angular momentum between two coupled like-spins whose resonance frequencies overlap.

The ^{1}H-^{1}H homonuclear dipolar interactions can be quite large. For example, the value of D_{12} for *geminal* protons is ~33 kHz. Such interactions result in very broad lines in the ^{1}H spectrum because of the large value of γ for protons and the presence of large number of ^{1}H spins in any given molecule. The intra- and inter-molecular interactions give rise to line-widths which are of the order of 100 kHz for ^{1}H.

On the other hand, because of their low γ, the homonuclear interactions for ^{13}C and ^{15}N are relatively small. For example, for two directly bonded ^{13}C the dipolar coupling is ~3 kHz, while the ^{13}C-^{15}N one bond interaction is ~0.9 kHz. Thus, the contribution to line-widths, due to homonuclear interactions of the spins such as ^{13}C and ^{15}N, are less than 5 kHz. Further, because of the low natural abundance of these two nuclei, the probability that one may have two ^{15}N or two ^{13}C bonded to each other is relatively small. The labelled molecules are however, exceptions.

2.5 Quadrupolar Interaction ($\mathcal{H}_Q$)

For spins I > 1/2, it is necessary to introduce an additional term in the Hamiltonian, arising from the quadrupolar interaction (Q). The effect of this interaction on the relaxation rates has been discussed in Chapter 1. It may be noted that the quadrupole coupling constants depend on the symmetry and the magnitude of electronic charge distribution, at the site of the nucleus. The electric field gradient tensor V (like the CSA tensor) can be represented by components V_{xx}, V_{yy} and V_{zz}, such that $V_{zz} \geq V_{yy} \geq V_{xx}$. In the case of non-spherical charge distribution, the Hamiltonian for such an interaction is given by:

$$\mathcal{H}_Q = e^2qQ\,[\,3I_z^2 - I(I+1) + \eta/2\,(I_+^2 + I_-^2)]\,/\,[4I(2I-1)], \qquad 9.10$$

Here, eQ representing the quadrupole moment of the nucleus, V_{zz}= eq and $\eta = (V_{xx} - V_{yy})/V_{zz}$. It may be noted that for systems with cylindrical symmetry η= 0, while in structures with spherical symmetry, this interaction vanishes. Thus, even if the quadrupole moment is large, the interaction term may not be dominant in such situations.

From the view-point of applications to biological systems, the most useful spin with quadrupolar moment is ^{2}H. This is both due to its low Q, and the fact that ^{2}H provides information on certain aspects of molecular conformation and dynamics, which are not accessible by ^{1}H NMR. The C-D bond, which is most extensively used to provide biological information, has a cylindrical symmetry (η= 0). To the first order, it leads to an interaction given by e^2qQ. In fact, there is also a second order and more complex term, which is important for nuclei that have a larger Q. In solids, such a term causes additional line broadening, and persists even when the first order term has been averaged by methods discussed below. For ^{2}H, the second order term is small and can be ignored for all practical purposes.

As seen from Equation 9.10, because of the $3\cos^2\theta$-1 dependence of the first order quadropolar term, the powder pattern of ^2H resonance will show a Pake doublet, with widths depending on the quadrupole coupling constant. The values of quadrupole coupling constants for ^2H lie between 100 and 300 kHz. For the C-D bond, ^2H has quadrupolar coupling constant is around 128 kHz.

In partially oriented systems such as liquid crystalline media, the quadrupole coupling constants exhibit a residual coupling behaviour. One example of such a situation is biological membranes in the liquid crystalline phase. In such a situation, the $\langle 3\cos^2\theta-1 \rangle$ term is replaced by an order parameter (S_{CD}= ½ $\langle 3\cos^2\theta-1 \rangle$). The line separation arising from the residual quadrupole coupling constant is given by:

$$\Delta\nu_Q = 3\, e^2qQ\, S_{C\text{-}D}/(2h) \qquad 9.11$$

In addition to ^2H, extensive use has been made of two other quadrupolar spins in biological systems. One is ^{14}N, which has been used to monitor changes in the concentration of chemical groups having spherical symmetry. Typical examples are NH_4^+ and NO_3^-, which have found applications in the study of nitrogen assimilation in plant sciences. The other is ^{23}Na which has been used to study ion transport of Na^+. The applications of such spins will be discussed in appropriate sections in later chapters.

2.6 Scalar Coupling [$\mathcal{H}_J$]

One may compare the interactions discussed above with the term in the Hamiltonian arising from scalar spin-spin couplings ($\mathcal{H}_J$). This term is same both in solution and solid-state. However, the magnitude of scalar interactions is much smaller than any of the interactions discussed above (Figure 9.1). The J coupled interactions are less than 0.2 kHz for ^1H-^{13}C, 0.05 kHz for ^{13}C-^{13}C and 0.01 kHz for ^1H-^1H. The smaller values of J couplings may not be observed in high-resolution solid-state NMR spectra due to larger line-widths. However, larger J couplings may be observed and can be used for structural studies.

3. STRATEGIES FOR OBTAINING HIGH-RESOLUTION NMR SPECTRA IN SOLID-STATE

The intra- and inter-molecular interactions, arising from factors discussed above, lead to line-widths in solids which are of the order of several kHz. The spectra have very poor S/N ratio. It is clear that not much information can be obtained from such broad spectra. Techniques have been developed recently, whereby sharp lines can be obtained in powder samples, and meaningful information can be obtained in biological materials. This requires use of techniques to suppress interactions contributing to the widths of NMR lines in solid-state.

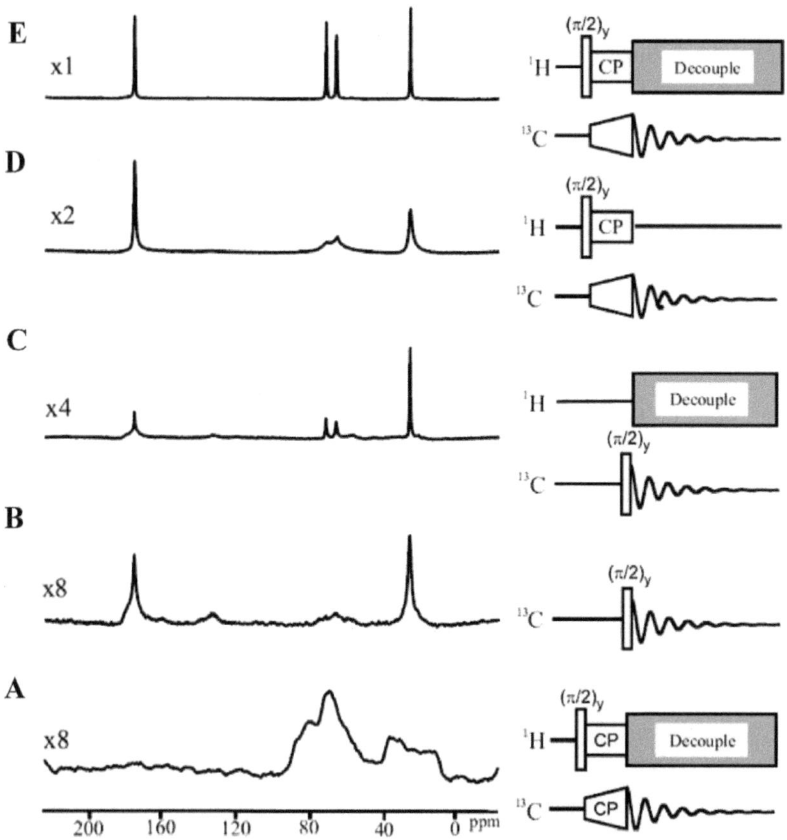

Figure 9.4: The [13]*C NMR spectrum of threonine (powder) using different techniques employed to obtain high resolution/sensitivity solid-state NMR: (A) normal spectrum; (B) MAS; (C) MAS with decoupling; (D) CPMAS without decoupling; and (E) CPMAS with BB decoupling. Note that the broad-band* [1]*H decoupling removes the dipolar interactions of the observed spins with protons and only the CSA term survives. Also, notice the sensitivity enhancement with the CP in panel D and E (Thanks to P.K. Madhu for recording these spectra).*

Note that apart from the magnetic properties of the nuclei, two distinct components contribute to different terms in the spin Hamiltonian. One is the spatial part, which depends on the orientation of the molecules (i.e. the term $3\cos^2\theta$-1). The second is the one that depends on the spin states of the interacting spins (terms such as I_zX_z). This provides two independent ways with which the spins can be manipulated to achieve narrower lines. One is to use motional effects on the orientation part of the Hamiltonian, for example, by using mechanical rotors. This is analogous to what is achieved by the isotropic tumbling motion of molecules in solutions. The second approach is to manipulate the spin part of the

Hamiltonian through suitable pulse sequences to average such interaction, at least, partially. An analogue to this approach in solution NMR is the spin-decoupling (Figure 9.4).

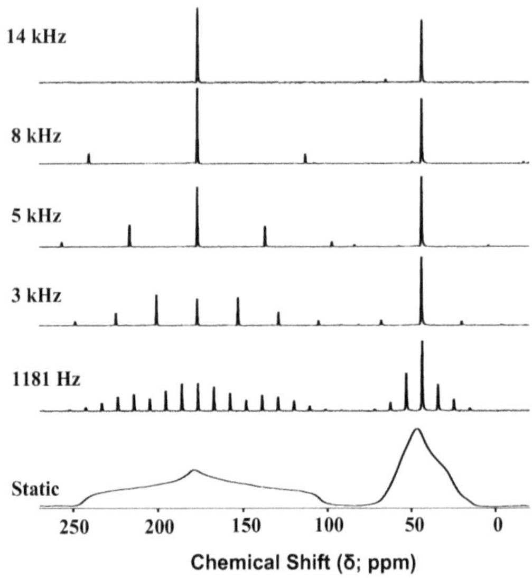

Figure 9.5: ^{13}C NMR spectrum of uniformly labelled L-glycine powder at different MAS frequencies. At lower rotational speeds, spinning side bands appear separated from the lines due to isotropic chemical shifts by multiples of MAS frequency. At higher spinning speeds, one gets a sharp resonance corresponding to the isotropic chemical shift. (Thanks to P.K. Madhu for recording these spectra).

Methods based on motions (such as molecular tumbling in solutions or magic-angle spinning using rotors in solid-state, discussed below), act on the spatial part of the Hamiltonian. This becomes time dependent. This approach therefore treats the average laboratory frame Hamiltonian under the effects of motions. In the methods based on the use of RF pulses, one manipulates spins, such that the time evolution of the Hamiltonian changes. When both types of methods are used simultaneously, there is a correlation between the two. Therefore, there may be a need for achieving a synchrony between the pulse sequences and MAS for certain types of experiments. The theoretical description of these aspects is beyond the scope of this book. We shall discuss only the basic concepts necessary to understand biological applications of high-resolution NMR of solids.

3.1 Magic-Angle Spinning (MAS)

Recall that at a magnetic field of 10 T, the ^{13}C CSA term is of the order of 20 kHz. Also, if the molecule can be oriented such that the angle θ is equal to 54.74° then the second term in Equation 9.4 becomes zero and only the isotropic term in the chemical shift survives. It is difficult to orient the molecules such that the vector $\mathbf{r}_{IX}$ is oriented at the magic angle. However, one can achieve a similar effect by rotating a polycrystalline sample at an angle of 54.47° relative to B_0. When the sample is spun, the CSA terms for crystallites in the powder sample become time-dependent, since the rotor motion causes time dependent variation in the term ($3\cos^2\theta$-1). The net result is that each crystallite experiences the average isotropic chemical shift, and each site in a molecule contributes to a single resonance, with splittings due to resolvable scalar couplings (Figures 9.4 and 9.5).

A complete averaging of the CSA interaction requires spinning at rates higher than the CSA. At lower spinning speeds, one observes spinning side-bands (Figure 9.5). These can be a nuisance, but have been used to reconstruct CSA patterns. Side bands can be suppressed by using pulse sequences such as TOSS (total suppression of side bands). The basic scheme of such a sequence uses four π pulses. However, such sequences may lead to a loss in sensitivity.

MAS also averages to a large extent, other interactions containing orientation terms such as the homonuclear dipolar couplings involving rare spins, the heteronuclear dipolar interactions in solids and the quadrupolar couplings. As seen in Figure 9.5, fast rotor speeds help to remove spinning side bands. Development of rotors which allow spinning at very high rates has therefore made a major impact on high-resolution NMR in solids. Currently, rotors are available which allow spinning at rates of around 80 kHz. Such spinning rates can average out several of the interactions encountered in solid-state. In view of their importance, rotors with higher rotational speeds are under development. In fact, the field of high-resolution solid-state NMR has seen rapid strides from the development of better- and higher-speed rotors.

3.2 Heteronuclear Dipolar Decoupling (DD)

The second approach for obtaining sharp lines in solids is to suppress heteronuclear dipolar couplings by manipulating the I_zX_z term in the Hamiltonian. This is possible if the orientation of the abundant spin I can be flipped continuously between the states α and β, while observing the spin X. Experimentally this is achieved by using a strong RF in a continuous-wave (CW) mode on the nucleus I. This technique is called broad-band spin-decoupling as has been discussed earlier (Chapter 3). Before the advent of multi-dimensional NMR, scalar-decoupling using CW broad-band irradiation was very popular for simplifying ^{1}H and ^{13}C spectra, in solutions. The decoupling of heteronuclear dipolar couplings is based on the same principle. However, while 5 watts of power is sufficient to remove J couplings (which are typically less than 200 Hz), dipolar decoupling requires much higher RF power. Also note that the method cannot be used for homonuclear decoupling as in this case, any perturbation of one spin (A) will also perturb the second spin (X).

One of the more widely used decoupling methods today is the *two-pulse phase modulation* (TPPM), which breaks the CW decoupling power into a series of phase alternating π pulses. The phases of the π pulse is alternated between ϕ and -ϕ (the value of ϕ depends on the spinning speed and can vary from 10°-70°).

3.3 Cross Polarization (CP)

A combination of MAS and DD techniques enables one to obtain NMR spectra in solid-state, for spins such as ^{13}C, ^{15}N and ^{13}P, with a resolution comparable to that obtained in solutions. However, the problem of low sensitivity of such spins has to be overcome. The inherent reasons for the low sensitivity are the low natural abundance of ^{13}C and ^{15}N and the low gyromagnetic ratios of these spins. During direct detection, high relaxation times in solids for such nuclear spins require the use of long acquisition times. Recall that this problem is common with solution-state NMR.

Techniques such as INEPT, refocussed INEPT and HSQC, where magnetization is transferred from ^{1}H to a less abundant spin and is then back transferred to ^{1}H have been discussed in Chapter 3. These methods allow one to exploit the higher sensitivity of ^{1}H. An analogue of such an experiment in solid-state is called cross polarization (CP). A particularly effective technique for CP is to use the Hartmann-Hahn approach (Figure 9.6). The Hartmann-Hahn technique uses two continuous RF fields. One corresponds to the abundant spin I (^{1}H) and the other to the nucleus X, which is to be detected. For example, if the intention is to detect ^{13}C resonance on a 500 MHz spectrometer, then one uses a RF field of amplitude B_{1H}, at a frequency 500 MHz for the abundant spin (^{1}H) and amplitude B_{1C} at 125 MHz, for the rare spin (^{13}C). Both spins I and X will rotate independently, at rates which are determined by the amplitude of the RF fields and the respective gyromagnetic ratios γ_H and γ_C of the two spins. Energy transfer between the two spins will take place if the following condition is satisfied:

$$\gamma_H B_{1H} = \gamma_C B_{1C} \qquad 9.12$$

This is the so called Hartmann-Hahn condition. When the nutational frequencies of the spins I and X are the same, an energy conserving dipolar contact is established between the two-spins. The magnetization transfer from the abundant spin ^{1}H (I) to the rare spin X (^{13}C or ^{15}N) can occur through the flip-flop mechanism. This can be visualised in double rotating frame, in which the spacing between the states α and β is equal for the two spins. It may be emphasised that in the laboratory frame, I and X have different Larmor frequencies.

In the CP pulse sequence shown in Figure 9.6, a π/2 pulse brings the ^{1}H magnetization on to the xy plane. The RF fields are then applied both on the ^{1}H and the ^{13}C for a time period τ_{CP} (called the *contact time*) during which the two spin systems exchange energy. Typical mixing times are 100 μsec to 10 msec, depending on the strength of the heteronuclear dipolar interaction (i.e. whether the observed ^{13}C has directly bonded protons or the distant ones). Finally, the ^{13}C RF field is turned off and a decoupling RF field at the ^{1}H frequency is applied to decouple ^{1}H. The maximum enhancement under the Hartmann-Hahn condition is equal to γ_I/γ_S. This is about 4 for ^{13}C and 10 for ^{15}N. Modification of the basic CP sequence through multiple CP contacts can provide further enhancement in sensitivity.

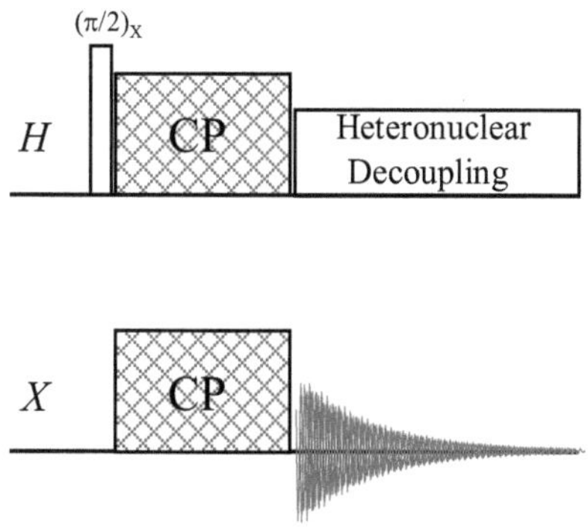

Figure 9.6: The pulse sequence used in a Hartmann-Hahn CP experiment.

Due to homonuclear dipolar-couplings between protons, the ^{1}H relaxation rates are much faster than that of the rare spins ^{13}C and ^{15}N. Therefore, many more scans can be acquired in a given time using CP, as compared to the situation in the direct detection of ^{13}C and ^{15}N spins.

3.4 ^{31}P NMR

As discussed earlier, even though ^{31}P is 100% abundant, it is usually present only in a small number of biological molecules. With the exception of ADP and ATP, there is usually only one phosphorus atom in each phosphorous containing biological molecule. The homonuclear ^{31}P-^{31}P interactions are therefore usually absent and ^{31}P signals can be detected in the same way as those discussed for ^{13}C and ^{15}N. The phosphate group has a large CSA. This property has been used for studying orientation of the phosphate group. Areas where solid-state ^{31}P NMR has found applications in biological systems are model membranes and nucleic-acids. Solid-state NMR has also been used to detect metabolites such as ATP, ADP, phospho-monoesters, inorganic phosphate and other phosphorus containing compounds in biological materials which have solid-like properties, such as hard tissues and bones.

3.5 Spectra of Dilute Spin ½ Nuclei

The combination of high speed MAS, CP and high power DD provides good quality spectra of compounds containing spins such as ^{13}C and ^{15}N. The high-resolution spectra recorded in solid-state for these two spins, have been used extensively in biological studies.

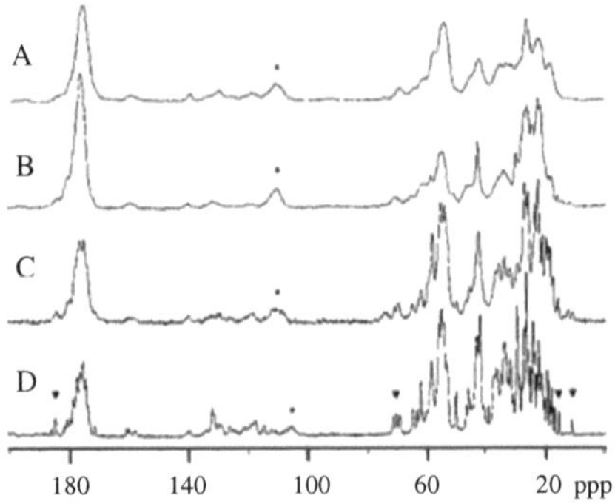

Figure 9.7: 1D ^{13}C CPMAS spectra of uniformly labelled α-spectrin SH3, using spinning speeds of 13 kHz at 18.8 T spectra (A to C), and at 17.6 T for (D). The samples were prepared by four different protocols. Sample (A) lyophilized from aqueous low-salt buffer; (B) after adding some water to (A); (C) lyophilized from a solution containing PEG-8000 and sucrose; (D) precipitated from $(NH_4)_2SO_4$ rich solution. J. Pauli, B. Rossum, H. Forster, H.J.M. Groot, and H. Oschkinat. J. Mag. Res. 143, 411-416 (2000). (Reprinted with permission from Elsevier).

3.6 ^1H NMR

The large homonuclear dipolar couplings involving protons cannot be removed efficiently by the rotor speeds currently available. Further, it is not possible to use CW decoupling, since perturbing one spin will cause a similar effect on the other homo-spin. One of the ways to reduce ^1H-^1H coupling is through selective deuteration. When the two protons are in close proximity (as is the case for *geminal* or *ortho* protons), deuteration of one of the two, considerably simplifies the ^1H spectrum in solid-state. One then has to deal only with pairs of protons in the molecule having relatively larger inter-atomic distances. This leads to sharper signal. The ^1H-^1H pairs thus observed in the spectrum have relatively smaller dipolar coupling constants.

Sophisticated techniques using special pulse sequences and multi-dimensional NMR have been developed for achieving homonuclear decoupling. These sequences form a group of pulse trains called *combined rotation and multiple pulse spectroscopy* (CRAMPS). A number of such methods have been developed. It is difficult to deal with the ever increasing literature on this subject. Multiple pulse line narrowing is based on the application of series of pulses with suitable delays, such that in the rotating frame, the spin vectors can be averaged suitably and results in the decoupling of protons. One of the approaches is called WAHUHA (or WHH-4, named after its inventors, Waugh, Huber and Haeberlen). This technique uses a four-pulse sequence such that the magnetization is aligned sequentially along the x, y and z axes. It can be shown that this pulse sequence toggles the direction of the local fields between the three axes, leading to an elimination of the homonuclear dipolar coupling. However, the overall chemical shift is scaled by a factor of $1/\sqrt{3}$ in this experiment.

A useful alternative to WAHUHA is the Lee-Goldburg sequence (LG), which operates such that the effective RF field in the rotating frame is inclined in the direction of the magic angle. Several decoupling super-cycles have been proposed to improve the performances of the above basic sequences. These pulse sequences utilise repeated use of WAHUHA, or suitable modifications of the LG approach. Selective excitation of certain MQ coherences also allows one to obtain simpler ^{1}H spectra.

In spite of these developments, ^{1}H resonances in solid-state spectra are much broader than those obtained in solutions. The smaller dispersion in chemical shifts further limits the information that can be obtained from ^{1}H spectroscopy of biological materials in solid-state.

3.7 Heteronuclear and Homonuclear Correlation Experiments

Another approach to obtain information on protons is to use heteronuclear correlation (HETCOR), between ^{1}H and ^{13}C or between ^{1}H and ^{15}N. During the t_1 period of such 2D experiment, homonuclear ^{1}H-^{1}H decoupling is achieved with the use of multiple-pulses or LG kind of schemes. Magnetization is transferred to the rare spins by CP and is detected during the t_2 period. HETCOR relies on the dipolar interactions and correlates spins which are at close distances to each other. This may be compared to the HSQC/HMQC experiments in solutions, which utilise J couplings. The ^{1}H line-widths are still large in such experiments (~1 ppm), but the method has found wide applications in studies of biological molecules in solid-state.

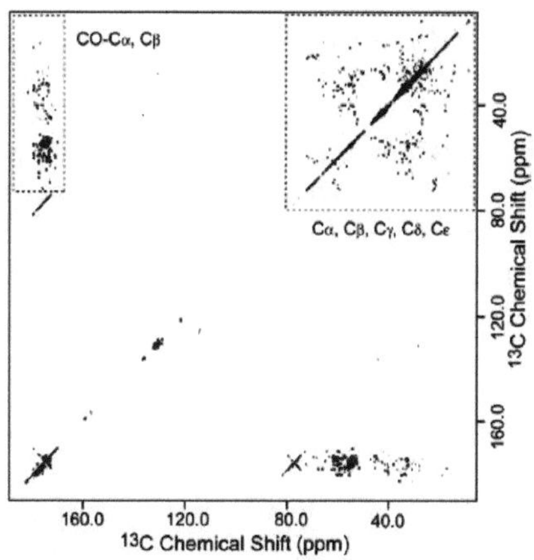

Figure 9.8: A 2D CP MAS ^{13}C-^{13}C spectrum of a microcrystalline sample of ubiquitin obtained on a 800 MHz spectrometer for protons. T.I. Igumenova, A.E. McDermott, K. Zim, R.W. Martin, E.K. Paulson, and A.J. Wand. J. Am. Chem. Soc., 126, 6720-6727 (2004). (Reprinted with permission from American Chemical Society).

A typical example of the quality of homonuclear 2D spectra, which can be obtained using CP-MAS is shown in Figure 9.8 for a 71 amino acid residue protein ubiquitin. Moderate proton decoupling fields and MAS speeds of 20 kHz were used.

The so called dipolar assisted rotational resonance (DARR) technique was used for one-bond transfer between ^{13}C spins. Most of the carbon resonances have line-widths in the range 0.5-0.8 ppm, which arise mainly from homonuclear couplings.

There are several other experiments related to 2D heteronuclear correlations techniques, which allow correlation of chemical shift of spin X with the heteronuclear dipolar couplings D_{IX}. A particularly useful technique for studies of proteins is called PISEMA (polarization inversion spin exchange at the magic angle). This technique allows 2D spectra comprising of ^{1}H-^{15}N dipolar coupling and the ^{15}N chemical shifts, and is discussed later.

3.8 ^{2}H NMR

The gyromagnetic ratio γ of ^{2}H is 6.51 times lower than that of ^{1}H and its natural abundance is 0.0156%. This results in a much lower sensitivity for ^{2}H resonance. This value is around 1.5×10^{-6} relative to that for ^{1}H detection. Since the electronic structures of the two isotopes in a particular molecule are very similar, most of the NMR interactions are also of similar nature. However, there are the isotope effects, and scaling due to different γ ratios. Thus, the ^{2}H chemical shifts are scaled down to 2-3 ppm. The scalar and the dipolar couplings between ^{1}H and ^{2}H are 15% of the corresponding ^{1}H-^{1}H couplings, while the scalar ^{2}H-^{2}H couplings are scaled down to 42%.

The ^{2}H relaxation time is dominated by the quadrupolar coupling constants. Under isotropic molecular motions, the values of R_1 and R_2 are given by:

$$R_1 = R_2 = 3/8\ (1 + \eta^2/3)\ [e^2qQ\ (h/2\pi)^{-1}]^2\ \tau_c \qquad 9.13$$

In previous chapters, we have used some of the properties of ^{2}H, for improving information in ^{1}H NMR. Therefore, the question arises as to why one needs to worry about direct detection of ^{2}H NMR? Some of the advantages of ^{2}H NMR have been discussed already. There are additional sources of information in terms of quadrupole coupling constants and relaxation rates. In principle, quadrupolar coupling can be reduced by MAS, if rotors with sufficient speed are available.

A convenient method to obtain ^{2}H signals is to use deuterium quadrupole echo spectroscopy using a pulse sequence shown in Figure 9.9. The quadrupole moment of ^{2}H facilitates its relaxation and good quality spectra can be obtained in a relatively short time.

Applications of ^{2}H NMR in biology are fairly widespread. The most important property of ^{2}H NMR is its sensitivity to internal motions in molecules. For motions which are fast on the NMR time scale ($< 10^{-7}$ sec), the pattern of the line shapes change in a predictive manner. For example, in cases of rotating methyl group, flips of aromatic rings, free diffusion of phenyl rings and two-site hops, the orientation of the C-D vector with respect to the axes of internal motion is at different angles. This information enables simulation of the distinct line shape patterns in each case. For intermediate rates of motions (10^{-3} to 10^{-7} sec), the line-shape can be calculated much in the same way as discussed in Chapter 1 for chemical exchange between two sites. The kinetic parameters can be determined by simulation of the line shapes. Another application of ^{2}H NMR is in the study paramagnetic reaction centres in proteins.

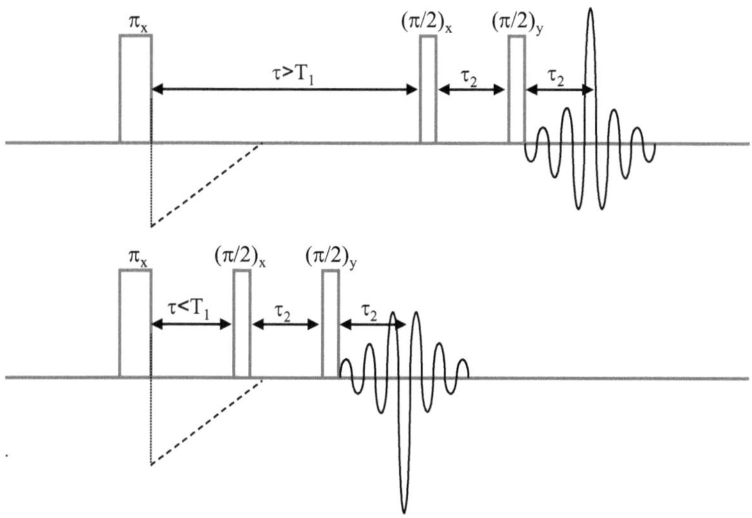

Figure 9.9: The pulse sequence for deuterium quadrupole echo spectroscopy. Inversion recovery quadrupole echo T_1 experiment.

3.9 Measurement of Internuclear Distances in Solid-State: Dipolar Recoupling

The dipolar couplings and the CSA contain information on internuclear distances and orientations of molecules. This information is lost when one uses MAS and DD, even though one obtains narrow lines. A number of homo- and hetero-nuclear techniques have been developed whereby, such information can be reintroduced, selectively and/or partially. Such methods are collectively known as re-coupling techniques.

Pulse sequences for homo- (such as ^{13}C-^{13}C) and hetero-nuclear (e.g. ^{13}C-^{15}N) dipolar recoupling and for CSA recoupling have been developed recently. Recoupling sequences are important not only for quantitative distance information but also for transfer of spin-polarization transfer in multi-dimensional spectroscopy in solid-state.

The re-coupling methods involve manipulation of the dipolar Hamiltonian using RF pulses. These pulses are designed to interfere with the spatial modulation of the dipolar interaction in synchrony with the sample rotation (MAS). For example, a two π-pulse technique called rotational-echo double resonance (REDOR), measures the heteronuclear dipolar coupling between isolated pairs of labelled nucleus such as ^{13}C and ^{15}N. A frequency selective (FS) extension of REDOR, allows recoupling of a single ^{13}C-^{15}N dipolar interaction in a multi-spin systems. Another method, commonly used for reintroducing homonuclear dipolar couplings under MAS is the rotational resonance (RR). When the isotropic chemical shift between two nuclei (Δ) is such that it is an integral multiple of the rotational speed ω_r (i.e. $\Delta = n\omega_r$), a reduced form of the homonuclear dipolar coupling reappears. Line-shape simulation can be used to calculate the magnitude of dipolar interaction.

A number of other sequences with different acronyms such as DRAMA, HORROR, MELODRAMA, RFDR and C7 have been developed for homonuclear and heteronuclear re-coupling. Each of these methods has their own strength and weakness. The more advanced methods are less sensitive to variations in RF field strengths or resonance offsets and have other advantages. Symmetry principles guide in the derivation of pulse sequences that recouple certain selected interactions. Such frequency selective recoupling techniques are useful in studies of multiple labelled samples. By using such techniques, ^{13}C-^{15}N distances up to ~5 Å and ^{13}C-^{31}P distances of up to ~7.5 Å can be measured. Thus, the range of distances, which can be measured using heteronuclear dipolar couplings, is larger than those obtained in solutions using NOESY.

Polarization transfers driven by dipole-dipole interaction have been used to obtain sensitive 2D ^{13}C-^{13}C and ^{13}C-^{15}N spectra of proteins. Figure 9.10 C shows the aliphatic region of a 2D ^{13}C–^{13}C NMR spectrum of $A\beta_{1\text{–}40}$ fibrils (uniformly ^{13}C and ^{15}N labeled at I31, G33, M35, G37, and V39), obtained at a 14.1 T magnetic field with a MAS of 22.3 kHz. The sample contained about 8 mg of lyophilized fibrils. The total data acquisition time is 9.5 and 17.5 hr, respectively. Dipolar recoupling has been applied during the 2.15 msec mixing period that separates the two spectral dimensions using the fpRFDR technique. As is evident from the figure, strong cross-peaks connect ^{13}C NMR chemical shifts of ^{13}C–labeled sites in each of the labeled amino acid residues, which form independent networks of coupled spins. Figure 9.10 D shows 2D ^{15}N-^{13}C NMR spectrum of the same sample, with a MAS of 11.14 kHz. Conditions have been used in these experiments so as produce strong cross-peaks between the chemical shifts of backbone ^{15}N and $^{13}C^{\alpha}$ spins belonging to the labeled residues.

Another class of methods which help in obtaining constraints on the ϕ and ψ dihedral angles in proteins are called tensor correlation techniques. These techniques make use of the orientation dependences of the dipolar couplings or the CSA from the ^{13}C-^{13}C correlated spectra. The double quantum NMR spectroscopy is also useful to obtain such geometrical information.

3.10 Reintroduction of CSA

At lower rates of spinning, the intensities of the side bands contain information about the CSA. The chemical-shift components δ_{11}, δ_{22}, and δ_{33} can be measured through an analysis of line shapes in a crowded spectrum. Through the use of 2D techniques, it is possible to spread overlapping CSA patterns in the second dimension. Methods such as PASS (phase adjusted spinning side bands) depend on this principle. Several other multi-pulse sequences have been developed to re-introduce CSA.

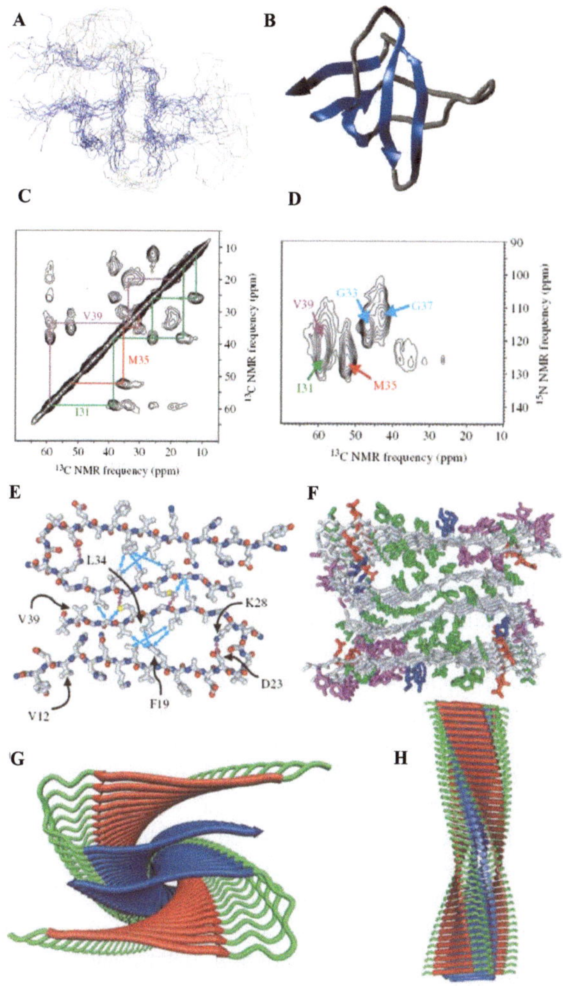

Figure 9.10: 3D structure of the α-spectrin SH3 domain as derived from solid-state MAS NMR data. (A) Stereo view of twelve lowest-energy structures, representing the fold of the SH3 domain. (B) The X-ray structure is shown for comparison with the structure shown in this panel. F. Castellani, B. van Rossum, A. Diehl, M. Schubert, K. Rehbein and H. Oschkinat. Nature, 420 98-102 (2002) (Reprinted with permission from Nature Publishing Group) (C) 2D ^{13}C–^{13}C and (D) ^{15}N–^{13}C NMR spectra of $A\beta_{1\text{–}40}$ fibril.(C) Chemical shift assignment paths for I31, M35, and V39; (D) ^{15}N–^{13}C cross-peak assignments (E-H) Proposed structure for the protofilament in Aβ1-40 fibrils. (E) All-atom representation of a pair of peptide molecules. Purple double-headed arrows indicate contacts established by measurements of ^{15}N–^{13}C dipole–dipole couplings. (F) Average structure of the same. (G and H) Cartoon representations, with residues 12–21 in red and residues 30–40 in blue. R. Tycko. Q. Rev. Biophys. 39, 1-55 (2006). (Reprinted with permission from Cambridge University Press).

4. PROTEIN STRUCTURE DETERMINATION USING SOLID-STATE NMR

Solid-state NMR is fast emerging as a powerful tool not only for studies in structural biology, but also for studying supramolecular assemblies. It may be noted that the line-widths in solid-state NMR do not depend on the molecular weight. Therefore in principle, unlike in solutions, there is no limit for the size of the system that can be investigated. The sample can be powdered material, which may be hydrated or dry. Both, fibrous or amorphous materials can be studied. Oriented solids can be used with advantage. Solid-state NMR can be used to obtain information on molecular orientation in materials such as biological membranes. Dynamic information using solid-state techniques covers a wider time-range as compared to that in solutions.

Solid-state ^{13}C and ^{15}N NMR studies usually require around 5 μmol quantities of labelled molecules. The precise quantity depends on the complexity of the measurements. The S:N ratio in solid-state NMR is higher at lower temperatures due to increased spin polarisation, probe circuit Q factors and lower thermal noise. Thus, the development of solid-state probes working at lower temperatures is likely to be very helpful.

The advantages discussed above, make solid-state NMR a powerful technique for studies in solid-like biological materials. In the next section, application of solid-state NMR techniques to protein structure determination is described. The work on tissues and other materials is described in subsequent chapters.

The work that exemplified the application of solid-state MAS NMR to proteins was that of the group of Oschkinat. They presented a solid-state 3D structure (Figure 9.10 A) that satisfactorily described the β-sandwich fold of the 62 amino acid residue long α-spectrin Src-homology 3 (SH3) domain. Almost complete ^{13}C and ^{15}N resonance assignments for a micro-crystalline preparation of the protein were made, which formed the basis for the extraction of 286 inter-residue ^{13}C-^{13}C and six ^{15}N-^{15}N restraints. These restraints were derived from proton-driven spin diffusion (PDSD) 2D NMR spectra of biosynthetically site-directed, labelled samples obtained from bacteria grown using [1,3-^{13}C]glycerol or [2-^{13}C]glycerol as sole carbon sources. The observed long-range distance correlations were up to 7 Å. This seminal work paved the way for the 3D structure determination of amyloids, small membrane proteins or receptor–ligand complexes.

4.1 Membrane Proteins

It has been estimated that almost 30% of the proteins coded by human genome are membrane proteins. There are difficulties in obtaining well ordered crystals of membrane proteins or to solubilise them in aqueous solutions. Therefore, very few structures based on X-ray diffraction of membrane proteins are available. Solution NMR studies are also not possible since these proteins give broad spectrum. Solid-state NMR methods are therefore valuable in studies of such systems.

Membrane proteins adopt unique 3D structures in their functional environment consisting of phospholipid bilayers. The peptide residues are immobile on time scales longer than 10^{-3} sec, as compared to time scales of around ~10^{-6} to 10^{-9} sec present in solutions. Therefore, one of the important applications of solid-state NMR spectroscopy has been in the determination of structures of membrane bound peptides and proteins in non-crystalline environment. This knowledge is important for understanding several important biological functions. In particular, studies on biological membranes, including aspects such as structures of membrane bound proteins, receptor activation, transport of molecules across membranes, membrane fusion and transport have provided valuable information.

The lipid bilayers in which the proteins are embedded, play an important role in modulating biological properties of membrane proteins. Approaches based on static and macroscopically oriented samples of proteins in the lipid bilayers have proved useful for certain class of proteins. At the same time, MAS based techniques have been used to probe structure of membrane bound proteins, as well as non-membrane proteins in solid-state. The orientation of the proteins in lipid assembly is also an important parameter in their function. Such problems can be tackled using MAS.

Solid-state methods yield structural constraints for local backbone conformation as well as the overall orientation and dynamics of polypeptides interacting in the model membrane environment. Oriented samples in model phospholipids environment can be prepared with ease. The two structural aspects mentioned above, can be probed separately within the same NMR protocol. Most proteins such as trans-membrane segments of ion channels, proteins involved in transport and G-protein coupled receptors are aligned in a direction, which is close to the normal to the phospholipid bilayers. NMR is particularly sensitive to changes in such an orientation. Membrane proteins generally adopt α-helical structures, though peptide assemblies also occur in multi-stranded β-assemblies. Both these secondary structures can be characterised by solid-state techniques.

Because of technical problems in ^{1}H spectroscopy, a large amount of information on biological systems has been derived from ^{13}C and ^{15}N NMR. The information on ^{1}H is generally derived indirectly, using heteronuclear 2D methods with the ^{1}H chemical shifts on one of the axes and that of the rare spins on the other. Other indirect methods used for information on ^{1}H utilize *multiple-quantum* coherences. The accuracy with which molecular orientation can be detected depends on the molecular and overall symmetry. Signal-to-noise considerations are of critical importance and the availability of ^{13}C and ^{15}N labelled compounds is a pre-requisite for a good S:N ratio. Both uniformly and selectively labelled compounds have been used with advantage. The spectral resolution even with the new developments is still much lower than that obtained in solutions.

There are two distinct approaches for solving structures of membrane proteins. One is based on the static oriented samples, using an approach known as PISA wheels. This method utilises ^{15}N CSA and ^{15}N-^{1}H dipolar coupling in oriented samples, and does not employ techniques such as MAS, and spin-decoupling. The second is based on high-resolution MAS multi-dimensional NMR data. It uses information on the chemical shifts and inter-atomic distances estimated from dipolar couplings. A combination of the two approaches has also been used. The two approaches have been discussed below along with some illustrative examples.

4.2 Polarisation Inversion at Magic Angle: PISA Wheels

Polarization Inversion Spin Exchange at Magic Angle (PISEMA) allows observation of the directly bonded ^{1}H-^{15}N dipolar coupling along with the ^{15}N chemical shifts in peptides and proteins oriented in lipid bilayers samples. Information on the two parameters is expressed in the form of a 2D spectrum. These experiments are done in the static mode, using uniformly or selectively ^{15}N labelled proteins. Because of the slow motions of peptide backbone in lipid environment, the dipolar couplings and the CSA are not averaged. The N-H bond length can be assumed to have a fixed value (1.07 Å) and therefore the strength of the dipolar-coupling constant (D_{IX}) can be calculated. The principal values of the ^{15}N chemical-shift tensor are also fairly well known. So is the case with the molecular orientation of the chemical-shift tensor, with respect to the N-H bond.

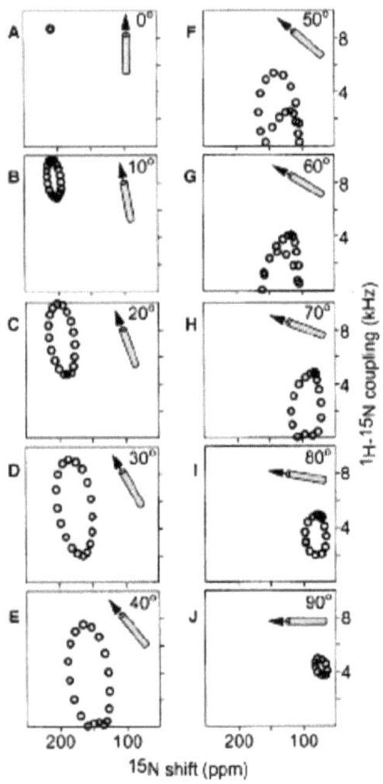

Figure 9.11: Simulated PISEMA spectra for a 19 residue α-helix at various helix-tilt angles (θ) relative to bilayer normal. Values used for the principal values of the N-H chemical-shift tensor (in ppm) used are: σ_{11}= 64, σ_{22}= 77, σ_{33}= 217. The angle between σ_{33} and the NH vector is taken as 17° and the N-H bond distance as 1.07 Å F.M. Marassi and S.J. Opella. J. Mag. Res. 144, 150-155 (2000). (Reprinted with permission from Elsevier).

Figure 9.11 shows a simulated PISEMA spectrum of an ideal α-helix at various tilt angles (θ) of the helix relative to the bilayer normal. When the helix axis is parallel to the bilayer normal, all N-H bonds have identical orientation relative to the direction of B_0 and a single peak is seen in the 2D spectrum. The N-H vectors in this case are almost parallel to the helix axis. Therefore, the resonance frequencies approach their maximum values. At other values of the angle θ, N-H vectors of amino acid residues have different orientations. Consequently the spectral features of the peptide are dispersed. Since both the dipolar couplings and the CSA have fairly large values, the spectral features are well resolved, even without the use of MAS.

These *wheel like patterns* of resonances in the 2D spectra of transmembrane helices, provide secondary structures and topology of proteins embedded in phospholipid bilayers. These spectral patterns are referred as polarity index slant angle wheels (PISA). For helices with θ greater than 40°, some N-H bonds may acquire angles close to the magic angle leading to low dipolar couplings. Others may have orientations such that the dipolar coupling changes sign. PISEMA spectra do not distinguish between positive and negative dipolar couplings, leading to the *non-wheel* like patterns for θ in the range 40-60°, as illustrated in Figure 9.11.

PISA wheels are sensitive indicators of secondary structure elements, particularly the α helix and β-strands. Only a limited number of assignments of ^{15}N signals are sufficient to characterize the complete architecture of a membrane protein.

4.3 Structure Analysis using MAS Solid-State NMR

The use of MAS provides in well resolved multi-dimensional NMR spectra of immobilized protein samples. The conformation dependent chemical shifts for secondary structure analysis are also valid for proteins in solid-state. The backbone (φ, ψ) angles can therefore be estimated by an analysis of $^{13}C^{\alpha}$, $^{13}C^{\beta}$, $^{13}C'$ and ^{15}N chemical shifts using the program TALOS discussed in Chapter 3. Use can also be made of some of the distance constraints, particularly there between ^{13}C and ^{15}N using methods already discussed in Section 3.9 above.

4.4 Ion Channel Proteins

Gramicidin A is a major constituent of gramicidin, which dimerizes to form an ion transporting channel. It consists of 15 amino acid residues, having both L- and D-amino acid residues. The sequence is HCO-VGA**L**A**V**V**V**(W**L**)$_3$W-CONHCH$_2$-CH$_2$OH (D-amino acid residues are shown in bold). Gramicidin A can be easily incorporated into phospholipid bilayers. The details of its structure and dynamics of both the isolated peptide and that incorporated in lipids have been studied extensively. A variety of techniques and NMR parameters have been used such as ^{13}C and ^{15}N CSA, ^{2}H quadrupole couplings, as also as some of the solid-state NMR methods such as REDOR. Extensive studies have also been made on analogues of gramicidin A.

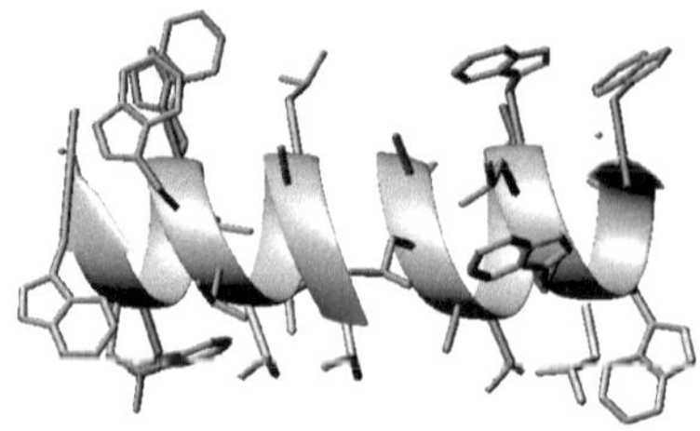

Figure 9.12: 3D structure of gramicidin (PDBID: 1grm)

The results establish a single-stranded, right-handed $\beta^{6.3}$ helical dimer as the structure of Gramicidin A, in multi-lamellar dispersions (Figure 9.12). The number of residues per turn of the helix has been estimated to be between 6 and 7. The backbone is stabilised by channel-stabilizing hydrophobic interactions. The large number of results from various groups using a variety of NMR parameters, unambiguously eliminate double-stranded helical dimers and left handed helical structures. This unusual peptide conformation is a result of alternating L and D residues. The peptide shows rapid rotations around the bilayer normal. Flexibility gradients along the peptide backbone have not been observed.

The 81 residue channel protein Vpu, is a typical member of membrane proteins. It has a long hydrophobic membrane spanning helix and two shorter amphipathic helices, which reside on the plane of the membrane. Vpu plays important functions as an accessory protein during the life cycle of HIV-1. The atomic resolution studies are important to understand the structural biology of HIV-1 and in the design of drugs with a different target other than those, which are in current use. Both the approach of high-resolution solid-state NMR and PISA wheels have been used to investigate peptides derived from Vpu. Peptide containing residues 2-30 of Vpu, shows that it is a trans membrane α-helix, with a kink. The structure of the trans-membrane domain may be responsible for the ion channel activity and may be associated with the role of the protein in its action on infected cells.

4.5 Bacteriorhodopsin (bR)

Bacteriorhodopsin (bR) is a 26 kDa protein. It is the only protein in the purple membrane of *Halobacterium halobium.* bR has seven transmembrane helices surrounding a retinal chromophore which is linked to Lys-216, and carries out light driven proton transport. This process involves photo-isomerization of the chromophore and involves an M intermediate during proton transfer reactions at the retinal chromophore. bR has been extensively studied using solid-state NMR techniques. The results have yielded valuable information about the structure and dynamics of the protein as a whole, and about the retinal chromophore, both in the ground state and in the M intermediate state.

In view of the large size of the protein (Figure 9.13), selective labelling techniques have been used to study the role of individual amino acid residues, such as Ala, Leu, Val in the structure and dynamics of bR. ^{13}C spectra of bR labelled at Lys (CO) carbons indicate that all the 36 Leu residues in intact purple membrane are immobile. Most of the Leu residues are found to lie in the transmembrane regions. Similarly, a ^{2}H investigation using bR labelled selectively at a number of amino acid residues, indicate a sharp component superimposed on a broad powder pattern. Thus, there are parts of bR which are mobile on the NMR time-scale, while others are immobile. Both the N- and C-terminals were found to be in random-coil conformation.

The *cis-trans* isomerization at C-13 position of the retinal in the M state has been followed using NMR. *R*otational *R*esonance (RR) technique has been used to measure distances and dihedral angles of chemical groups. Changes in the conformation during various stages of functional bR have been established.

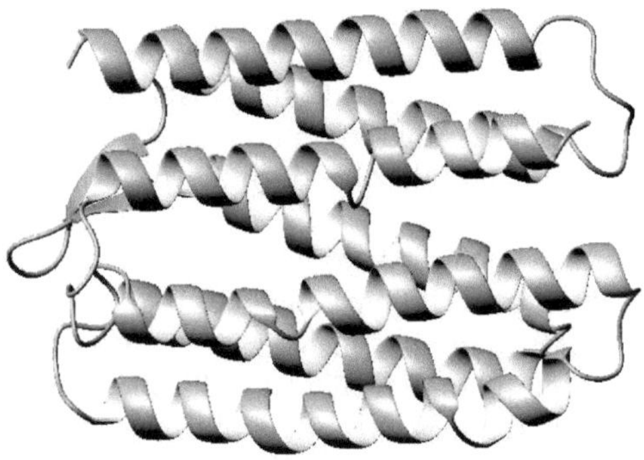

Figure 9.13: 3D structure of bacteriorhodopsin (bR; PDB ID: 1jgj)

4.6 Amyloid Proteins

Amyloid fibrils are formed by spontaneous aggregations of a variety of peptides and proteins. Amyloid fibrils are highly organised filamentous structures formed by the self assembly of polypeptides. The formation of such fibrils are associated with a number of protein deposition diseases, called amyloid diseases This class of diseases include Alzheimer's disease, type II diabetes, and spongiform encephalopathies or prion disease and Huntington's disease. Interest in amyloid fibres have also arisen from the realisation that amyloid fibrils appears to be a generic stable structure of polypeptides, along with the globular and monomeric unfolded states.

Electron microscopy and atomic force microscopy indicate that the fibrils are long, unbranched and have a diameter of ~100 Å. X-ray diffraction patterns indicate an extended β-sheet structure, the sheets running parallel and individual peptide strands oriented perpendicular to the long axis of the fibrils. The structures are stabilised by inter-chain hydrogen bonds roughly parallel to this axis. Obtaining supramolecular details of such structures are difficult because of insoluble nature of these fibres.

Considerable progress has been made in the past decade to understand the structures of amyloid fibres using solid-state NMR techniques. It has been possible to obtain full structural models including both the molecular conformation and supramolecular organization. These studies have provided valuable insights into the interactions that stabilise amyloid structures, mechanism of binding and inhibition of such fibrils by small molecules.

Solid-state NMR techniques have been applied to investigate fibrils formed by the 40 residue β-amyloid peptide involved in Alzheimer disease as well as that obtained from fragment 105-115 from the amyloidgenic protein transthyretin. In the later case, a large number of structural constraints has been obtained using MAS solid-state NMR, including thirty five ^{13}C-^{15}N distances which lie in the range of 3-6 Å. These results therefore provide structural details at atomic resolution both for the backbone atoms and the side-chains. Both studies show that amyloid fibrils adopt a parallel β-sheet structures and rule out previous suggestions of anti-parallel structures. The fibrillar form has a long range order generally associated with crystalline materials. These results provide an explanation for the unusual stability of such polypeptide assemblies, and are illustrative of the power of solid-state techniques for probing structural details which are on par with those obtained in solution NMR studies.

Extensive solid-state NMR spectroscopy studies on amyloid fibres have established several features of such fibres. Figure 9.14 shows the proposed structure for the protofilament in $A\beta_{1\text{-}40}$ fibrils. This structure is based on 2D ^{13}C-^{13}C and 2D ^{15}N–^{13}C data. In this Figure, conformationally disordered stretch of amino acid residues has been omitted. The long axis of the fibril is perpendicular to the page in panels A–C. The long axis is vertical and parallel to the page in panel D. The protofilament is a four-layered β-sheet structure with C2- symmetry about its long axis. Residues 10-22 abd 30-40 have β-strand conformations. These residues form two separate in-register, parallel β-sheets. In the Figure, blue double-headed arrows indicate side-chain–side-chain and side-chain–backbone contacts.

Figure 9.14 B shows average structure resulting from 10 independent molecular dynamics/energy minimization runs on a cluster of 12 peptide molecules, with interatomic distance and backbone torsion angle restraints dictated by the solid-state NMR data. Hydrophobic, polar, negatively charged, and positively charged side-chains are colored green, purple, red, and blue, respectively. A left-handed twist of 0.833°/Å is imposed, although direct experimental constraints on the twist, in agitated $A\beta_{1\text{-}40}$ protofilaments, are not yet available. The four-layered β-sheet structure is stabilized primarily by hydrophobic interactions in the core of the protofilament. Polar and charged side-chains are on the exterior, with the exception of oppositely charged K28 and D23 side-chains, which form salt bridges.

The fibrils structures show more general features of polypeptide conformations than was believed a few years back. The structures are favoured by all amino acid residues except Pro. They have a specific parallel or anti-parallel β-sheet structure with specific inter-strand hydrogen bonds. The fibrils have specific inter-sheet hydrogen bond interactions. Thus, these structures have well defined primary, secondary, tertiary and quaternary structures. Peptide sequences with long hydrophobic residues are oriented such that the hydrophobic interactions are maximised without introducing electrostatic repulsions. The parallel β-sheets are preferred over anti-parallel sheets. The later occur only in short peptides. This is because the parallel alignment maximizes favourable hydrophobic and other interactions.

4.7 Structure of the Coat Protein in fd Filamentous Bacteriophage Particles

Filamentous bacteriophages are important in prokaryotic biology. Bacteriophages are used in cloning and sequencing of DNA and generation of screening of peptide libraries. Approaches based on PISA wheels have been applied to study the structure of the fd coat protein. The NMR derived structure of this 50-residue protein differs from that determined by X-ray diffraction. The protein adopts a nearly ideal straight helix between the residues 7 and 38, followed by a kink. Another straight helix follows between the residues 39-49. The first five residues show a random-coil structure.

4.8 Collagen

Collagen is the most abundant protein in biological systems. It is a major protein component of connective tissues in animals. Collagen is a structural protein, containing the repeating unit (Gly-X-Y), with a high content of imino acids (at positions X and Y). There is a large percentage of Pro and Hyp residues among the repeat units at these two positions. Other residues such as Leu and Arg may also be present at these positions. The Hyp residues at position Y result from hydroxylation of Pro. Collagen imparts tensile strength to connective tissues such as skin, bone and tendon. Degeneration of collagen limits the shock absorbing capacity of the tissues and is responsible for arthritis. Mutation in the triple helix domain, particularly replacement of Gly, thus breaking the Gly-X-Y repeating pattern, leads to a variety of human disorders.

The collagen triple helix is a basic motif which is adaptable to a wide range of other proteins. The collagen triple helix is also present as a motif in a variety of proteins, particularly those involved in host-defence functions. Collagen triple helix domains have been found in a number of other organisms including bacteria and viruses.

The mechanical strength and biological functions of collagen arise from its rope-like structure (Figure 9.15). This structure consists of a left-handed supercoiled triple helix stabilized by inter-chain hydrogen bonds. Each polypeptide contains about 1000 amino acid residues. The overall length of each fibril is ~ 300 nm, with a diameter of ~1.5 nm. The CH_2 group of Gly residues are in the core of the triple helix, while the side chains of residues X and Y are on the exterior of the surface of the super-coiled triple helix. Mutual interactions among the side chains direct the assembly of collagen and are responsible for its structure.

A great deal of information on the structure of collagen, its hydrogen bonding pattern, hydration, side chain interactions and ligand binding have been obtained from theoretical consideration, fibre diffraction, NMR and electron microscopy. At least 27 distinct types of human collagens have been identified. They carry diverse functions in tissues and have distinctive mode of supramolecular organization.

A variety of solid-state NMR studies have been conducted on collagen. With the improvement of technology, new insights have been obtained on the structure and dynamics of this important molecule. Selective labelling at specific sites has been used to gain information on the 3D structure of collagen, and the low amplitude segmental motions present in this molecule. Studies based on ^{13}C and ^{15}N NMR has been used to determine conformational changes arising from the degree of hydration. Hydration has a strong effect on the amplitude of molecular motions. This may be due to the effect of hydration on the hydrogen bonded structure of collagen. The hydrogen bonding patterns are consistent with NMR hydrogen exchange studies on ^{15}N labelled residues.

Figure 9.15: Hydration and structure of triple stranded collagen (PDB ID: 1cgd)

Even though collagen triple helix is considered as a rigid rod, NMR and fluorescence studies indicate significant sequence dependent molecular motions in the backbone and side chains. This sequence dependent mobility is important for its biological functions. Dynamics of collagen is important for providing flexibility in the tissues. Cross polarization studies on ^{13}C labelled collagen indicate two types of molecular motions. There is a rod-like reorientational diffusion motion along the longitudinal axis of the molecule and a torsional reorientation. NMR studies indicate that the imino acid rings have angular fluctuations. Significant azimuthal motions around the helix axis have also been observed. Water is an integral part of collagen triple helix and the function of Hyp has been linked to the hydration network. The binding of collagen to various ligands such as integrins mediates cell adhesion.

The effects of mineralization and cross-linking, which can take place in collagen of connective tissues has been studied by labelling Gly residues with ^{13}C. Cross-links reduce the anisotropic backbone motions to a smaller extent as compared to mineralization. Side chain motions are only slightly hindered. These studies are important in the context of the biological functions of collagen in healthy and diseased tissues. The fluidity of surface domains allows collagen in soft tissues to distribute stress rapidly and uniformly. The backbone motions on the other hand help in the absorption of mechanical energy, when tissues experience tensional stress.

4.9 Other Fibrous Proteins

Silk is another structural protein, which has been investigated using solid-state NMR. Silk consists of a large percentage of Ala and Gly residues. The structure has also been studied using fibre-diffraction. The long stretches of alanine residues favour formation of β-sheet structures. The NMR results are generally consistent with those obtained by fibre diffraction.

4.10 Other Systems

With the emergence of high through-put screening technology in pharmaceutical industries, solid phase technique has become a routine technique in combinatorial chemistry. MAS methods can be used to analyse molecules on solid support during separation procedure. The method can also be used to monitor chemical reactions on solid surfaces.

Another area of research, which is finding applications in biological NMR, is the use of MAS for narrowing spectra of tissues. Such studies provide information on the metabolism of different tissues, and will be discussed in later chapters.

5. FUTURE OF SOLID-STATE NMR IN BIOLOGY

Solid-state NMR is fast developing into a powerful technique for studies of not only membrane proteins, but also a variety of other supramolecular structures, where neither the solution NMR nor X-ray diffraction methods are possible. There is no limit on the M_r of the supra-molecular assemblies, which can be studied by solid-state NMR. However, difficulties in interpretation of the data on complex systems coupled with relatively lower resolution, may limit the extent of information that can be derived. Such difficulties may be tackled through more extensive use of selective isotope labelling.

The diversity of the systems that can be studied by solid-state NMR is large. We have covered specific examples of applications to lipid bilayers, membrane proteins and fibrous polymers. However, the scope of such techniques is more extensive. The range of time-scale of motions, which can be studied by these techniques, are much wider than those amenable to other methods. The samples can be prepared with different levels of hydration. In the case of membranes, properly aligned samples can be used. All this augurs well for a bright future for this emerging field.

At the same time, there are several reasons why the area of solid-state NMR has not found such wide applications as the solution phase NMR research or the techniques of MRI and MRS. First, the commercial instruments with novel hardware advances became available only in the last decade. Even with the present day advances in rotor technology, sample filling and aligning the molecules, is not as straight forward, as preparing a solution of biological molecules. The quality of spectrum in solids depends critically on how the sample is prepared.

A structural biologist can easily get lost in the maze of solid-state pulse sequences, designed by pulse-engineers and given fancy names. No doubt the same situation exists in other areas of NMR research. However, in these cases, the various sequences are simple modifications or extensions of certain basic sequences. In solid-state NMR, the pulse sequences are more complex. One has to match the rotor speeds, with the timing of the sequences, and re-introduce some of the structural information, which has been removed in order to obtain sharp lines. Therefore, research in solid-state NMR in biological systems is done in relatively fewer laboratories.

With the improvement in technology, and greater awareness about the importance of such techniques in biology, the field will emerge as a potential tool in pharmaceutical industries and structural biology.

6. REFERENCES

6.1 Further Reading

E.R. Andrew, A. Bradbury, A. and R.G. Eades, *Removal of dipolar broadening of nuclear magnetic resonance spectra of solids by specimen rotation*, Nature **183**, 1802–1803 (1959).

I.J. Lowe, *Free induction decays of rotating solids*, Phys. Rev. Lett. **2**, 285–287 (1959).

A. Pines, M.G. Gibby, and J.S. Waugh, Proton-enhanced NMR of dilute spins in solids, J. Chem. Phys. **59**, 569–590 (1973).

A. Seelig, and J. Seelig, *Effect of a local anaesthetic on hydrocarbon chain order in membranes* Biochemistry **13**, 4839–4845 (1974).

J. Schaefer, and E.O. Stejskal, *Carbon-13 nuclear magnetic resonance of polymers spinning at the magic angle*, J. Am. Chem. Soc. **98**, 1031–1032 (1976).

A.H. Roufosse, W.P. Aue, J.E. Roberts, M.J. Glimcher, and R.G. Griffin, *Investigation of mineral phases of bone by solid-state phosphorus-31 magic angle spinning nuclear magnetic resonance*, Biochemistry **23**, 6115–6120 (1984).

R.L. Smith, and E. Oldfield, *Dynamic structure of membranes by deuterium NMR*, Science **225**, 280–287 (1984).

F.M. Marassi and S.J. Opella, *NMR structural studies of membrane proteins*, Curr. Opin. Struc. Biol. **8**, 640–648 (1998).

F.M. Marassi, and S.J. Opella, *A solid-state NMR index of helical membrane protein structure and topology*, J. Mag. Res. **144**, 150–155 (2000).

J. Pauli, B. Rossum, H. Forster, H. J. M. Groot, and H. Oschkinat, *Sample optimization and identification of signal patterns of amino acid side chains in 2D RFDR spectra of the α-spectrin SH3 domain*, J. Mag. Res. **143**, 411–416 (2000).

F. Castellani, B. van Rossum, A. Diehl, M. Schubert, K. Rehbein and H. Oschkinat, *Structure of a protein determined by solid-state magic-angle-spinning NMR spectroscopy, Nature* **420**, 98–102 (2002).

D.D. Laws, H.L. Bitter, and A. Jerschow, *Solid-State NMR Spectroscopic Methods in Chemistry*, Angew. Chem. Int. Ed. **41**, 3096–3129 (2002).

B.J. Van Rossum, E.A.M. Schulten, J. Raap, H. Oschkinat, and H.J.M. Groot, *A 3-D structural model of self assembled chlorophyll a/H_2O from multispin labeling and MAS NMR 2-D dipolar correlation spectroscopy in high magnetic field,* J. Mag. Res. **155**, 1–14 (2002).

B. Bechinger, and C. Sizun, *Alignment and structural analysis of membrane polypeptides by ^{15}N and ^{31}P solid-state NMR spectroscopy*, Concepts Magn. Reson. **18A**, 130–145 (2003).

S. Luca, H. Heise, and M. Baldus, *High-resolution solid-state NMR: Applied to Polypeptides and membrane proteins*, Acc. Chem. Res. **36**, 858–865 (2003).

P.T. Eles and C.A. Michal, *A DECODOR NMR study of backbone orientation in Nephila clavipes dragline silk under varying strain and draw rate*, Biomacromolecules **5**, 661–665 (2004).

A. Simmons, S.K. Sarkar, and L.W. Jelinski, *Solid-state NMR and Its Applications to Biological Research,* in *NMR Spectroscopy and Its Application to Biomedical Research*, S.K. Sarkar (ed.) pp. 345–380 (1996).

T.I. Igumenova, A.E. McDermott, K. Zim, R.W. Martin, E.K. Paulson, and A.J. Wand, *Assignments of carbon resonances for microcrystalline ubiquitin,* J. Am. Chem. Soc. **126**, 6720–6727 (2004).

B. Brodsky and A.V. Persikov, *Molecular structure of the collagen triple helix*, Adv. Protein Chem. **70**, 301–338 (2005).

D. Huster, *Investigations of the structure and dynamics of membrane-associated peptides by magic angle spinning NMR*, Prog. Nucl. Magn. Reson. Spectrosc. **46**, 79–107 (2005).

R. Tycko, *Molecular structure of amyloid fibril: insights from solid-state NMR*, Q. Rev. Biophys. **39**, 1–55 (2006).

6.2 Books and Reviews

C. A. Fyfe, *Solid-State NMR for Chemists*, C.F.C. Press, Guelph, Canada (1983).

K. Schmidt-Rohr, and H. W. Speiss, *Multidimensional Solid NMR and Polymers,* Academic Press (1994).

M.H. Levitt., *Spin Dynamics. Basics of Nuclear Magnetic Resonance*, John Wiley & Sons., Singapore (2001).

M.J. Duer (ed.), *Solid-state NMR Spectroscopy: Principles and Applications*, Blackwell Science (2002).

A. S. Ulrich, and A. Ramammorthy (eds), *Solid-state NMR on biological systems* (Special Issue), Mag. Res. Chem. **42(2)**, (2004).

H. Saito, I. Ando and A. Naito, *Solid-State NMR for Biopolymers: Principles and Applications*, Springer (2006).

CHAPTER 10

BIOMOLECULAR INTERACTIONS AND SUPRAMOLECULAR ASSEMBLIES

1. INTRODUCTION

So far our attention has been centered on processes which occur at the level of single molecule, within the environment of water or lipid bilayer matrix. However, most biological processes involve two or more macromolecules. The interaction among biopolymers therefore plays a dominant role in the regulation of biological processes. The effect of water and lipid bilayers on the properties of biological molecules has been discussed, already. The more complex biomolecular structures involve protein-protein, protein-nucleic acids and polysaccharide-protein interactions. Such interactions may lead to the formation of multi-molecular assemblies. Some examples of such subcellular assemblies are ribosomes, biological membranes, virus and chromosomes. In fact, most biological processes involve multi-molecular interactions. Such interactions may vary in time and space during biochemical reactions.

Understanding biomolecular interactions that stabilise such complex structures is viewed as the next phase of research in structural genomics and proteomics. Study of intermolecular interactions and complexes, therefore forms a crucial step towards understanding the biochemistry of cellular, plant and animal systems. This area is of great interest from the view point of structural biology. It is also critical for obtaining a basic understanding of aspects such as the control of biochemical reactions, biomolecular aggregation, molecular dynamics of biological systems, controlled binding and unbinding of ligand and receptors and several other aspects of biomedical research.

Supramolecular assemblies also play an important role in the formation of structural materials in complex biological systems. Examples of such assemblies are bones, muscles, biological membranes and other systems. While some of these assemblies may not have a precise functional role, such supramolecular structures are important for the health of living systems.

NMR investigations of biomolecular complexes have been limited. Such systems are complex and require considerable effort because of the larger sizes of supramolecular structures involved. Determination of NMR structures of large biological complexes is of recent origin and they are limited to molecular sizes of around 50 kDa. The advent of more sensitive techniques such as TROSY for macromolecular structure determination, availability of cryogenically cooled probes and spectrometers operating at higher fields, have immensely contributed for early leads in this very challenging area.

Much of the information on biomolecular interaction so far has come from X-ray crystallography. This is reflected in the fact that till date, the PDB holds NMR structures of only 129 protein-nucleic acid complexes as compared to 1598 complexes derived from X-ray crystallography. In several cases, model systems have been used in NMR studies to obtain information on more complex systems.

In this Chapter, application of both solution and solid-state NMR to study macromolecular interactions, formation and organization of macromolecular assemblies and related biological functions have been discussed. This subject has seen major developments both by X-ray and NMR methods, though the former has proved its superiority. One major advantage of NMR method is that it can be used also for weak bindings between two molecules. Weak and transient complexes are difficult to study by other physical techniques.

2. INTERMOLECULAR INTERACTIONS

The intermolecular interactions in biological system occur with the simultaneous involvement of several molecules. Weak molecular interactions play a key role in the formation of macromolecular assemblies. The formation of intermolecular complexes include van der Waal's, hydrogen bond, stacking of aromatic moieties, and water mediated hydrophobic and hydrophilic interactions. The stabilisation of intermolecular structures may include both enthalpy and entropy components.

3. NMR APPROACHES TO STUDY BIOMOLECULAR INTERACTIONS

NMR approaches to study interaction between biological molecules, follow a different pattern as compared to those which has been discussed in the study of individual structures of proteins and nucleic acids. Since the molecular weight of a complex is the sum of the molecular weights of the individual molecules involved, the resultant NMR spectrum of the complex has relatively broader lines. The use of sensitive methods therefore becomes imperative. Further, even though most of the biomolecular interactions take place through induced-fit, the main body of the structures participating in molecular assemblies do not change too much. Thus, it is desirable to obtain the structures of individual molecules before embarking on the structure determination of larger complexes or assemblies.

3.1 Experimental Considerations

Both labelled and unlabelled samples have been used to identify spectral features which change as a result of binding between two macromolecules. While studying proteins and nucleic acid interactions, it is essential that the two molecules are not contaminated by nucleases or proteases, since these two enzymes may cleave proteins and nucleic acids when the samples are mixed together.

The stoichiometry and the salt and pH conditions have to be suitably adjusted to obtain meaningful spectra and to avoid aggregation during the formation of the complex. The stoichiometry of the two macromolecules in the complex may not be 1:1. The exact composition of the two macromolecules has to be carefully adjusted to avoid complications due to unreacted species. These conditions can be usually monitored and optimized from 1D ^{1}H NMR or 2D [^{15}N-^{1}H] HSQC and biochemical assays.

For resonance assignments, one normally starts with the assignment of individual macromolecule. Generally, resonances arising only from regions involved in the interaction show large or appreciable changes in the chemical shifts. The identification of the perturbed resonances in a complex, follows the principles already discussed for individual macromolecules. However, there may be difficulties arising from unfavourable relaxation properties and poor chemical shift dispersions, particularly in larger complexes.

Novel methodologies such as TROSY and CRINEPT can overcome these problems to some extent. However, resonances are broadened in the presence of conformational exchange or other forms of internal motions. Quite often, the exchange broadened regions are the regions of interest. In such cases, even TROSY is not helpful and NOESY methods are indispensable. Spectral simplification may be achieved through the use of perdeuterated samples. Though, partial deuteration leads to lower sensitivity in ^{1}H detection, on ends up with considerable reduction in the individual line-widths.

3.2 Use of NOE and RDC

NMR derived constraints, such as NOEs and RDCs on the entire complex, have been used in several cases. However, for large molecules, estimating such constraints pose serious difficulties, because of spectral overlaps, lower chemical shift dispersion and larger resonance line-widths. Isotope labelling has resolved such problems to some extent. In the case of protein-nucleic acid systems, the NOE constraints are often limited to neighbouring units of nucleotides.

De novo structures of complexes may be derived using NOE and RDC. However, this may not be required provided that the structures of the two interacting molecules have been determined with good precision. Most of the structural changes take place on the interacting surfaces. Here also, a special attention should be paid to intermolecular constraints arising from the binding of the two molecules. In particular, RDC data provides valuable clues on the orientation of the two macromolecules. Such information on intermolecular interactions and chemical shift perturbations can be combined with docking procedures. In particular, RDCs are ideally suited for studies of complexes involving nucleic acids.

3.3 Use of Perturbations in Chemical Shifts (CSPs)

Quite often, the chemical shift perturbations (CSPs) take place at the interaction surfaces involved in binding. The CSPs, coupled with modelling and docking procedures, can help to characterise biomolecular complexes. Although such information is not adequate for *de novo* structure determination at atomic detail, the mapping of binding surfaces is a valuable tool for functional genomics and in directing biochemical and genetic experiments to probe functional aspects of proteins that bind to nucleic acids.

A convenient and fast way to follow complex formation is by monitoring 2D [^{15}N-^{1}H] HSQC of a ^{15}N labelled protein with the gradual addition of unlabelled protein or nucleic acid. In the case of slow exchange, one finds two sets of peaks corresponding to the bound and the free labelled protein. The addition of the other biopolymer can be continued till the free protein peaks arising from the labelled molecule disappear. In case of fast exchange, as is the case with weak complexes, the lines shift and reach a final position when the complex formation is complete. Once NMR assignments and the structure of one or both the molecules involved are known, this method provides a quick insight on the binding surfaces of the two molecules. The only draw back of the method is the perturbations in the chemical shifts due to conformational changes in the two molecules as a result of binding. This may occur in some cases, even on sites which are not directly involved in the binding.

In the case of protein-nucleic acid interaction, complex formation can be followed by looking at the spectrum of nucleic acids, during titration with a protein solution. In this case, the 2D [^{13}C-^{1}H] HSQC or HMQC spectrum of labelled nucleic acid has been used. The chemical shift changes of aromatic and anomeric ^{1}H and ^{13}C resonances, often allows one to follow the interacting surface. The strategies discussed above provide a good map of the surfaces involved in the interaction of the two molecules, simply from chemical shift perturbations. Selective isotope labelling of the possible binding surfaces has also proved valuable.

Extensive use is made in such studies of molecular modelling and docking procedure. The complex formed by the two molecules is modelled using the knowledge of binding surfaces obtained either from changes in chemical shifts or NOE and RDC data, arising from intermolecular interactions.

3.4 Relaxation Rates

Other techniques are based on changes in ^{15}N relaxation time T_2, following binding of a protein with other proteins and nucleic acids. The amino acid residues involved in the binding surface show large changes in T_2. This can be followed from the line-width analysis. Large increases in relaxation times of selected spins on the formation of a complex helps in identification of the groups involved in intermolecular interaction. Unlike chemical shift perturbation, the changes in relaxation times are not significantly influenced by conformational rearrangement during binding, but reflect the overall changes in molecular motions.

3.5 Use Paramagnetic Probes

The use of paramagnetic tags has also proved useful in studies of intermolecular complexes. As discussed earlier, such probes allow measurements of long-range interactions through broadening of signals and pseudo-contact shifts. In a complex, the binding regions are protected from such effects and can supplement information obtained by other techniques. Estimate of even a few distance constraints can have a significant effect on the derived structures. These techniques are particularly valuable for large molecules because of difficulties in obtaining NOE information. Use of lanthanides (Ln^{3+}) and site-specific labelling using a spin label such as PROXYL have proved useful in studies of intermolecular interactions. The use of constraints derived from paramagnetic probes can considerably improve the quality of structures of molecular assemblies and in deciding on the orientations of individual sub-units, in such complexes.

3.6 Cross-Saturation Method

With the advent of TROSY, there is yet another method to map interfacial residues in large complexes. This method is called *cross-saturation* or *saturation transfer*. It is based on the concerted use of deuteration of one of the molecules, coupled with TROSY detection and *cross-saturation* phenomena. The principle of the *cross-saturation* is illustrated in Figure 10.1. The method involves use of uniformly ^{2}H and ^{15}N labelled protein (protein II in Figure 10.1; preferably the one with lower proton density) which is complexed with unlabelled target molecule (protein I ; the one with higher proton density). Non-selective saturation of the aliphatic proton resonances of the unlabelled target molecule using a radio-frequency (RF) field, instantaneously saturates its aromatic and amide protons due to *spin-diffusion* effect. One may note that u-[^{2}H, ^{15}N]-labelled protein which is complexed with an unlabelled target molecule is not directly affected by the RF field. However, saturation effect is transferred from the unlabelled molecule to the doubly labelled molecule by *cross-relaxation* across the interface of the complex.

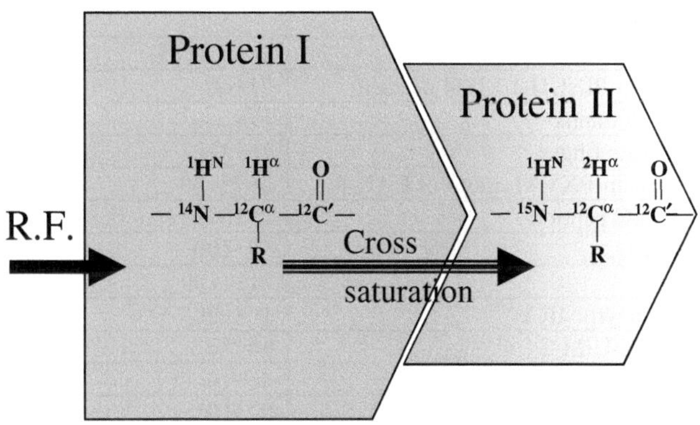

Figure 10.1: The principle of cross-saturation for determining the interface of large protein-protein complexes.

The reduction in the peak intensities due to such a *cross-relaxation,* allows the mapping of the binding interface with greater precision than the methods based on CSPs and ^{1}H-^{2}H exchange. Experimentally observed intensity changes from *cross-correlation* can be then used as constraints in the docking procedures. The *cross-saturation* method can be used to study not only protein-protein complexes but also other complexes such as protein-nucleic acids and protein-lipids.

3.7 Strategies for Structure Simulations

The protocols used in structural simulation of intermolecular complexes differ somewhat from those used for individual components. The complexes present additional problems not only because of their sizes but also because of the induced fit nature of complex formation. The strategies for structure determination in such cases often use predetermined structures of individual molecules, as a starting point. The molecules are then docked to each other by using intermolecular constraints such as NOEs between the two component molecules and optimization of the intermolecular interactions, in particular hydrogen bonds and electrostatic interactions. The optimised structures may be subjected to the usual tests for checking the validity of the acceptable structure(s).

4. PROTEIN-PROTEIN INTERACTIONS

Table 10.1 lists some of the protein-protein complexes studied by NMR and deposited in the PDB during the last five years. The number of such complexes is not large, but the number is increasing as more refined techniques become available to NMR spectroscopists.

Table 10.1: Protein-protein complexes studied by NMR during the last five years (Courtesy: Protein Data Bank)

Complex name	**Size (aa)***	**PDB ID**
NOE based complexes		
Enzyme $IIA^{mannose}$-HPr	2×(129+85)	1VRC
Nck-2 SH3 domain-PINCH-1 LIM4 domain	71+66	1U5S
CD3t-CD3δ ectodomains	178	1XMW
Ubiquitin-Npl4 zinc finger	76+31	1Q5W
HP1 chromo domain-PXVXL motif of CAF-1	2×75+30	1S4Z
ThKaiA108C-KaiC peptide	2×107+34	1SUY
CPB TAZ1-CITED2	50+100	1R8U
EIF4Ecap-eIF4G	213+100	1RF8
Enzyme IIA^{Glc}-Enzyme $IICB^{Glc}$	168+90	1O2F
HHR23A UBI-S5a-UIM-2	45+78	1P9D
CUE-Ubiquitin	46+76	1OTR
CITED2 TAD-p300 CH1	52+101	1P4Q
MMP-3-N-TIMP-1	168+126	1OO9
Chemical shift matching based complexes		
RPA32 C domain-SV40 T antigen domain	105+132	1Z1D
Plastocyanin-cytochrome f	105+254	1TU2
UbcH5B-CNOT4	147+52	1UR6
Atx1-Ccc2 copper transporting ATPases	73+72	1UV1/2

*aa: amino acids

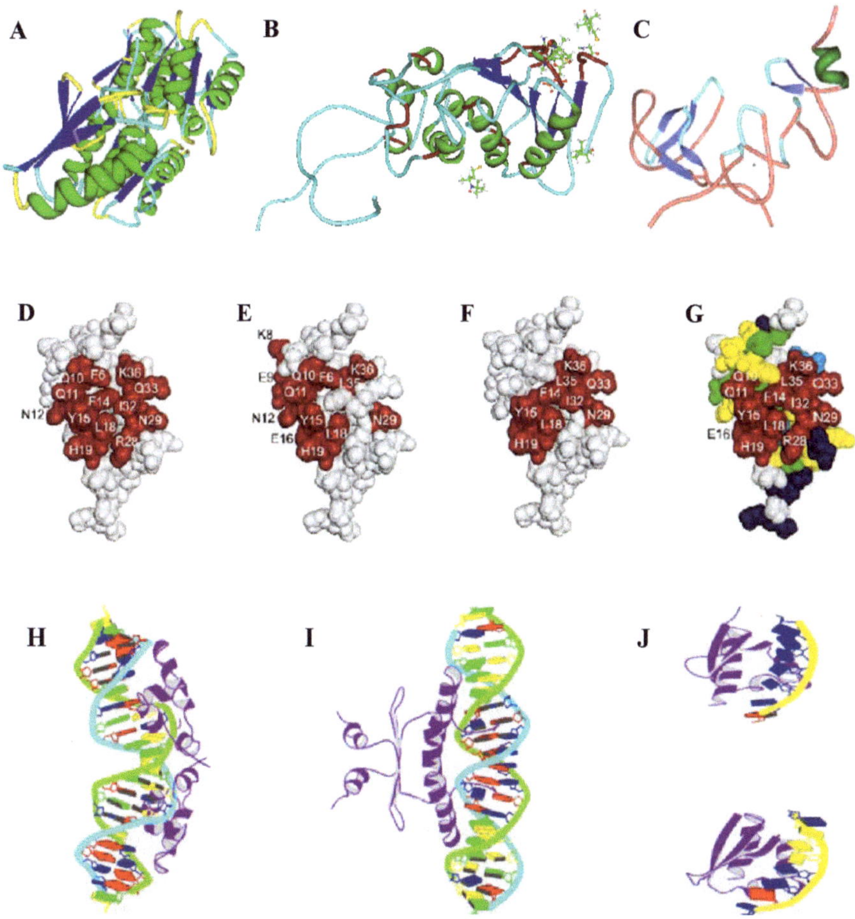

Figure 10.2: NMR derived structures of protein-protein complexes (A-C) and protein-DNA complexes (H-K) (Courtesy: Protein Data Bank): (A) The enzyme $IIA^{Mannose}$/HPr complex (pdb id-1vrc), (B) eIF4G/$eIF4E^{cap}$ complex (pdb id-1rf8), and (C) Nck-2 SH3 domain and PINCH-1 LIM4 domain complex (pdb id-1u5s). (D-G) Comparison of the binding interface in FB-Fc complex as derived by various methods (H. Takahashi, T. Nakanishi, K. Kami, Y. Arata and I. Shimada, Nature Structural Biology, 7 220-223 (2000); Reprinted with permission from Nature Publishing Group): (D) X-ray crystallographic studies; (E) CSP experiment; (F) ^1H-^2H exchange experiment; (G) cross-saturation experiment. (H) Structure of Lac rep. HP62-DNA (pdb id-1cjg); (I) MEF2A-20mer DNA (pdb id-1c7u); (J) FUSE bind Protein-DNA (pdb id-1j4w).

A number of such structures have been derived based entirely on NMR constraints, coupled with modelling. However, a fairly large number of complexes have been solved by partial NMR data based on chemical shift perturbations, followed by docking procedures. These also include structures of transient complexes. It may be noted that only a few of these depositions are larger than 200 amino acid residues. This reflects the fact that solving the structures of larger complexes by NMR is still a challenging task. The enzyme $IIA^{Mannose}$/HPr complex (Figure 10.2 A) has been solved based on intermolecular nOes and RDCs. The $eIF4G/eIF4E^{cap}$ complex (Figure 10.2 B) has been solved based on intermolecular nOes and the restraints derived from paramagnetic broadening effects. These are among the largest complexes solved by NMR. The structure of the Nck-2 SH3 domain and PINCH-1 LIM4 domain complex (with millimolar range K_d; Figure 10.2 C) reveals the uniqueness of NMR in studying weak complexes.

As an illustrative example, Figure 10.3 A and B show the results of *cross-saturation* experiment used in mapping of the binding interface of a large protein-protein complex formed between u-[^{2}H, ^{15}N]-labelled FB and the Fc fragment dissolved in 10% H_2O/90% ^2H_2O.

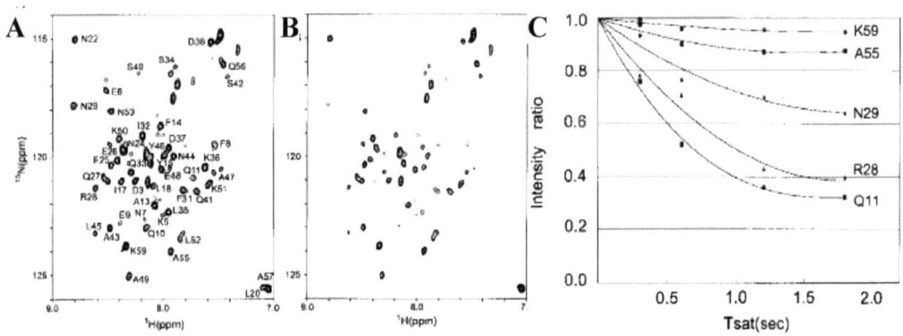

Figure 10.3: The [^{15}N-^{1}H]-TROSY-HSQC spectra recorded with 0.5–1.0 mM u-[^{2}H, ^{15}N]-labelled FB in complex with the Fc fragment taken in 10% H_2O/90% ^2H_2O: (A) without and (B) with irradiation. Measurement time was around 20 hrs. (C) Effect of the saturation time on the ratios of the cross-peak intensities originating from the backbone amide groups with and without irradiation. The saturation times were: 0.3, 0.6, 1.2 and 1.8 sec. H. Takahashi, T. Nakanishi, K. Kami, Y. Arata and I. Shimada, Nature Structural Biology, 7, 220-223 (2000). (Reprinted with permission from Nature Publishing Group).

A comparison of the binding interfaces of the B domain of a cell wall component of *staphylococcus aureus* (FB) and Fc fragment of immunoglobulin G (Ig) complex as determined by X-ray crystallographic study with that of *cross-saturation* experiment (Figure 10.2 D and G, respectively) reveals that they are identical. The corresponding contact residues overlap perfectly. The *cross-saturation* method provides direct information on through-space interactions between the two interacting molecules. It is generally more accurate as compared to structures derived by NMR methods that rely on CSP and H-D exchange (Figure 10.2 D-F).

5. PROTEIN-DNA INTERACTIONS

One of the most important and emerging fields in molecular biology is related to the interactions between the two functional biopolymers namely, proteins and nucleic acids. It is clear that several enzymes in the key biological functions take their *instructions* and *directions* from nucleic acids in order to carry out various biochemical reactions. Thus, several biological processes involving the two major macromolecules are governed by their inter-molecular interactions. The interactions are not limited to 1:1 complex between a specific protein and a particular sequence in nucleic acids. A group of proteins can operate cooperatively to control biological processes, such as those involved during the process of replication and transcription. Some of the DNA binding motifs have helped in understanding forces that stabilise protein-DNA complexes.

The interactions between two or more nucleic acid strands are well understood in terms of distinctive patterns of donor and acceptor groups involved in inter-base hydrogen bonding. However, both protein-protein and protein-nucleic acid interactions are much more complex. The protein-nucleic acid complexes involve organization of the side-chains of amino acid residues of a binding protein such that it leads to a specific interaction between the side chain of the protein and the bases of nucleic acids (or base pairs in a double stranded nucleic acid). In the case of protein-protein interactions, the molecular interaction involves amino acid residues of the two molecules.

Specificity of protein-nucleic acid complexes may be higher for complexes which involve two hydrogen bonds, such as those between Glu and A, and between Arg and G. One common element of such interactions is the structural adaptability of the two molecules. Variations in the sequence dependent local conformations of the two molecules occur only at or near the binding surface(s). The complexes are stabilised by the net effect of intermolecular interaction energy, deformation energies of the two molecules to reach an induced fit, dehydration near the binding surfaces to enable the binding and loses in the conformational and translation entropies as a result of binding.

5.1 Biological Importance

Protein-nucleic acid interactions constitute an important aspect of molecular biology. Several processes such as gene regulation, replication, recombination, transposition, processing and repair are governed by interactions between proteins and nucleic acids. Such interactions also play structural role in the formation of structures such as ribosome, nucleosome and chromosome. Such interactions form an integral part in the formation of viral particles.

The proteins involved in these processes, such as *polymerases, synthetases*, repressors, activators and histones are fairly large molecules. Such molecules often involve complex tertiary structures. The DNA-binding regions are often small, and interactions take place through selected base pairs in double-stranded structures, using the exposed sites in major and minor grooves, or through stacking interactions by partial unfolding of the base pairs. On the other hand, protein-RNA structures involve much larger domains of binding regions. The enzyme activities involved in such processes may also be complex. For example, DNA polymerase-I has three different catalytic activities (polymerase, 3′-5′ exonuclease and 5′-3′ exonuclease), which are carried out by different domains of the protein.

5.2 Nature of Protein-Nucleic Acid Interactions

While the rules for interactions between two strands of nucleic acids are well understood in terms of WC base-pairing, those for protein-nucleic acid interactions are still in an emerging state. However, it is clear that while the functional rules for interactions between nucleic acids can be understood in terms of alphabets designating the bases [e.g. A:T(U) and G:C during replication and transcription], those for protein-nucleic acid interactions require reading of whole words (for example, triplet code during translation or 4-6 letter words in recognition by restriction enzymes) or whole sentences (e.g. in the case of aminoacylsynthetase and tRNA) on the surface of nucleic acids by the interacting proteins. It is possible that simpler representation of protein-nucleic acid interactions may emerges which may involve a small number of simple patterns of protein-nucleic acid complexes. Such investigations may ultimately lead to an understanding of general recognition codes in protein-nucleic acid interactions.

The protein-nucleic acid interactions usually take place through non-covalent interactions. The most important of these are Coulombic interactions between positively charged amino acid side-chains (Arg, Lys, His) and the negatively charged phosphate backbone. Amino acid residues, such as Arg, Glu, Asp, Asn, Gln, are involved in specific hydrogen bond interactions with the nucleic-acid bases and backbone. Nucleic-acid base pairs can be accessed in duplex form *via* the grooves of ds-DNA and loops and bulges in RNA. His and aromatic amino acid residues, Trp, Tyr, and Phe participate in stacking with nucleic acid base-pairs. Such interactions may require partial unwinding of the duplex structures to accommodate protein side-chains. Hydrophobic and van der Waals interactions and salt bridges can also contribute to the stability of the protein-nucleic acid complexes, though these are less specific.

5.3 Structure Calculation of Protein-Nucleic Acid Complexes

Only certain regions of structural moieties present in the two macromolecules are involved in protein-nucleic acid interaction. Perturbations in chemical shifts and relaxation rates are good indicators of the binding topologies. Generally, only the binding regions show large changes in the structure on complex formation. Additional geometrical constraints can be obtained from NOE between the protein and nucleic acids. Due to shorter T_2 of both ^{1}H and ^{13}C in protein-nucleic acid complexes, 2D NOESY experiments with X-filters have an advantage over the 3D experiments. As usual, scalar couplings, NOE, RDC and changes in chemical shifts are used to obtain structural information on the binding pattern.

One of the convenient strategies for obtaining structure of intermolecular complexes, is to start with the individual structures of the protein and nucleic acid. The two molecules can then be docked, using one of the structural algorithms discussed in Chapter 4. However, the fact that such interactions may take place through induced fit, rather than lock and key arrangements, causes problems. Some of these interactions may be quite long-range. Therefore, it is advisable to adapt these algorithms such that the conformations of both the protein and the nucleic acid are allowed to vary. Such computations are demanding in view of the limited number of experimental constraints that may be available for the complex.

5.4 Examples of Protein-DNA Interactions

There have been extensive biochemical and biophysical studies to understand such interactions. Insight into these interactions has been achieved from studies involving both X-ray crystallography and NMR spectroscopy. The first 3D structures for DNA binding proteins such as λ– and *cro*- repressors were obtained from X-ray crystallography in 1980s. In 1985, *Lac* repressor head-piece, another DNA binding protein, turned out to be the one of the first proteins, whose structure was determined by NMR spectroscopy. This was followed by a low resolution 3D structure of the complex between *lac* repressor head-piece and its operator (Figure 10.2 H). All these DNA binding proteins contained a common helix-turn-helix motif as the main DNA binding domain (Figure 10.2 H). During the last two decades, several other DNA binding motifs have been characterised (Figure 10.2 H-J). These include helix-loop-helix proteins, β-sheet DNA binding proteins, zinc fingers and leucine zippers.

The success of NMR, in this respect, is primarily because the DNA binding domains of the individual proteins arc rclatively small. Concerted use of various biochemical and biophysical methods along with the NMR structural studies provides more insights into the structure-function relationships of protein-nucleic acid complexes. Table 10.2 lists some of the protein-DNA complexes studied by NMR and deposited in the PDB. As is evident from the list, the number of protein-DNA complexes studied is not large. However, but the number is increasing as more refined techniques become available to NMR spectroscopists. In the following paragraphs, two such complexes are discussed.

5.4.1 Myocyte Enhancer Factor 2A (MEF2A)–DNA Complex

The myocyte enhancer factor 2 (MEF2) proteins comprise a group of transcription factors that play an important role in myogenesis and morphogenesis of muscle cells. The MEF2 factors bind specifically to a conserved A-T rich DNA sequence in the control regions of the majority of muscle specific genes. This includes several genes that encode myogenic basic helix-loop-helix (bHLH) proteins. There are four members of the mammalian MEF2 family, which are identified as MEF2A, MEF2B, MEF2C and MEF2D. Residues 1-85 of these proteins contain core DNA-binding domain. This stretch has been further divided into two sub-domains: a MADS-box (1-58) and a MEF2S domain (59-73).

Table 10.2: Protein-DNA complexes studied by NMR (Courtesy: Protein Data Bank)

Complex name	**Size (bb/nt*)**	**PDB ID**
Oct1-Sox2-Hoxb1 DNA(ternary complex)	167+88+19bp	1O4X
MarA-α-CTD RNAP-DNA(ternary complex)	132+84+20bp	1XS9
Cdc13 CTD-ssDNA	199+11nt	1S40
Dimeric lac headpiece-non-specific DNA	2×82+18bp	1OSL
Core NFATC1-DNA	178+24bp	1A66
Mutant Antennapedia –DNA	68+28bp	1AHD
Lac Repressor HP62-DNA	62+22bp	1CJG
TRP Operon Repressor-DNA	107+40bp	1CO0
Ethanol regulon transcriptional factor/DNA	65+20bp	1F4S
Erythroid transcription factor GATA-1/DNA	60+16bp	1GAT
Human SRY/DNA	76+16bp	1HRY
Methyl-CpG Binding Protein/DNA	75+12bp	1IG4
Sex-determining region y protein/DNA	85+28bp	1J46
FUSE binding protein/DNA	174+29bp	1J4W
Heterogeneous nuclear ribonucleoprotein K/DNA	89+10bp	1J5K
Nonhistone chromosomal protein 6A/DNA	93+30bp	1J5N
Dead ringer protein/DNA	139+30bp	1KQQ
Lactose operon repressor/DNA	62+46bp	1L1M
Steroid hormone receptor ERR2/DNA	98+26bp	1LO1
MYB/DNA	109+32bp	1MSE
Homeobox protein VND/DNA	77+32bp	1NK2
Transcription factor Oct-1/Transcription factor SOX-2/DNA	167+88+38bp	1O4X
Transcription factor IIIA/DNA	92+30bp	1TF3
GAGA-FACTOR/DNA	54+22bp	1YUI
Pituitary homeobox 2/DNA	68+26bp	1YZ8
Homeotic bicoid protein/DNA	68+26bp	1ZQ3
Steroidogenic factor 1/DNA	102+30bp	2FF0
CalC/DNA	155+22bp	2GKD
AREA/DNA	66+26bp	5GAT
Protein (integrase) mutant (C57A)/DNA	169+26bp	1B69
Protein (integrase) mutant (C57A)/DNA	69+26bp	1TN9
Nucleocapsid protein 7/DNA	42+5bp	1BJ6
HMG-I(Y)/DNA	25+24bp	2EZE
Telomeric repeat binding factor 2/DNA	63+26bp	1VFC

*bp: base-pair; nt: nucleotide

The MADS-box family members possess distinct DNA-binding specificities. For example, the MEF2A recognizes the consensus sequence CTA(A/T)$_4$TAG to form a 33 kDa protein-DNA complex. The NMR derived solution structure of the complex (Figure 10.2 I) reveals that the recognition and specificity between the protein and DNA are achieved through interactions between the MADS-box and both the major and minor grooves of the DNA. The observed differences in the protein-DNA contacts seen in the MEF2A-DNA complex and the related MADS-box transcription factors SRF and MCM1, provide a structural basis for the modulation of sequence specificity and DNA bending. Further, the discrepancies in the 3D structures of the MEF2S domains in MEF2A and the corresponding domains in SRF and MCM1 account for the absence of cross-reactivity of MEF2A with other proteins that interact with these transcription factors.

5.4.2 Far-upstream Element-Binding Protein (FBP) Bound to Single-Stranded DNA

Recognition of DNA deformations which are induced by stress generated, during transcription, tightly controls gene regulation. The FUSE (far-upstream element)-binding protein (FBP) is a regulator of c-myc expression. It contains four K homology (KH) repeats separated by linkers of varying lengths. The minimal ssDNA binding domain comprises of KH3 and KH4.

The solution structure of a complex between the KH3 and KH4 domains of FBP and a 29-base long ssDNA from FUSE has been determined by NMR. The structure of the complex (Figure 10.2 J) reveals that the KH domains recognize the two sites, which are 9-10 bases in length and are separated by 5 bases. The KH4 binds to the 5′ site while the KH3 binds to the 3′ site. The central portion of each site comprises a tetrad of sequence 5′ d-ATTC for KH4 and 5′ d-TTTT for KH3. Dynamics data show that the two KH domains bind to ssDNA as articulated modules. The binding provides a flexible framework which facilitates recognition of the transient, moving targets.

6. RNA STRUCTURES AND THEIR INTERACTION WITH PROTEINS

As discussed earlier, RNA plays a key role in protein synthesis. The specificity in RNA-protein interaction is critical for gene expression and its regulation. However, the role of RNA molecules in other biological processes, including catalysis, has been realized in recent years. The expanding knowledge on the forms and functions of RNA in biological processes has brought about a very active interest in their studies by different structural tools. During the last decade, there has been an explosion in new structures of RNA and their complexes with proteins. Most of this information has come from X-ray crystallography. NMR has also contributed to the structures of relatively smaller systems.

Such studies have provided valuable insights into how RNA recognise protein or DNA in individual cases. However, the general principles for such interactions are not very well understood. The RNA induced conformational changes are common. Such interactions also change the dynamics of the two macromolecules involved in interaction.

RNA molecules assume complex 3D structures and have several specific roles. However, unlike proteins, RNA structures assume lesser number of structural motifs. Such structural elements are linked and influenced by tertiary interactions. Docking such structures with proteins or other nucleic acids is quite demanding since the conformational structures of both the interacting components have to be varied to achieve an induced fit. The alternative approach of simultaneously allowing all components to fold together, under the experimental constraints, avoids any bias in conformational search. However, this is quite demanding.

6.1 Structural Motifs in RNA Structures

Structural studies of RNA and RNA-protein complexes are vital for understanding a variety of biological processes. These include gene expression and regulation. Recent studies on RNA structures have thrown valuable light on the secondary and tertiary motifs in RNA structures and their role in their function, folding and evolution. RNA structures are generally modular. In addition to their relevance in protein synthesis, studies on RNA structures have importance in RNA engineering. Applications of RNA design are important in several diverse areas, such as approaches to nanotechnology, drug-design, anti-sense and ribozyme technology and non-coding RNA stretches in genomic sequences.

Well defined structural motifs are an integral part of RNA structures. These motifs are stabilised by long-range intramolecular interactions. Double stranded regions are dispersed with motifs such as pseudoknots, ribose zippers, hairpin loops, coaxial helices and tetraloop-tetraloop receptor interactions. The long range interactions in RNA may be mediated by interactions with water, ligands, metal ions particularly Mg^{+2}, or other macromolecules. Thus, the structure and function of RNA is much more challenging and requires knowledge of various inter-molecular interactions which mediate such interactions.

6.2 Introduction to NMR of RNA

RNA structures are much more complex and difficult to tackle by NMR techniques, as compared to studies on DNA and proteins. Further, most RNA in their functional states are bound to proteins and DNA. This adds to the complications in NMR studies. Therefore, most of the NMR studies have been restricted to studies on model systems using well defined nucleotide sequences. Such information has been a valuable complement to that derived from X-ray crystallography on larger structures. Progress in this direction has resulted from refinement in existing techniques and development of new methods in NMR for studying molecules with higher M_r and complexity.

The main structural motif for RNA is double-stranded A-helix. However, alternative forms of helical structures, hairpin loops junction and internal loops, and other secondary structures are frequently found. As in the case of DNA, each base-pair in the double helical RNA stretches give rise to at least one hydrogen bonded imino proton resonance [G(1NH) or U(3NH)]. Such resonances appear in the 10-15 ppm region. This spectral region provides a quick method to estimate the number of hydrogen bonded base-pairs. Since, the rise per residue (h) along the helix axis ranges from 2.6 to 3.4 Å, NOEs are observed between the adjacent imino protons of the sequential base pairs all along the helical axis. This enables assignment of almost all the imino ^{1}H resonances, even in large molecules such as tRNA.

The RNA structures are also stabilised by non-canonical hydrogen bonds and long-range interactions. Such interactions give rise to unique structural motifs, which have highly specific 3D structures. The structure are dictated by base sequences. Some of these secondary structures are known to perform role in recognition, interaction and in enzymology. Examples of such motifs include tetraloop, sarcin-ricin loop, T-loop and the kink turn. Such motifs may have been conserved during evolution. The structures may perform specific functional roles.

The biologically important RNA segments are much larger than the corresponding DNA segments used in model studies discussed, previously. This leads to lower dispersion of chemical shifts of protons. The labile nature of imino protons, particularly those which are not involved in hydrogen bonding, also complicates the assignments. Besides, RNA molecules are generally more extended than globular proteins of similar sizes. This leads to slower overall tumbling times, and shorter transverse relaxation times. This aggravates spectral overlaps for the NMR spins present in RNA. At the same time, RNA molecule fold into specific 3D shapes which have significant fraction of their secondary structures in a non-duplex form. This diversity in structures provides a greater dispersion in chemical shifts.

Solving the 3D structure of RNA involves knowledge of base pairs, base triples, loops, pseudo knots, and other secondary and tertiary interactions. The secondary structure elements in RNA are often stabilised by Mg^{+2}. In view of such differences and difficulties, NMR investigations have concentrated on smaller and thermodynamically stable sub-domains of large biologically important RNAs. Such domains are put together to model the larger RNA molecules. Assuming that the sub-domain structures are conserved in larger RNA, model structures of biologically important RNAs can be built, through a combination of NMR and modelling approaches. One can use multi-dimensional NMR experiments to overcome problems of resonance overlaps to a large extent.

It is relatively easy and economical to prepare uniformly ^{13}C and ^{15}N labelled RNA, using biochemical methods. DNA-dependent RNA *polymerases* such as T7-RNA-*polymerase* have been used extensively for this purpose. The precursors for biosynthesis are appropriately labelled nucleotide tri-phosphates. The *polymerase* requires DNA template, which is provided as linearized plasmids or ds-DNA. This method is more convenient and less expensive than the chemical approach based on phosphoramidite chemistry. The samples are purified by polyacrylamide gel electrophoresis under denaturing conditions. The labelled compounds are of utmost importance in studies of RNA-protein interactions.

6.3 Transfer RNA (tRNA) and its Interaction with Aminoacyl-tRNA Synthetase

One of the basic steps in protein synthesis is decoding of the sequence on mRNA and the insertion of appropriate amino acid residue in a growing polypeptide chain based on the three nucleotide long codon. Therefore, tRNA interaction with aminoacyl tRNA synthetase is one of the most important examples of protein-nucleic acid interaction. tRNAs are involved in the sequential recognition of codons on mRNA and subsequent insertion of specific amino acid residue in a growing polypeptide chain during the process of translation. Each tRNA molecule has a specificity for the amino acid residue that it inserts. Thus, each of the tRNA present in the cell cytoplasm has two functions:

(i) Recognition of the three letter code on mRNA using its anticodon triplet and getting attached to ribosome.

(ii) To accept the incoming amino acid using the 3′-terminal adenosine, and carry it to the so called A-site (acceptor site) of peptidyl transferase, a ribosomal centre where the new peptide bond is formed.

The aminoacylation of the 3′-terminal adenosine occurs in the cytoplasm and requires energy, which is provided by ATP. The ribosome catalyses the peptide transfer from the donor peptidyl-tRNA to the acceptor aminoacyl-tRNA. Aminoacyl-tRNA synthetases are responsible for genetic code expression through a tRNA aminoacylation at the 3′ end of tRNA. The reaction follows a two step process. It is accompanied by conformational changes in both molecules with mutual adaptation for cognate complexes.

tRNA are relatively small molecules. Their typical M_r is ~ 25 kDa, and are usually composed of 73-93 nucleotides. The tRNA possess a terminal phosphate at its 5′ end and a free hydroxyl group at the 3′ end. tRNA are single-stranded ribonucleic acids with a complex folding. In addition to A, G, C and U, several invariant and semi-invariant purines and pyrimidines are present in tRNA. The polynucleotide chain of tRNA folds back on itself to generate double-stranded helical regions which are interspersed with single-stranded loop regions which adopt hairpin or bulge-like conformations.

Holly proposed a model for tRNA called the *clover-leaf model* using G:C and A:U base pairing complementarity (Figure 10.4 A). It was shown that almost half of the nucleotides are base-paired and form four helical segments. The 3′ end has unpaired CCA terminal and acts as the acceptor end for amino acid and the growing polypeptide chain. In addition, there are three loops and an extra arm. These are so called DhU loop, the anticodon loop, and the TΨC loop. The DhU loop contains a dihydrouracil base. The anticodon loop contains the trinucleotide sequence complementary to the trinucleotide codon on mRNA and leads to the selectivity for inserting the right amino acid in the growing polypeptide chain. The TΨC loop contains an unusual nucleotide pseudo-uridine (Ψ). Each of these loops is associated with their corresponding double helical regions called *stem*. The acceptor stem at the 3′ terminus is the site of aminoacylation. In addition, there is a variable loop whose length depends on individual tRNA. These loop regions are responsible for specifically picking the correct aminoacyl-tRNA synthetase, which allow activation and incorporation of the right amino acid.

The 3D structures of tRNA established in recent years, has confirmed the essential features of clover-leaf model. The overall structure is L-shaped. The conformation is stabilised by divalent cations. The A:U and G:C WC base pairs contain a single imino proton [U(3NH) and G(1NH)], that participates in the inter-base hydrogen bonding. The first attempt to observe these imino ^{1}H resonances was

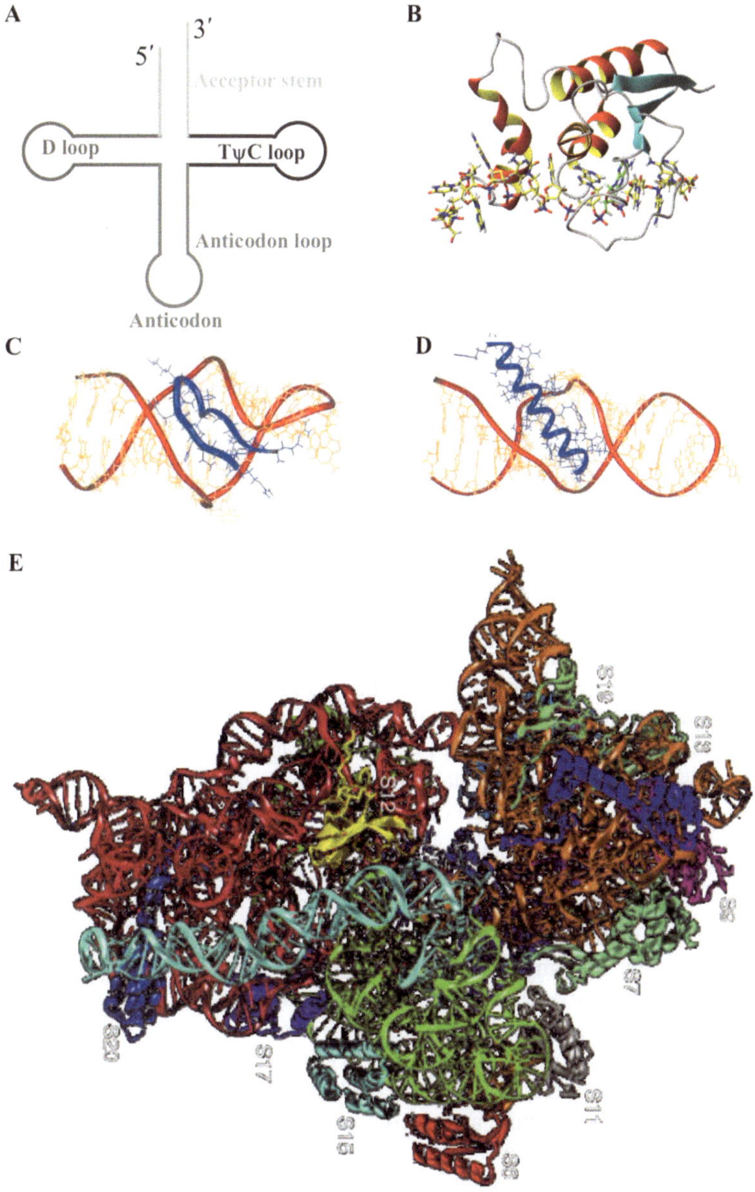

Figure 10.4: (A) Clover-leaf model of tRNA. (B-D) NMR derived RNA-Protein complexes: (B) SF1-RNA (PDB ID-1k1g) (C) BIV TAR-tart (PDB ID-1zbn); (D) HIV RRE-rev (PDB ID-1etf) and (E) Crystal structure of the ribosome (PDB ID-1GIX).

by the group of Shulman, in 1970. The high resolution 220 MHz ^{1}H spectrum of yeast tRNA^{Phe} in H_2O, exhibits several resonances in the down-field region between 10 and 15 ppm. These have been assigned to the imino protons belonging to the A:U and G:C base-pairs. Several of these resonances were resolvable with line widths of the order of 50 Hz at room temperature. These spectral features expectedly broaden out at higher temperature following thermal unfolding of the tRNA. This study, for the first time, demonstrated that NMR spectroscopy can monitor each individual base pair present in molecules which are as large as tRNA. It also provided support for the clover leaf tertiary folding of the tRNA.

In recent years, spectra recorded using highly purified tRNAs have provided excellent resolution and have enabled assignments of imino protons resonances involved in secondary and tertiary base pairing. For example, Reid and co-workers observed a number of well-resolved imino ^{1}H resonances in the NMR spectrum of *E.coli* tRNA^{Val}. By spectral simulation they could account for the existence of 27 base paired imino protons. From the fact that *E.coli* tRNA^{Val} contains twenty WC base pairs, it has been concluded that the extra 7 slowly exchanging protons correspond to tertiary interactions involving imino protons. The presence of tertiary interactions was independently confirmed by several groups based on NMR studies on *E.coli* tRNA and yeast tRNA^{Phe}.

The structure of tRNA has revealed several motifs, which led to certain common principles for RNA structures. Several of the non-canonical hydrogen bonded base pairs and triads discussed earlier, have been found in both tRNA and rRNA. The hairpin turns occur as abrupt changes in the direction around one or two nucleotides has been discussed with DNA loops. Coaxial stacking occurs frequently both in tRNA and rRNA. The use of ^{13}C and ^{15}N labelled tRNA has enabled good assignment strategies and has established NMR as a useful probe for studying structure, dynamics and interaction of tRNA.

6.4 Ribosomal RNA

Biosynthesis of proteins, leading to the formation of peptide bond, is the most important reaction in living systems. This reaction takes place on multimolecular assembly called ribosome (Figure 10.4 E). In functional sense, ribosome is the most important supra-molecular assembly as it is responsible for protein synthesis. Ribosome is a large ribonucleo-protein particle with a M_r of 2.5 MDa in bacterial systems. It consists of three molecules of ribosomal RNA (rRNA) and more than 50 proteins. RNA and proteins are present in approximately 2:1 ratio in this supramolecular assembly. The bacterial ribosomes are composed of three large RNA and over 50 proteins that are rapidly and efficiently assembled into the molecular machine responsible for protein synthesis.

The large and small subunits of the ribosome of *E. coli* are known as 50S and 30S, as characterized respectively, from their rates of sedimentation in a centrifuge. These subunits are held together to form a 70S scaffold for protein synthesis. The peptidyl transferase centre is located on the 50S subunit. Ribosomal RNAs 5S, 16S and 23S characterised into these domains not only provide the structural scaffolding for the binding of mRNA and tRNA, but also play key roles in protein synthesis. Biochemical and structural studies have shown that peptidyl centre is composed of RNA only, though contributions by residual proteins is possible. Thus, ribosome is the largest known RNA catalyst. A fairly detailed picture of the catalytic mechanisms in this process is now available, following x-ray crystallographic studies.

6.5 Protein-RNA Interactions

Knowledge of protein-RNA interactions is an integral part in understanding the function of RNA. Most RNAs are bound or are transiently associated with one or more proteins. Unlike DNAs, which often bind to proteins in a duplex form, interactions for RNA use bulges in duplex regions, loops of a hairpin or single-strand regions. Both the protein and the RNA conformations change in the complex.

Several classes of structural domains in RNA binding proteins have been identified. Both NMR and X-ray crystallography have been used for determining protein-RNA structures. However, in view of the complexities and difficulties in studying such structures, relatively few NMR structures are available in the PDB data bank. NMR however has an advantage over crystallography, as the isolated protein or RNA domains may be unstructured, which are amenable to NMR studies (Figure 10.4 B-D).

6.6 NMR Studies of Protein-RNA Interactions

Table 10.3 lists some of the protein-RNA complexes studied by NMR and deposited in the PDB. As in the case of protein-protein complexes, the number of protein-RNA complexes is not large. However, the number is increasing as more refined techniques become available to NMR spectroscopists. In the following paragraphs, three such complexes are discussed.

6.6.1 JDV Tat–BIV TAR Complex

An arginine-rich peptide from the Jembrana Disease Virus (JDV) Tat protein binds bovine immunodeficiency virus (BIV) TAR RNAs in two different binding modes. The affinity for BIV TAR is even higher than the cognate BIV peptide. The NMR structure of the JDV Tat–BIV TAR high-affinity complex (Figure 10.4 C) reveals that the C-terminal residues in JDV Tat are involved in a network of inter- and intramolecular hydrogen bonding and stacking interactions that stabilize the unstructured peptide into a stable β–hairpin in the complex and simultaneously form a base triple in the RNA. The study also reveals that amino acid residues in the C terminus of the JDV peptide enhance RNA-binding affinity. This mechanism of co-stabilization, which is a common observation in protein-RNA complex formation, highlights how intramolecular interactions can enhance the binding affinity of two flexible partners. Another important observation is that the conformation of the complex of BIV TAR is nearly indistinguishable from that of HIV TAR. This throws light upon how side-chain moieties in the JDV peptide may be used in developing high-affinity HIV TAR binding molecules.

6.6.2 HIV-1 Rev peptide-RRE RNA complex

Another arginine-rich Rev peptide from human immunodeficiency virus type-1 (HIV-1) (DTRQARRNRR RRWRERQRAA AAR) binds to stem-loop IIB of the Rev-response element (RRE) RNA. The peptide which adopts an α–helical conformation binds in the major groove of the RNA near a purine-rich internal loop. NMR solution structure of the HIV-1 Rev peptide-RRE RNA complex (Figure 10.4 D) based on the nOe data reveals that several arginine side chains make base-specific contacts, and the lone aspargine residue contacts a G-A base pair. Further, the phosphate backbone adjacent to a G-G base pair adopts an unusual structure that allows the peptide to access a widened major groove. It is interesting to note that RNA elements as small as TAR and RRE can provide such specific 3D frameworks for proteins to fold. This complex highlights the importance of RNA structure for protein recognition.

Table 10.3: Protein-RNA complexes studied by NMR (Compiled from: Protein Data Bank)

Complex name	**Size* (nt)**	**PDB ID**
Tandem zinc finger domain of TIS11d-AU-rich ss RNA	70+9nt	1RGO
Nucleocapsid protein-MIv encapsidation signal RNA	56+101nt	1U6P
Nucleocapsid protein-AACAGU	56+6nt	1WWD
Nucleocapsid protein-UUUUGCU	56+7nt	1WWE
Nucleocapsid protein-CCUCCGU	56+7nt	1WWF
Nucleocapsid protein-UAUCUG	56+6nt	1WWG
RNase III (Rnt1P)-dsRNA	90+32nt	1T4L
Phage P22N Peptide-Box B/RNA	19+15nt	1A4T
Bovine immunodeficiency virus TAT PEPTIDE/RNA	17+28nt	1BIV
Ribosomal protein L25/RNA	94+37 nt	1D6K
U1A/RNA	101+22 nt	1DZ5
REV PEPTIDE/RNA	23+34nt	1ETF
HU ANTIGEN C/RNA	174+10nt	1FNX
Transcription antiterminator licT/RNA	55+29nt	1L1C
36-mer N-terminal peptide of the N protein/RNA	35+15nt	1QFQ
Nucleolin/RNA	175+21nt	1RKJ
Transcription factor IIIA/RNA	87+55nt	2HGH
Nucleoprotein p10/RNA	56+6nt	1WWD
U1A 102/RNA	101+30nt	1AUD
60S ribosomal protein L30/RNA	104+33nt	1CK5
HTLV-1 REX PEPTIDE/RNA	16+33nt	1EXY
Nucleolin RBD12/RNA	175+22nt	1FJE
Nucleoprotein p10/RNA	56+103nt	1U6P

*nt: nucleotides

6.6.3 Splicing Factor 1-RNA complex

Splicing factor 1 (SF1), which belongs to the STAR (signal transduction and activation of RNA) family of proteins, specifically recognizes the intron branch point sequence (BPS; UACUAAC) in the pre-mRNA transcripts. NMR structural characterization of the SF1-RNA complex (Figure 10.4 B) shows that the KH-QUA2 region (which is a highly conserved region called *'Quaking homology 2'*, located at the COOH terminal to a KH domain) of SF1 adopts an enlarged KH (hn RNP K) fold, which is necessary and sufficient for BPS binding. Such recognition involves numerous STAR-specific amino acids and is consistent with mutations that affect RNA binding *in vitro* and splicing. The SF1-BPS complex structure suggests how SF1 may facilitate spliceosome assembly.

7. SUPRAMOLECULAR ASSEMBLIES

Supramolecular assemblies are responsible for several structural features of living systems. They are also involved in the catalysis some of the more complicated biological processes as illustrated in the case of protein synthesis on ribosome. In this section, some of other assemblies have been discussed.

7.1 Bone

The bones contain $CaPO_4$ as a mineral phase. Even though X-ray diffraction has been used for study of bones, the small size of the mineral crystallites in bone, provide very broad and diffuse reflections.

The exact chemical and structural nature of the solid phases of the mineral deposits in chicken bone from species of various ages, have been studied by MAS ^{31}P NMR. The spectra of powdered bone samples, have been compared with those of the synthetic samples, such as brushite (BRU; $CaHPO_4.2H_2O$), crystalline hydroxyapatite [$Ca_{10}(PO_4)_6(OH)_2$], poorly crystalline hydroxyapatite (PCHA), monetite (MON; $CaHPO_4$), octacalcium phosphate (OCP) [$Ca_8H_2(PO_4)_6.5H_2O$], carbanotohydroxyapatite (CA; HPO_4^{--}) and amorphous calcium phosphate (ACP). Most of these compounds can be differentiated in solid state ^{31}P NMR studies based on their isotropic and anisotropic chemical shifts. The spectral changes were also observed as a function of the age of the animal and the age of the bone mineral.

The result based on spectral simulation suggest that the major $CaPO_4$ mineral in bone are type B CA containing about 5-10% CO_3^{2-} and 5-10% HPO_4^{2-} groups in BRU configuration. These are average values and the amount of CA increases with age and maturation. The results are consistent with the chemical analysis of the samples and indicate the possibilities of analysis of components of biological materials, present in solid-state. Following the demonstration of these early solid state NMR studies of the bones, the technique has been applied extensively to study the contribution of different forms of phosphate materials in the bone, and their alteration during ageing and disease processes.

7.2 Skeletal Muscle

Unlike some of the other tissues studied by the techniques of MRI and MRS (Chapters 11-13) muscle in human body show solid-state NMR properties. This manifests in the MRS studies of skeletal muscles. For example, CSA and dipolar coupling effects due to anisotropic motional effects are visible in ^{1}H resonances of several metabolites. MAS has been shown to narrow the resonance lines of metabolites in tissues. This is another area where, *in-vitro* and *in-vivo* NMR has potential biomedical applications.

7.3 Large DNA and its Complexes

Solid-state techniques provide the means to look at large DNA and their complexes. Unlike in solutions, ^{1}H and ^{13}C NMR is not very helpful in view of the broad resonances. However, ^{31}P has been used to determine the conformation and dynamics of phospho-di-ester backbone. There have been several NMR reports on DNA fibres oriented in the A- , B- and C- forms of DNA. The changes in the conformation as a function of the degree of hydration have been extensively studied.

^{15}N has been incorporated in biosynthetically grown DNA. The ^{15}N-^{1}H dipolar coupling has been used to study the protonation and to localize hydrogen bonds. Selective deuteration has been used to determine base-tilt angles and anisotropic motional amplitudes as a function of hydration and temperature. Such experiments also enable estimation of the sugar pucker. While these experiments are important, the possible areas of future research are the studies on histone bound DNA and protein-DNA complexes.

7.4 Chlorophyll

A typical example of the quality of 2D spectra obtainable using techniques discussed earlier, and potential applications is shown in Figure 10.5 for a self-assembled Chlorophyll a/H_2O aggregate. Chl-a is the green pigment involved in harvesting of light and its conversion into chemical energy by plants. It self assembles in hexane in presence of water. The molecular arrangement in this aggregate, represents a paradigm for a special type of photo-synthetic light harvesting system, and has importance in artificial photosynthetic research.

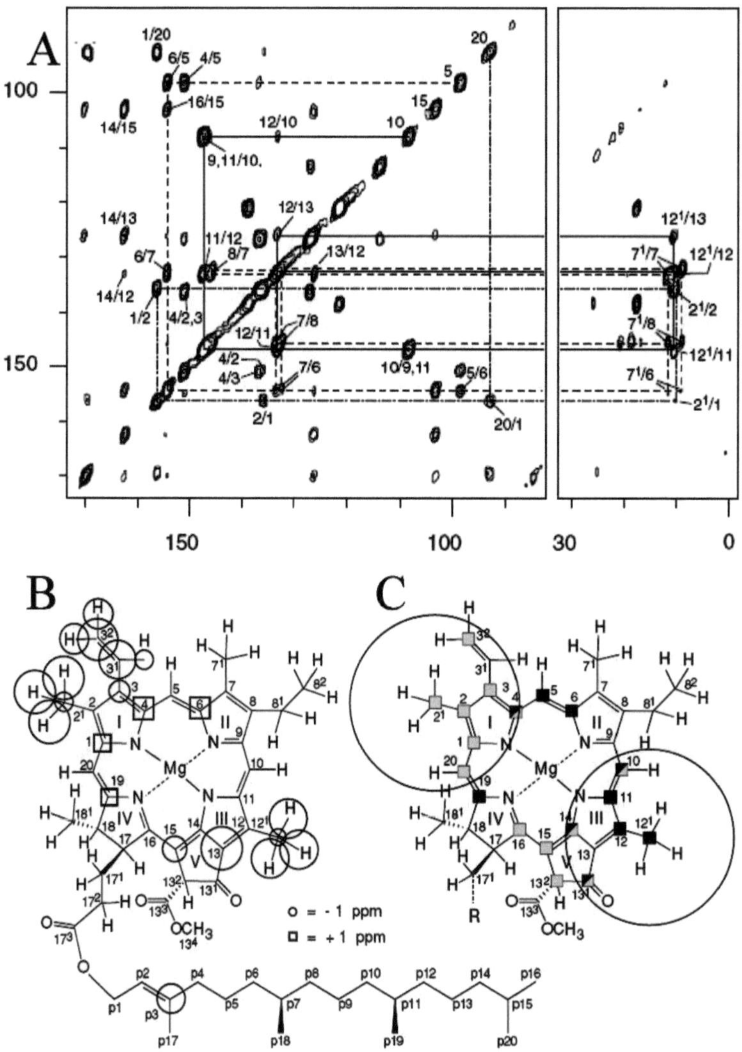

Figure 10.5: (A) 2-D MAS NMR homonuclear (^{13}C-^{13}C) dipolar correlation spectrum of ^{13}C-enriched Chl a/H_2O aggregates at 14.1 T. A spinning rate of 13 kHz has been used. The correlation networks are shown for the assignment of the 2^1, 7^1, and 12^1 signals. Solid lines connect the 10-11-12-12^1-13, dashed lines the 4-5-6-7-7^1-8, and dot-dashed lines the 20-1-2-2^1-3. (B) The numbering scheme and changes in shielding constants as a result of aggregation. Circles represent upfield shifts, while squares represent downfield shifts. B. J. van Rossum, E. A. M. Schulten, J. Raap, H. Oschkinat, and H. J. M. de Groot. J. Mag Res., 155, 1–14 (2002). (Reprinted with permission from Elsevier).

The complete assignment for Chl-a/H_2O aggregates, have been reached through ^{13}C-^{13}C dipolar correlations (Figure 10.5 A). Long range ^{1}H-^{13}C dipolar correlation and the ^{1}H and ^{13}C aggregation shifts have been used to establish the stacking of Chl-a molecules in the aggregate. A 3D model has been proposed, which matches the models inferred from electron microscopy and X-ray powder diffraction, but not with the one proposed based on neutron diffraction.

8. FUTURE OF NMR STUDIES ON SUPRAMOLECULAR SYSTEMS

NMR has become a very powerful tool for study of macromolecular interactions. Much of these advances have come because of the novel and sensitive techniques for study of larger molecular systems. It is expected that with the advent of high resolution solid state NMR and new developments in the techniques for solution NMR, new windows will be available to study this relatively difficult area.

9. REFERENCES

9.1 Further Reading

M.R., Wormald, and R.A. Dwek, *Glycoproteins: Glycan presentation and protein-fold stability*, Structure 7, R155–R160 (1999).

H. Takahashi, T. Nakanishi, K. Kami, Y. Arata and I. Shimada, *A novel NMR method for determining the interfaces of large–protein complexes*, Nature Structural Biology 7, 220–223 (2000).

T.I. Igumenova, A.E. McDermott, K. Zim, R.W. Martin, E.K. Paulson, and A.J. Wand, *Assignments of carbon resonances for microcrystalline ubiquitin*, J. Am. Chem. Soc. **126**, 6720–6727 (2004).

J. Israelachvili, *Differences between non-specific and bio-specific, and between equilibrium and non-equilibrium interactions in biological systems*, Quar. Rev. Biophysics **38**, 331–337 (2005).

G. Paillard and R. Lavery, *Analysing protein-DNA recognition mechanisms*, Structure **12**, 113–122 (2004).

9.2 RNA Structure

R.W. Holley, J. Apgar, G.A. Everett, J.T. Madison, M. Marquisee, S.H. Merrill, J.R. Penswick and A. Zamir, *Structure of a ribonucleic acid*, Science **147**, 1462–5 (1965).

P.J. Lukavsky and J.D. Puglisi, *RNAPack: an integrated NMR approach to RNA structure determination*, Methods **25**(3), 316–32 (2001).

K. Bondensgaard, E.T. Mollova and A. Pardi, *The global conformation of the hammerhead ribozyme determined using residual dipolar couplings*, Biochemistry **41**, 11532–11542 (2002).

V. Ramakrishnan, *Ribosome structure and the mechanism of Translation*, Cell **108**, 557–572 (2002).

B. Furtig, C. Richter, J. Wohnert and H. Schwalbe, *NMR spectroscopy of RNA*, Chembiochem **4**, 936–962 (2003).

M.P. Latham, D.J. Brown, S.A. McCallum and A. Pardi, *NMR methods for studying the structure and dynamics of RNA*, Chembiochem **6**, 1492–1505 (2005).

P.J. Lukavsky and J.D. Puglisi, *Structure determination of large biological RNAs*, Methods Enzymol. **394**, 399–416 (2005).

H.F. Noller, *RNA structure: reading the ribosome*, Science **309**, 1508–1514 (2005).

9.3 Protein-Nucleic Acid Interactions

J.L. Battiste, H. Mao, N.S. Rao, R. Tan, D.R. Muhandiram, L.E. Kay, A.D. Frankel and J.R. Williamson, *Helix-RNA major groove recognition in an HIV-1 rev peptide-RRE RNA complex*, Science **273**, 1547–1551 (1996).

K. Huang, J.M. Louis, L. Donaldson, F.L. Lim, A.D. Sharrocks and G.M. Clore, *Solution structure of the MEF2A-DNA complex: structural basis for the modulation of DNA bending and specificity by MADS-box transcription factors*, EMBO J. **19**, 2615–2628 (2000).

M. Allen, L. Varani and G. Varani, *Nuclear magnetic resonance methods to study structure and dynamics of RNA-protein complexes*, Methods Enzymol. **339**, 357–376 (2001).

Zhihong Liu, Ingrid Luyten, M.J. Bottomley, A.C. Messias, S.H. Molango, R. Sprangers, K. Zanier, A. Kramer, M. Sattler, *Structural Basis for Recognition of the Intron Branch Site RNA by Splicing Factor 1*, Science **294**, 1098 (2001).

D.T. Braddock, J.M. Louis, J.L. Baber, D. Levens and G.M. Clore, *structure and dynamics of KH domains from FBP bound to single-stranded DNA*, Nature **415**, 1051–1056 (2002).

G. Varani, Y. Chen and T.C. Leeper, *NMR studies of protein-nucleic acid interactions*, Methods Mol. Biol. **278**, 289–312 (2004).

A.M.J. Bonvin, R. Boelens and R. Kaptein, *NMR analysis of protein interactions*, Curr. Opinion Chem. Biol. **9**, 501–508 (2005).

V. Calabro, M.D. Daugherty, and A.D. Frankel, *A single intermolecular contact mediates intramolecular stabilization of both RNA and Protein*, PNAS **102**, 6849–6854 (2005).

H. Wu, D.L. Finger and J. Feigon, *Structure determination of protein/RNA complexes by NMR*, Methods Enzymology **394**, 525–545 (2005).

M.V. Rodnina, M. Beringer and W. Wintermeyer, *Mechanism of peptide bond formation on the ribosome*, Q. Rev. Biophysics **39**, 203–225 (2006).

9.4 Books

W. Saenger, *Principles of Nucleic Acid Structure*, Springer Verlag, Heidelberg, 315 (1984).

T.M. Jovin, K. Rippe, N.B. Ramsing, R. Klement, W. Elhorst, and M. Vojtiskova, *Structure and Methods, Vol. 3: DNA and RNA* (R.H. Sarma and M.H. Sarma, eds.), Adenine Press, Schenectady, New York, 155 (1990).

V.A. Bloomfield, D. Crothers, I. Tinoco and J. Hearst, *Nucleic Acids: Structures, Properties and Functions*, University Science Books (2000).

D.L. Nelson and M.M. Cox, *Lehninger Principles of Biochemistry*, 3rd ed., W.H. Freeman & Company (2000).

B. Lewin, *Genes*, 7th edition, Oxford University Press (2000).

L. Stryer, J.M. Berg and J.L. Tymoczko, *Biochemistry*, 5th edition W.H.Freeman & Co Ltd. (2002).

CHAPTER 11

MAGNETIC RESONANCE IMAGING

1. INTRODUCTION

Magnetic resonance imaging (MRI) provides a spatial map of the distribution of a specific NMR active nucleus in a particular compound within the volume of interest (VOI). Because of the reasons of sensitivity, one mostly uses ^1H resonance of water in imaging. MRI is one of the latest modalities in clinical imaging and represents a continuing revolution in medical technology for human health. MRI provides detailed images of human body with unprecedented soft tissue contrast and sub-millimetre resolution and that too in few seconds to minutes. Images of any desired planes, either orthogonal (perpendicular to B_0) or oblique (at an angle with respect to B_0), can be obtained by manipulation of the magnetic field B_0 using magnetic field gradients and RF pulses. Tissue anatomy, pathology, metabolism, flow and diffusion are all amenable to non-invasive evaluation by MRI.

In this Chapter, the basic concepts of MRI have been introduced. How the volume selection is done and how a region of interest (ROI) is picked up? How the image is constructed and how its contrast is manipulated? This is followed with a description of applications of MRI in clinical and pharmacological sciences. The focus will be on ^1H images using water in living systems. MR images using other nuclear spins and MRS are discussed in Chapter 12.

2. PRINCIPLES OF NMR IMAGING

In MRI, one generally makes use of a single molecule and observes a specific resonance in the three dimensional space, for example in a body organ. An obvious choice is the ^1H resonance from H_2O. As discussed in Chapter 4, water comprises ~70% of the body weight. This corresponds to almost 77 M ^1H density, considering that the molecule has two protons. Thus, water in cells gives rise to a strong signal. Therefore, the problem of sensitivity is not demanding while imaging water.

Water in living system interacts with other macromolecules, cell surfaces and tissues. It can flow freely in several parts of the body. However, it may have a restricted and anisotropic motion in other parts. This leads to significant differences in relaxation times of water in different parts of the body. On the other hand, several pathological processes modulate the behaviour of water, though in a non-specific way. The density of ^1H (water) varies in different parts of the body. Besides, different parts of the body have different levels of endogenous paramagnetic substances, such as deoxy-haemoglobin. Thus, though the water resonance is a single line, many factors modulate its NMR properties and provide contrast in MR images. MRI of water provides anatomic details with exquisite resolution. In some cases, a resolution better than 100 μm has been reported.

Lipids and fats are the next group of abundant molecules. The ^1H resonance in this case is a superposition of several methylene protons. Hence, the NMR signal is relatively broad. Fat signals have also been used for imaging. Care is taken to avoid interference between the ^1H signals arising from fat and water.

^{31}P resonances in ATP and PCr have been used in MRS. Other spins used in MRI and MRS include ^7Li, ^{19}F, ^{23}Na and ^{129}Xe. The later have been used for specialized applications, and have not yet become routine clinical tools.

The typical magnetic fields used in clinical MRI investigations are 0.2 to 3 T for studies in humans. Higher magnetic fields are sometimes used for specialized human investigations, and are routinely used in animal studies. Although the S/N ratio of image increases with magnetic field, the contrast in proton images may not necessarily improve with higher fields. It may be even be lower as one uses very high magnetic fields. The magnet bore sizes are much larger (typically 100 cm) compared to those used for molecular studies. The subjects can lie supine in the magnetic field. Larger bore size leads to higher magnetic field in-homogeneities over the VOI. An integral part of MRI scanners are gradient coils, which are positioned within the magnetic bore to generate magnetic field gradients. These additional fields can be altered in space to modify B_0 in a precise manner. Typical field gradients used in MRI are 10-60 mT/m.

Major considerations in biomedical imaging is safety of subjects, their comfort, image contrast, spatial resolution, ability to carry out reproducible *non-invasive* scanning, and ability to relate observed differences with pathological conditions. A small percentage of patients feel claustrophobic in a closed magnet environment, even though no noticeable hazard to health is observed at magnetic fields used. The RF power is also kept at fairly low level. Though radiofrequency is among the softest of electromagnetic radiations, it can cause local heating because of its absorption by water and fat. This can cause tissue damage, particularly when stronger RF fields are used.

The steps involved in MRI can be summarized as follows:

(i) The position (**r**) of the selected nucleus (say ^1H in H_2O) is spatially localized (*encoded*) using magnetic field gradients (G) along orthogonal directions, such that different voxels (volume elements in 3D) of the VOI, experience different magnetic fields and therefore, have different resonance frequencies.

(ii) A slice in a particular direction in the organ is chosen using a field gradient orthogonal to the plane of the slice of interest.

(iii) Phase and frequency encoding are used to distinguish signals from different points (picture elements or pixels) in a selected slice to get a 2D image.

(iv) The time domain signals are weighted to highlight a suitable NMR property, such as proton density (ρ), T_1, T_2, diffusion, etc. of the molecule under investigation. This is achieved by using a specific pulse sequence.

(v) The signals in the time-domain are Fourier transformed to obtain the required intensity information of individual pixels in the selected field of view (FOV).

(vi) The frequency representation of the selected slice thus obtained is translated into a description of location representation.
(vii) The intensity and properties of the NMR signal in the 3D space is displayed as an image. The image reconstruction techniques used for this purpose are similar to those employed in CT.

3. SPATIAL LOCALIZATION OF THE NUCLEUS TO BE IMAGED

To differentiate resonances from different regions of the organs, one uses magnetic field gradients [**G(r)**]. The gradients are superimposed on the static magnetic field ($\mathbf{B_0}$) during preparation, excitation or acquisition phases. Thus different parts of the organs are at slightly different magnetic fields. The ^{1}H frequency of a single molecule such as water is position dependent, according to the Larmor equation.

$$\nu = (\gamma/2\pi)\,[\boldsymbol{B_0} + \boldsymbol{G(r)}] \qquad 11.1$$

In Equation 11.1, **G(r)** is the additional contribution due to gradient field at a point **r** (x, y, z). The gradients vary linearly along the three orthogonal directions (x, y and z axes). The resonance frequency (v) reflects the spatial position from which the NMR signal arises. The gradient coils are located within the magnet bore. A current is passed through the coils to modify the field in a spatially controlled fashion. Gradient coils can be switched on and off at will at different times during a pulse sequence. The x, y and z gradients add vectorially and can be switched on selectively to get *axial* (head to trunk direction), *sagittal* (anteroposterior direction) or *coronal* (top to bottom) images. To obtain an oblique plane, all the three gradients are excited simultaneously (Figure 11.1).

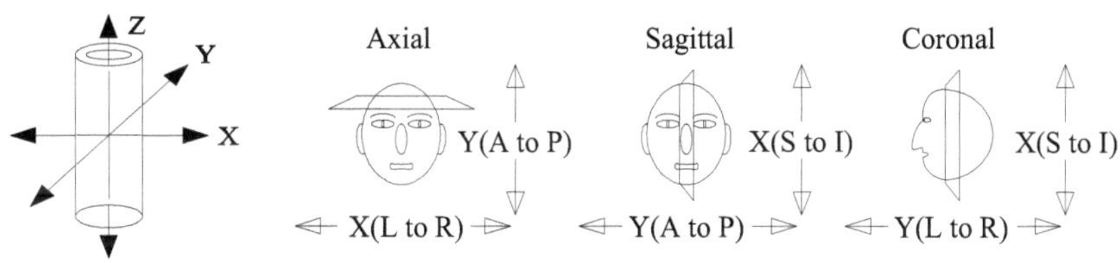

Figure 11.1: The ability to obtain images at any desired angle is an important advantage in MRI, compared to other imaging techniques. MR images in sagittal, axial and coronal, or any other direction can be obtained without changing the position of the subject.

The subject usually lies horizontally (*supine*) in the magnetic field. The z-axis defines this direction. Traditionally, the x-direction is chosen along the horizontal axis across the subject. The G_x gradient enables the spatial selection along this axis (Figure 11.1). The y-axis is the anterior-posterior direction (vertical direction). A fast RF pulse sequence with a large FOV is used to see the position of the organ of interest. Using this image, the ROI is determined and a smaller FOV and appropriate scan planes are chosen. The MR slice plane is determined by the gradient coils. The subject's position is not changed unlike the case for CT scan. The slice angle in MRI is not limited by physical factors, such as CT gantry tilt. This is an important advantage in MRI, since images with any orientation (Figure 11.1) can be obtained simply by the right choice of direction of the magnetic field gradients.

3.1 Slice Selection

Field gradients are used to differentiate ^{1}H at different positions in space. This is usually done by first selecting a slice in the body. This is followed by localizing of pixels (2D picture elements) in the plane of the selected slice. The digitized MRI is thus composed of a 2D matrix of elements or pixels. The z-axis is along the direction of the magnetic field (B_0), which is usually chosen as the longitudinal direction in which the subject is placed (Figure 11.2). Thus, the z-gradient (referred as G_z or G_s) changes the main magnetic field along the head to feet direction of the body. The use of G_z thus allows selection of a slice of the organ for imaging.

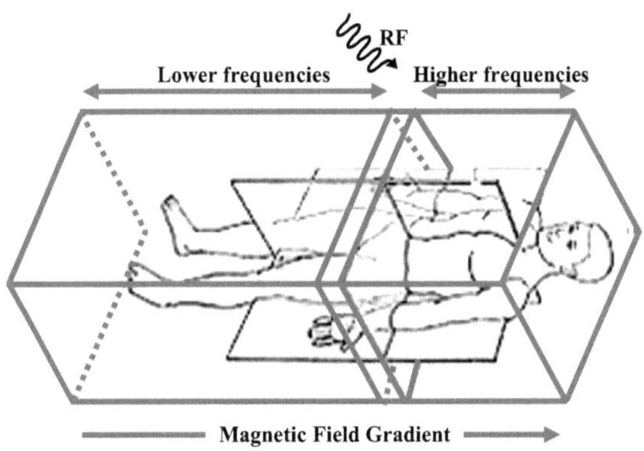

Figure 11.2 Selection of a slice of human body using a field gradient (G_z) parallel to the axis of the magnetic field B_0.

The effect of G_s is to change resonance frequency linearly along Z axis. It therefore allows dividing the object into planes perpendicular to z-axis. One can visualise that the subject is partitioned into narrow slices, such that the magnetic field and hence the Larmor precessional frequency of protons within each slice are approximately the same. Only if the frequency of the applied RF pulse matches the Larmor frequency, the protons within the selected slice are excited. Protons outside the slice have their Larmor frequency either higher or lower than the frequency of the exciting pulse. A different slice can be chosen by changing the excitation frequency.

The RF pulse has a spread of frequencies, called the bandwidth. The pulse will excite protons whose resonance frequencies matches the bandwidth of the pulse. As a result, in the presence of G_s, the RF pulse selects a slice of the chosen organ with a finite thickness. The thickness of the excited slices depends on the strength of the slice selection gradient and on the bandwidth of the RF pulse. If the RF bandwidth is kept constant (which is generally the case in most MRI instruments), then steeper field gradients will produce thinner slices. Conversely, with a weaker G_s and the same RF bandwidth, a thicker slice is selected. In this case a larger number of spins will be excited.

Typically, the gradient fields used are a few mT/meter. The slices used in human imaging are 1 to 5 mm in thickness. Thinner slices provide better spatial details of smaller body structures, such as cervical disk and facial nerves. However, the effective volume is reduced. This leads to a lower S/N ratio. Hence larger scanning time has to be used to image thinner slices. Higher magnetic fields help in shortening of the scan time while higher filed gradients are required for the selection of thinner slices.

An inter slice gap is used in multi slice imaging. Multi-slice imaging can be used in interleaved or continuous fashion. Interleaved slices help to avoid degradation of the image quality due to cross talk between two consecutive slices. There is a risk of missing a thin lesion in the skip area due to introduction of gap. Interleaving of slices is commonly used to avoid skip areas at the expense of the scan time.

3.2 Sensitive-Point Method

Having chosen a plane using G_s, the next step is to localize pixels in the orthogonal plane. Clearly, by using gradients in three orthogonal planes, it is possible to associate a particular magnetic field, and hence a unique ^{1}H resonance frequency with each volume element (voxel) in the three dimensional space defined by (G_x, G_y, G_z) and the RF pulse width. The location of the voxel is thus defined uniquely by the resonance frequency of the signal associated with each voxel.

It is conceptually useful to introduce the principles of MRI through a *phantom* experiment consisting of water filled in three tubes in the ratio of 1:2:3 and placed in a specific geometrical design (Figure 11.3). In presence of only the z-gradient, all water molecules in a given *x-y* plane will show a single signal with a frequency determined by the resonance condition $2\pi\nu = \gamma(B_0+G_s)$. Using radio frequency pulse of narrow bandwidth together with Z-gradients selects a slice orthogonal to z-axis (x-y slice). Water molecule present in the selected slice will have same frequency. Application of two additional gradients (x and y), which are orthogonal to each other as well as to z-axis, makes the resonance frequency different at each point in the selected x-y plane.

The image is reconstructed by combining information contained in these projections by image reconstruction algorithms. The images are usually displayed as signal intensities with different shades of grey or by using different colours.

Addition of suitable exogenous molecules to tubes 2 and 3 changes the relaxation behaviour of water in different tubes so as to correspond to the *slow, medium* and *fast* tumbling rates, as defined by regions A, B, and C in Figure 1.12 (Chapter 1). The relaxation times T_1 and T_2 of water are also different in these tubes. For example, if the water in tube 1 has both T_1 and T_2 corresponding to pure water, the corresponding water resonance is therefore sharp, but once excited, the water molecules will return to equilibrium at a slow rate. Thus, if short repetition times (TR) are used in multi-sequence detection, the longitudinal magnetisation will not recover completely and thus results in lower intensities. Likewise, the resonance for tube 3 has shorter T_1 and T_2 will be broader and has lower amplitude. Therefore, relatively it results in higher signal intensity in shorter TR. As the signals from the central tube (tube 2) have T_1 and T_2 in between the signal intensity will also be in intermediate range. Thus, not only the ^1H signals of H_2O from the three tubes are separated in frequency space, but also their NMR properties become distinguishable by appropriate use of the gradients.

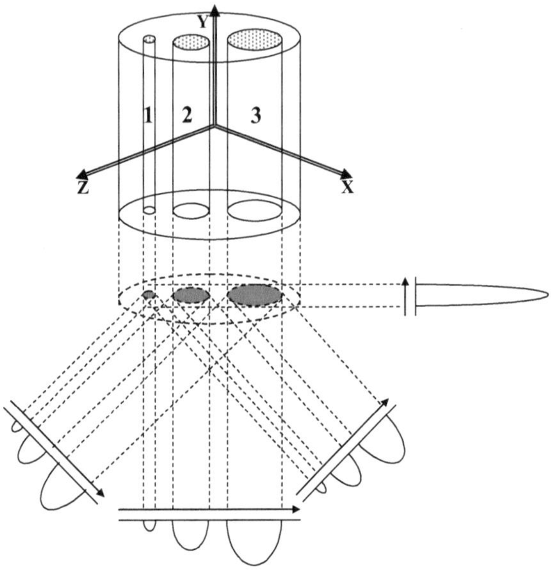

Figure 11.3: A phantom consisting of 3 tubes filled with water, placed in a geometrical arrangement as illustrated. The amount of water is different in different tubes and so are the relaxation rates. By application of field gradients, projection of the object in different directions can be obtained. An image is reconstructed using standard algorithms.

3.3 Two-Dimensional Fourier Imaging

While the methodology discussed above is conceptually simple, in practice it takes a long time to obtain images. For a 256×256 matrix, 10 slices, 4 scans, and TR of 500 msec, the total time required is 364 hours. Thus, an obvious limitation of such sensitive point technique is the excessive time required to construct an image.

A number of alternative schemes such as sensitive-line, sequential plane and 3D zeugmatography have been reported. Of these, the most commonly used method in commercial instruments falls in the category of *Fourier Imaging.* It exploits properties of Fourier transforms as conceived by Ernst and his coworkers. Because of the alteration or *twisting* of phase used for spatial localization, this method as modified for imaging is also called *spin warp*. The steps involved in this method are as follows:

3.3.1 Slice Selection

The first step in the 2D FT method for MRI is activation of a slice-selective pulse G_s (Figure 11.2) along the z-axis as discussed earlier. All ^1H spins in the selected slice have same Larmor frequency. When a RF pulse of a narrow bandwidth and matching frequency is applied, protons belonging to this slice alone are excited.

3.3.2 Phase Encoding

The next step is to spatially code the signal in the transverse plane. This is achieved by two separate processes called *phase* and *frequency* encoding. This enables one to obtain information from individual pixels within the selected slice. For phase-encoding, a short magnetic field gradient (referred as G_ϕ or G_y) (Figure 11.4) is applied immediately after the slice selective pulse. The gradient G_ϕ essentially causes a linear change in the magnetic field as a function of position along the y-axis, and hence influences the precession frequency of the spins. As soon as G_ϕ is switched off, all spins resume their initial precessional frequency, but with a change in their individual phases. The phase difference (Φ_{net}) depends on the position and the duration for which such a field gradient has been applied.

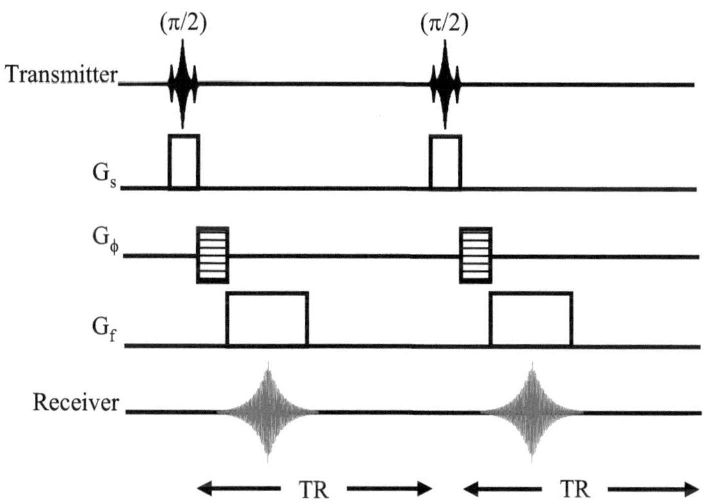

Figure 11.4: Phase and frequency encoding in a 2D Fourier imaging.

$$\Phi_{net} = \Phi_{initial} + \gamma G_{\phi} t \qquad 11.2$$

Repeating this process while incrementing the value of G_{ϕ} and keeping the time fixed will result in multiple signals with different phase encodings. This introduces a linear relationship between Φ_{net} and G_{ϕ}, which can be translated into spatial position along the y-direction.

Phase encoding is equivalent to choosing strips from the selected slice. Each strip gives rise to a signal identified by the phase, which is picked up in the next step, called frequency encoding. Typically, 128, 256 or 512 phase encoded steps are chosen depending on the resolution desired. The range of φ_{net} is 0 to 2π. This is equivalent to 0 to π in the positive direction and 0 to $-\pi$, in the negative direction.

3.3.3 Frequency Encoding

Frequency coding is used to differentiate individual pixels tagged earlier with the same phase. It thus picks up a point from the strip encoded with phase encoding. This encoding is achieved by applying a field gradient (referred as G_f or G_x) during the detection (*read-out* period) of the NMR signal, which is in the time domain. During the detection period, spatial position along the x-axis is encoded into the frequency content of the signal, arising from the frequency-encoding gradient. This axis is called the frequency-encoded direction. Because the signal is *read* during the time t_x (the time during which G_f is present), it is also called readout gradient. The signal is a function of time t_x. It is measured in a series of pixels (typically 128, 256 or 512 depending on the resolution desired).

A combination of phase and frequency information allows creation of a grid in which each pixel is defined by its phase and frequency code. This results in a data set containing signal amplitudes, the size of which is determined by the number of points in the frequency and phase directions. For example, we can start with 256 phase encoding steps, followed by 256 readout points in each step, leading to a 256×256 data matrix. This determines the number of pixels and hence the spatial resolution in the selected slice.

The grid of raw data is called the k-space. The signal in the k-space is defined by two time variables, t_x and t_y. It may be noted that following existing literature, we have used x, y and z to define axis along which different gradients have been applied. The choice of the coordinate system is rather arbitrary and may have no relationship with the directions of the planes, which define the subject's orientation (i.e. axial, sagittal, coronal). It is therefore preferred that the nomenclature G_s, G_{ϕ} and G_f be used to define the three axes for spatial, phase and frequency coding, respectively.

3.3.4 Fourier Transform

The time domain signal $S(t_x)$ sampled in the presence of readout gradient (G_x) is given by:

$$S(t_x) = \int A(x,y) exp[i\gamma \int_0^{t_x} G_x(t)\, dt] dx \qquad 11.3$$

Here A(x, y) is the amplitude of the NMR signal at the point (x, y). When a second gradient G_y is applied, the experimental data is a series of phase-encoded signals in two time domains $S(t_{x,}, t_y)$. This is related to the amplitudes of the signal at various points A(x, y) by the relation:

$$S(t_x) = \int A(x,y) exp[i\gamma \int_0^{t_x} G_x(t)\, dt + i\gamma \int_0^{t_y} G_y(t)\, dt] dx\, dy \qquad 11.4$$

Fourier transformation of this data along both the frequency and phase encoding directions provides the required signal amplitudes A(x, y) for each pixel in the chosen 2D plane (slice) of interest. These pixels collectively form the MR image.

One generally obtains signals with a specific phase angle and time average the read-out signal with several scans using G_f. This is followed by the first FT. The whole process is repeated by using a different phase gradient (G_ϕ). The second FT is done on such a data set obtained with all the phase encoded signals. Thus one gets a digital image of the object under study, in terms of the amplitude of the NMR signal arising from a small voxel [the volume around a point (x, y, z)] as selected by the gradients (G_x, G_y, G_z) or (G_ϕ, G_f, G_s)].

As discussed earlier, time-averaging of the data dictates the S/N ratio. The total imaging time for each slice will thus depend on the number of signals with different phases, which are acquired. Thus for an image slice of 256×256 pixels, only 256 signals are acquired. This leads to a reduction in time by a factor of 256 compared to the sensitive point method. The total scan image can be achieved in about an hour. However, it is still a long time for most of the applications. Hence, there has been a need for even faster methods.

3.3.5 Multi-slice Imaging

The technique described above, produces image from a single selected slice at a time. There is considerable amount of time between each *readout* period and the next excitation. During such relaxation delay, it is possible to perform additional excitation and obtain images of other slices of interest. Thus, the time required to image several slices is only slightly more than the time required to image a single one. This is basically the principle behind the so called multi-echo multi-planar (MEMP) technique, which remarkably reduces the total scan time.

3.4 k-Space

It was realised quite early in the theory of MRI that Equation 11.4 can be recast into spatial phases or spatial frequency by simple substitutions:

$$k_x = \gamma \int_0^{t_x} G_x(t)\, dt; \quad and \quad k_y = \gamma \int_0^{t_y} G_y(t)\, dt \qquad 11.5$$

In this equation, the raw data, which consist of the arrangement of the transverse magnetization over the sample in space, is recast in terms of spatial phases k_x and k_y. In fact, k_x and k_y reflect changes in phase per unit length in the x and y directions. Both k_x and k_y express the *phase advance* or *phase retardation* that the spins experience per unit length in the x- and y- directions. If x and y are expressed in cm, then the units of k_x and k_y are cycles/cm. The spatial frequency (phase axes) k_x and k_y, constitute a grid of time domain raw data, which is the Fourier inverse of the image.

Figure 11.5: MRI data is acquired in the form of intensities as a function of phases along the x and y axes. MRI pulse sequences fill the k-space by moving in a trajectory through changes in phases caused by applications of magnetic field gradients along the two directions. The image is obtained through a double FT of the signal in the k-space. (Thanks to N.R. Jagannathan for providing this figure).

The raw data acquired by MRI (Figure 11.5) peaks around the origin of the k-space. This is because the points near the centre have the least phase dispersion, whereas at the ends there is maximum amount of dispersion. The large amount of de-phasing removes the low frequency components and only the high frequency component are picked up by the *readout* pulse. The low frequency components are associated with the contrast information and are located near the centre of the k-space. The high frequency components provide the resolution, which are in the peripheral areas.

The concept of k-space has an analogy with the reciprocal space in diffraction. It plays a key role in the theoretical treatment and physical understanding of MRI. It is important to note that unlike other methodologies, the resolution of the MRI is not dictated by the wavelength of the radiation employed (in the case of ^{1}H NMR at 1.5 T, this is of the order of 5 meters), but by the resolution in the k-space.

Different imaging protocols fill-up the k-space in different ways with difficult trade offs in scan time, S/N ratio, resolution, and imaging artefacts. One can conceive several ways to achieve this task. The trajectory used for mapping the k-space is not relevant. Therefore several ways to achieve this task can be conceived. This gave rise to different protocols (discussed in the following section) to fill up the k-space in MRI. The consequences of different methodologies used are reflected on the speed of acquisition and the quality of image.

4. PULSE SEQUENCES IN MRI

We shall now discuss some of the pulse sequences used to excite and detect signals from various points with in a selected VOI, such that they provide biological information on the object to be imaged. Just as in the case of NMR spectroscopy of biological molecules, hundreds of sequences have been designed for MRI. Every year, several new sequences are added to the arsenal. Thus, there are continuous additions to the family of acronyms for pulse sequences used. However, most of these sequences fall into two families: spin-echo (SE) and gradient recalled echo (GRE). Techniques based on *reading* single Fourier-encoded echo are referred to as single-echo, as compared to multi-echo methods, which utilize multiple echoes with different Fourier encoding. Methods using a single excitation, followed by filling of the k-space with multiple phase-encoded echoes are referred to as single-shot techniques. Certain techniques in this group require preparation of the longitudinal magnetization to improve contrast.

A single pulse sequence, which uses a $\pi/2$ pulse followed by detection of the free induction decay (FID) is ideal for obtaining proton density. For several reasons, it is rarely used in MRI. First, it is difficult to record the signal immediately after the strong excitation pulse due to *ringing* in the receiver coil, which may last for several milliseconds. Secondly, since the bore of the MRI magnet is much larger than the ones used in high-resolution molecular studies in solution, the magnetic field in-homogeneity over the sample volume is much larger. Consequently de-phasing of the signal is dominated by $T2^*$, which has to be compensated.

The choice of a pulse sequence in MRI often depends on the information desired, the organ under investigation, and limitations on the scan time. Several general criteria have to be met in each case. Most importantly, the whole process of excitation and signal detection should be fast. The pulse sequence should be designed such that it can provide best contrast primarily by using relaxation rates or some other property of the nucleus under study. The two critical factors, contrast and speed may not go hand in hand. Hence, a compromise between the two has to be reached. Finally, it should be possible to correlate the information in the images with the biochemical behaviour of tissues in VOI. Several sequences satisfy these criteria.

There are two features of excitation and detection of signals in MRI, which are distinctly different from those used in spectroscopy of biological molecules. The excitation $\pi/2$ pulses used are rectangular in shape in most of the spectroscopic pulse sequences. A rectangular pulse shape in the time-domain has a conjugate frequency-domain transform called the *sinc-function* modulated waveform (sinc pulse), which has largest amplitude at the frequency of the RF. Such pulse-widths (t_p) are generally of the order of 10 μsec duration, with a constant amplitude and phase. Such pulses excite nuclear spins non-selectively over a band-width of $1/2t_p$ (200 kHz). They were called non-selective or hard pulses. The amplitude or duration of the pulse defines whether it is $\pi/2$ or π pulse.

In MRI, the application of such a non-selective $\pi/2$ pulse with a magnetic field gradient in the x direction, rotates only some of the spins which are in a plane perpendicular to the x axis by 90°, while the rest of the spins within the selected slice have a B_1 either less or more than that required for a $\pi/2$ rotation. Hence, the selected spins do not actually constitute a slice. To overcome this problem, in MRI, the RF pulse used for exciting the spins is shaped in the form of a *sinc pulse* such that one has uniform excitation of spins over the entire selected slice. A *sinc pulse* shape in the time-domain has a conjugate frequency-domain transform in the form of a square distribution. Note that a *sinc pulse* has infinite number of side lobes on either side of the central lobe which decrease in their amplitude. Hence, in practice, such *sinc pulses*, which are usually of the order of 0.5 msec duration, are truncated. As in the case of non-selective pulses, the duration of the *sinc pulse* defines the range of excitation frequencies, which is related to the desired slice thickness. The nuclei resonating within the selected slice contributes to the MRI signal.

Further, unlike in spectroscopic studies wherein one needs to separate chemically shifted resonances and coupling constants, in MRI, one needs to look at only a single resonance, that is of water and obtain its properties such as relaxation rates. Therefore, it is desirable to use pulse sequences such as spin-echo, which highlight such properties.

4.1 Spin-echo (SE)

The most common pulse sequence used in MRI is spin-echo (SE). It is also called conventional spin-echo (CSE, to distinguish it from other members of the SE family) or an advanced version of the basic spin-echo sequence (Figure 11.6). In this sequence, spins in slice is excited by a $\pi/2$ pulse, which tilts the equilibrium magnetization into the transverse plane. The transverse magnetization thus created is dephased for a time delay of TE/2, before it is subjected to a refocussing π pulse (Chapter 2). The echo signal, which appears after yet another time delay of TE/2 is detected by the receiver.

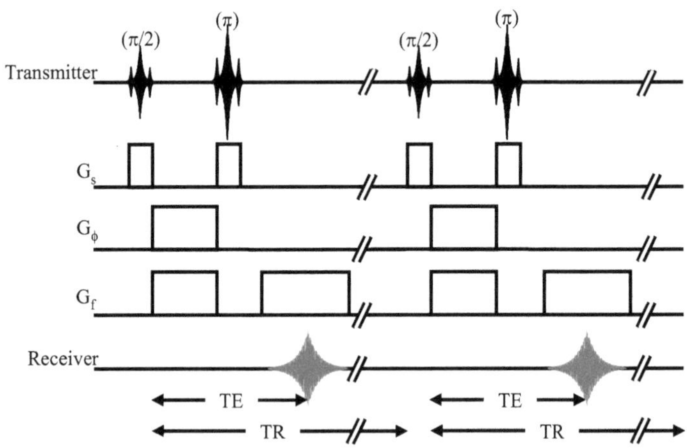

Figure 11.6: Spin-echo pulse sequence with slice (G_s), phase (G_ϕ) and frequency encoding gradients (G_f) used in MRI.

The SE scheme integrated with all the spatial encoding pulses is shown schematically in Figure 11.6. The slice selection is done during the excitation pulse. Phase encoding is achieved by the gradient pulse immediately after the $\pi/2$ pulse and frequency encoding is done during the acquisition of the echo signal. For each slice, the raw data (k-space) is a matrix having intensity at various points in the phase-frequency encoded plane as a function of time. The 2D FT of this time domain signal provides the signal amplitude of different voxels selected by the use of the three gradients. The signal intensity in a SE sequence decays as exp (-TE/T2).

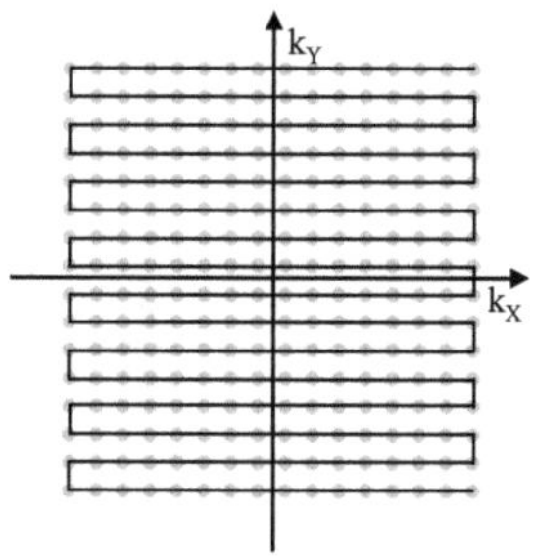

Figure 11.7: The schematics of the acquisition of points in the k-space in a SE.

To acquire multi-slice images, RF pulses are repeated with different RF frequencies to excite slices centred at different positions as defined by G_s. In this SE sequence, the k-space is filled by sampling N_p lines (where N_p is the number of phases), each having N_f data points (number of points sampled along the frequency axis) (Figure 11.7). To improve the S/N ratio, the pulse sequence is repeated with a certain repetition time set to TR. The detected signal is a function of both TE and TR and is proportional to $\rho\{1\text{-}\exp(-TR/T_1)\}\ \{\exp(-TE/T_2)\}$, where ρ is the proton density.

4.2 Half Fourier Transform Imaging (HFI)

The phases in the G_ϕ gradient can be varied between 0 and 2π. If we have 256 (N_p) steps spanning the range of 2π, then in the phase encoding direction of the k-space the pixels, can be conceived of as ranging from -128 to 127. The first $N_p/2$ points have phases distributed between $-\pi$ and 0. The remaining points have phases between 0 and π. The points with the negative phase gradients have the same information as those with positive gradients. Thus, only half of the steps are required to produce the entire image. In other words, there is a symmetry in the k-space and the point (k_x, k_y) has the same information as ($-k_x$,$-k_y$). This symmetry can be used to reduce the scan time by half. In practice, one encounters phase distortions, which destroy this symmetry. This is corrected by adding a few lines after crossing the centre of the k-space.

4.3 Inversion Recovery (IR)

The pulse sequence for the IR experiment in MRI is similar to the one discussed in Chapter 1, except that the read pulse $(\pi/2)_y$ is replaced by the SE pulse sequence (Figure 11.8). The first π pulse inverts the magnetization to the –z-axis. During the period between the first π and $\pi/2$ pulses (TI) period the inverted magnetization undergoes partial relaxation of longitudinal magnetization before the SE pulse sequence is applied. All the RF pulses used in this experiment are applied with the slice selective gradients (Figure 11.8). The phase encoding gradient is applied between the $\pi/2$ and π pulses of the SE sequence. Further, the frequency encoding gradient is applied immediately after the π pulse. The intensity of the signal thus depends on TI, and T_1 of the spins in tissue.

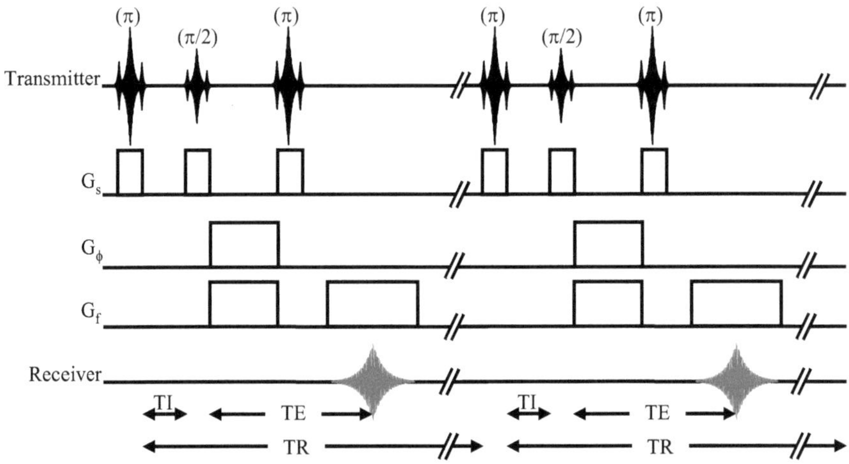

Figure 11.8 Inversion recovery pulse sequence for MRI.

IR-SE provides excellent contrast between various tissue protons (water, fat and blood), which differ in their T_1 values. For example, the grey and white matters in the brain can be distinguished with excellent contrast in MR images recorded with IR sequences. An important property of the IR signal is that as the value of TI is increased, the longitudinal magnetization goes from a negative value, through zero (*null point*) on its way to $M_z(0)$. The *null point* is different for different tissues, depending on their T_1 value. Thus, while signal from one set of molecules is zero, the other set may give a visible signal.

One of the important applications of this concept is for removing the fat signal. Fat has a T_1 of 200-260 msec at 1.5 T. If a value of 140-150 msec is set for TI, before the SE sequence is applied, then the M_z component arising from the fat is close to zero. It does not contribute to the signal. Signals arising from water in the tissues, however, have finite intensities as their T_1 values are generally longer. This forms the basis of the short T_1 inversion recovery (STIR) technique. STIR is usually used for diagnosis of fat rich regions such as liver, spine etc.

Alternatively, choosing long TI values allow magnetization from fluids having very long T_1 values to reach the null point at the end of TI. This forms the basis of the FLAIR (Fluid Attenuation by Inversion Recovery) sequence. FLAIR can be pre-inserted in a T_2-weighted SE sequence to delineate focal lesions, while suppressing signals from fluids.

4.4 The Need for Fast Imaging Techniques

The conventional techniques discussed above are satisfactory for several types of clinical applications. However, there are situations when image acquisition has to be done at a much faster rate. Examples are the imaging of abdomen, lungs and heart, where breathing motion is a problem. To overcome the problem of breathing motions one would like to have the scan of several slices with in a single respiratory cycle or heart beat. Functional brain imaging, where one observes changes in cortical signal, following neuronal activity, also requires faster imaging techniques. The imaging speed it depends not only on the data acquisition time, but is also dictated by technical considerations such as signal generation by RF pulses, contrast considerations and performance of the gradient system.

The task of designing a strategy for fast imaging is related to the task of covering k-space as efficiently and as rapidly as possible. We shall now discuss some of the methodologies used for fast and ultra-fast imaging.

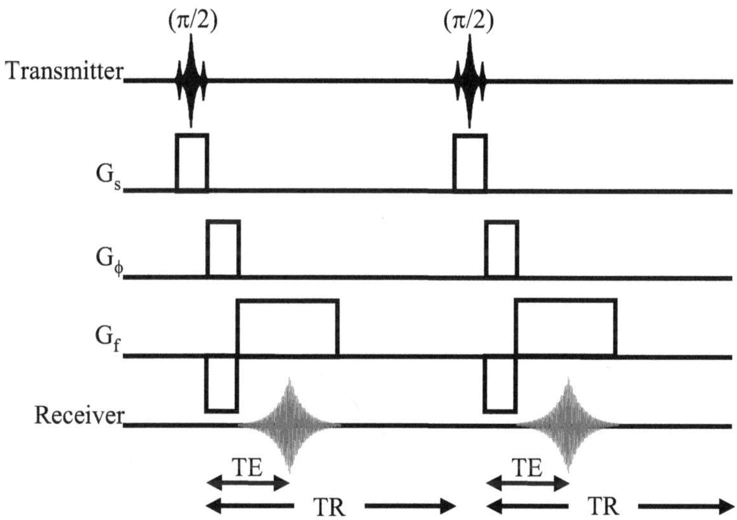

Figure 11.9: The pulse sequence for Gradient Echo (GRE).

4.5 Gradient Echo (GRE)

After the first $\pi/2$ pulse in a SE experiment, the magnetization is in the transverse plane and all spins have phase coherence. If a uniform field gradient G_x for a time t is applied, then the spins will accumulate a phase given by:

$$\phi = \phi_{initial} + \int_0^t \gamma G_x \, dt \qquad 11.6$$

When, the second term (time integral of G_x) becomes zero, the transverse magnetization acquires its original value, regardless of the position. If all spins are in phase at time t = 0, then they will regain the phase at a time t, such that $\int_0^t \gamma G_x dt = 0$. Thus, a simple reversal of G_x refocuses the spins. This forms the basis of GRE pulse sequence (Figure 11.9).

Unlike in the case of SE sequence, the amplitude of signal in a GRE experiment is modulated by T_2 and not by T_2. This is because the spatially fixed spin de-phasing processes such as magnetic susceptibility are not refocussed by gradient reversal. One of the major advantages of GRE sequence is that it does not use a π pulse for refocussing. It is therefore not necessary to use excitation pulse as a $\pi/2$ pulse. The GRE experiment is usually performed using a smaller flip angle (α), which can be much smaller than the $\pi/2$ pulse used in SE experiment. This allows M_z magnetization to recover faster, avoids saturation and hence enables the use of shorter TR.

GRE was one of the first fast-scan MRI pulse sequence, introduced under the acronym fast low angle shot (FLASH). The technique works with the steady state longitudinal magnetization, avoiding build-up of transverse steady state. A different variant is called fast imaging with steady precession (FISP). In this case, de-phasing done for spatial encoding is re-phased after each Fourier line, and a steady state build-up of transverse magnetization is utilized. This leads to a good contrast for long T_2^* situations. Techniques under the FISP family are gradient recalled acquisition into steady state (GRASS) (also known by other acronyms, such as FAST).

A periodic sequence of multiple pulses can be employed in GRE, at time spaces, which are much smaller than the T_1 or T_2 of the tissue. Depending on the refocussing behaviour of the preceding pulse, a variety of sequences in the GRE family have been devised. It may be emphasized that in all steady state multi-pulse sequences, it is required that the time-integrals of G_s, G_ϕ and G_f are brought to zero before the next train of pulses is applied. Any residual magnetization seriously affects the quality of images and can lead to undesirable *ghost* signals.

For short TR, and TR << T_2, the transverse magnetization can be shown to be:

$$M_{xy} = M_0 \sin\alpha \,/[\,(1+ T_1/T_2\,) + \cos\alpha\,(1- T_1/T_2\,)] \qquad 11.7$$

Thus, the signal intensity does not depend on TR, and constant signal amplitude can be maintained even at very short repetition times. GRE can be incorporated with IR preparation. The 3D volume coverage can be achieved in a few minutes.

4.6 Turbo-Spin-Echo (TSE)

In the sequences discussed above, TR is much longer than TE. Only one line in k-space gets encoded for each period of TR. However, if the objective is to obtain T_2 -*weighted image*, then one can obtain more than one echo within the same TR period.

One of the techniques used to reduce imaging time in the SE frame-work is TSE. This is based on the principle of CPMG sequence introduced in Chapter 1. The fact that $T_1 > T_2$ in body tissues, can be utilised to use a train of refocussing π pulses within a sequence, and obtain multiple echo signals. The difference however, is that each echo is phase-encoded differently. Thus, following a single excitation of a slice, several images belonging to different TE can be obtained (Figure 11.10) within a single (or a few) TR periods to obtain T_2-weighted images. These views are mapped in the same k-space. This idea was originally named as rapid acquisition with relaxation enhancement (RARE), since the echoes enhance the contrast due to T_2 relaxation. The number of refocussing pulses used is called the echo train length (ETL) or the TURBO factor. The sequence is called TURBO-SE sequence (TSE) or fast-spin-echo (FSE). The reduction in measurement time is proportional to the number of echoes utilized.

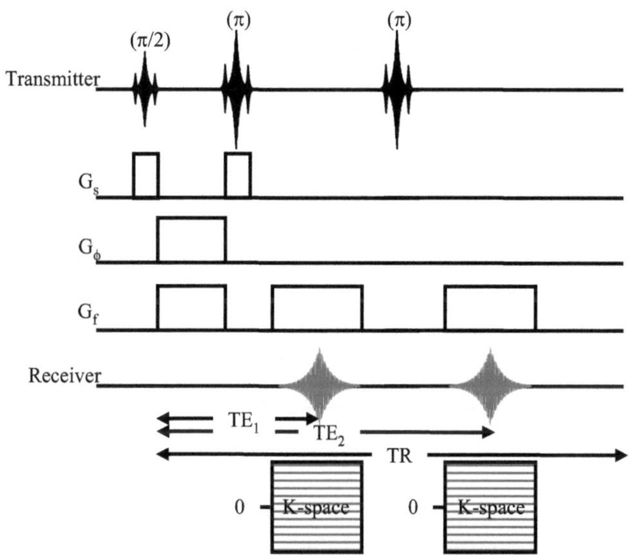

Figure 11.10: Pulse sequence for a TSE with two echoes and filling of the k-space.

An obvious advantage of TSE over the CSE is a significant reduction in scans time. The contrast properties are however, different. TSE images carry some T_1 effect and they are less sensitive to magnetic susceptibility. Thus, the regions containing fat appears brighter and haemorrhage region appears less dark. The larger RF power deposition in the TSE sequence causes a lower limit to echo spacing. The technique requires high-speed and self-shielded gradient technology to generate multiple echoes. TSE is routinely used to minimize respiratory artefacts as images can be obtained within *breath-hold* time.

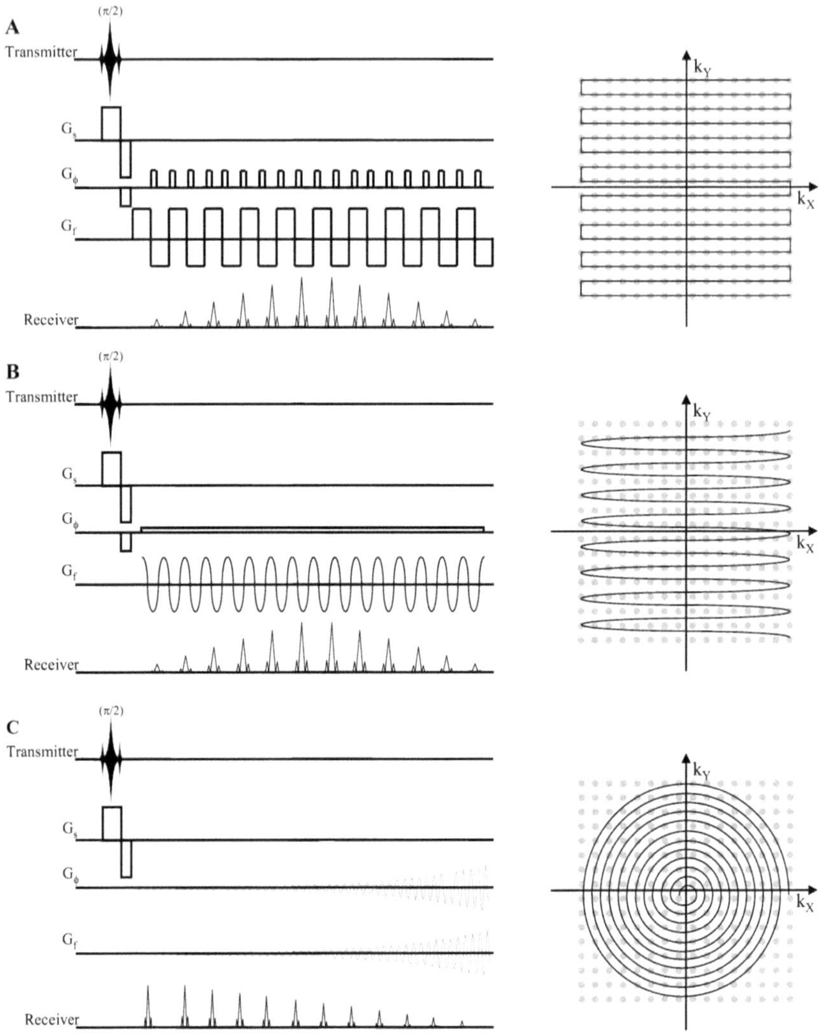

Figure 11.11: Sequences and the filling of the k-space, using different methodologies in single shot techniques; (A) rectilinear, (B) sinusoidal and (C) spiral.

When TSE is combined with HFI, the technique is called HASTE (*H*alf-Fourier *a*cquired *s*ingle sho*t* spin-*e*cho). HASTE can be used in conjunction with an inversion pulse (HASTIRM sequence). To minimize loss of T_2 dependent signal loss, such applications are normally performed such that centre of the k-space is acquired early in the echo train. The HASTE technique shows very decent image quality.

4.7 Single-Shot Techniques: Echo Planar Imaging (EPI)

One of the most useful approaches to obtain fast image is to use a single excitation of the slice followed by continuous readout through the k-space using GRE pulse sequences. The echo-planar imaging (EPI) (Mansfield, 1977) essentially belongs to this class of imaging technique. It enables image data acquisition in times less than 100 milliseconds. EPI is one of the important approaches for rapid scanning.

EPI maps the multiple phase-encoded lines of the k-space with a single RF excitation. It utilizes a clever manipulation of the encoding gradients for frequency (G_f) and phase (G_ϕ). At least three different approaches have been described in literature, whereby the k-space can be filled (Figure 11.11). In the original EPI sequence, the whole k-space was filled using multiple gradient echoes created by an oscillatory gradient pulse. In a more general approach, G_f is inverted after each line-scan, while G_ϕ is applied very briefly in *flips* after completion of a line. Inversion of a gradient inverts the scan direction in the k-space. The strong gradient G_ϕ is rapidly switched to obtain very fast echoes and G_ϕ is flipped as narrow pulses which concide with the zero crossing of G_f. The gradients are programmed to act in synchrony generating a zig-zag route (Figure 11.11 and 11.12). The NMR signal is continuously sampled during the k-space transverse. The encoding gradients are manipulated such that the gradient G_x is inverted after each line-scan, while G_y is applied very briefly after the completion of each line.

More recently, spiral EPI has been introduced. In this technique, an oscillating gradient for phase encoding and the direction of readout causes acquisition pathway in the form of spiral through the k-space. Here, gradients introduced in both directions leads to somewhat faster images. Thus, the behaviour of images obtained using EPI and spiral scanning are different.

EPI can be implemented in either the SE or the GRE mode. For example, when a π pulse is inserted between the initial excitation pulse and the signal sampling, it results in SE-EPI. An extension of this technique for 3D imaging is echo volume imaging (EVI). The speed limit of EPI experiment depends on the total acquisition time, maximum gradient amplitude and the switching times.

EPI images have found wide applications. Such images are the method of choice for several types of investigations. For this, the instrument should be equipped with strong gradients, free of eddy currents and with rise and fall times of a few milliseconds. Fast ADC is required, which can handle repeated gradient echoes. Short acquisition time for each line requires large receiver bandwidth. Today, the state-of-the-art EPI can collect a series of images in less than 100 msec.

EPI enables several biological processes to be followed in real time. However, it has several limitations. The S/N ratio is poor. Each line has to be digitized within a short time such that the entire k-space is spanned before the decay of the transverse magnetization. Spatial resolution is limited. Fat image appears as ghost. The sensitivity of the gradient readout to the susceptibility effects results in image distortions. Nevertheless, EPI is very useful in situations where fast imaging is required. Examples are: type of studies in real time; functional MRI; cardiac imaging; diffusion and perfusion studies; and bolus tracking. In several situations, this class of technique is the only alternative in spite of its limitations. Quite often, to supplement the information such single shot images are compared with high quality images obtained by other techniques.

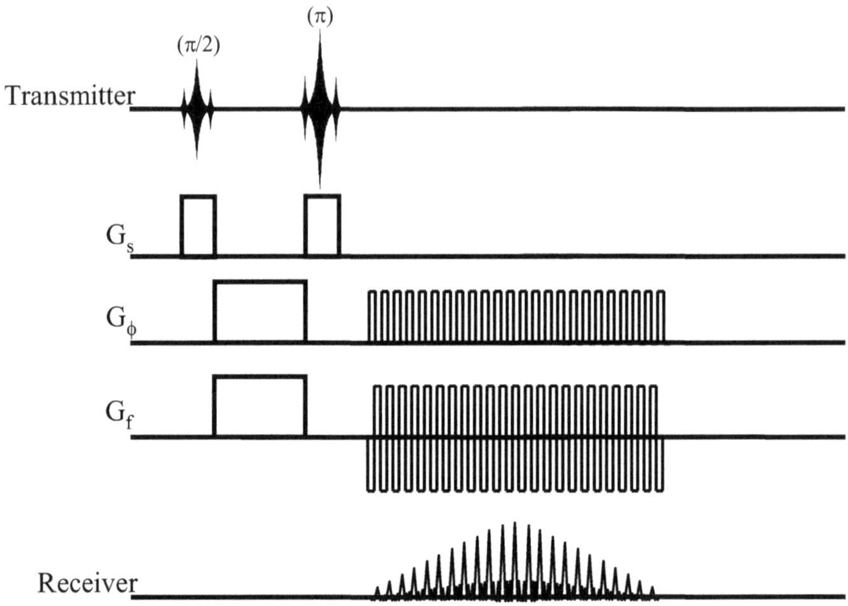

Figure 11.12: Single shot echo-planar imaging sequence showing how it spans the k-space.

4.8 Parallel Imaging

Strong gradients, RF pulses and their fast switching in sequences such as EPI can produce neuromuscular stimulation, deposit high level of RF energy and result in heating of the tissues. A common feature of these sequences is that they acquire data in a sequential fashion. The imaging speed is therefore limited by the maximum switching rate compatible both with the machine capability and the patient safety. These factors limit the image acquisition time to about 10-20 msec per slice.

Parallel imaging is a new paradigm, where several data points are acquired simultaneously, rather than sequentially. Thus, parallel imaging uses multiple detectors. Each detector provides distinct component of the spatial information. For this purpose, spatial encoding by gradients can be replaced by an array of coils, which is coupled with appropriate reconstruction strategies. Therefore, several lines in the k-space are sampled simultaneously. This increases the speed of imaging without increasing the gradient switching rate or RF power deposition in the tissue.

Two such proposed methodologies are: simultaneous acquisition of spatial harmonics (SMASH) and sensitive encoding technique (SENSE). The gain in imaging speed is related to the number of receiver coils. Parallel imaging can reduce the image acquisition times to less than 10 milliseconds.

4.9 Three Dimensional (3D) Imaging

The 3D methods involve excitation and acquisition of data from a VOI instead of thin 2D slices. A relatively thick slab is initially excited. This volume is then partitioned into a number of thin slices by phase encoding in the z-direction as well. Reconstruction of data requires use of a 3D algorithm. The 3D imaging methods use weaker G_s. Therefore thinner sections (around 1 mm) of contiguous slices can be scanned. In this sense, 3D imaging has an advantage over multi-slice 2D imaging. However, imaging times are long. It is not possible to use conventional SE pulses, which have large TR. GRE sequence using smaller flip angles and shorter TR (typically less than 30 msec) allow 3D data to be collected in a few minutes. 3D GRE imaging has found applications in musculoskeletal systems and in MR angiography.

4.10 Motion Suppression Techniques

In several MRI modalities the scan times are relatively large. The movement of subject during the period of imaging can introduce artefacts. Methods for removing artefacts due to motion of the subject are divided into three categories: restraint, sedation, and breath holding. Breath holding is practical only if scan times are reduced to an order of seconds. Respiratory gating used in conjunction with a mechanical device can be used for data acquisition during a phase of respiratory cycle. The phase encoding steps may be timed according to the respiratory motion, which is the basis of a technique called respiratory-ordered phase-encoding (ROPE). Systematic motions can be taken care by changing the shape of the gradient with time (gradient-moment nulling). Finally, the use of ultra-fast imaging techniques discussed above, can overcome such motional artefacts.

4.11 Scan Speed

It is clear that apart from the comfort of the subject, motion related issues demand fast scanning techniques. The time taken (T) for obtaining a SE image can be written as:

$$T = N_p \times TR \times N_{ex} \qquad 11.8$$

Where, N_p is the number of lines in the phase encoding axis, TR is the pulse repetition time and N_{ex} is the number of excitations used for signal averaging. For 256 phase encoding lines, repetition time of 500 msec, 4 excitations and 10 slices, this amounts to 85 minutes. This is relatively high. Scan time can be reduced by reducing N_p, which reduces the spatial resolution, but leads to a higher S/N ratio.

Lower values of N_{ex} decreases S/N ratio, though avoids some of the motional artefacts. Multi-slice acquisitions can bring the scan time within acceptable limits.

Use of multi-echo or parallel imaging technique considerably decreases the scan time. Faster sequences such as GRE can reduce TR to almost 20-200 msec. Ultra-fast GRE and EPI can acquire a single slice within 30 msec. However, field in-homogeneity and susceptibility artefacts degrade the quality of such images. Parallel imaging allows speeds less than a millisecond.

Recent developments in pulse sequences address to one or the other aspect of these problems. Each technique has its own limitation and has different requirement for hardware. Therefore, the choice of which sequence to use is dependent on a number of considerations. NMR technology has advanced such that the speed of EPI and other ultra-fast technique is set not by the instrumental limitations but by the physiology (neuro-simulation) of the subject. There is involuntary nerve stimulation due to eddy currents where higher strengths of field gradients are used. Secondly, there is heating of the organs due to RF deposition. The absorption rate depends on the frequency of RF, the tissues excited and the size of the patient. The RF absorbance by the human body reaches a maximum at around 3T magnetic field. The limits of safe RF exposure vary for different parts of the body. Thus the patient safety and comfort limits scan speed.

4.12 Receiver Coils

We have so far discussed issues relating to spatial encoding and excitation. Another important factor in MRI is receiver coils which pick up the NMR signal(s). Most modern day scanners have separate transmitter and receiver coils. Appropriate receiver coils can be placed in close proximity to ROI to optimise the signal. Three basic types of coils are used. These are surface, volume and body coils. Surface coils are used to image structures such as spine. The coil contacts part to be examined and can study smaller FOV. These are ideal for small structures such as optic nerve. Large joints such as the knees are best imaged by using coils, which surround them. To scan a large volume such as spine, a phased array system consisting of several individual coils arranged in line are used. Volume coils are necessary for examining deeper structures such as brain. The volume coils are close to the part under examination. Both surface and volume coils move with the subject. Body coils are incorporated within the magnet bore and are used to image relatively larger parts, such as chest and abdomen.

4.13 Display

The resolution of the image depends on the number of phase- and frequency- coded signals collected. The FT data is in digital form and provide intensity A(x, y) at different points of the body. The digital information is converted into analog form using a DAC and can then be represented on a 2-D device such as a video screen. The coordinates of the various pixels represent anatomical position. The brightness represented in different shade of grey or other colours used for visual presentations show the signal intensity. The numerical data are usually stored on DVD or CD. Data acquired by 3D-FT or interleaving can be used to produce a 3D image. These are useful in surgery. The rapid cinematic presentation of multiple contiguous sections can facilitate integration of information over several sections. Cine presentations are particularly useful for dynamic cardiac studies.

4.14 Future

It may be recalled that the large number of pulse sequences available for MRI can be classified in one of the two families SE (e.g. CSE, IR, STIR, FLAIR, FSE, HASTE) or GRE (FLASH, EPI). One can also use hybrids of the two, such as gradient and spin-echo (GRASE). These can be applied as single echo or in a multi-echo format, with or without magnetization preparation. There have been continuous efforts to improve the quality of images and to reduce the scan time using pulse engineering. The limits of fast imaging are also dictated by S/N ratio and patient comfort. Using parallel imaging, images may be obtained in μs times. Hyperpolarization offers the possibility of increasing S/N ratio several folds and provides a solution to imaging of ^{13}C and other less abundant nuclear spins.

5. TISSUE CONTRAST BASED ON RELAXATION RATES AND ^{1}H DENSITY

One may recall that in MRI one often looks at water molecule, which gives a single line in ^{1}H spectrum. Water signals are obtained from different pixels of the object. If the properties of water are the same in different regions thus scanned, then the image will have uniform intensity and will not provide us with any biochemical or clinical information. To obtain an image contrast, one uses a property of water, which is different in different pixels and can thus provide with means to introduce contrast in images. At least ten potential parameters have been used to obtain contrast. These include proton density (PD), relaxation times (T_1, T_2 and T_2*), T_1 in rotating frame ($T_{1\rho}$), chemical shift (δ), susceptibility (χ), flow, diffusion (D), perfusion and magnetization transfer. Thus unlike other imaging protocols, which usually work using a single property, NMR provides us with a number of parameters for achieving contrast. Each one of these reflects different aspects of anatomy and biochemistry of the tissues under investigation.

Most MR images utilize differential values of PD, T_1 and T_2 of water for different tissues (Table 11.1). Properties of water may also differ in different tissues as a result of flow, bulk susceptibility and diffusion. Use of paramagnetic contrast reagents to obtain T_2*-weighted images is also a valuable technique. Several of these properties may differ in healthy and pathological tissues. This forms the basis for diagnosis of diseases using MRI. To proceed further, it is desirable to have an idea about the magnitudes of these quantities in different tissues and organs. Table 11.1 lists water relaxation times in some tissues, organs and body fluids. By a clever manipulation of pulse sequences, one of these properties can be highlighted (called *weighting*) to obtain a contrast in the images. One can thus obtain what are called *T_1-weighted, T_2-weighted, ρ-weighted* or *diffusion-weighted* MR images. Thus, there are a variety of ways in which a particular problems relating to anatomy, physiology, pathology or biochemistry of tissues and organs can be handled. In this section, we shall discuss PD, T_1- and T_2-weighted images.

Table 11.1: Typical values of PD (in %), T_1 and T_2 (in msec) in various tissues.

Tissue	**PD**	**T_1**	**T_2**	**T_1/T_2**	**T_1 and B_0**
Fat	90	180	90	2.00	1.50
Grey matter	69	520	90	5.78	2.03
White matter	61	390	90	4.33	2.23
Oedema	86	900	77	11.68	
CSF	100	2000	300	6.67	
Blood	90	800	180	4.44	
Spleen	92	480	80	6.00	2.15
Liver	91	270	50	5.40	2.39
Water	100	2500	2500	1.00	1.00
Cardiac muscle	100	600	34	17.64	2.30
Skeletal muscle	100	600	40	15.0	2.63

The reported values of the above parameters vary significantly from one reference to another. These should be taken as typical numbers. In the last column, the field dependence of T_1 is indicated as fractional changes when the magnetic field is increased from 0.5 to 1.5 T. The T_1 and T_2 values are at 1.0 T.

5.1 Proton Density (PD)

It is useful to re-emphasise that only mobile protons contribute to signal in MR images. Large macromolecules, such as proteins and nucleic acids and water strongly bound to such molecules do not give detectable signals. Likewise, protons in solid like materials such as bone-structures may contain ^1H, but these are immobilized and do not provide detectable signals. There are also situations such as in the lungs, where due to profound magnetic field in-homogeneities, T_2^* is short and ^1H signals are too broad to be detected. Therefore, the contribution to the total signal strongly depends on the mobility of ^1H. For example, PD may vary only by a few percent in soft tissues. However, the contribution to the measured ^1H MR signal tends to vary from a few to 30%.

The largest contributions to proton density come from water present in body fluids such as cerebrospinal fluid (CSF), and urine. Soft tissues such as kidney and spleen, liver, grey and white matter in brain, and articular cartilage, also contribute to PD-weighted images. In a number of pathological conditions, there is an increase in PD. Oedema formation is one of the most common conditions, which leads to an increase in PD. Other situations, which lead to higher ^1H densities, are infection, inflammation, demyelination, tumours, cysts and haemorrhage. Decrease in PD may be seen in scar formation, fibrosis, certain types of tumours, membrane formation, and calcination. Several types of bone pathology (i.e. fracture and tumour) are reflected in MRI due to an increase in signal arising from haemorrhage or oedema.

Though calcification is poorly visible in SE images, use of GRE pulse sequences can help better enhance detection arising from susceptibility effects.

5.2 Relaxation Times T_1 and T_2

As compared to ^1H density, both T_1 and T_2 in a tissue show wider variations (Table 11.1). Recall that the ^1H relaxation times for water can be influenced by the viscosity of the medium, its state (i.e. whether it is free or bound), presence of other molecules (particularly paramagnetic molecules) and also the magnetic field employed. In Chapter 1, the basic principles of relaxation times were discussed. The magnetic field component arising from the intra-molecular ^1H-^1H dipole-dipole interaction in water molecule is relatively high (~0.5 mT). This changes in amplitude and direction as the water molecule moves and rotates. The correlation time (τ_c) of pure water is $\sim 10^{-11}$seconds. However, in cells and tissues, the motion is modulated by several factors. Though for pure water T_1 and T_2 are equal (about 2500 msec), the values of both T_1 and T_2 in tissues are significantly lower.

A typical voxel used in water images is 1 mm^3. In this volume there are a very large number of cells and each cell has compartments. Typically, the composition of cells is around 80% water, 5-15% proteins and 2-6% lipids. The remaining mass consists of small molecules and ions. There is a constant interaction between these molecules and water, which changes relaxation times of water.

The tissues are chemically heterogeneous. Therefore, water experiences a variety of environments and interacts with a number of molecules. Cell and tissue structures are divided into multiple compartments. This restrict flow and diffusional motion of body fluids. Water is present in hydration layers of proteins and other molecules. There is continuous exchange between bound and free water in cells. The protein to water ratio can change. However, even a 1% change in the relative composition of the two may lead to 10% change in relaxation times of water. Besides at the macro level, the relaxation rates of water also depend on the molecular weights of the proteins. High molecular weight proteins are more efficient in promoting relaxation. Water protons exchange with side-chain groups in proteins such as hydroxyl and amide protons, which act as efficient conduits for relaxation. The effect of chemical exchange is particularly important for T_2 relaxation. The contributions due to chemical exchange are pH dependent. For example, the observed increase in relaxation time in exercising muscle is ascribed to a change in pH and osmolarity within muscle cells, which increase contribution to relaxation due to chemical exchange. The tissues also contain ions, small molecules and lipids, which interact with water, and therefore affect relaxation time.

Certain tissues have unique properties, which determine their relaxation behaviour. For example, due to the presence of paramagnetic melanin, melanotic melanoma lesions have shorter T_1 and T_2^*. In liver, the presence of high surface area offered by endoplasmic reticulum for binding water and the presence of paramagnetic substances lead to shorter T_1. In certain tissues, restricted motion of water leads to shorter T_2 and hence weaker signals. Most VOI contain heterogeneous relaxation environment and thus several relaxation times may exist.

Water relaxation rates in cells and tissues are average of the factors discussed above. T_2 is almost invariably much smaller than T_1. Typical values of T_1/T_2 are in the range 2-18 (Table 11.1). In fact, because of the restricted motion of water and its association with biological structures and macromolecules, the relaxation rates follow the behaviour of slowly tumbling molecules. Even for body fluids, which have significantly faster motions, values of T_1 and T_2 are significantly lower than those for pure water and depend on the environment. For example, T_1 of cerebrospinal fluid (CSF) is several times higher than for white matter.

While T_2 values are somewhat flat, the values of T_1 tend to increase significantly with the magnetic field (Table 11.2). This increase depends on the type of tissue. For several tissues, the increase may be more than 100% by a ten fold increase in the field. However, T_1 relaxation of fat does not change as much as that of other tissues. Thus, the choice of B_0 also has an important bearing on the choice of parameters of pulse sequence and on the image contrast. Scan parameters, which produce good contrast at one field, may give poor images at another. Thus, the nature and parameters of pulse sequences should be carefully chosen to obtain a good contrast.

Table 11.2: Typical values of proton T_1 and T_2 (in msec) in various tissues at 1.0 and 1.5 T.

Tissue	T_1(msec)		T_2(msec)	
	1.0 T	1.5 T	1.0 T	1.5 T
Fat	240	260	85	80
Grey matter	810	920	100	100
White matter	680	780	90	90
CSF	2500	3000	1400	1420
Blood	525	600	260	260
Muscle	730	860	45	50

T_1 and T_2 form the primary basis of contrast in imaging. Because of disease or other factors such as age, these relaxation rates may change. This forms the basis for clinical work. The changes can be very large. For example, the neonatal brain contains 92% water, which drops to 83% in two years. There is a 300% decrease in T_1 due to this change.

5.3 Achieving Contrast

Let us now discuss how to achieve weighted images using the SE pulse sequence as an example. In tissues, since we usually have $T_2 < T_1$, the signal intensity for short TE in a SE sequence can be written as a product of three terms (as placed in square brackets):

$$S = [C\,\rho M_z(0)]\,[1 - \exp(-TR/T_1)]\,[\exp(-TE/T_2)] \qquad 11.9$$

In a SE experiment, there are two experimental parameters, TR and TE. These can be manipulated to highlight one of the three terms in the right hand side of above equation. The first term has a constant C, which depends on the instrument configuration, the initial longitudinal magnetization [$M_z(0)$] and the proton density. At short TE and long TR the last two terms reduces to unity, only proton density term survives. The images under these conditions are PD-weighted.

The second term which contains T_1 term, indicates the return of longitudinal magnetisation to equilibrium after excitation in the previous steps in multi-pulse experiments. It shows that the magnetization attains a steady state value, which is lower than $M_z(0)$, unless TR is much higher than T_1. Use of short TR and short TE, will highlight this term.

Table 11.3: Typical values of TE and TR for obtaining T_1-,, T_2- and PD-weighted images. The actual numbers are chosen from estimated values of relaxation times.

Weighting	TE	TR
T_1	$<<T_2$	$<T_1$
T_2	$\sim T_2$	$>T_1$
PD	$<<T_2$	$>T_1$

The third term is the T_2 term. It represents decay of the transverse magnetization and thus the decay of the signal intensity with the echo time, TE. The use of long TR and TE will highlight the T_2 effect. An interesting point that emerges from this equation is that the effect of increase in the two relaxation times on signal intensity act in opposite direction. For a given TE, an increase in T_1 leads to lower signal for the same TR (saturation effect), while that for T_2 leads to higher intensity.

Several diseases are associated with increase in water content in the tissue, and a concomitant increase in T_1 and T_2. If pulse parameters are chosen such that both T_1 and T_2 make equal contributions to the image contrast, then the two effects tend to cancel each other and it may be difficult to distinguish between diseased and healthy tissues. From the equation, it can be seen that shorter TR and shorter TE, places greater weight to T_1. The T_2-weighted images can be obtained by using higher TE and longer TR. For proton density, use of a higher TR (no saturation) and shorter TE (good signal intensity) is required (Table 11.3). Considering the typical values of relaxation parameters for different tissues, one arrives at the ball mark for obtaining T_1-, T_2- and PD-weighted images as given in Table 11.3.

A choice of TR and TE thus enables one to obtain images, which are weighted to enhance the relative contributions of PD, T_1 or T_2. Generally, diseased tissues appear less bright than normal tissues in T_1 -weighted images, while in T_2 -weighted images such tissues appear brighter than normal tissues. Tumours and other lesions have slightly higher proton densities. However, these differences also influence T_1- and T_2-weighted images. Thus, the best way to obtain a good PD -weighted image is to minimize relaxation effects by choosing long TR and short TE.

It may be observed that due to the first term in Equation 11.9, when large value of TR is chosen, the effect of TR on T_1 -weighted image is less, since all signals are fully relaxed. Thus, lower TR should be used for obtaining T_1 contrast, such that saturation effects are visible. However, very short TR leads to low S/N. Maximum contrast is obtained when TR is somewhat less than the T_1 of the tissue under study. For most tissues, a value of TR in the range 300 to 800 msec is useful. TE should be minimized to reduce the effect of T_2 weighting, which arises from the second term. Since TE cannot be put as zero, some degree of T_2 weighting in T_1 images is inevitable. The choice of TR >> T_1 eliminates the T_1 effects and the SE signal is completely dominated by ρ and the T_2 term in Equation 11.9. However, the use of large values of TR increases scan times to prohibitively large values. A compromise is therefore reached by allowing some mixing of the T_1 effect. A higher mixing is however self-defeating as T_1 and T_2 effects act in opposite directions in clinical situations. The choice of TR is also dependent on the field strength since T_1 increases with the B_0.

Table 11.4: Typical values for pulse parameters to achieve contrast in MRI. Weighting of spin-echo images as a function of repetition times (TR) and echo times (TE).

	Repetition time (TR)		
	Short (<300 msec)	Intermediate (300-1000 msec)	Long (>2000 msec)
Short TE (<40 msec)	T1 weighted	SD and T1 weighted	SD weighted
TE (>80 msec)	T1 and T2 weighted	SD, T1 and T2 weighted	T2 weighted

5.4 Lipid Signals

The ^{1}H spectra of tissues from several organs contain strong but broad signals from lipids, in addition to the signal from water. The lipid signals are due to superimposition of methylene protons and are approximately 3.5 ppm upfield from the water signal. Because of the difference in the resonance frequencies, lipid and water signals alternate in and out of phase after the application of a $\pi/2$ excitation pulse. The two signals will be out of phase after a time (TE_0) which is given by:

$$TE_0 = 1/(2\nu_0\delta) \qquad 11.10$$

Obviously, TE_0 depends on the magnetic field since $\nu_0\delta$ is proportional to B_0. For the two resonances, which are 3.5 ppm apart, the signals will be out of phase after 3.7 msec at 1.0 T, 2.2 msec at 1.5 T and 1.1 msec at 3 T. If the two molecules are in comparable concentrations in a voxel, then the cancellation effects can be used for imaging. Such a situation occurs at interfaces between subcutaneous fat and produces a dark marking between tissue planes.

5.5 Susceptibility Effects: Diamagnetic Interfaces

The net magnetization induced by the magnetic susceptibility (χ) of the medium is given by χB_0. The net effect of diamagnetic or paramagnetic materials on the internal magnetic field of the tissue depends on the differences in magnetic susceptibility as well as on the geometry of the region. Changes in the magnetic field arising from differences in magnetic susceptibility can be used for obtaining contrast in imaging. Often, susceptibility effects can also produce artefacts in images.

The contrast due to diamagnetic substances is pronounced at interfaces, particularly at the air-tissue or the bone-tissue interfaces. The shape of the body itself (e.g. water surrounded by air) can produce non-homogeneous fields. Certain pathological conditions such as haemorrhage and calcification can be detected using susceptibility effects. It may be pointed out that the susceptibility effects change T_2^* and therefore the gradient echo (GRE) sequences are more useful for such images rather than CSE. The susceptibility effects are larger at higher magnetic fields.

5.6 Paramagnetic and Ferromagnetic Materials: Contrast Enhancement (CE)

Both T_1 and T_2 change significantly if paramagnetic or ferromagnetic materials are present in the organ. These may be endogenous compounds such as deoxy-haemoglobin or breakdown products of haemoglobin. For example, the values of T_1 and T_2 for liver are lower than expected from its wet nature and may be associated with the presence of compounds of iron. Effects of paramagnetic species are also seen in haematomas, where deoxy-haemoglobin, methaemoglobin, haemosiderin and free iron decrease T_1 and T_2. In fact, deoxy-haemoglobin is a very important endogenous contrast reagent and has opened the whole field of functional MRI.

One can also introduce paramagnetic substances exogenously so as to act as contrast enhancement (CE) reagents. It is useful to improve contrast using exogenous compounds in a variety of clinical situations. The most commonly used CE reagents are ligands of paramagnetic metal ions. From clinical point of view, the agents may be grouped in terms of their biological distribution in the body, as being intravascular, extracellular or organ specific. These properties in turn, depend on the nature of the chelating agent (i.e. whether it is hydrophilic or hydrophobic, charged or neutral etc.). The mechanism of CE is therefore governed by several factors: the type of contrast agent (paramagnetic, super-paramagnetic or ferromagnetic), its absorption behaviour and equilibrium concentration in the tissue, compartmentalization in various patho-physiological spaces and the imaging protocol used.

Paramagnetic molecules such as chelates of Gd^{3+} and Mn^{2+} have been used extensively as CE agents. The paramagnetic ions are chelated to carefully selected ligands so as to reduce their toxicity and improve excretion. One of the commonly used reagent is (Gd-DTPA), a chelate of Gd^{3+} with diethyl-amine-triamino-pentaacetic acid. The basic pharmacology of some of these reagents is well understood. After intravenous injection, these agents circulate within vascular system and do not cross the normal blood-brain barrier in healthy subjects. However, they may do so in case of clinical abnormalities. The chelates are excreted through the kidney in the urine.

The effect of paramagnetic reagents on relaxation rates is well understood and has been discussed in Chapter 3. The unpaired electrons behave like tiny magnet and the fluctuating magnetic fields caused by tumbling motions causes exchange of energy between the nuclear spins and the free electron in the lattice. Gd chelates produce a decrease in both T_1 and T_2, but the relative changes are different. The overall CE effects on the images are complex and depend on dose, time of administration, tissue type and tissue pathology, besides the pulse sequence used. The effects are also dependent on the magnetic field. The CE by paramagnetic compounds has proved valuable in several clinical applications. These include small acoustic neuroma, tumour recurrence, distinguishing tumours from oedema and in meningiomas. In malignant tumours, the results parallel with those seen by using iodinated contrast agents in CT. A recent development of CE is in its use in various forms of chemical shift imaging, designed to suppress signal from lipids.

The other class of compounds used are super-paramagnetic and ferromagnetic iron oxide particles (MIOPs). Such reagents alter T_2^* by inducing local field gradients in the cells. MIOPs are metabolised into the iron pools of the body and are excreted slowly. The particles cause local magnetic field inhomogeneity resulting in shorter T_2. Several formulations such as super paramagnetic iron oxide (SPIO) and USPIO have been used as T_2 contrast reagents for enhancing liver, spleen and bone marrow lesions, arising from their selective absorption. The effects are pronounced. However, careful examination of the adverse clinical side effects of CE agents is essential, before these are introduced for clinical applications. At the same time, dynamic Contrast Enhancement MRI (CEMRI) has provided a new protocol for the diagnosis of several diseases in humans. Development of new contrast reagents is an active field of research in MRI.

5.7 Other Imaging Parameters

In addition to NMR properties discussed above, MRI based on a number of other parameters is in the process of development. For example, $T_{1\rho}$ (T_1 in the rotating frame) provides information similar to that obtained with T_1 and T_2 but is more useful for obtaining information on large molecules. Tissue orientation can also have substantial effect on signal intensity. These occur typically when tissues have an oriented fibre structure and a highly ordered structure such as in tendon. When such a structure is oriented at the magic angle relative to the field axis, the signal shows a dramatic increase. This causes bright areas in portions of curved tendon and ligaments, such as in rotator cuff and knee.

6. IMAGE CONTRAST BASED ON FLOW, DIFFUSION AND PERFUSION

In recent years, it has become practical to use properties associated with the movement of body fluids for contrast in MRI. The movement of fluids produces important changes in NMR signal intensity. Situations of interest include blood, CSF, fluid collections such as urine in bladder etc. A typical example is investigation of flow in blood vessels. With advances in MRI technologies, such techniques have proved valuable for probing cardiovascular anatomy and diseases. The general principles of these methods are based on:

(i) Behaviour of the fluid remaining in the slice between excitation pulses.
(ii) Fluid that is washed out.
(iii) Fluid that is moved in.
(iv) Phase-changes introduced by magnetic field gradients.

6.1 Magnetic Resonance Angiography (MRA)

An understanding of the influence of flow of blood on MR images has led to development of magnetic resonance angiography (MRA). The vascular anatomy is generated by the protons in the flowing blood, which produce a large signal against the background of weak signals from stationary tissues. In practice, one uses the so called *bright blood* technique, wherein the signal from the flowing protons is enhanced relative to that due to stationary protons. As a result, both arteries and veins show up as hyper-intense signals (Figure 11.13).

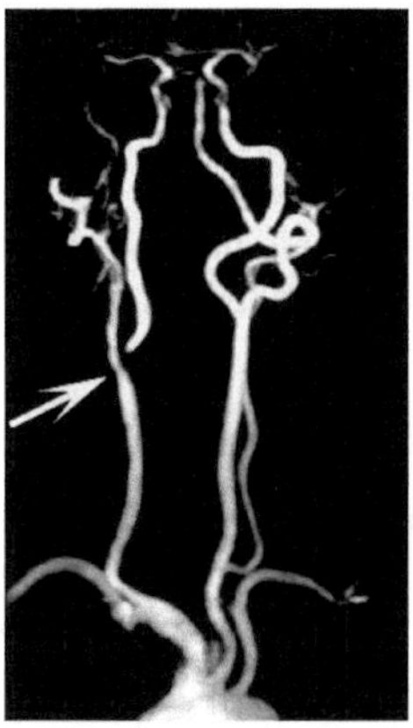

Figure 11.13: Contrast-enhanced 3D magnetic resonance angiography showing stenosis at the origin of right internal carotid artery. (Thanks to N.R. Jagannathan for providing this figure).

In the time of flight (TOF) technique, a series of thin slices are obtained using GRE technique. A TR value much shorter than the tissue T_1 and moderate flip angles are used. During image acquisition, protons in stationary tissues are partially saturated by the RF pulses. However, ^1H in the flowing blood are continuously entering and exiting the volume receiving RF pulses. These are therefore fully magnetized. Thus, the signal from flowing blood is stronger. The signals are brightest at the entry point of the slice, as it has not been excited by the RF pulse. The degree of saturation depends on TR, α, and flow velocity. TOF may be run in 2D-sequential or 3D-volume acquisition mode. 3D TOF provides best in-plane resolution, but suffers from saturated out signal due to the large number of excitation pulses. Magnetization transfer techniques can be incorporated to enhance detection of small vessels.

Phase-contrast (PC) angiography depends on alteration of the phase of precessing protons as they move along a magnetic field gradient, relative to spins, which are stationary. The amount depends on the direction and velocity of flow relative to the applied gradient and its parameters. A phase shift is induced in the moving protons through flow encoding gradient pulses between the excitation and the read out pulse. A second signal acquisition is made with reversed polarity of the encoding pulses. On subtracting the two acquisitions, stationary protons give zero signals, while moving protons give strong signals. It is possible to obtain quantitative information on the flow volume. Owing to their directional sensitization, PC methods enable separation of flow in individual directions. The technique also provides a better background suppression compared to TOF method. However, PC requires longer scan time, since four separate acquisitions are required to measure the three flow components. A prior knowledge of the maximum flow velocity is necessary to ensure that a proper sensitizing gradient pulse is used.

MRA has proved valuable for evaluating vascular diseases. New approaches for measuring blood flow have expanded the role of MRI in the diagnosis of cardiac abnormalities and in studies of renal blood flow. MRA enables determination of flow direction and velocity. SE pulse sequence enables visualisation of blood flow within the vascular system and its disruption due to pathology such as blood clot. Several blood-flow related pathologies such as aneurysms, arteriovenous malformation and stenoses can be detected using MRA.

6.2 Diffusion-weighted Imaging (DWI)

Diffusion of water in tissues and cells is a result of Brownian motion. Diffusion is a random motion and is dictated by the thermal energy. Free water and that in body fluids diffuse faster than the water molecules bound to large molecules or when it is present in cells and tissues. The extent of such a translational motion is also modulated by the underlying structure, which can limit the diffusion rate and the direction. For example, diffusion of water in white matter in the brain is highly anisotropic as compared to grey matter or CSF. This directionality arises from myelin fibres in white matter, which prevents motions in directions perpendicular to the fibre axis.

The mean translational distance travelled by a water molecule is described by the Einstein equation for diffusion:

$$\langle L \rangle = \sqrt{2Dt} \qquad 11.11$$

Here, $\langle L \rangle$ is the average distance travelled by a molecule in time t and D is the diffusion coefficient (which is expressed in cm^2/sec). D is a tensor quantity having diagonal components D_{xx}, D_{yy}, D_{zz} (which denote diffusion in three orthogonal planes). The off-diagonal elements such as D_{xy} denote correlation of motion in various directions. D depends on the nature of environment of the imaged molecule such as water. The displacements due to diffusion are small on the time-scale of

MRI experiments. For example, D for grey matter is about 10^{-5} cm^2/sec. In a typical time of 40 msec, the average displacement along one direction is approximately 9 μm.

The random motion of water molecules in tissues due to Brownian motion is distinct from the coherent motion in flow. Therefore, a different strategy is used to obtain contrast based on diffusion. A pair of bipolar field gradient pulses of amplitude G and a time duration δ are added on both sides of the π pulse in the SE sequence (Figure 11.14). The interval between the gradient pulses is denoted by Δ. The purpose of these pulses is to *phase tag* the diffusing spins, which is in a way similar to the use of a radiotracer. During the first pulse, protons experience a gradient that is dependent on their location (a_1). The second pulse produces a phase shift. This depends on the position a_2 at this time (called spin *un-tagging*). Since there is a π pulse between the two gradient pulses, the net phase shift (ϕ_{net}) is given by:

$$\phi_{net} = \gamma G\delta\ (a_2 - a_1) \qquad 11.12$$

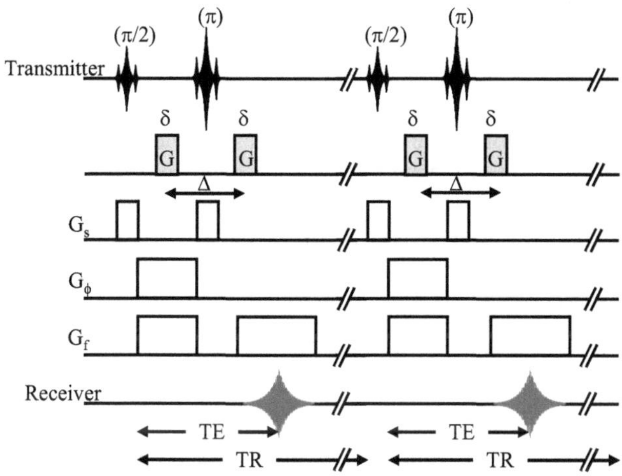

Figure 11.14: Basic SE sequence with the diffusion gradient applied along the slice-select gradient. Δ is the diffusion time and G denote gradient pulse.

For static spins, $a_1 = a_2$. The gradient pair therefore has no effect on phase. For spins undergoing diffusion, there is a net phase change, which depends on the history of spins during the time Δ, between the two gradient pulses.

The macroscopic magnetization arising from the vector sum of individual spins having different phase shifts results in an attenuation of the signal amplitude. This is given by:

$$\ln (S_g/S_0) = -b_g D;\ \text{with}\ b_g = -(\gamma G\delta)^2(\Delta - \delta/3) \qquad 11.13$$

$$\textit{when } \delta \textit{ is small compared to } \Delta, \quad S_g/S_0 = exp\,[-(\gamma G\delta)^2 \Delta D]. \qquad 11.14$$

Here, S_g and S_0 are signal intensities, with and without the application of gradient pulses and b_g is a gradient factor, which depends on G and δ. In most commercial spectrometers, δ may not be negligible as compared to Δ and undesirable background gradients may be present from thc imaging sequence.

The strategy for measuring diffusion coefficients is to obtain images with different gradients G, while keeping other parameters constant. However, the value thus obtained for an *in-vivo* tissue is not the same as the measured diffusion coefficient of water in a test tube. In the former case, there are contributions effects such as bulk flow and tissue vibrations caused by systolic arterial excursion. The *in-vivo* measured values are therefore called apparent diffusion coefficient (ADC). Since D is a tensor, at least three measurements with diffusion gradients applied orthogonally are required to measure the trace of ADC.

Diffusion-weighted MRI provides unique information on the state of molecular translational motion of water. Diffusion can provide information on cellular density, heterogeneity and anisotropy. The diffusion coefficient is slower in intracellular compared to intercellular compartments. DWI is the only non-invasive method for observing diffusion processes. It provides information on microstructure, tumour grade and local tissue architecture, which is an early indicator of abnormality. Generally, increased water diffusion leads to an increase in the signal. However, during the early phases of stroke or ischaemic injury, the diffusion may be retarded, leading to an attenuation of the signal. DWI has been applied to a variety of clinical studies such as intracranial lesions, including acute cerebral infarcts, tumour and demyelinating diseases. The technique has proved to be more valuable than T_1-weighted images in the detection of myelin. For detection areas with acute ischemia and infarction it has proved to be more sensitive than T_2-weighted images. DWI is one of the most sensitive indicators of early ischaemic damage in the brain.

6.3 Perfusion

Perfusion is a microcirculation process, which measures delivery of nutrients to the body tissues. It plays a critical role in vertebrate physiology. Perfusion imaging is a useful technique in nuclear medicine and its potential in clinical practice is already established. MRI has been used to exploit motional sensitivity of images using flowing blood as an endogenous marker. Several MRI techniques have been developed for measuring perfusion. The more useful ones use contrast agents and measure changes in T_2^*, due to passage of the agent through the tissue microvasculature. Diseased state can lead to changes in microvasculature of tissues. For example, tumour vessels have larger diameter and pores and lack contractile properties. Perfusion imaging is directly related to tumour microvascular perfusion. It provides a direct index on tissue blood volume. MR techniques offer original insights into the problem of perfusion, which are not addressed by conventional methods.

6.4 Magnetization Transfer (MT)

As pointed out earlier, only mobile protons are observed in a ^1H density measurement. In several types of tissues, there can be substantial amount of water, which is not observed directly, simply because T_2 is small (typically < 0.1 msec) and the lines are broad. However when excited, such molecules can transfer their magnetization to mobile water, thus decreasing the signal intensity. This effect is called magnetization transfer.

Tissues having only mobile water (for example CSF, blood and urine) are unlikely to show magnetization transfer. Signals from tissues having short T_1 (fat, Gd enhanced images) are also not affected. However, there are regions, such as heart and skeletal muscle, skin, and articular cartilage, where the effect of magnetization transfer can be quite large. To avoid direct saturation of the tissue signal, magnetization transfer pulses are applied non-selectively, at an offset of typically 1.5 kHz from the ^1H resonance frequency. Therefore magnetization transfer pulses use high RF power and are applied repeatedly. This leads to a high RF power deposition on the tissues. Magnetization transfer based contrast has found applications for intracranial MRA. This improves blood-brain contrast enabling smaller vessels to be seen. It is also helpful in characterising degree of demyelination in plaques.

7. FUNCTIONAL MRI (fMRI)

The main function of haemoglobin is to carry oxygen to tissues and organs. Haemoglobin can exists in four different states, depending on the spin state and oxidation states of Fe. The two forms that are important in biological functions are the oxy- and deoxy- forms of Fe^{2+} containing heme group. The oxygenated form of haemoglobin is diamagnetic while deoxy-haemoglobin is paramagnetic (Chapter 4). Blood therefore acts as an endogenous contrast reagent. The consequences of change in magnetic properties of blood led to the development of a new and emerging field in NMR called functional MRI (fMRI).

7.1 Principle of fMRI

In this technique, one measures the blood oxygenation level dependent (BOLD) changes in the water signal in brain, which arise with changes in neuronal activity, following a change in brain state. Examples are stimulus or while performing a particular task. MRI studies of brain function rely on secondary signals, which link changes in neuronal activity with alterations in glucose metabolism (glucose and oxygen consumption), perfusion (blood flow and blood volume) and blood oxygenation. Functional MRI (fMRI), has emerged as a powerful tool for studying brain function by detecting small changes in signal amplitude associated with neuronal activity. It uses techniques similar to conventional MRI. In fact, the same equipment is used both for conventional MRI and fMRI.

It is known that to meet the larger demands for oxygen and other substrates, the neuronal activation in a region of cortex results in an increase of blood flow to the local vasculature. The increase in blood flow exceeds the amount needed value. The increase in the corresponding level of oxygen extraction is lower. This leads to venous hyper-oxygenation and results in a reduction in the paramagnetic deoxy-haemoglobin in the excited area. There is consequently an increase of the T_2^*-weighted MR signal amplitude. Thus, deoxy-haemoglobin acts as an endogenous contrast-enhancing agent. The value of T_2^* may vary by around 30 msec as a result of oxygen effect. The changes in the signal amplitudes are around 1%. This however, varies depending on the strength of the magnetic field. The BOLD effect is more pronounced at higher fields.

The magnitude of the BOLD signal depends on several physiological parameters such as changes in the blood flow, volume within the tissue, and change in local oxygen tension. Therefore, unlike in the case with positron emission tomography (PET), fMRI does not report absolute changes such as units of flow. Further, as neurons become more active, there is a time delay of several seconds between a stimulating event and the corresponding BOLD response. The images may differ even when there is no change in brain activity due to noise. Therefore, fMRI studies are used to compare sets of images during different conditions.

The ability to directly map the brain structure and organization associated with functional activities has opened a new frontier in neurological studies. One of the advantages of fMRI is that it is non-invasive. It offers repeated studies of subjects, provides sub-mm resolution, yields temporal resolution of about 1 second and allows flexibility in design of cognitive paradigms.

7.2 Methodology

Pulse sequences which are best-suited for fMRI studies are high speed EPI and FLASH. Both are sensitive to changes in T_2 and T_2^*, which in turn are coupled to concentrations of deoxy-haemoglobin. Advances in gradient coil technology have made it possible to obtain cross sectional images in 50-100 msec and permit multi-slice recording of the entire brain in a few seconds. Since BOLD contrast depends on changes in magnetic susceptibility, it is enhanced by acquiring signals at prolonged echo times and higher magnetic fields. Use of higher magnetic fields gives better contrast in fMRI. Scanners using magnetic fields higher than 2 T are usually preferred.

For obtaining fMRI images, the subject is placed in an MRI machine where different kinds of stimuli can be used or the subject is asked to do specific task. MRI changes are then recorded. BOLD signal intensity increases have been documented for several functional activities such as motor tasks, visual, auditory, speech, language related activities and sensorimotor cortical processing. In view of the lower level of changes in signals coupled with the noise arising from the coils, results are usually interpreted by comparing images acquired under different conditions. Multiple recordings help in signal averaging and provide more reliable data. For example, in the block design of experiments, the stimuli are presented as alternating short runs (blocks) of several seconds duration. For visual-stimulus, the subject may be presented with a brightly lighted board followed by a dark screen. Such blocks may be repeated several times and images may be recorded for several slices. The voxels affected by the stimulus show alteration in the BOLD effect. In this way the visual cortex can be mapped. Voxels not affected by stimulations provide a steady set of MRI image. Some variance may arise due to noise not related to the task, for example, due to respiration, cardiac pulsations, patient motion, respiration and coil and gradient related noise. Statistical analysis of the data helps in checking the reliability of the data.

Figure 11.15 A shows the BOLD images in several situations. It illustrates how the MRI technique can help in identifying activation centres in the brain and in pathological situations.

7.3 Applications of fMRI

The possibility of observing brain function and identification of processing locations for different types of body functions has opened up a new frontier for understanding brain organization and for neurological studies. Functional MRI has found applications in basic neuroscience as well as in clinical situations.

Brain functions are coordinated interactions of components of brain systems that are spatially distinct but are connected in functional network. The location of such areas has been explored. Areas in the brain associated with motor tasks, such as finger tapping, movement of toes, elbows and tongue, visual processing, word generation, speech perception, language learning, spatial memory and several other functions have been documented. Studies have been developed for identification of different traits and pathologies. These are built on tasks used routinely in neurophysiology. One example is the application of the so called *Stroop test* to fMRI. This test is among the psychological test that may help discriminate subjects with depression and disorder.

Several biomedical centres have developed protocols for fMRI mapping of motor, auditory, visual and language areas for evaluation of the brains of patients (Figure 11.15). The functional images can be superimposed on anatomical images, which have better contrast, to help in the positional specification and foci within the brain. The data can be integrated with image guided neuro-surgical procedures. For example, fMRI helps in identification of epileptic zone and assessment of memory functions. Localization of cerebral origin of seizures in patients is possible and functional images can be used to facilitate surgery. fMRI is finding increasing applications in studies of patients recovering from stroke and in understanding losses in cortical functions in degenerative diseases such as Alzheimer. Another important area is examination of subjects diagnosed with schizophrenia. It has been shown that during a verbal working memory task, normal subjects show good activation in the prefrontal cortex, while the patients show lower and short-lived activation. fMRI has been found useful in a number of other psychiatric disorders such as anxiety, depression and attention deficit.

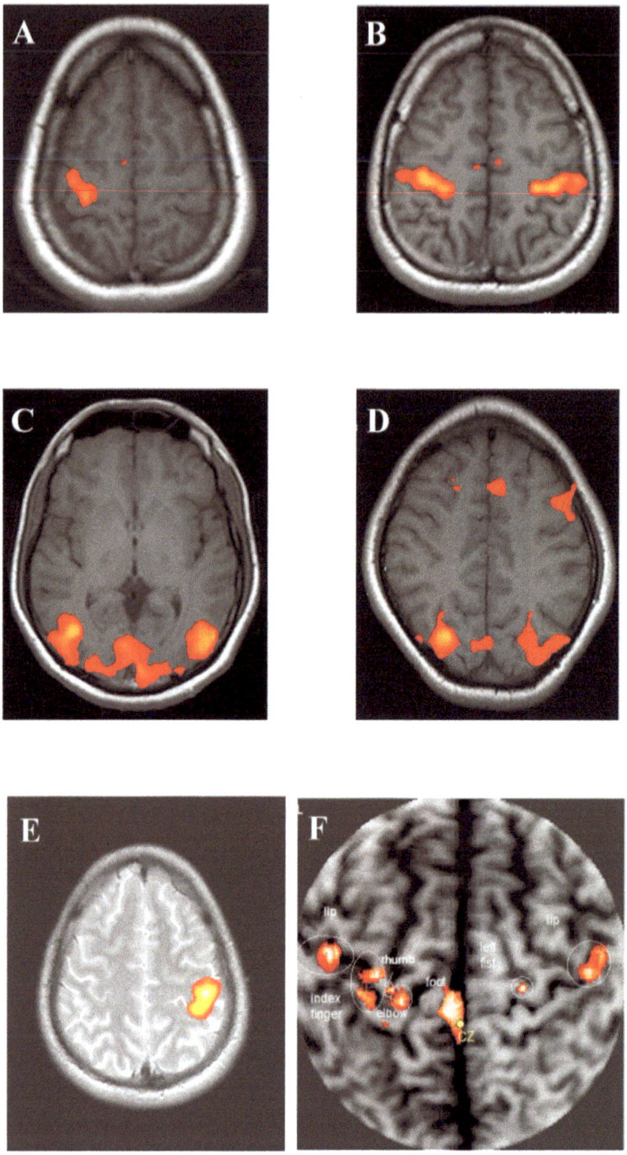

Figure 11.15: Functional MR image of motor cortex function of (A) Left hand finger tapping, (B) both hands hand finger tapping, (C) visual, (D) working memory, and (E) right hand finger tapping (superimposed on T_1-weighted MR image). (F) BOLD images of a healthy volunteer showing different activation centres in human brain. (Thanks to N.R. Jagannathan for providing the panels A-E and to W. Grodd for the panel F of this figure).

The effects of therapy and learning have also been successfully achieved. A very important area is the study of neuro-development and learning disorders in children. In particular, when applied to children, the safety and non-invasive nature of the technique are important factors. fMRI provides an accurate and painless method for mapping critical brain functions and may play an important role in clinical applications in future.

7.4 Limitations and Future Outlook

fMRI is one of the most important emerging areas of NMR. The images provide changes in the state of brain, effects of learning and neuronal changes. The effect of various medical treatments can be monitored. It is likely to play a major role in management of patients.

The current limitations of fMRI over conventional MR images are mainly due to weak signals. The use of fast imaging techniques reduces the spatial resolution but increase signal amplitude by an order of magnitude. fMRI requires multiple sampling of brain responses. The temporal resolution is limited by the nature of hemodynamic response. The reliability is reduced when there are subject motions or physiological variations. The effect of variations in blood level of substances, foods and hormones are yet to be understood. At this stage we do not understand several aspects of this phenomenon at a quantitative level. Some of these problems may be solved with the development in MRI technology.

An important area for future development is the application of data analysis techniques and modelling to interpret fMRI data. Some of the areas where fMRI may play an important role are the behaviour patterns, physiological measures, co-variations among images obtained for various tasks and complex task patterns where several areas of brain may be involved collectively. The techniquc may have an important role in evaluating benefits of learning strategies and pharmacological intervention and social sciences in general.

8. MRI AS A CLINICAL TOOL

MRI has become a useful tool in medical diagnostics. One of the earliest applications was in diagnosing abnormalities of the *c*entral *n*ervous *s*ystem (CNS). Thereafter, MRI has been applied to clinical studies of all human organs. A detailed discussion on all documented uses of MRI in medicine is beyond the scope of this book. However, salient applications are highlighted and discussed in this section. For further details, the readers are referred to more advanced clinical reviews and book on the subject given in the list of references.

8.1 Central Nervous System (CNS)

Human brain is one of the most extensively studied organs by MRI. This is both due to its complexity and due to its vital role in health. Clinical situations arising from haemorrhage and infarction stroke, tumour, trauma, haemorrhage, inflammation, congenital conditions, degenerative diseases etc. can be routinely studied using CSE or GRE techniques. For vascular details, MRA and DWMRI have been used. Functional aspects are studied using fMRI. MRI is more sensitive in most abnormalities compared to CT. However, CT is a better protocol for characterization of calcification, fractures and abnormalities in joints.

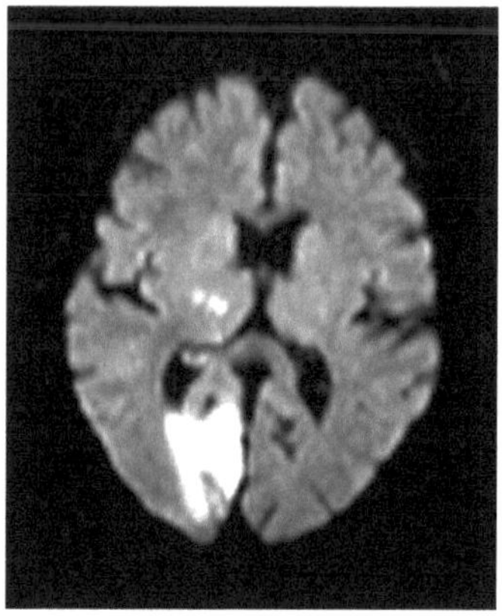

Figure 11.16: Diffusion-weighted MR image of patient with stroke in the right occipital region of the brain. (Thanks to N.R. Jagannathan for providing this figure).

8.1.1 Ischaemic Stroke

Stroke is a common life-threatening neurological disorder, which occurs as a result of loss of blood supply to brain. It is the single largest cause of disability among the survivors. Age, diabetes and hypertension are important risk factors for stroke. Other factors include smoking, high blood lipids and obesity. Treatment and patient management for stroke-afflicted patients, still pose a challenge. MRI has helped in detecting early changes and in understanding the pathophysiology behind stroke. While T_1- and T_2-weighted images have limited success, DWI is widely used in stroke research and in directing the treatment of acute ischaemic stroke (Figure 11.16). Regions of abnormal water movement can be detected in the early stages after the onset of ischaemia. Regions of reduced water movement appear hyper-intense. DWI has also been used for evaluation of therapeutic efficacy of neuro-protective agents and for evaluation of new drugs.

8.1.2 Haemorrhage Stroke

The sensitivity of MRI to biochemistry of blood makes it possible to explore different stages of haemorrhage. Images are affected by a number of processes related to haemorrhage such as properties of various chemical forms of haemoglobin and its breakdown products as well as those of the red blood cells. Fresh arterial blood filling haemorrhage has long T_1 and T_2. This shows up as bright in T_2- and dark in T_1-weighted images. With time, the blood in haemorrhage is deoxygenated leading to shorter T_2 and T_2^*. This makes the afflicted areas appear dark on T_2 and T_2^* (particularly in GRE) images. As the cells lyse, this effect disappears. Haemoglobin gets converted to methemoglobin, which allows much better access to water. As a result, T_1 is decreased giving brighter T_1 weighted images. Finally, iron is broken down and this leads to dark T_2 images. Thus, it is possible to explore the natural history of haemorrhage. The susceptibility effects are smaller in TSE images. Conventional SE is a better alternative to detect hemosiderin containing lesions. In comparison with the CT scan, the effect of haemorrhage is related only to the concentration of iron.

8.1.3 Neurological Diseases

Epilepsy is a common neurological disorder. It is characterized by recurrent epileptic seizures. MRI has increased the ability to detect causes of seizure disorders, plan medical or surgical therapy, and helps in the prognosis of disorders and therapy. Computer image guided surgery allows exact implantation of intracranial electrodes, resection of abnormal structural or functional imaging foci. fMRI has proved particularly valuable evaluating patients for surgery.

Alzheimer's Disease (AD) is a chronic neuro-degenerative disorder. It is characterized by progressive deterioration of memory and higher cortical functions. It ultimately leads to total degradation of intellectual and mental capabilities. MRI and MRS have contributed significantly to increase our knowledge of these degenerative processes. It has helped in detection of earliest changes in AD and to differentiate AD from other forms of dementia.

Parkinson is another disease, which has attracted attention by researchers interested in this disease. Because of the involuntary movements of the patients, there are difficulties in patient management, without using sedatives. Therefore it has not found wide-spread use in clinical diagnostics. However, this is an important area of clinical research.

8.2 Musculoskeletal System

Due to excellent soft tissue contrast, major progress has been made in the application of MRI in articular cartilage, knee, hip, wrist and shoulder. T_2-weighted characteristics are different for fat, muscles, ligamentous and fibrous structures, marrow diseases, nerves and blood vessels. MRI has therefore become a versatile and indispensable modality for musculoskeletal systems (Figure 11.17). The basic requisites for optimised MRI are high spatial contrast and sensitivity for the visualisation of free water. Detection of pathology is often hindered by a strong signal from neighbouring adipose tissue. Fat suppression is therefore desirable. STIR technique is therefore, frequently used. Contrast agents are used for detection of inflammation. Accurate assessment of osteoarthritis can be achieved using CSE, FSE and GRE techniques. Fat suppression improves contrast between cartilage and fatty marrow. Conventional MRI has 50-95% accuracy for detection of lesion.

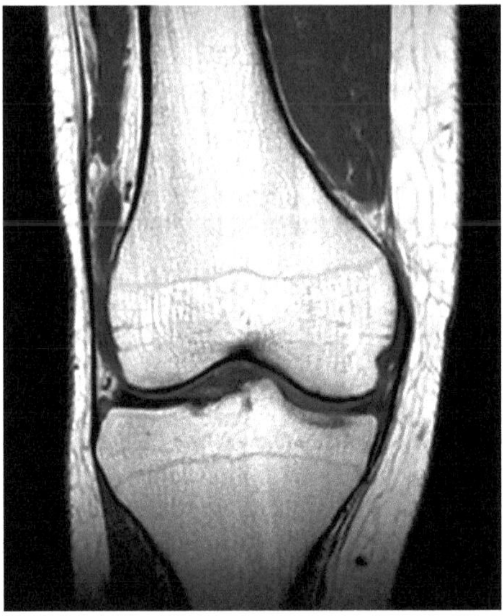

Figure 11.17: T_1-weighted coronal MR image of knee of a volunteer. (Thanks to N.R. Jagannathan for providing this figure).

8.3 Heart and Cardiovascular System

Cardiovascular MRI provides information about anatomical and functional assessment of heart, myocardial viability and coronary artery disease. To avoid artefacts from heart motion, electro-cardiogram (ECG) or pulse triggering is used. Pulse sequences which couple ECG and heart beats have been developed and MRI compatible equipments are available. It is possible to acquire high quality images using breath-hold technique. The advent of GRE has enabled dynamic evaluation of the heart beat. Atherosclerotic plaques, diseased vessel walls and surrounding tissues can be characterised. Presence of thrombus or calcification can be identified. It is possible to evaluate cardiac morphology, function and perfusion in a clinical setting, and to obtain information on mycocardial viability and coronary anatomy. MRI of flow has evolved as a powerful technique to evaluate vascular diseases. The technique enables determination of flow velocity, direction and profiles within the vessel. This has provided new tools to study cardiac abnormalities, renal blood flow, arterio-venous malformation, aneurysms, venous angiograms and abnormal/aberrant vessels.

8.4 Breast

Breast cancer is one of the leading causes of cancer-related deaths in women. An early detection is therefore crucial to reduce mortality and improve clinical outcome. Mammography, which takes advantage of differential attenuation of X-rays by normal and abnormal breast tissues, is usually the first line of imaging modalities. However, even when performed optimally, its sensitivity is between 70-90%.

Initial attempts to use MRI for breast cancer were not very encouraging. However, the use of contrast agents through intravenous injection has proved valuable. Malignant tumours usually show a fast uptake and high intensity in dynamic *c*ontrast *e*nhanced *MRI* (CEMRI). The surrounding tissues show normal image. MR techniques have significant impact in detection of breast cancer in particular in multifocal and multicentric disease that was previously unsuspected (Figure 11.18).

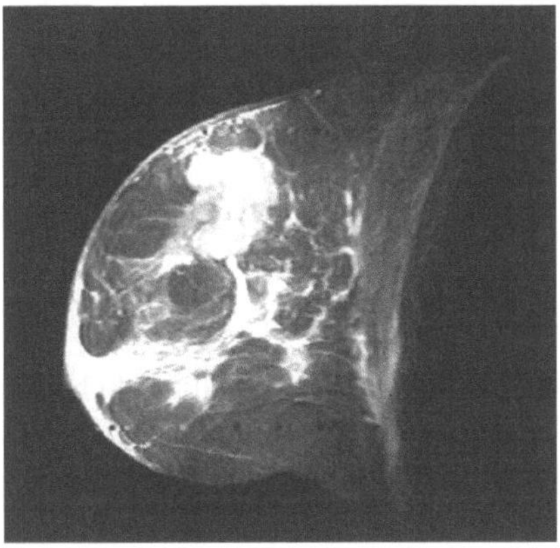

Figure 11.18: STIR image of breast of a patient suffering from infiltrating ductal carcinoma. (Thanks to N.R. Jagannathan for providing this figure).

The successful protocols in breast imaging use surface coils for enhancement of spatial resolution dynamic contrast. Most cancers show a steep rise in the signal typically within first 2 to 3 minutes of the injection of the contrast agent. This is followed by a plateau and then a wash out. On the other hand, most benign lesions either do not show enhancement or show a continued enhancement with time, followed by a slow wash out. These patterns are related to the micro-vascularity changes in malignant tissues. Techniques based on diffusion, perfusion and flow have also shown promise. Breast MRI is also applied in silicone based implant imaging. Application of 3D imaging sequences allows contiguous mm level resolution over the entire VOI.

In recent years, MR techniques have significant impact in detection and staging of breast cancer in particular in multifocal and multicentric disease that was previously unsuspected. Several review articles have been published on the application of MR techniques in breast cancer. Dynamic CEMRI permits differentiation of malignant from benign lesions. The reported sensitivity of CEMRI for breast cancer detection is as high as 94-100%. However, its specificity varies between 37% and 97%. To improve the specificity of MR breast imaging, several strategies that focus on the enhancement kinetics have been developed. A model based method that optimizes the spatial resolution by using only three time points and yet characterizes tumour heterogeneity in terms of microvascular permeability and extra-cellular fraction has been used. By integrating the morphologic and kinetic information obtained with MR imaging, higher specificity may be achieved. MR techniques such as diffusion and perfusion have been applied to breast cancer.

Thus, the breast MRI has reasonable sensitivity and specificity. In some cases, it detects contra lateral breast cancer that is missing in mammography and clinical examination. However, it is felt that the added sensitivity of MRI of contra lateral breast cancer came at a high cost in terms of false positive results and over-diagnosis. Recent American Cancer Society (ACS) guidelines specify that breast MRI screening should only be used with certain group of high-risk women, such as BRCA mutation carriers and those with a family history of breast cancer. The guidelines do not recommend MRI for women at average risk, those with very dense breasts or those who already have had breast cancer, including ductal carcinoma in situ (DCIS). The ACS also says that breast MRI should be used with conventional mammography as part of an annual screening regimen in certain women.

8.5 MRI in Pregnancy

In a pregnant women MRI is the imaging modality of choice as it avoids the risk of radiation exposure to the developing foetus unlike CT. Many countries and hospitals do not recommend imaging during pregnancy. However, in special cases MRI has been rated as a valuable complement in treatment and clinical decisions during pregnancy.

8.6 Genitourinary System

Applications of genitourinary system using T_2-weighted images using FSE techniques are somewhat hampered by contribution from fat. GRE images have helped in detection of pelvic neoplasms and pelvic floor movement. MRI is useful in evaluation of renal masses.

8.7 Gastrointestinal System

Introduction of TSE has revolutionized protocols for gastrointestinal system. Images can be obtained by fast imaging techniques within a breath hold. High T_2 weighting is achieved with long echo trains and fluid filled cavities give hyper-intense appearance for fluid-filled cavities. GRE along with T_1-weighted imaging is a standard protocol for liver, pancreas and abdomen. These advances have opened a new area called magnetic resonance cholangiopancreatography (MRCP).

8.8 Prostate Cancer

Prostate is one of the leading cancers in males. The development of prostate cancer does not present any specific symptoms, which makes detection of prostate cancer difficult. Currently used methods such as digital rectal examination, transrectal ultrasound, prostate specific antigen (PSA) and sextant biopsy for detection of prostate cancer, have limited efficacy. Technique such as MRI, CEMRI, MRS and

DWI with endorectal coil have shown tremendous potential in diagnosis of prostate cancer. MRI provides multiplanar images with excellent depiction of prostatic zonal anatomy. Currently, MRI has low specificity for diagnosis of prostate cancer. It is only useful for advance stage for detection of extra capsular extension, seminal vesicle invasion, nodal and bone marrow metastasis. CEMRI technology has been reported to improve tumour assessment. DWI is yet another technique which has shown promise in differentiation of malignant and normal tissues.

8.9 Interventional Imaging

Due to restricted patient access in conventional MR machines, MRI was not considered as a guidance tool during interventional procedures. Recently, open access magnet have been designed which make it possible to use MR guided surgical procedures. Improved patient access allows intervention in any desired plane. Drainage procedures and percutaneous biopsy are currently the easiest procedures performed under MRI control. Susceptibility artefact provided by non-magnetic needles, provide contrast for visualisation of the instrument. Typically, biopsy needles are displayed as dark areas surrounded by bright rim. For drainage procedures, MR compatible sets are available. Bone and soft tissue lesions can be accessed by MR guided interventions. Deeply embedded soft tissue lesions are more amenable to MRI rather than CT or ultrasound guided procedures. The clinical use and efficacy of such procedures are under development.

8.10 Patient Comfort and Precautions

One of the sources of discomfort to subjects arises from the narrow confines of the magnet bore, isolation of the magnet room and knocking sound of the scanner associated with field gradients. Magnet safe ear plugs and ear phones are provided, if the acoustic noise exceeds 85 dB. Magnetic field compatible, patient monitoring systems (for body temperature, heart and respiratory rates, blood pressure and ECG), ventilation support devices and equipment for anaesthesia are commercially available.

As discussed earlier, the heating and neuronal simulation for fast imaging techniques use strong RF and fast gradient pulses. These problems are more pronounced at higher magnetic fields. A good check on *in-vivo* body temperature is the chemical shift for water resonance. Most countries have set limits for the maximum magnetic fields to be used in clinical MRI. However, these vary from one country to another.

While no adverse effects have been reported for MRI, a common hazard is the attractive force exerted by the NMR magnet on ferromagnetic materials. Due diligence and training is therefore essential for the operation and maintenance of NMR laboratories. While most orthopaedic hardware, hemostatic clips and medical implanted devices may be safe, patients with certain type of ferromagnetic implants (intrauterine devices) and pacemakers cannot undergo MRI.

9. REFERENCES

9.1 Further Reading

R. Damadian, *Tumor detection by nuclear magnetic resonance*, Science **171**, 1151–1153 (1971).

A. Kumar, D. Welti, and R.R. Ernst, *Fourier zeugmatography*, J. Mag. Res. **18,** 69–83 (1975).

P.C. Lauterbur, *Image formation by induced local interactions: Examples using NMR*, Nature **242**, 190–191 (1973).

R. Damadian, M. Goldsmith, and L. Minkoff, *NMR in Cancer, Fonar Image of live human body*, Physiol. Chem. Phys. **8**, 97–108 (1977).

W.S. Hinshaw, P.A. Bottomley, G.N Holland, *Radiography thin section of the human wrist by nuclear magnetic resonance*, Nature **270**, 722–723 (1977).

P. Mansfield and A.A. Maudsley, *Medical imaging by NMR*, Br. J. Radiol. **50**, 188–194 (1977).

P. Mansfield, and A.A. Maudsley, *Planar spin imaging by NMR*, J. Mag. Res. **27**, 101–119 (1977).

R.C. Hawkes, G.N. Holland, W.S. Moore and B.S. Worthington, *NMR tomography of Brain: A preliminary clinical assessment with demonstration of pathology*, J. Comput. Assist. Tomography **4**, 577–586 (1980).

L. Esserman, D. Wolverton, N and Hylton. *Magnetic resonance imaging for primary breast cancer management: Current role and new applications*, Endocr-Relat. Cancer **9**, 141–153 (2002).

S. Sinha and U. Sinha. *Functional magnetic resonance of human breast tumors. Diffusion and perfusion imaging.* Ann. NY Acad. Sci. **980**, 95–115 (2002).

9.2 Books and Reviews

D. Le Bihan, *Diffusion and Perfusion Magnetic Resonance Imaging*, Raven Press, New York (1995).

I.P. Arlert, G.N. Bougertz, and G. Manchal, *Magnetic Resonance Angiography*, Springer, Berlin (1996).

H. Higgins and A. Helms (eds), *Magnetic Resonance Imaging of the Body*, 3rd Edition, Lippincot-Raven, Philadelphia (1997).

E.R. Danielsen and B. Ross, *Magnetic Resonance Spectroscopy: Diagnosis of Neurological Diseases*, Marcel Dekker, New York (1999).

W.R. Nitz, *Magnetic resonance imaging: acronyms and clinical applications*, Eur. Radiol. **9**, 979–997 (1999).

C. Westbrook, *Handbook of MRI Technique*, 2nd Edition, Blackwell Science Ltd (1999).

M.T. Vlaardingerbrock and J.A. Denboer, *Magnetic Resonance Imaging*, Springer (1999).

V. Kuperman, *Magnetic Resonance Imaging: Physical Principles and Applications*, Academic Press (2000).

C.T.W. Moonen and P.A. Bandetti (eds), *Functional Magnetic Resonance Imaging*, Springer-Verlag, Heidelberg (2000).

R.B. Buxton, *Introduction to Functional Magnetic Resonance Imaging*, Cambridge University Press (2001).

N.R. Jagannathan (ed.) *MR Imaging and Spectroscopy in Pharmaceutical and Clinical Research*, Jaypee Brothers (2001).

R.H. Hashemi, W.G. Bradley Jr., and C.J. Lisanti, *MRI: The Basics*, 2nd Edition, Lippincott Williams and Wilkins (2004).

N.R. Jagannathan (ed.) *Biomedical Magnetic Resonance: Proceedings of the International Workshop*, Jaypee Brothers (2005).

N.R. Jagannathan (ed.) *Recent Advances in MR imaging and Spectroscopy*, Jaypee Brothers (2005).

P. Raghunathan, *Magnetic Resonance Imaging and Spectroscopy in Medicine: Concepts and Techniques*, Orient and Longman, Hyderabad (2007).

CHAPTER 12

STUDY OF METABOLISM: CELLS AND TISSUES

1. INTRODUCTION

Recent advances in molecular biology have led to a new appreciation of the complexity of the living state. Techniques in molecular genetics and structural biology have identified genes and proteins involved in several biochemical reactions. Previous Chapters have discussed how NMR can be used to study the structure and dynamics of individual proteins, nucleic acids and organized supra-molecular assemblies such as membranes and protein-nucleic acid complexes. The information enables one to understand their function at an individual level and under *in vitro* conditions. This provides physico-chemical explanations for several properties, through a reductionist approach, where complex reactions are broken down into individual components.

However, even in the most basic biological processes, hundreds of reactions and enzymes are involved. Understanding of biological phenomenon such as bioenergetics, biosynthesis, molecular transport, changes arising from disease, behaviour, evolution, development and stress, require a bottom-up approach. How does one utilize the two sets of information discussed above to understand complex but highly coordinated processes involved in the biochemistry of living systems? Why do individual organs behave differently, though they have common genes? How can NMR help us in this endeavour? These aspects can be dealt at the level of cells and tissues, as well as at the level of whole organs and individual living systems.

NMR has the unique capability of revealing metabolic changes in healthy and diseased individuals at the level of cells, tissues and whole organs. In the next three Chapters, discussion on some of the NMR approaches to bridge the gap between the properties of individual molecules, and the function and the behaviour of living systems have been presented. This Chapter concentrates on the basic metabolic processes in living systems and NMR studies on cells and tissues.

1.1 Cell-metabolism is the Bridge between Proteomics and Function

Metabolic pathways consist of a series of coupled chemical reactions, which are catalysed by enzymes. The cells control these processes through regulation of genetic expression, feedback mechanisms and signalling pathways. The demand on metabolism changes as a result of changes in the environmental factors (an area called xenobiotics). The rates of reactions in metabolic pathways are regulated to meet the requirements of organisms.

The non-invasive nature of NMR offers several advantages over conventional analytical techniques in *in vivo* studies of biochemical process. Most analytical methods are capable of detecting only specific molecules for which the technique has been designed. NMR is not biased towards a particular compound. It does not require precise selection of analytical conditions to obtain quantitative information from endogenous metabolites and xenobiotics. NMR can simultaneously detect molecules which are expected, unexpected or which are difficult to assay using standard chemical methods. Therefore, NMR has been extensively applied to study biochemistry and physiology in cells and living tissues.

Valuable information on body metabolism can also be obtained through the analysis of body fluids. The only limitation is related to the low sensitivity of the method, which at this stage of technology, limits detection levels in cells and tissues for ^1H to 10 μM concentration of metabolites. This limit may be pushed down with the availability of cryoprobes and other technical advances. In particular, several applications of cryogenic probes have been reported in the analysis and study of metabolic pathways.

1.2 Measurement of Products of Metabolism

Metabonomics is a recent term coined, for an emerging area devoted to the understanding of the behaviour of cellular systems and organs at a molecular level and aberrations in their function as a result of physiological or genetic modification, diet, differences arising due to gender, physical and biochemical stress, toxicity and changes during cell development and differentiation. Alternatively, a term metabolomics is used to cover studies on analysis of changes in intra- and extra-cellular metabolites in simple biological systems. On the other hand, metabonomics is generally used for a dynamic multi-parametric metabolic response of living organisms to genetic modifications, and patho-physiological response. The applications of these techniques cover several important areas of life sciences. NMR of cells, tissues, body fluids and whole organs has therefore become an important area of research in biological and medical sciences and has contributed immensely to metabonomics.

The total number of compounds synthesised by the genome may vary from about 600 for a small organism such as *Saccharomyces cerevisiae* to about 200000 in plants. Their concentrations vary from pM to mM ranges. The concentrations depend not only on genomics and proteomics of the living system but also on environmental factors. For example, the concentration variations in humans of the metabolites are influenced by factors, such as sex, diet, life style and diet habits, diurnal variations, health or disease and environmental factors. The concentrations of various metabolites can be monitored in several body fluids such as urine, serum, plasma and cerebrospinal fluid. NMR is an important technique for such analysis. When used in conjunction with other physical techniques such as mass spectroscopy (MS) and under controlled conditions, important conclusions can be made on the relationship between metabolism and external factors.

2. METABOLIC CYCLES AND PATHWAYS

Cell metabolism is a highly intricate network of coupled biochemical reactions. Metabolism links genes and proteins to higher-level biological functions. Each metabolic pathway consists of a series of individual chemical reactions, which are linked to each other. These include biosynthetic pathways, which usually require energy and result in complex biological molecules such as proteins, nucleic acids, hormones, neurotransmitters, and lipids. In fact, as discussed in Chapter 4, bacterial cells have been be used to prepare large amount of labelled proteins, taking advantage of biosynthetic pathways. Nucleic acids and fatty acids are likewise synthesised by energy requiring pathways.

Most biological molecules are in a highly reduced state. The energy and reducing power for such reactions is provided by oxidative pathways, which break molecules such as glucose, amino acids, and fatty acids. To understand later discussions, some of the major metabolic pathways are recapitulated.

Adenosine triphosphate (ATP) is the universal currency of energy. It is generated during bioenergetic pathways. ATP helps in maintaining ionic balance between the two sides of biological membranes through active transport, in signalling pathways and in signal amplification. It is a key molecule for the biosynthesis of proteins and nucleic acids, for which it provides energy through coupled reactions. ATP is synthesized by phosphorylation of adenosine diphosphate (ADP). The reaction is reversibly catalysed by ATP synthase. Another energy rich compound is phosphocreatine (PCr). It serves as energy reserve for ATP in muscles and brain. PCr is generated from ATP by transfer of the phosphate group to Cr. Under normal conditions, an equilibrium is maintained between the levels of ATP and PCr.

Thus, there are two sets of metabolic pathways: one helps to synthesize molecules required for life (*anabolism)*; the second, produces energy and building blocks for synthetic-pathways (*catabolism)*. Several of these pathways operate simultaneously and act in a highly coordinated fashion, such that each one is able to sense the status of others. This achieves optimal control and function in living systems.

The synthetic and degenerative pathways are distinct. Depending on the needs of the system, the one required at a particular stage is active, while the other one is deactivated. This contributes to metabolic control, and allows an integration of metabolism. Metabolic regulation is achieved through a number of factors. These include allosteric interactions in proteins, concentration levels of enzyme, cofactors and hormones, covalent modification and compartmentalisation in cells. In mammalian systems, each organ (e.g. brain, muscle, adipose tissue, kidney and liver) has a different pattern of metabolism. This is a result of differential gene expression in different organs.

2.1 Glycolysis

The basic function of catabolism is to form energy rich molecules (mostly ATP). Catabolism also serves to generate reductive power for the living system, in the form of NADPH. It generates building blocks (amino acids, ribose etc) for biosynthesis. Energy stored in the form of fatty acids, amino acids, lipids or sugars is released during their oxidation, and is utilized to form ATP and/or PCr.

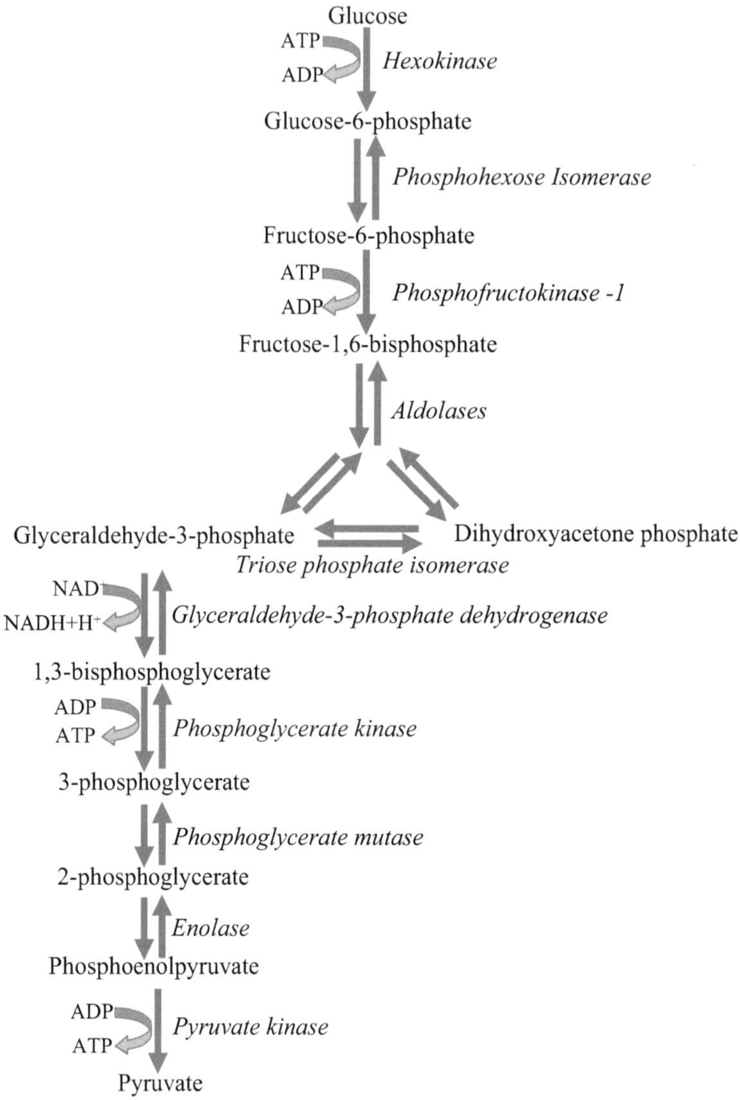

Figure 12.1: The steps and enzymes involved in glycolysis.

There are three major stages in energy producing metabolism. The first is glycolysis, during which one molecule of glucose is metabolized into two molecules of pyruvate through a series of ten sequential reactions (Figure 12.1). The net result is the reduction of two molecules of nicotinamide-adenine dinucleotide (NAD) to NADH, and production of two molecules of ATP. Glycolysis provides ATP and carbon skeletons for biosynthesis. These reactions take place in cytosol.

The 3D structures of the enzymes involved in various steps of glycolysis have been solved by X-ray crystallography and/or by NMR. A great deal is known about these pathways, structure-function relationships of the enzymes involved, and how these reactions are controlled.

Glucose (Glc) is the most important molecule in sustaining metabolism and for providing precursors for living systems. Certain organs such as brain do not have their own source for Glc. It has to be transported to brain continuously. Certain organisms survive entirely on glycolysis. Further, under anaerobic conditions (lack of oxygen) glycolysis may be the only energy source. Since NADH has to be regenerated under such conditions, pyruvate is reduced to lactate or fermented to alcohol or some other molecule. Simultaneously, NADH is oxidized back to NAD.

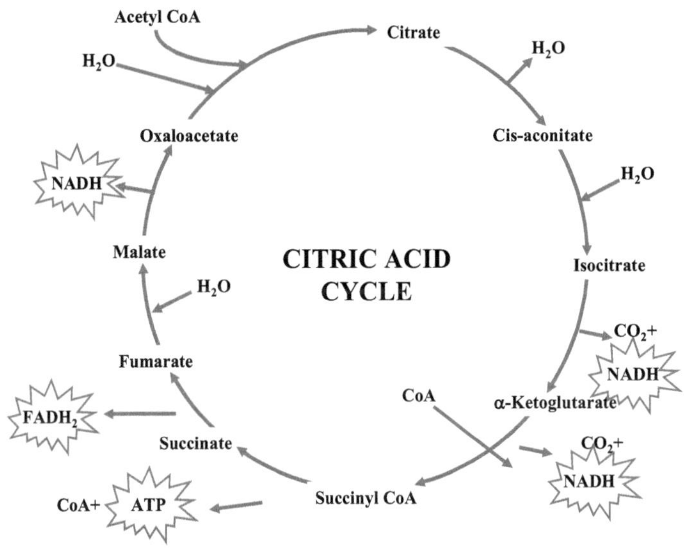

Figure 12.2: Kreb's cycle, also called citric acid or tri-carboxylic acid (TCA) cycle consists of a series of reactions. They occur in a cyclic fashion, ultimately resulting in the oxidation of substrates and reduction of coenzymes.

2.2 Citric Acid (Kreb's) Cycle

In the presence of oxygen, pyruvate is oxidized by tri-carboxylic acid (TCA) cycle to CO_2 and H_2O and releasing energy in form of ATP. The acetyl group is transferred to coenzyme A (CoA), forming acetyl CoA, which enters the Kreb's cycle and is further metabolized. This process is also called as the citric acid cycle.

Molecules containing two carbon atoms are oxidized to CO_2 and H_2O. Simultaneously, one molecule of GTP is produced, and one molecule of coenzyme FAD and three molecules of NAD^+ are reduced to $FADH_2$ and NADH, respectively. These reactions take place inside mitochondria, and utilize carbohydrates, amino acids and fatty acids in the form of acetyl coenzyme A (CoA). The series of steps occur in a cyclic fashion such that oxaloacetate is regenerated (Figure 12.2).

2.3 Oxidative Phosphorylation

The final stage called oxidative phosphorylation occurs in the inner mitochondrial membranes (Figure 12.3). During the electron transport chain, the reduced coenzymes are oxidized back. The energy released is stored in the form of 34 molecules of ATP. Efforts to understand different stages of this reaction, and the mechanism of enzymes involved in oxidative phosphorylation, have met with limited success. The catalytic processes in these reactions require large macromolecular assemblies, and take place on biological membranes. This adds to difficulties in studying molecular mechanism involved in such reactions.

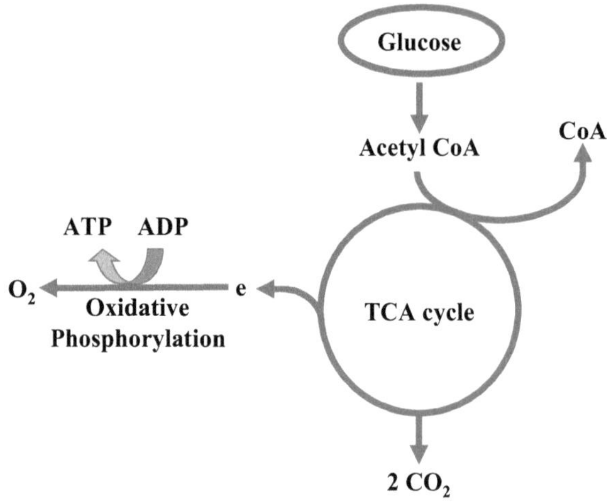

Figure 12.3: Series of reactions during oxidative phosphorylation lead to oxidation of reduced coenzymes. The large amount of energy released, is used for the synthesis of ATP molecules.

2.4 Pentose Phosphate Pathway, Gluconeogenesis and Glycogen Synthesis

Glucose is involved in two other pathways, which occur through an intermediate, glucose 6–phosphate (G6P). One is the pentose phosphate pathway. This is a two-step process that takes place in cytosol. The first step results in the production of two molecules of NADPH from $NADP^+$ for the purpose of reductive biosynthesis. Simultaneously, one molecule of ribose 5-phosphate is generated, which may be used for the synthesis of nucleotides. In the next step, the system decides whether the five carbon phosphorylated sugars are to be converted to introduce ribose for catabolism or to generate it for biosynthesis.

The second pathway decides on the synthesis or degradation of glycogen, depending on the energy needs of the body. Glycogen is a branched polymer, and is a store house of Glc. Depending on the energy requirements, either the degradation or the synthesis of glycogen is activated. The two processes are mediated through hormone regulation. Because of the suspected role in diabetes, glycogen synthesis has been extensively studied. Although glycogen is a high molecular weight molecule, it provides fairly sharp NMR signals.

Glucose is synthesized in liver and kidneys from lactate, glycerol and amino acids. Pyruvate serves as an entry point, which is carboxylated to oxaloacetate in mitochondria. Fructose 1,6-biphosphate and fructose 6-phosphate, are the key intermediates in glucose synthesis. The reaction is catalysed by fructose 1,6-biphosphatase. Glycolysis (degradation) and gluconeogenesis (synthesis of Glc from non-carbohydrate precursors such as Lac, propionic acid, Gln and Ala) act, and are usually regulated in a reciprocal fashion.

2.5 Fatty Acid Synthesis and Degradation

Acetyl CoA also acts as an entry point for fatty acid synthesis. The synthesis takes place in cytosol through the carboxylation of acetyl CoA. This leads to an activated intermediate, malonyl CoA, which provides the route to add two carbons to an acyl carrier protein. Depending on the needs of the body, fatty acids are synthesised or degraded in mitochondria. Fatty acid metabolism plays a key role in providing energy for organs such as liver, heart and skeletal muscle.

2.6 Amino Acid Metabolism

Carbohydrate metabolism may not provide sufficient energy in situations such as fasting or diabetes. In such cases, energy is derived through amino acid metabolism whereby L-α amino acids are converted to α-keto acid, and formation of L-Glu by amination of α-KG. The α-keto acid enters TCA cycle resulting in CO_2, H_2O, and NH_3. Several amino acids such as Ala, Thr, Gly, Ser, Cys, Phe, Tyr, Lys, Arg, His, Glu, Gln, and Pro can enter TCA cycle. The process involves production of ammonia, a toxic compound, which is transported to liver in a non-toxic form where it is converted into urea.

2.7 Integration and Control

The energy pathways are controlled at several stages. Three important molecules are involved at the interface between the synthetic and degenerative pathways: Glc-6-P, pyruvate and acetyl CoA (Figure 12.4).

Phosphofructokinase (PFK) is the most important enzyme in glycolysis. A high level of ATP and citrate inhibits PFK. AMP acts in the reverse fashion. Glucolysis stops when there is an excess of ATP and Glc is stored in the form of glycogen. Gluconeogenesis and glycolysis are reciprocally activated. The conversion to NADPH and ribose 5-phosphate in pentose phosphate pathway are controlled by the level of $NADP^+$. Acetyl CoA carboxylase is stimulated in presence of large amounts of citrate. Thus, the presence of large amount of ATP and acetyl CoA activates fatty acid synthesis. The need for ATP activates degradation of fatty acids.

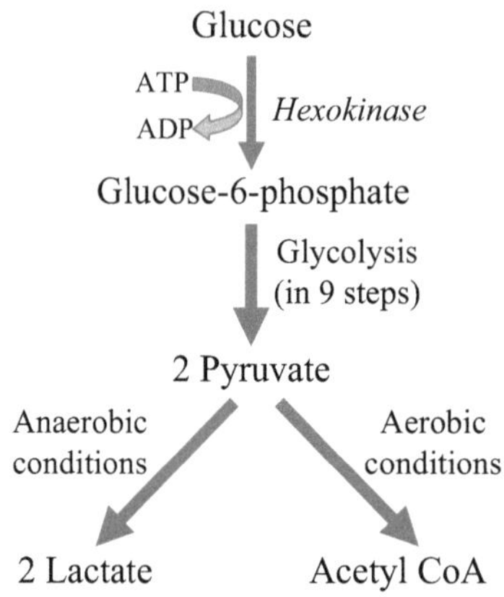

Figure 12.4: The integration of metabolic pathways is largely controlled at three key points, and involves glucose 6-phosphate, pyruvate and acetyl CoA.

2.8 Metabolic Profile and Requirements of each Body Organ is Different

Body organs provide additional means for regulation of catabolism and anabolism. Each living being has a unique genome, which is derived from the fertilization of spermatozoa (carrying paternal gene), and egg (the maternal gene). Cell differentiation during the development phase leads to different organs in the body. Each organ has a unique metabolic pattern, which decides its function.

Metabolically, brain is one of the most active organs. In a normal human, Glc is the main fuel for brain. With only 2% of the body mass, brain consumes 20% of Glc and O_2 of the total human body requirements. The main function of brain is to process information. It requires a continuous source of Glc, which is its sole fuel, except during crisis. Thus, Glc and O_2 have to be transported to brain on a continuous basis. Most of the metabolic energy is used to maintain Na^+-K^+ concentration across the cell membrane and to synthesize neurotransmitter and to propagate nerve impulses across neurons.

Liver is a key organ for metabolic activity. It is responsible for providing Glc and other fuels to brain and muscles. It plays a central role in regulation of lipid metabolism, for providing building blocks for biosynthesis, and detoxification by converting NH_3 into urea. Liver mobilises glycogen and carry out gluconeogenesis.

Muscles can use Glc, fatty acids and ketone bodies as fuels. Muscles also have a large store of glycogen.

A close link exists between the metabolism of liver and muscles. The two organs share the metabolic load of the body. Adipose tissue is responsible for synthesis, storage and release of tri-acylglycerols. It receives fatty acids from liver. In muscle and brain, another energy rich compound is PCr. When sufficient O_2 and fuels are available to facilitate ATP synthesis, abundant amount of PCr is produced. During energy crisis (hypoxia, ischemia or cell damage) PCr serves to maintain ATP levels to allow cellular function. However, if ATP is not synthesized and PCr stores are depleted, some of the body-functions can be impaired, either temporarily or permanently.

The main function of kidney is to produce urine, decontaminate the body and reabsorb glucose. Urine is in fact, one of the most common body fluid studied using NMR. Besides other metabolites, the metabolism in kidney produces trimethylamine (TMA), and trimethyl-amine-N-oxide (TMAO).

Above functions may be disturbed if there is a change in the chemistry of the body. For example, in a well fed human, insulin stimulates the formation of glycogen, triacylglycerols and proteins. A low level of Glc in blood stimulates generation of Glc through glycogen degradation, gluconeogenesis and triacylglycerol hydrolysis. The picture may be modified in different organs in situations such as disease, starvation, heavy exercise and parasitic infection. In human and animals, it is not only the genes of the host that decides the control of these reactions, but one also has to consider the bacterial systems present in the body (particularly the gut bacteria), and parasites (in parasitic diseases). The genomes of such organisms are different from human body and may disturb controls of the metabolic pathways.

Study of metabolism therefore provides vital information, which is important not only to understand basic biology and the function of organs but also in diagnosis of diseases. The chemistry of individual organs can be studied using NMR at different levels: body fluids, cells, tissues and whole organs.

2.9 ^{31}P NMR and Metabolism

^{31}P is the nucleus of choice for studying bioenergetics in cells and organs. Insights on several aspects of bioenergetics can be obtained by ^{31}P NMR from the brain of a normal subject. A typical spectrum is shown in Figure 12.5. Such a spectrum provides quantitative information on the levels of different phosphorus metabolites expected in the muscle.

A major advantage of ^{31}P NMR is that this spin is present in only a few key molecules. It provides information on the central metabolic pathways, which are vital for the energy state of cells and organs. The problems of assignments are not severe. In fact, some of the earliest studies on MRS of whole organs were reported using ^{31}P NMR spectroscopy.

The ^{31}P NMR spectrum typically consists of 6-7 peaks (Figure 12.5). The three phosphates in ATP (α, β, and γ) contribute to unique signals in the spectrum. The peak corresponding to β-ATP is sensitive to bound Mg^{+2}, and can be used to measure Mg^{+2} concentration. The peaks due to AMP and ADP are usually weak as their concentrations are small. However, these signals have higher intensities in situations where the cells have been starved. The α-ATP occasionally contains a shoulder corresponding to $NADP^{+}$/NADPH. The signals due to sugar mono phosphate (PME) arise from hexose 6-phosphates such as glucose 6-phosphate (G6P) and fructose 6-phosphate (F6P), as well as AMP and IMP. Peaks labelled as phosphocholine (PC), and phospho-diesters (PDE), are likewise superimposition of a number of related compounds such as glycerophosphocholine (GPC) and phosphoethanolamine (PDE). While in chemical studies, 85% H_3PO_4 (or phenylphosphine) is usually recommended as an external reference for ^{31}P NMR spectroscopy, the spectra in living systems are calibrated with respect to the internal signal from PCr, which does not depend on pH. It may be noted that the intra- and inter-cellular pH may be different resulting in two P_i signals.

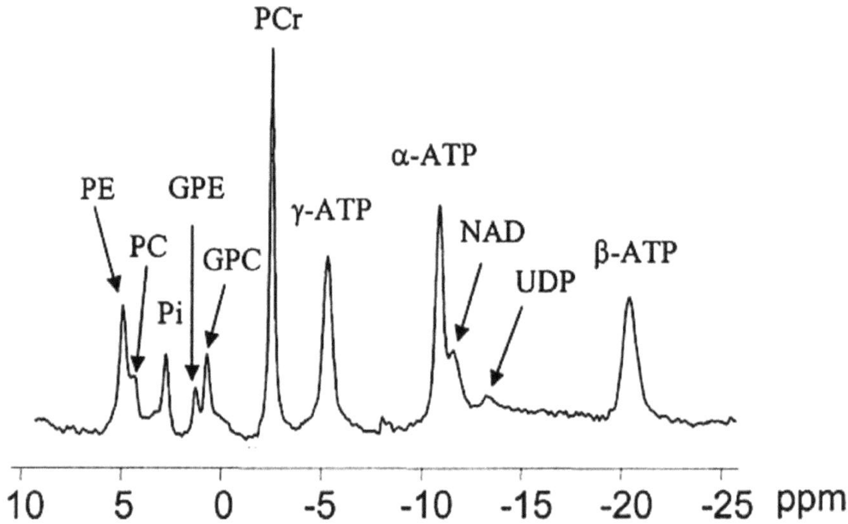

Figure 12.5: Surface coil localized ^{31}P NMR from human brain acquired by using 7T NMR imager/spectrometer. H. Lei, X.H. Zhu, X.L. Zhang, K. Ugurbil and W. Chen. Magnetic Resonance in Medicine 49:199–205 (2003). (Reprinted with permission from John Wiley & Sons).

The quantitative measurement of *in vivo* ^{31}P spectroscopy allows measurements of a number of metabolic parameters, such as oxidative phosphorylation potential and the free energy for ATP hydrolysis. These parameters reflect different aspects of the control and regulation of energy metabolism, ion transport and muscular contraction. The peak positions and their characteristics may be different for a patient suffering from deranged metabolism.

The inorganic phosphate (P_i) shows resonance at around 5 ppm. This signal is a time-average of different forms of the tribasic acid H_3PO_4. Under physiological conditions, the two major species are $H_2PO_4^-$ and HPO_4^{-2}, with a pK_a value of 6.75.

The chemical shift of P_i is therefore sensitive to pH and moves to high fields with acidic pH. These changes have been calibrated over a wide pH range and serve as key indicators of changes in pH arising from metabolism in cells and organs. Typical applications include metabolic changes due to lack of oxygen in the tissues following damage or heavy exercise as well as measurements of difference between intra- and inter-cellular pH.

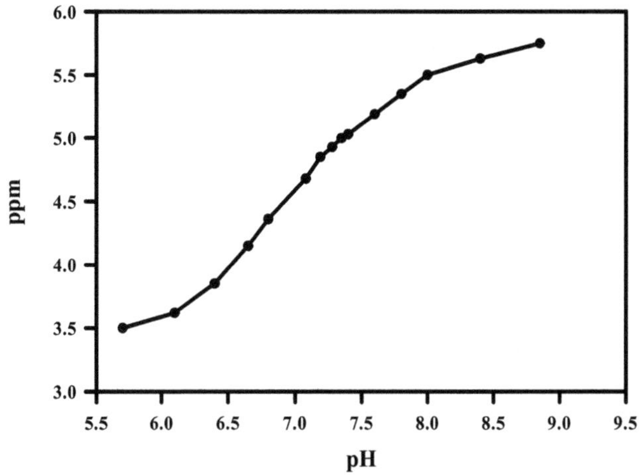

Figure 12.6: The chemical shift of inorganic phosphate as a function of pH in cells. (Thanks to S. Srinivastava for providing this figure).

2.10 ^{1}H NMR Spectroscopy

^{1}H spectra of tissues, or organs, normally show two strong signals, corresponding to molecules present in molar concentrations: a broad resonance due to methylene protons lipids and fats, and a relatively sharper signal due to water. Even these two signals can provide very useful information, to probe metabolic changes in healthy and diseased organs.

To observe ^{1}H signals from molecules present in mM concentrations, the above two resonances have to be suppressed. As discussed earlier, only mobile nuclei are observed in NMR. In cells, one has almost 10-15% mass in the form of proteins. Several of the large macromolecules and molecules embedded in membranes give relatively broad signals. Even for relatively mobile proteins, the spectral features are widely dispersed, making it difficult to observe their individual spectra. One does not observe nucleic acids as their concentrations are small, molecular weights are large, and consequently resonances are broad.

Thus, *in vivo* NMR signals in cells, tissues and whole organs are mainly due to small molecules such as amino acids, nucleotides, precursors or products of metabolic pathways, and other endogenous and exogenous molecules. In a typical ^{1}H and ^{13}C spectrum from cells and tissues almost 30-100 molecules can be observed, simultaneously. On the other hand, in *in vivo* spectroscopy from whole organs, less than ten molecules are usually observed due to considerable broadening of spectral features.

2.11 ^{13}C NMR Spectroscopy

Spectra from cells can be observed in natural abundance (1.1%) of ^{13}C. These spectra provide levels of different metabolites and their precursors. However, in studies of metabolic pathways, it is often useful to introduce ^{13}C label in one of the precursors (or an exogenous compound). The flow of ^{13}C from a specific substrate can be followed through the appearance of the corresponding ^{13}C signals in other compounds arising from metabolism of substrates. The rates of the reactions in the metabolic pathways can be followed in real time, by measuring the appearance of ^{13}C in different metabolites. Such measurements of turnover rates are valuable in measuring *in vivo* kinetics during steady-state conditions where ^{13}C concentration of end product(s) increase(s), while their total concentration of intermediates unchanged. For example, labelled Glc has been used to measure rates of glycogenesis and gluconeogenesis simultaneously in the human liver. These *in vivo* experiments show high rates of both processes, and have provided insights into role of glycogen.

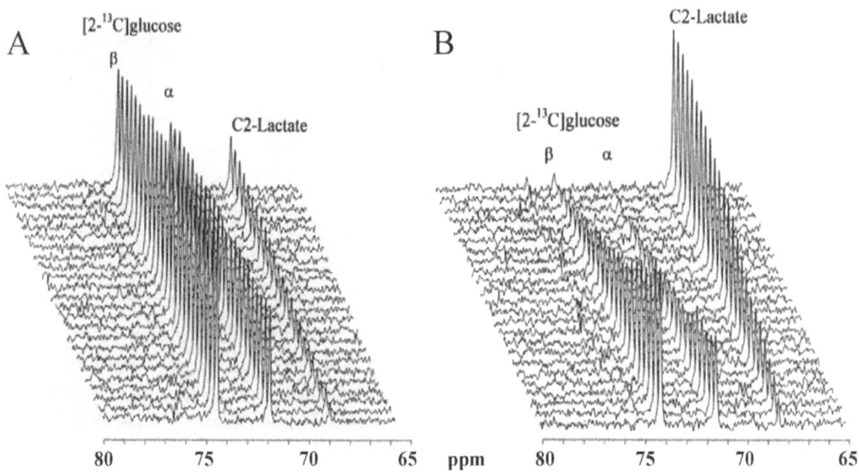

Figure 12.7: ^{13}C NMR spectra as a function of time recorded from (A) normal erythrocytes and (B) an approximately 2% of erythrocytes infected with the malarial parasite – Plasmodium falciparam, incubated with [2-^{13}C]-glucose. Each spectrum was recorded in ~22 min. The first spectra were recorded 42 and 72 min, after addition of labelled glucose, respectively. (Thanks to H.M. Sonawat for providing this figure).

As an illustrative example, Figure 12.7 shows ^{13}C NMR spectra of normal erythrocytes and erythrocytes infected with the malarial parasite – *Plasmodium falciparam*, incubated with [2-^{13}C]-glucose. The spectral features change as a function of time. One can notice the rates of utilization of glucose and concomitant production of [2-^{13}C]-lactate, which are more in the case of erythrocytes infected with malarial parasite as compared to the normal erythrocytes. This data suggests that the metabolism of infected erythrocytes increases to meet energy requirements of parasite.

3. STUDIES OF BODY FLUIDS

The genetic code of a living system is ultimately reflected in its metabolic behaviour. Genomics and transcriptromics are primarily responsible in deciding the expression of proteins in an organ, at any given time. According to the physiological state of the system, different genes are expressed and the corresponding m-RNA molecules are produced. Post translational modifications to the proteins translated from mRNA can cause additional variations. However, the combination of genomics, transcriptromics and proteomics is not sufficient to decide the metabolism of a particular organ. Metabolite concentrations can change rapidly in response to signals such as changes in metal ions (such as Na^{+}, K^{+} and Ca^{+2}), diseases, toxins, drugs, food etc. Somewhat more gradual changes are observed in case of viral and bacterial infections, and deficiencies in the immune system. Diseases such as cancer, AIDS and neurodegenerative diseases can bring about major local changes in the body systems. Age, gender, and the physiological state of the body can also influence metabolic profiles.

3.1 Metabolomic Analysis using Biological Fluids

Use of NMR based metabolic approach is fast becoming a powerful tool to study animal function and behaviour, both in the healthy and diseased state. It has also become a tool in the development of new biologically active compounds and for tests on toxicology of such molecules. Figure 12.8 illustrates an overview of the approach.

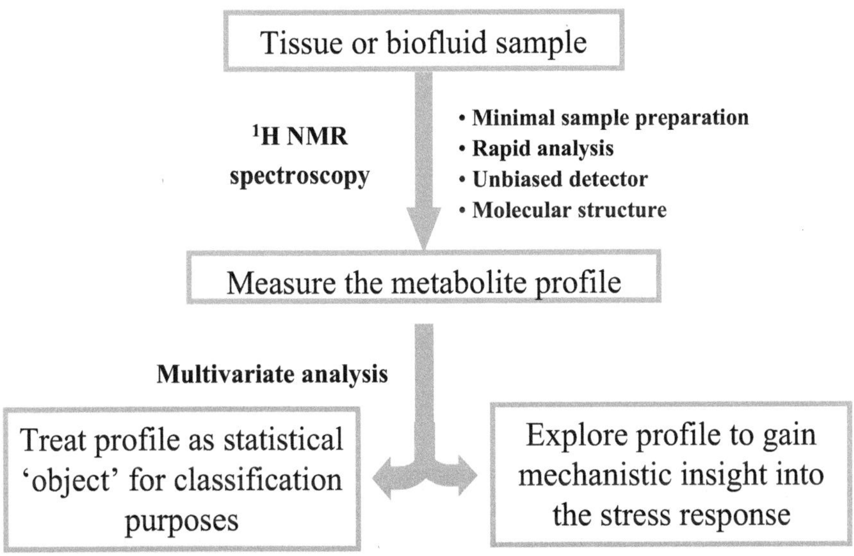

Figure 12.8: A typical protocol for metabonomics approach.

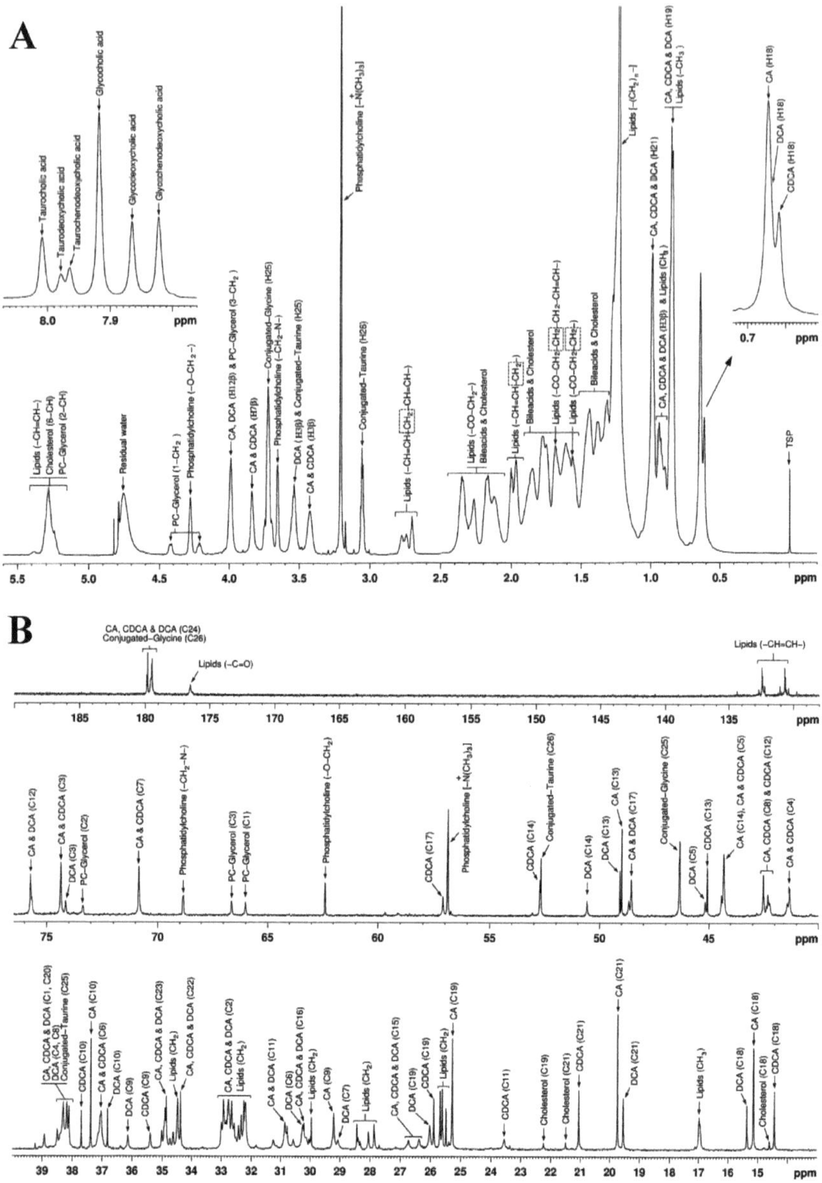

Figure 12.9: (A) ^{1}H and (B) ^{13}C NMR spectra of human bile recorded at 800 MHz. G.A. Nagana Gowda, Omkar B. Ijare, B.S. Somashekar, Ajay Sharma, V.K. Kapoor, and C.L. Khetrapal. Lipids 41, 591-613 (2006). (Reprinted with permission from American Oil Chemists Society).

Body fluids such as urine, stool, abdominal fluids, serum, seminal fluid and milk can be obtained easily and studied by NMR, without much effort. However, studies have also been conducted on a large number of other fluids such as cerebrospinal fluid (CSF), amniotic and synovial fluids, which are obtained by catheterization or needle biopsy. Usually, minimal preparation of the sample is needed, and spectra of fluids can be observed directly using a high resolution spectrometer.

Unlike body fluids, in the case of biopsy samples, it is advisable to store the sample under liquid nitrogen conditions to prevent metabolic changes, till one is ready to record NMR. The analysis is non-destructive, non-selective, and takes only a few minutes. The spectra are usually complex as they contain signals from several metabolites. Use of higher magnetic fields and edited 2D techniques offer major advantages over 1D spectroscopy.

For example, human bile being a complex mixture of numerous metabolites provides highly overlapping and complex spectra requiring the need for the highest field NMR, especially for quantification of individual components. Figure 12.9 A and B show ^{1}H and ^{13}C NMR spectra of human bile recorded at 800 MHz, respectively, which have been used for such characterization. As is evident from this Figure, one can fairly have well-dispersed resonance lines. Such spectra are amenable for straight forward assignment of signals, analysis and quantification.

High-field, high-resolution NMR spectra provide valuable information on the metabolic profile of the organs from which the body fluids. All low M_w molecules and other mobile metabolites can be studied, simultaneously. Abnormal metabolism is reflected in the form of changes in the levels of metabolic products or by appearance of new compounds. A metabolic perturbation at the level of genomics, transcriptomics or proteomics, can have an indirect effect on at least one of the NMR detectable metabolites. The metabolic profile also shows excretion of exogenous compounds (such as drugs or toxins), as well as the metabolic products of such compounds. While arriving at general conclusions for human related studies, several samples may be required to take care of environmental and genetic diversity.

Other methods, which have been used extensively for metabolic studies of body fluids are HPLC and GC-MS. The data obtained from NMR can be coupled to these techniques, for obtaining greater insights.

3.2 ^{1}H and ^{13}C Chemical Shifts of Common Metabolites: Assignments

Unlike in the case of proteins and nucleic acids, it is not possible to make use of sequential assignments for the large number of spectral features present in a sample of body fluids. However, each metabolite gives its own finger print in the total spectrum, in the form of its unique chemical shifts and coupling constants. A dictionary of chemical shifts serves as an aid to the assignment in the complex spectra from body fluids, cells and tissues. As a prerequisite to metabolic studies, it is important to have knowledge of the chemical shifts of the molecules present in body.

Tables 12.1 and 12.2 show the ^{1}H and ^{13}C chemical shifts for some of the common metabolites, present in body fluids, cellular suspensions and organs. The chemical shifts of some of these compounds are sensitive to pH, ionization state of the molecule, ionic concentrations and temperature. This should be taken into account during analysis. The chemical shifts and the multiplet behaviour of each spectral feature provide important clues on the molecules present in the mixture. The assignments can be further confirmed by using spiking method. A small amount of suspected compound is added to the sample. Observed spectra before and after the addition are compared. Since the amount of each metabolite is different, spectral simulations using standard chemical shifts helps in data processing and quantification.

Table 12.1: ^{1}H and ^{13}C (shown in brackets) chemical shifts of resonances arising from molecules usually present in cells, tissues and body fluids. (Courtesy S. Srivastava and H.M. Sonawat).

Molecule	C1H	C2H	C3H	C4H	C5H	C6H
GPC	3.64 (63.73)	3.77 (73.31)	3.94 (67.76)	4.31 (60.64)	3.67 (67.16)	3.20 (55.13)
Hyp	3.36 (34.66)	2.64 (56.62)				
Creatine		3.92 (55.01)		3.03 (38.13)		
m-Inositol	3.53 (72.32)	4.05 (73.40)	3.53 (72.32)	3.61 (73.67)	3.27 (75.53)	3.61 (73.67)
s-Inositol	3.32 (75.81)					
L(+)Lactate		4.09 (69.73)	1.30 (21.21)			
α-Ala		3.78 (51.71)	1.46 (17.26)			
β-Ala		3.26 (47.86)	1.26 (9.39)			
Arg		3.76 (55.80)	1.92 (28.55)	1.64 (24.94)	3.23 (41.69)	
Asp		3.89 (55.37)	2.67, 2.79 (37.59)			
Gln		3.78 (55.48)	2.11 (27.94)	2.44 (31.95)		
Glu		3.71 (55.41)	2.04 (27.84)	2.37		
Gly		3.55 (42.61)				
His		3.99 (55.69)	3.18,3.25 (28.50)			
Ile		3.67 (60.79)	1.96 (36.81)	1.45, 1.25 (25.43)	0.92, 0.99 (12.1, 15.8)	

Leu	3.75 (55.85)	1.69 (40.76)	1.70 (25.02)	0.94, 0.95 (21.83, 23.05)	
Lys	3.74 (55.77)	1.84 (30.80)	1.47 (22.36)	1.70 (27.30)	3.02 (40.23)
Met	3.85 (55.22)	2.12 (30.60)	2.63 (29.89)	CH_3 2.12 (15.01)	
Pro	4.12 (62.45)	2.26, 2.34 (30.05)	2.00 (24.80)	3.33, 3.41 (47.23)	
Tyr	3.95 (57.30)	3.17 (36.57)		7.16(2,6)	6.91(3,5)
Thr	3.58 (61.50)	4.26 (67.30)	1.31 (20.70)		
Ser	3.83 (61.44)	3.84 (57.58)			
Val	3.59 (61.61)	2.27 (30.30)	0.98, 1.03 (17.80, 19.07)		
NAA	2.00	4.38	2.67, 2.48		
TMA	2.88				
TMAO	3.55				
CHO	4.05	3.50			

GPC: Glycerophosphorylcholine; NAA: N-acetyl aspartate; TMA: Trimethylamine; TMAO: Trimethylamine N-oxide.

3.3 Suppression of Water and Other Undesirable Signals

To obtain good ^{1}H spectrum from molecules present in body fluids in low concentrations, it is essential to suppress the signal from water and fats. In previous Chapters, several methods have been discussed for saturating H_2O signals. Essentially, the same methods have been used for studying body fluids, cells and tissues. For example, a low power selective pulse can be used to saturate the water signal. Alternatively, one may use jump and return (JR) experiment, DEFT or double inversion recovery. Another problem that may arise in the analysis is due to broad overlapping signals from macromolecules and lipids. This problem is particularly acute in fluids rich in proteins and lipids. In such cases it is advisable to use CPMG spin echo methods to eliminate the broad resonances due to such molecules, which generally have shorter relaxation times. These methods also help to attenuate the water signal.

3.4 Multi-dimensional NMR and Quantification

A large amount of information can be derived from simple 1D ^{1}H NMR of body fluids. Indirect detection using INEPT helps in picking up ^{13}C resonances, in natural abundance. However, the use of 2D methods offers obvious advantages. Extensive use of ^{1}H-^{13}C hetero-nuclear methods has been made. Another nucleus, which offers somc advantage, is ^{15}N. However, at present, applications of this nucleus to study body-fluids, are somewhat limited. In recent years, multiple-quantum filtered COSY, phase sensitive COSY and other versions of homo- and hetero-nuclear correlated spectroscopy have been used for assignments and resolution of J-correlated multiplets from different metabolites. Both HMQC and HSQC experiments have been used extensively. Development of pulsed-field gradient (PFG) technology has also provided a great impetus to hetero-nuclear spectroscopy of cells and tissues.

Thus, several techniques discussed for macromolecular studies are finding place in both *in vivo* and *in vitro* studies. Spectral editing techniques are also extremely useful. Quantification of metabolites is based on the fact that the intensity of an NMR line is proportional to the total amount of a compound in the sample. Since metabolic disturbances change only quantitative features, there is a growing need to obtain accurate quantitative information, particularly in relation to normal metabolic patterns from healthy volunteers. Some methods in use for protons for such analysis are integrated areas, use of external capillary with a control sample, internal endogenous or exogenous marker, and reference to internal water ^{1}H concentration.

For ^{13}C spectra, low flip angles and large repetition times have to be used, to avoid saturation and to obtain reliable quantitative information. Most ^{13}C spectra are obtained using broad-band ^{1}H decoupling. The spectral intensities should be corrected for nuclear Overhauser effects.

3.5 Multi-variant Statistical Analysis

Two different approaches have been used for the analysis of body fluids and relating special features to changes in metabolic fluxes. In one, the aim is to assign as many signals as possible making use of multi-dimensional NMR. The results are often complemented by linking NMR to on-line chromatographic separation techniques or parallel mass spectroscopy and NMR analysis. A quantitative estimate of the metabolites is thus achieved.

Since spectra of bio-fluids are complex, it may not be possible to assign all the peaks. The complexity in metabolic changes arising from genetic differences, dietary habits, disease, age and environmental variations require studies on a large number of volunteers. This results in hundreds of spectral determinants. Several of these biochemical changes occur in parallel and need not necessarily produce predictable results. Assignment, selection and quantification of NMR data are time consuming. To obviate such problems, statistical methods have been developed. These methods need not necessarily use spectral assignments but treat the spectral data as multi-parameter features. In this approach, the assignment of peaks is ignored and a multivariate analysis is carried out to compare sets of spectra. Thus, one makes use of simple combination of groups of metabolites, rather than a single molecule. The applications of NMR to the new and emerging field of metabonomics and analysis of metabolic fluxes, complements the results obtained from mass spectroscopic analysis of extracted materials from plants and animals.

A variety of methods such as linear discriminant analysis, neural network, pattern recognition and genetic algorithms have been developed. The NMR data can be digitized into narrow windows of integrated spectral regions (typically around 0.05 ppm) to reduce the number of data points and to allow for small pH changes.

3.6 Applications of Metabonomics

NMR analysis of body fluids offers great potential for understanding diseases in humans. Several genetic diseases result in the accumulation of intermediates proximal to the step influenced by a defective enzyme. This is reflected in changes in the spectrum of serum, urine and other body fluids. Studies dealing with metabolic changes include several areas including diseases (such as genetic, renal, diabetes, neurological, cancer, AIDS), physiological status (including hormonal variation), therapeutic intervention (e.g. renal transplant) and toxicity. Applications of such approaches include detection of metabolic disorders, biochemical basis of drug and xenobiotic metabolism, organ damage, toxicological processes and drug-design. Metabonomics is an emerging field and the literature is growing very fast. We shall outline some representative examples to indicate the power of this approach.

Urine is one of the fluids, which is easily available and can be handled conveniently. The ^{1}H spectrum of a healthy subject contains signals from DMA, TMAO and hippurate, in addition to other usual molecules. Renal damage alters levels of some of these metabolites. The spectra of patients following renal transplantation are considerably different. Acute graft rejection results in increased level of TMAO. The rejection of kidney can be detected much earlier by NMR of urine as compared to kidney biopsies.

Due to high variability from one subject to another in the urine samples, spectra are usually normalised to unit area. Other fluids may not require such normalization. Statistical analysis provides a classification in spectral changes as a result of disturbances in metabolic functions. It also indicates which parts of the spectra are driving changes in the classification. This is referred as feature selection. Such studies help to link the observed changes to specific molecules.

Next to urine, blood is a fluid, which can be obtained with very little pain to humans. It also contains vast amount of information. For example, certain disorders of mitochondria, can cause accumulation of lactic acid and lowering of serum pH. Absolute concentrations of triglycerides, phospholipids, total and free cholesterol, total protein, very low and low density lipoproteins can be estimated from the line shape analysis of ^{1}H NMR. The assays can be done in a matter of hours as compared to the laborious biochemical techniques.

Coronary heart diseases (CHD), is a major cause of mortality and morbidity. It affects as many as 30% of the individuals under the age of 70. Based on clinical studies, several environmental and biochemical risk factors have been associated with CHD. High levels of cholesterol in triglyceride-rich lipoprotein particles (very low density lipoproteins (VLDL) and low density lipoproteins (LDL)) and lower levels of cholesterol in high density lipoprotein (HDL) particles are associated with high risks of CHD. Conventionally, one directly measures the amount of cholesterol carried by these lipoproteins through measurements of cholesterol or apo-lipoproteins. ^{1}H NMR provides a direct analysis of lipoprotein subclasses of different sizes as a basis of quantification. Such analysis has provided better understanding of the risk profiles related to lipoprotein heterogeneity and for LDL lowering therapeutical procedures. Pattern recognition and ^{1}H NMR based metabonomics using human serum allows correct diagnosis, not only of the presence but also of the severity of CHD. The proposed diagnostic assay may prove as a simple replacement for the invasive and expensive angiographic procedures which are currently employed for CHD.

Unlike urine and serum, cerebral spinal fluid (CSF) can be obtained only from subjects for whom lumber puncture is done for other pathological tests. The metabolic profile of CSF provides a window for the central nervous system (CNS). The NMR spectrum of CSF provides a powerful strategy in the diagnosis of neurological diseases. Several markers of neuronal damage have been identified in ^{1}H NMR. These include, neuron specific molecules (enolase, creatine, citrate, inositol), metabolic changes following nerve ischemia (through molecules such as Ala, Phe, Tyr, Lys), products of anaerobic metabolism (lactate, acetate and ethanol), and molecules resulting from the damage of membranes of cells (e.g. choline and products of lipid per-oxidation). For example, elevated level of m-inositol is a good indicator for Alzheimer disease. In degenerative dementia, there is a significant increase in Leu and Ile levels. There are significant changes in the metabolic profile of CSF in patients suffering from multiple sclerosis.

Perchloric extracts (PCA) of tissues also provide useful materials for studies of metabolism. The ^{1}H and ^{13}C multi-dimensional experiments on PCA extracts enable quantitative measures of various metabolites present in tissues under various conditions.

4. CELLULAR NMR SPECTROSCOPY

Recall that the non-invasive nature of NMR makes it an ideal tool for *in vitro* applications to study macromolecules in aqueous solutions and solid state studies of macromolecular assemblies. NMR can also be used in a non-destructive manner to study suspension of cells in suitable growth media. It has become a well established tool for *in vivo* observation of metabolites and metal ions. Metabolic fluxes can be measured in bacterial and even certain kinds of mammalian cells. More recently, the technique has been used to obtain information on the structure and dynamics of over-expressed macromolecules in cells. This area is sometimes called *"In-Cell NMR"*.

4.1 Technical Aspects of NMR in Cells

As discussed earlier, NMR can provide information only on *mobile* nuclear spins. This is simply because if the rotational correlation times are longer than 100 μs, then the broadening of the resonance signals can lead to their complete disappearance. Thus, the viscosity of intra-cellular medium can be a crucial factor in detection of cellular molecules. Experiments based on fluorescence and NMR show that the viscosity of cytoplasmic medium is only about two times that of free water. This implies that it should be possible to detect most of the molecules present in soluble form in cells (by NMR). However, binding of molecules to large cellular components such as membranes and other macromolecules decreases mobility causing serious line broadening and potential loss of signals. In such cases, it may be necessary to disrupt the cells and obtain information on the metabolites in cell free extracts (CFE).

The main challenges for *in-cell* NMR studies are to produce samples with good homogeneity and to keep the cells alive and metabolically active during the course of experiments. Both the cell survival-rate and the homogeneity of the sample increase with lower concentration of cells. However, this is at the expense of sensitivity of the detected signals. The quality of spectrum of tightly packed cells is lower, possibly because of non-uniform distribution of cells and accompanying inhomogeneity of such samples. In a typical *protocol,* the cells are harvested from a growth medium and suspended in suitable medium. During re-suspension, care has to be taken so that the cells do not lyse. The cell suspension is made as homogeneous as possible avoiding macroscopic clumps of cells.

Most cell lines require well regulated medium including buffered pH, constant supply of nutrients and oxygen and sterile conditions to survive and to retain their activity during the long periods of time over which the NMR experiments are performed. The survival rates depend on the cell type. For example, insect cells are very delicate. Bacterial cells survive longer when grown and re-suspended in commercially available growth media. Use of ^2H_2O can also affect cell metabolism and their survival rates. For example, bacterial cells survive longer in deuterated medium.

The cells can remain active for longer periods of time and without alteration in their metabolic profile, if the NMR tubes are modified to allow constant exchange of nutrient media and oxygenation. Among the various possibilities to achieve these conditions, gel thread methods are the simplest and closely approximate biological situations. By using suitable perfusion media, cells can be kept alive for several days in a NMR tube. Special bio-reactors have been designed for this purpose. Temperature is an important parameter for cell growth and survival. The medium can be exchanged at a constant temperature, such that it is ideal for the cell survival and growth (i.e. 37^0C for human cells) and by insulating the transfer lines. Immobilized cells experience metabolic restrictions in the form of reduced diffusion of metabolites and ions and selected absorption on the gel matrix. Most of these modifications require use of 10 mm NMR tubes, with the concomitant problems of homogeneity. Also lack of cryogenic probes for such sample tubes restrict use of exchange media.

Thus, the study of cell metabolism under aerobic conditions requires special modifications in the probe and care in sample handling. On the other hand, glycolysis, and inactive cells can be analysed by simply putting the sample in a 5 mm NMR sample tube, making some provision to prevent cells from settling down. Shimming may be difficult with cell slurry. Shimming may therefore be done on a dummy sample with similar solvent conditions.

Metabolic studies can also be made by freezing the cell activity at a particular time, and preparing a perchloric acid cell free extract. Extracts give very sharp NMR signals. Even tissues can be powdered and extracted with per chloric acid for this purpose.

4.2 Studies on Spermatozoa

Several aspects of the type of information that can be obtained from cellular NMR can be illustrated from the reported studies conducted on mammalian spermatozoa. There are several reasons for choosing these cells. These are the paternal cells in reproductive biology of mammals. The cells are readily available in large quantities from several sources of animals and humans. The metabolic behaviour of spermatozoa has wide implications in understanding basic reproductive biology, clinical health care, artificial insemination, fertility control and animal husbandry. Spermatozoa survive on glycolysis and fructolysis under anaerobic conditions. These cells do not form lumps or settle in NMR tube. The conclusions thus arrived are quite general and are representative of the type of studies that can be made on other cell-lines.

The overall structure of spermatozoa can be divided into three parts. The head contains the paternal chromatin. It has a cap like cover called acrosome and contains certain hydrolytic enzymes, which are released during fertilization with the egg. The mid-piece contains mitochondria and contributes to most of the NMR signal. The tail is responsible for the movement of spermatozoa (*motility*) in the seminal fluid. Motility and fertilizing ability are the major markers of spermatozoan activity, which can be influenced by a variety of physical and chemical factors.

The male reproductive organ epididymis acts as an efficient storage system and enables spermatozoa to mature. There are three main parts of epididymis: caput is the birth place of spermatozoa, corpus where they are in semi-matured stage and cauda where they are fully matured. During passage through epididymis, spermatozoa undergo major morphological and biochemical changes. The final stages in the cycle of activity of spermatozoa are capacitation, a process where spermatozoa are activated and the acrosomal reaction where the cap region delivers hydrolytic enzymes so that it can fertilize with the egg. These processes take place in the female reproductive tract. Changes in the biochemistry of spermatozoa in these steps have been studied using NMR and other physical techniques.

4.3 Identification of Low Molecular Weight Compounds in Cells

^{1}H and natural abundance ^{13}C NMR allows direct detection of the molecules present in the cells, which can be picked up through the use of 2D techniques. The wealth of information in 2D DQF-COSY and HSQC spectra is obvious from Figure 12.10. For a particular cell-line, the composition and concentration of metabolites changes dramatically, not only from species to species, but also within samples from the same group of animals. The signals and levels of several amino acids such as Ala, Arg, Asp, Gly, Glu, Ile, Leu, Lys, Met, Pro, Thr, Val, molecules such as Lactate (Lac), m-inositol, glycerophosphorylcholine (GPC), hypotaurine and aromatic compounds, such as uridinediphosphoglucose (UDP) can be identified and their concentrations measured. The relative concentrations of these molecules change as a result of maturation. Two unexpected molecules detected by NMR in spermatozoa are β-Ala and hypotaurine, which have been reported for the first time in goat spermatozoa (Figure 12.10).

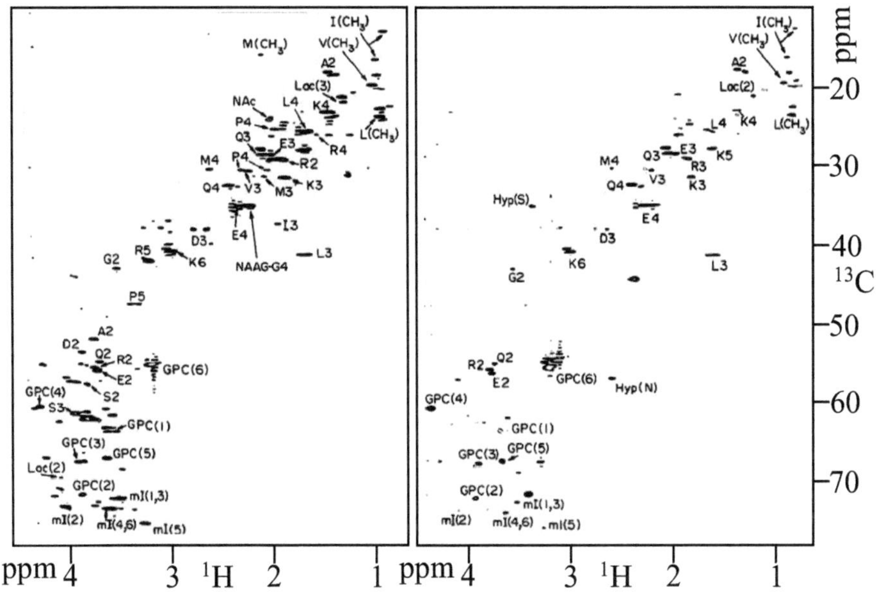

Figure 12.10: 2D ^{13}C-^{1}H HSQC spectra of spermatozoa from: (A) caput (least matured) and (B) cauda region (fully matured cells). Major changes in the levels of different metabolites during the cell maturation can be picked up. In particular note that hypo-taurine, which is a key molecule involved in cell maturation, is switched on once the cells are in a matured stage. S. Srivastava and G. Govil. 5, 1039-1057, 2001. (Reprinted with permission from Bentham Science Publishers).

4.4 Biochemical Changes during Cell Maturation, Modification and Differentiation

When spermatozoa cells are born in caput region, one sees mainly ADP signals with very little contribution from ATP. ATP levels increase as the cells mature. There are also major changes in the levels of several other compounds during cell maturation as spermatozoa move from caput to cauda region (Figure 12.10). The total inositol (m-inositol+s-inositol) decreases with cell maturation, while the relative concentration of GPC and amino acids Arg, Glu and Gln increases. The signal for the unusual amino acid hypotaurine, which is important for sperm survival capacitation and fertilization, is weak in caput region.

4.5 Glycolysis in Cells

The overall reaction in glycolysis under anaerobic conditions can be represented by:

$$Glc + 2\,ADP + 2\,P_i \;= 2\;\; Lac + 2\,ATP + 2\,H^+ + 2\,H_2O$$

Metabolic pathways can be followed using ^{13}C labelled substrates as part of the nutrients. For example, when cells are fed on [1-^{13}C]Glc, the label appears at the corresponding carbon atoms in metabolites resulting from normal glycolysis. For example, a decrease in the Glc signal and a concomitant build up of signals from [3-^{13}C]Lac can be monitored in real time (Figure 12.11). Thus, the rate of glycolysis can be followed in several ways: a decrease in the [1-^{13}C]Glc signal or an increase of [3-^{13}C](Lac) signal, with intermediate metabolites containing labelled carbons; using ^{31}P NMR to detect ADP and ATP levels and intra and inter cellular pH; or ^{1}H NMR for detecting intermediate metabolites.

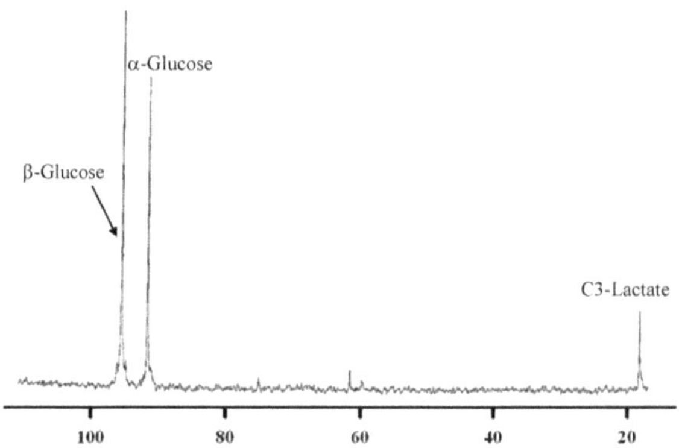

Figure 12.11: ^{13}C NMR of cells undergoing glycolysis, using ^{13}C labelled Glc. Note that the levels of intermediate metabolites allows the reaction to be followed in real time. The two Glc signals are due to co-existing α and β anomers of Glc. (Thanks to S. Srivastava for providing this figure).

The effect of various external factors on glycolysis can be determined using NMR. In the case of spermatozoa, the energy released in glycolysis and trapped in the form of ATP is needed for its biochemical requirements, maturation, capacitation, motility and fertilization. Cells, which are dormant (for example, when stored under frozen conditions or starved) have a low level of ATP and higher level of ADP and P_i. When starved cells are fed on Glc, the ATP level builds up. Inactive cells do not consume Glc and there is no increase in ATP/ADP ratio. Such cells also do not show motility or fertilizing capability. Such studies have obvious advantages in studies on fertility and artificial insemination.

4.6 Effect of Exogenous Compounds on Metabolism

Glycolysis can be used as an example to understand the effect of exogenous compounds on the metabolic pathways in cells. Such studies are of potential value in drug discovery and in study of inhibition or activation of cell growth.

L-Arg is an important molecule in sperm metabolism and is known to increase spermatogenesis. Administration of Arg to oligospermic and asthenospermic patients leads to both an improvement in sperm count and in motility. NMR studies have shown that the presence of Arg increases Glc and Fru consumption by cells with a concomitant increase of Lac production and a decrease in pH. It also reduces damage due to the presence of ionizing radiation and reverses impairment caused by glycolytic inhibitors (potential contraceptives). Irrespective of the nature of induction of per-oxidation, L-Arg is found to reduce the extent of lipid per-oxidation in a concentration dependent manner. Both L-Arg and α-tocopherol act synergistically in preventing lipid per-oxidation. L-Arg has a high degree of specificity in its catalytic activity as structurally related amino acids such as L-ornithine, L-Lysine and nitro-arginine do not act as activators. NMR studies have also established that Arg stimulates nitric oxide synthesis in spermatozoa, which is a key factor in the enhancement of metabolic activity and plays a role in preventing lipid per-oxidation. It has been concluded that the mechanism of action of L-Arg is primarily through increased biosynthesis of nitric oxide in spermatozoa.

4.7 In-Cell Studies of Macromolecular Structure and Dynamics

As discussed earlier, there are several methods for studying conformation, function and dynamics of biological macromolecules under *in vitro* conditions. However, the role of a macromolecule inside a cell is determined by its location and the 3D structure present inside the cell, which may be different from the structure under *in vivo* conditions. It may appear that given the large number of narrow and intense resonances from small molecules inside the cell, it may be difficult to look at signals of biopolymers, against the background of small molecules. It has been possible to use genetic tricks to over-express molecules of interest with ^{15}N, ^{13}C and ^{19}F labels in live-cells and to obtain decent ^{1}H-X HSQC spectrum. With the use of cryoprobes, intracellular concentrations of around 200 μM of the over-expressed proteins and other biological molecules have been detected.

The objective of such studies is to obtain conformation and dynamics of biopolymers in their natural environment. Indeed it has been shown that the bacterial protein FlgM is unfolded under *in vitro* conditions, but has an ordered structure in *Escherichia coli* cells. One reason for this behaviour is molecular crowding. Alternatively, interaction with other molecules in the cell can change the structure and dynamics. For this reason, studies of *in-cell* protein structures have attracted attention by several groups. The other technique, which has been used for this purpose is fluorescence spectroscopy. However, fluorescence has limited applications as it cannot provide detailed information on 3D structures.

The main trick used in the NMR studies of large molecules in cells is to selectively label the protein (or other molecules of interest) such that the signals from the label spins are much above the background signals from small molecules. In fact, the conditions under which over expression is conducted, is the most crucial factor in influencing spectral quality and information content of the resulting spectra (see Figure 12.12, for an example of in-cell HSQC spectrum). Use of ^{15}N labelled medium can normally put the isotope at all nitrogen sites in the cell. However, through clever genetic manipulations, it has been possible to over-express selective proteins. A more potential tool is to use ^{15}N labelled amino acids such as Lys, His and Arg in the growth medium and thereby selectively label (or unlabel) proteins during synthesis in cells. These labelled proteins can provide valuable information about the backbone and side chain conformations and dynamics at the selective sites.

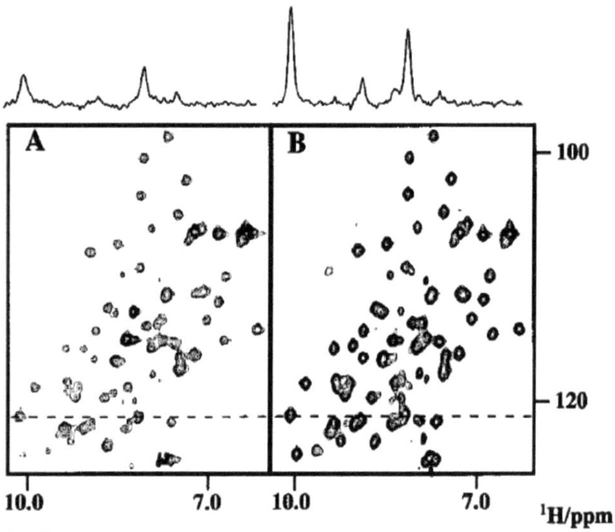

Figure 12.12: ^{1}H-^{15}N HSQC in-cell spectrum of a protein Nmer A. (A) Sample grown on ^{15}N-labeled M9 minimal medium and (B) sample grown on ^{15}N labelled and deuterium medium. Z. Serber, R. Lewidge, S.M. Miller and V. Dotsch. J. Am. Chem. Soc. 123, 8895-8901 (2001) (Reprinted by permission from The American Chemical Society).

The use of ^{13}C based approach has several advantages over ^{15}N labelling and provides valuable information about side chain conformations and dynamics. Protons in the C-H bond do not exchange with cellular medium, unlike amide protons. Methylene and methyl groups in protein side chains can be detected with high sensitivity due to larger number of protons attached to such side-chains. These groups have fast internal rotations and the lines are relatively sharper. However, the greater abundance of carbon in small molecules also gives rise to higher background during the metabolism of cells using ^{13}C labels in the growth medium. Again, selective labelling procedures can be helpful. For example, the use of ^{13}C labelled methionine has been used for the expression of calmodulin for selective labelling of Met. Ala is another useful amino acid for selective labelling.

An interesting label is ^{19}F, since there are almost zero background signals for this isotope. ^{19}F can be introduced through chemical modification in the building blocks of macromolecules, before their incorporation in proteins. Many interesting examples exist in literature, describing incorporation of ^{19}F in proteins, with little effect on their structure and function.

Even though *in-cell* studies of biological molecules are important and can provide valuable information on *in vivo* activity of proteins and other molecule, it cannot replace *in vitro* experiments discussed earlier, for the complete 3D structure determination. Only a limited number of experiments can be performed under *in vivo* conditions. The conditions are less stable and lines are much broader. The relative insensitivity of the NMR technique demands a high concentration of the over-expressed molecule in the cytoplasm, to a level as high as almost 5% of individual protein out of the total soluble proteins. The high cellular density in the NMR tube leads to lack of oxygen, thus switching the cells to anaerobic metabolism. This influences the intracellular pH.

A combination of the two lines of studies is likely to be very fruitful for obtaining detailed information on structure and function of biological molecules.

5. STUDIES OF TISSUES USING SOLID-STATE NMR TECHNIQUES

Tissues differ from biological fluids and cells in that they are solid like materials. Further, cells in freshly excised tissues are metabolically active. Thus, *ex vivo* tissue is not really *dead*. Therefore, it is desirable that tissues are frozen at liquid nitrogen temperature, after removal from the host, and a suitable protocol employed to thaw them, before obtaining the spectra. If proper precautions are taken, then one can obtain *metabolic snapshots*, of the metabolic state of the tissue at the time that the sample was removed from the body.

As discussed earlier, one approach to obtain information from tissues is to grind them and prepare cell free extract using perchloric acid. However, even when conducted at low temperatures such procedures can produce artefacts due to a disturbed metabolism. However, solid state NMR techniques can be used for studies of intact tissues. Recently, high-resolution solid-state techniques have been applied to study tissues as well as for *in-cell* NMR studies. Using Magic Angle Spinning (MAS), well resolved spectra are obtained. A direct observation on tissues avoids artefact introduced by extraction, though the lines are broader and less resolved than those from cells and cell free extracts. In fact, MAS based analysis of tissues has been developing into a powerful technique for obtaining high resolution spectra, particularly intact (hard) tissues from animals (for example, sections of brain or kidney tissues) as well as plants (roots, leaves).

As an illustrative example, Figures 12.13 A and B show the ^{1}H MAS-NMR spectra of intact rat heart tissue and its extract, respectively. Likewise, Figures 12.13 C and D show the ^{1}H MAS-NMR spectra of rat heart mitochondria and its extract, respectively. Notice that the spectra from the intact cells (Figure 12.13 B and D) are dominated by lipid signals, which are absent in the cell, extracts (Figure 12.13 B and D). On the other hand, signals from several low molecular weight metabolites are seen in the spectra of extracts rather than that from intact cells, suggesting that these metabolites are in highly restricted environments in intact cells.

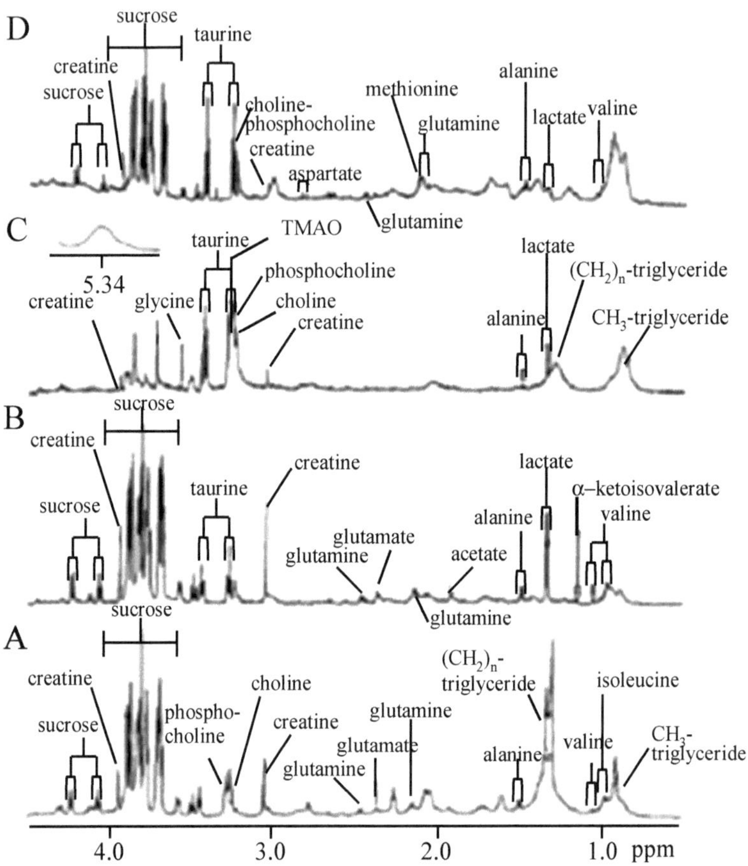

Figure 12.13: 600 MHz ^{1}H MAS-NMR spectra of (A) an intact rat cardiac tissue, (B) an extract of rat cardiac tissue, (C) rat heart mitochondria, and (D) an extract of rat heart mitochondria. W.H. Dunn, N.J.C. Bailey and H.E. Johnson. Analyst 130, 606-625 (2005). (Reprinted by permission from Royal Soc. London).

There are several problems in using solid-state NMR techniques for such samples. The filling of rotors used in solid state NMR is a time-consuming process. Only a small amount of material can be placed in the rotors. There are possibilities of degradation of the sample, even if the entire process is done at low temperatures. In the case of cells, survival time in solid-state rotors is another problem. In heterogeneous systems, magnetic susceptibility gradients are generated at the interfaces of regions with different magnetic properties, which may cause line-broadening. Fortunately, the susceptibility of water and cytoplasmic media is very similar. This may not be the case for molecules residing near the membranes. This is combined with the fact that the currently cryogenic high-resolution magic angle probes have not been developed. Thus the sensitivity of NMR for such studies is much lower than that for conventional *in vitro* systems.

There is also the problem that a metabolically active tissue would consume available oxygen and become hypoxic during NMR measurements. Nevertheless, NMR studies of tissues, is an emerging area and may find applications in several areas of biological and clinical sciences. One such emerging area is the study of genetically modified cells with knocked out genes, using animal models. The metabolism of such cell lines can provide important information on gene function and in drug development.

6. REFERENCES

6.1 Further Reading

H. Antti, E. Holmes and J. Nicholson, *Multivariate Solutions to Metabonomic Profiling and Functional Genomics*, http://www.acc.umu.se/~tnkjtg/Chemomatrics/Editorial (2002).

D.I. Hoult, S.J. Busby, D.G. Gadian, G.K. Radda, R.E. Richards, and P.J. Seeley, *Observation of tissue metabolites using ^{31}P nuclear magnetic resonance*, Nature **252**, 285–287 (1974).

P.A. Srere and J. Ovadi, *Enzyme-enzyme interactions and their metabolic role*, FEBS Letters **268**, 360–364 (1990).

M. Ala-korpela, A. Korhonen, J. Keisela, S. Horkko, P. Korpi, L.P. Ingman, J. Jokisaari, M.J. Savoloinen and Y.A. Kesalniemi, *^{1}H NMR-based absolute quantitaion of human lipoproteins and their lipid contents directly from plasma*, J. Lipid Res. **35**, 2292–2304 (1994).

Z. Argov, M. Lofberg and D.L. Arnold, *Insights into muscle diseases gained by phosphorus magnetic resonance spectroscopy*, Muscle and Nerve, **23**, 1316–1334 (2000).

D.E. Bugay, *Characterization of the solid-state: Spectroscopic techniques*, Adv. Drug Deliv. Rev. **48**, 43–65 (2001).

Z. Serber, R. Lewidge, S.M. Miller, and V. Dotsch, *Evaluation of parameters critical to observing proteins inside living Escherichia coli by in-cell NMR Spectroscopy*, J. Am. Chem. Soc. **123**, 8895–8901 (2001).

J.T. Brindle et al., *Rapid and non-invasive diagnosis of the presence and severity of coronary heart disease using ^{1}H-NMR-based metabonomics*, Nat. Med. **8**, 1439–1443 (2002).

J.D. Otvos, E.J. Jeyarajah, and W.C. Cromwell, *Measurement Issues related to lipoprotein heterogeneity*, Am. J. Cardiol. **90S,** 22i–29i (2002).

O. Corcoran, and M. Spraul, *LC-NMR-MS in drug discovery*, Drug Discov. Today **8**, 624–631 (2003)

Y.-L. Chung, M. Stubbs, J. Griffiths, *Metabolic Profiling in Tumors by in Vivo and in Vitro NMR Spectroscopy* in *Metabolic Profiling: Its Role in Biomarker Discovery and Gene Function Analysis*, G.G. Harrigan and R. Goodacre (eds.), (2003).

H. Lei, X.H. Zhu, X.L. Zhang, K. Ugurbil and W. Chen, *In vivo ^{31}P magnetic resonance spectroscopy of human brain at 7 T: An initial experience*, Magn. Reson. Med. **49**, 199–205 (2003).

M.R. Viant, E.S. Rosenblum and R.S. Tjeerdema, *NMR-based metabonomics: A powerful approach for characterizing the effects of environmental stressors on organism health*, Environ. Sci. Technol. **37**, 4982–4989 (2003).
W.H. Dunn, N.J.C. Bailey and H.E. Johnson, *Measuring the metabolome: current analytical technologies*, Analyst **130**, 606–625 (2005).
M. Mehta, H.M. Sonawat and S. Sharma, *Malaria parasite-infected erythrocytes inhibit glucose utilization in uninfected red cells*, FEBS Letters **579**, 6151–6158 (2005).
G.A. Nagana Gowda, Omkar B. Ijare, B.S. Somashekar, Ajay Sharma, V.K. Kapoor, and C.L. Khetrapal, *Single-step analysis of individual conjugated bile acids in human bile using ^1H NMR Spectroscopy*, Lipids **41**, 591–613 (2006).

6.2 Books and Reviews

M.P. Deutscher (ed.), *Guide to Protein Purification*, Methods Enzymol. **182**, 1–894 (1990).
J.M. Berg, J.L. Tymoczko and L. Stryer, *Biochemistry*, 5th International Edition, W.H. Freeman and Co., (1995).
U. Holzgrabe, *NMR Spectroscopy in Drug Development and Analysis*, Wiley, (1998).
A.L. Lehninger, *Principles of Biochemistry*, W.H. Freeman and Co. (2000).
S. Srivastava and G. Govil, *Applications of NMR to the study of cells and body fluids*, Current Organic Chemistry **5**, 1039–1057 (2001).

CHAPTER 13

MRS STUDIES OF METABOLISM IN ANIMALS AND HUMANS

1. INTRODUCTION

One of the most spectacular advances of modern science is the ability to monitor metabolic processes in intact organs of animals and humans. *In vivo* spectroscopy of body organs is perhaps the most challenging and fruitful outcome of the development of NMR. This emerging area is called (nuclear) Magnetic Resonance Spectroscopy (MRS). The technique of MRS has provided new insights on the biochemical changes, which drive the function of complex systems such as human brain and the effects of environmental factors, genetics and disease on metabolic processes in complex living systems. It has provided a platform for discovery of biologically active molecules. More recently, MRS has been extended to plant systems, discussed in Chapter 14.

Obviously, it is more meaningful to study parts of the larger organs at a time, instead of looking at the whole organ. The first step in the study of metabolism in intact organs is therefore to select a volume of interest (VOI) using special techniques. This process is called spectral localization. The NMR spectrum is obtained from such a VOI. At today's level of NMR sensitivity, one uses volumes which are of the order of a fcw ml. The NMR spectrum from the VOI provides chemically shifted resonances of different metabolic compounds in the organ. Their quantitative analysis provides vital information of the function of the cells with in the VOI. MRS has added a new dimension to NMR, since the concentrations and synthetic rates of individual molecules in the VOI of specific organs such as liver, brain and muscles can be measured. Such studies provide insight into the biochemical activity of organs.

Each organ consists of several types of cells. Each group of cells has a distinct function. The MR signals from a VOI in an organ are sum total of the signals from millions of cells, with multiple cell types. The resonance lines are fairly broad. Consequently, the detection limits in whole organs are much lower as compared to excised cells and tissues. Metabolites which can be detected in such studies are limited by their broad resonances, and the associated problems of sensitivity and resolution. Therefore, only a limited number of metabolites can be detected. The interpretation is complicated due to multiple cell types. Nevertheless, *in vivo* spectroscopy of organs provides a means to probe the biochemistry of living systems. The relative concentrations of the metabolites provide information on metabolic activity of the organ. It also provides information on the biochemical differences between normal and diseased cells. The response of therapeutic modalities and effects of medical treatment can be monitored.

An early lead to MRS was provided by ^{31}P NMR studies of intact muscles. In recent years, resolved ^{1}H signals from N-acetyl aspartate (NAA), Glu, Gln, Cho, N-acetyl aspartyl glutamate (NAAG) and other metabolites have been detected in human brain and other organs using ^{1}H NMR. Through the use of ^{13}C-^{1}H correlation techniques, several other metabolites have been identified. Because of reasons of sensitivity, the studies are often limited to the more abundant nuclear spins such as ^{1}H and ^{31}P.

2. TECHNIQUES FOR DETECTING MR SIGNALS

Techniques of MRI and MRS are mutually complementary. However, there is one basic difference between the two. In MRS, one needs a highly homogeneous field over the VOI during signal acquisition in MRS. Therefore, field gradients cannot be used during signal acquisition. As discussed earlier, in MRI one usually observes a single molecule (H_2O). Its resonance is separated in the 3D space using magnetic field gradients. On the other hand, in MRS the resonances of several different metabolites are observed simultaneously from the selected VOI. The dispersion is achieved by the chemical shifts of different spins. Consequently, the pulse sequences used in MRS are much more complex.

Since the objective of MRS is to detect metabolites which are present in relatively lower concentrations, water suppression pulses have to be used for ^{1}H NMR studies. Similarly, resonances from fat have to be suppressed. Such complex pulse sequences can produce several undesirable phase-coherences. Therefore, spoiler pulses have to be regularly applied in MRS to avoid artefacts in the detected signals.

A number of methodologies have been developed for MRS. These include the use of surface coils, rotating frame zeugmatography, Topical Magnetic Resonance (TMR), Depth Resolved Spectroscopy (DRESS), Fast ROtating Gradient Spectroscopy (FROG), Image Selected *In vivo* Spectroscopy (ISIS), Stimulated Echo Acquisition Mode (STEAM), Point RESolved Spectroscopy (PRESS), and Chemical Shift Imaging (CSI). While some of these techniques are of historical value, extensive use has been made of surface coils, DRESS, ISIS, CSI, STEAM and PRESS in studies of *in vivo* spectroscopy. A major difference among these techniques arises from the way that the spectral localization is achieved. This is usually done by manipulation of the magnetic field B_0 or the RF field B_1. The use of 2D FT spectroscopy has also been made in some cases, though due to large line-widths the additional information derived is limited.

2.1 Spectral Localization

To obtain meaningful results from MRS, it is essential that the volume from which the signal originates is carefully selected and identified. The signals from the VOI should be obtained with the best possible sensitivity and with minimal contributions from other regions. The technique for selecting the VOI from the whole body organ is called *localization.* Volume localization is in fact, the first step in MRS. Several techniques have been proposed for this purpose. However, only a limited number of methods have proved useful in clinical and basic studies.

Historically, for the purpose of spectral localization, the use of surface coil was introduced way back in 1980. The surface coil consists of one or two copper loop of wire and it is positioned on or under a subject's organ, close to the anatomical volume of interest. The coils are positioned such that the major component of the B_1 RF field generated by the coil is orthogonal to B_0. Choice of a suitable geometry of the coil provides a rough spatial isolation in such experiments. Surface coils can be used both for RF transmission and for the detection of NMR signals. The problem with such coils is that the *localization* is not very precise. This is primarily because of field inhomogenity of RF coil. Metabolites close to the coil surface are detected with better sensitivity, while distant regions give poor signals. Other difficulties include inhomogeneous transverse magnetic fields and the problems associated with the location and assessment of the VOI. However, surface coils have been used with advantage with certain protocols in MRS, to achieve volume localization. The method is still used to obtain spectra from surface and superficial tumours.

Another technique, which depends on the use of surface coils, is the rotating frame spectroscopy. It employs gradient B_1 fields of a transmitter coil, such that B_1 decreases with the distance from the coil. The spins therefore experience a distance dependent flip angle. This technique is used for spatial coding along the axis of the surface coil. Rotating frame spectroscopy has been used to obtain spectra from various depths along the axis of the surface coil. However it lacks lateral discrimination. The method also suffers from contamination of signals from surrounding layers of tissue outside the VOI.

Introduction of the technology of controlled magnetic gradients for MRI, has led to development of several other localization techniques. One of the earlicst one is Topical Magnetic Resonance. This technique uses non-linear magnetic field gradients to generate a spherical region of homogeneous magnetic field. Surface coils can be used to transmit and receive signals from the selected VOI. The subject has to be realigned and careful shimming has to be done, such that the profiled field lies in the VOI.

In modern MRS techniques, magnetic field gradients in either the B_0 or the B_1 fields are used to achieve localization. The methods which are most commonly used in clinical settings are the STimulated Echo Acquisition Mode (STEAM), Point RESolved Spectroscopy (PRESS) and Image guided *In vivo* Spectroscopy (ISIS). Both STEAM and PRESS are highly effective for volume localization. However, each of the two techniques has their relative advantages and disadvantages. These sequences acquire data from a Single Volume (SV). Multi-Voxel (MV) spectroscopic methods have also been developed. These techniques acquire spectra from several voxels simultaneously and allow Chemical Shift Imaging (CSI), or Spectroscopic Imaging (SI).

2.2 Water Suppression

In ^{1}H MRS, detection of signals of small concentrations of metabolites against the background of the strong water signal (and sometimes the fat signal) is a major problem. Therefore, an important requirement for *in vivo* ^{1}H spectroscopy is suppression of resonances from water and lipids and fat. In Chapter 2, we have discussed methods for suppression of strong signals. The most common method for water suppression used for *in vivo* ^{1}H NMR studies, is to use a train of water targeted pulses. This is commonly known as CHEmical Shift Selective water saturation (CHESS). These pulses which excite a limited band of RF frequencies (typically around 60 Hz) are placed before using the localized volume sequence. CHESS usually consists of three frequency selective pulses (WS1, WS2, WS3), which are used to excite a narrow band of frequencies around the water signal, along with crusher gradients to eliminate water signal in the transverse plane. The other metabolites in the selected region are excited, at a time when the water magnetization is beginning to recover and is not available for excitation. The efficiency of water suppression depends on the field homogeneities and on T_1 of water in the tissues.

2.3 Depth Resolved Surface Coil Spectroscopy (DRESS)

One of the early methods for avoiding spatial contribution from unwanted regions of an organ was the use of Depth REsolved Surface coil Spectroscopy (DRESS; Figure 13.1). In this, a shape selected RF excitation pulse (B_1) is applied in the plane of the surface coil, in presence of a slice selection G_s gradient. This gradient is applied in a direction perpendicular to the coil surface, for providing depth selection. The volume observed is approximately a thick slab, which is located at the intersection of the excited slice and the sensitive volume of the coil. The sensitivity of the spectrum reduces with the distance from the cylindrical axis of the surface coil, and is directly related to the depth of the selective pulse. There is a significant loss of signal for metabolites with shorter T_2, due to a mandatory delay of a few msec between the excitation and detection.

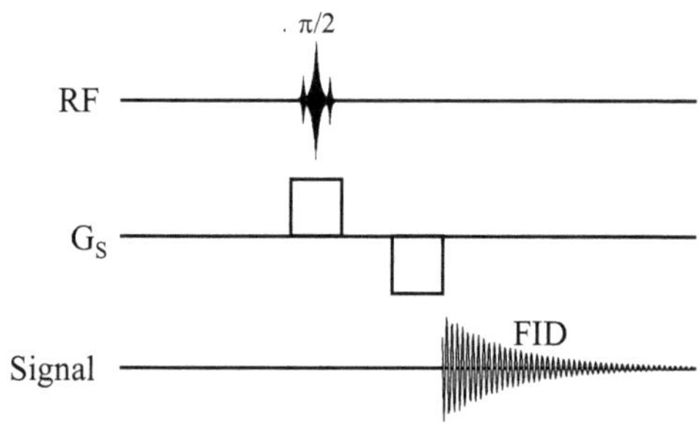

Figure 13.1: Pulse sequence used for DRESS. The RF pulse is applied in presence of magnetic field gradient (G_s).

2.4 Image Guided Protocols

Several steps are involved in the current methodologies for obtaining meaningful ^{1}H spectra from body organs. Most MR localization methods use MRI to guide the selection of VOI for doing spectroscopy. Therefore, a typical spectroscopy protocol includes an image obtained using MRI as a guide for volume selection. This is followed by global and voxel shimming over the selected voxel and then by the water suppression protocol. The remaining steps involve excitation of the VOI, followed by detection of the signal in the time domain. Application of FT provides the spectrum in the frequency domain from the VOI.

2.5 Image Selected in vivo Spectroscopy (ISIS)

The method involves a series of measurements having different preparation schemes (Figure 13.2 B).

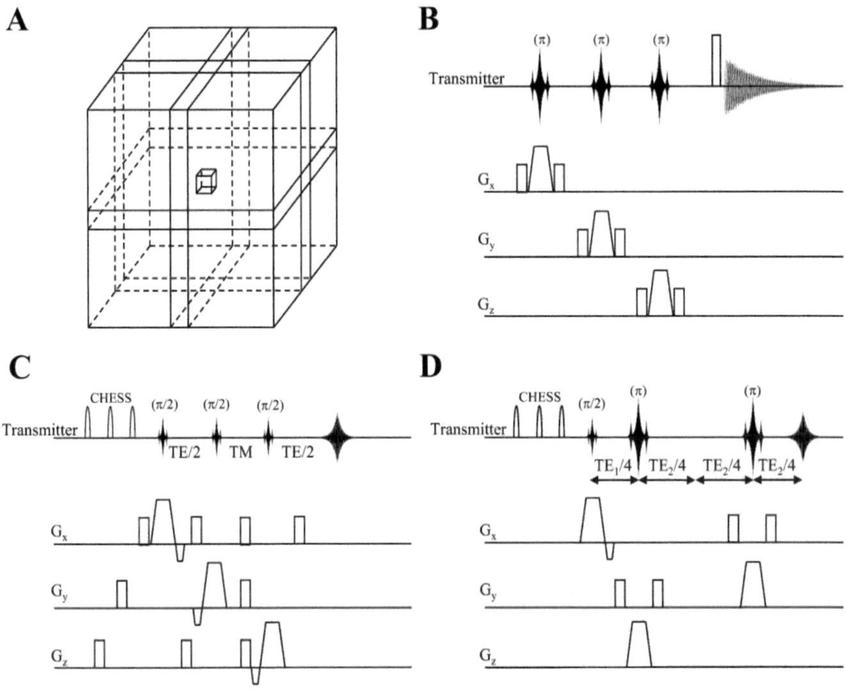

Figure 13.2: (A) In MRS, a voxel is selected. This is usually defined by the intersection of three orthogonal planes by applying field gradients in orthogonal directions. Some of the important pulse sequences used in MRS of organs are: (B) ISIS, (C) STEAM and (D) PRESS. The three pulse CHESS sequence has been used in C and D to saturate water resonance.

During the preparation period, selective inversion (π) pulses are applied in the presence of gradient pulses, along the orthogonal directions to the three axes, x, y and z. Application of selective π pulses in a sequence, allows selection of a cube at the intersection of the planes that are orthogonal to x, y and z axes, respectively (Figure 13.2 B). The preparation period is followed by a time delay to allow for eddy currents to decay. The longitudinal magnetization is sampled by an excitation pulse, which is followed by acquisition of the FID. The excitation pulse itself can be a simple $\pi/2$ pulse, or a sequence of pulses designed so as to achieve measurements of $T_1$and T_2, water suppression, or spectral editing. To achieve signal from a VOI, a set of eight acquisitions is performed, each consisting of a different combination of inversion pulses. Addition of the eight FID signals, results in a constructive averaging of signals from inside the VOI, while destroying the signal from outside this region. This technique has been discussed in Chapter 3, under the name phase cycling. The size and position of the VOI can be varied by a judicious choice of the overall profile of the pulse scheme. Surface coils are employed with ISIS to improve S/N ratio.

ISIS is T_1-weighted because the longitudinal magnetization recovers during the period between the preparation period and the excitation pulse. There is no T_2-weighting, since the signal is acquired just after the excitation pulse. Thus, the signal loses due to T_2 and J modulations are minimal. ISIS is therefore a method of choice in situations where metabolites have short T_2.

2.6 Stimulated Echo Acquisition Mode Spectroscopy (STEAM)

STEAM involves successive excitation of three slices through the use of three selective $\pi/2$ pulses. Each one of these pulses is applied in the presence of mutually orthogonal x, y and z gradients. The localized VOI is at the intersection of the three slices, chosen by the gradient pulses, as indicated in Figure 13.2 C. After the first two pulses, half of the resonating spins are in the transverse plane; the other half are in the longitudinal plane. There is a mixing period TM following the two pulses. The stimulated echo signal of our interest is detected following a time interval TE/2, after the application of the third pulse. During the mixing period TM, the phase prepared longitudinal magnetization decays according to T_1. The magnetization is brought into the transverse plane by the third pulse. This magnetization generates the STEAM signal. The other half of the magnetization, which is in the transverse plane, is de-phased further by the TM crusher gradient. It does not contribute to the stimulated echo. Thus, only one half of the spin-magnetization in the VOI is translated into the STEAM signal.

The STEAM sequence is conceptually simple. However, the three RF pulses used in this technique, not only generate the stimulated echo, but also lead to a number of undesirable spin coherences. For example, following the third pulse, four echo signals are formed. One is at TM, which arises from the second and third pulses. The second is at TE/2-TM due to the first two pulses. A third one is at TE/2 +TM, and this arises from the first and third pulse. The one of interest is the stimulated echo at the end of the second TE/2 period. Therefore, spoiler pulses have to be used during the TE/2 and TM periods to suppress undesirable coherences.

STEAM has excellent localization properties in single shot. Hence, it circumvents motion related problems, when it is used in single shot experiments. Use of $\pi/2$ flip angles for the three pulses is not crucial. Therefore, lower angles (α) can be used, as long as each of them is less than $\pi/2$. The advantages of using smaller flip angles are the speed of acquisition and the lower power deposition in the VOI, which varies as α^2.

STEAM has been widely used for ^{1}H NMR studies of organs. The information content of the spectrum can be manipulated by choosing the value of TE. Use of long TE (typically 100-300 msec), helps in detecting signals from a number of useful metabolites, such as Cho, NAA, Cr and PCr. However, the use of high TE values lead to spin-dephasing of signals which have multiplet resonances due to the J-couplings. Such signals can be detected by using short TE (typically less than 30 msec).

2.7 Point Resolved Spectroscopy (PRESS)

PRESS uses a double SE pulse sequence and is an extension of the 1D DRESS technique for localization in the 3D space (Figure 13.2 D). In this technique, three mutually perpendicular slices are excited by one $\pi/2$ and two π, frequency-selective RF pulses. These pulses are applied in conjunction with field gradients applied in orthogonal directions.

The first $\pi/2$ frequency selective pulse is applied in the presence of a magnetic field gradient G_x. This selects a slice in the yz plane. Magnetization is allowed to evolve for a time period $TE_1/4$. The first selective π refocussing pulse, in conjunction with a gradient in the z-direction, is applied next. This generates the first echo signal at a time $TE_1/2$. The echo signal is allowed to dephase for a time $TE_2/4$. At this stage, the spins are refocussed by a second RF pulse applied in an orthogonal direction (i.e. the y-axis). The final echo signal appears after a time t (where, $t = TE_1/2 + TE_2/2$). This second SE signal results from the spins in a volume element, which is defined by the intersection of the three excited slices, created by the field gradients. These gradients are applied on orthogonal directions.

Application of this highly asymmetric pulse-sequence drastically reduces problems arising from eddy currents, as the time-interval between the last switched gradient and acquisition is maximized. As in the case of STEAM, spoiler gradients are applied directly following the volume selection gradients. This destroys unwanted coherences and contamination from the volume outside the VOI.

2.8 Comparison of STEAM and PRESS

The techniques of STEAM and PRESS are highly efficient methods for volume localization. The two techniques differ in the nature of the echo signal. In PRESS the complete magnetization is focussed into the echo signal, while in STEAM only part of the magnetization results in useful signals. The techniques have been widely used for single-volume studies. However, each method has its positive and negative points. Since STEAM employs short echo times, it has an advantage while detecting ^1H resonances of metabolites, which have short T_2 relaxation and J modulation. More efficient water suppression can be achieved because CHESS pulses can be placed not only during the preparation period, but also within the sequence, without affecting the echo time used. Use of smaller flip angles helps in lower RF deposition. During the mixing period, the relaxation is only due to T_1 relaxation, because the magnetization is along the longitudinal axis, while the T_2 relaxation occurs only during the TE periods. Thus, by using short TE, very little signal is lost. In comparison, in the PRESS technique, T_1 and T_2 relaxations occur through out the period. A major disadvantage of STEAM is a loss by a factor of two, in the signal intensity. The volume localization sequence is highly susceptible to motion, diffusion and homo-nuclear couplings. STEAM has been applied extensively for ^1H NMR spectroscopy of organs.

The advantages of PRESS over STEAM include a gain in signal intensity by a factor of two. The volume localization is less sensitive to the motion of the subject, the J couplings, and effects arising from eddy currents. For long echo times, PRESS is preferred over STEAM. However, the more efficient water suppression techniques cannot be used with PRESS. Both T_1 and T_2 relaxation occur throughout the pulse period. PRESS is not suitable for metabolites having short T_2.

2.9 Multiple Volume Spectroscopy (MVSI or CSI)

Single volume MRS can be extended to multi-voxel spectroscopy. This methodology is also referred to by acronyms such as Spectroscopic Imaging (SI), Chemical Shift Imaging (CSI), or Magnetic Resonance Spectroscopic Imaging (MRSI). The methods are based on phase encoding of spatial position to obtain spectroscopic information in a single experiment from adjacent volumes, spread over a large VOI.

The spatial information is introduced in such techniques through the phase coding of the B_0 field in one or more dimensions. The pulse and gradient sequences used in such experiments are similar to those used in MRI. However, since the final signals should contain information on the chemical shifts, data is collected in absence of a read-out gradient. Depending on the sequence used for phase coding, spectra can be resolved in n dimensions (resulting in 1- , 2- or 3-D SI), where n is the number of dimensions used in phase encoding. The resultant signal obtained after FT in the n+1 dimension is an n+1 dimensional map. This consists of n-spatial and one spectroscopic dimensions. The principles of volume selective excitation with SI are the same as those for single-volume spectroscopy, except that the defined volume is normally a large slab. Both PRESS and STEAM protocols have been incorporated in SI, as is illustrated in Figure 13.3.

Spectroscopic Imaging (SI) combines features of both MRI and MRS. Spectral patterns or metabolite intensities can be represented either as spectra from adjacent voxels, or one can obtain distribution pattern of a particular metabolite. The SI information has been used in understanding basic features of metabolism and disturbances of metabolism in clinical situations. A major advantage of SI is its ability to provide spectroscopic information in the 3D space of the organ under study. Spectra from several locations are observed in a single step. A major limitation arises from the fact that only a limited number of phase encoding steps are feasible, as compared to those in normal imaging acquisitions. This results in decreased localization and higher contamination from volumes outside the VOI. The requirement for shimming of the magnetic field over large volumes and long acquisition times are some of the other limitations of SI.

SI sequences based on EPI have been designed, which are relatively fast. Gradient induced eddy currents and field inhomogeneity creates problems in suppression of water signal. Methods have been developed, which combine the advantages of both the SV and the SI localizations, as is shown in the Figure 13.3.

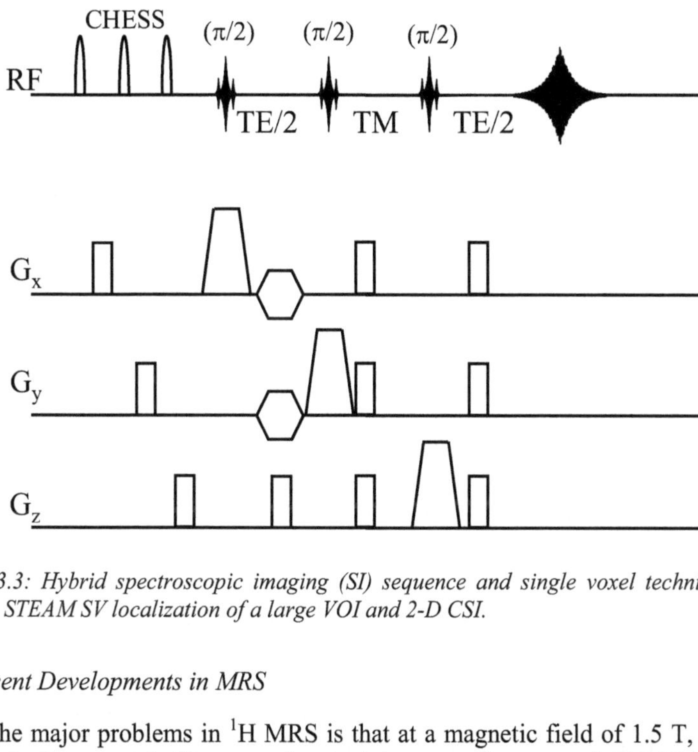

Figure 13.3: Hybrid spectroscopic imaging (SI) sequence and single voxel technique that combines STEAM SV localization of a large VOI and 2-D CSI.

2.10 Recent Developments in MRS

One of the major problems in ^{1}H MRS is that at a magnetic field of 1.5 T, most of the signals from metabolites are clustered in a small range of 400 Hz. Higher magnetic fields help to obviate this problem. However, this has to be weighed against the safety and comfort of the person under study.

The use of improved radiofrequency coil technology, particularly the use of phased array coils is helpful in the improvement of sensitivity. Parallel imaging approaches such as *sensitivity encoding* are also under development. Another new frontier in MRS is the use of two dimensional NMR methods. However, a major problem in introducing these techniques to human systems is the length of time that a person has to be kept in the NMR machine.

2.11 Detection of Metabolites using 2D NMR

One of the difficulties in studies of whole organs is the extensive overlap in spectral regions of interest. In recent years, the use of homo-nuclear 2D techniques such as COSY, has found applications in studies of whole organs. It may be noted that the broad lines in ^{1}H NMR of whole organs and the larger instrument times needed, makes it difficult to use multidimensional NMR on a routine basis in animal studies. However, in recent years, techniques such as COSY have been applied to detect overlapping resonances. For example, Figure 13.4 shows a 2D JPRESS spectrum in the peripheral zone of a healthy volunteer for the detection of prostate metabolites. The resonances from polyamines such as spermine and spermidine have been resolved using 2D spectroscopy from the overlapping peaks due to Cr and Cho.

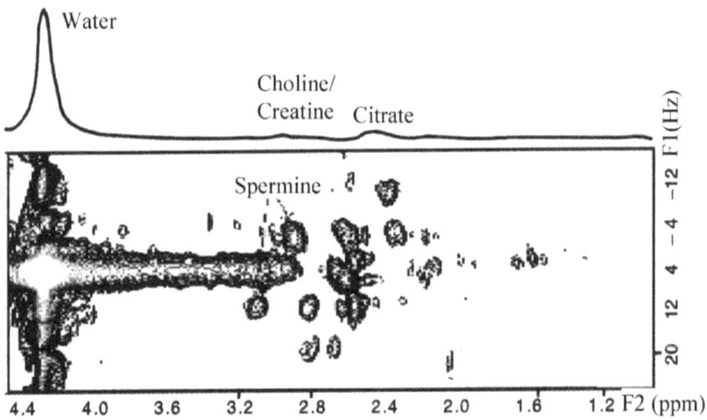

Figure 13.4: An illustrative 2D JPRESS spectrum. Volume dimensions $1\times1\times2$ cm^{3}, TR/TE = 2 sec/30 msec. A pelvic phased array (PPA) assembly with a rectal coil were used for "receive". (Thanks to M.A. Thomas for providing this figure).

3. UNDERSTANDING THE CHEMISTRY OF BRAIN THROUGH MRS

The bioenergetic pathways in brain are significantly different from those in other body organs. There are several pathways, which are unique to brain (Figure 13.5). There are different types of cells in brain, each of which has different functions. Unlike some of the other organs, it is difficult to obtain tissues and cells from brain. Consequently, the metabolic patterns of brain are significantly different from other organs. NMR has made important contributions to the basic understanding of the metabolic pathways in the brain and the levels of different detected molecules in healthy and diseased systems.

Considering the importance of brain in human activity, and the ability of NMR to explore both its anatomical structure and the function, it is the most extensively studied human organ. Important basic information on the metabolic pathways in brain has been obtained using rat brain as a model system. NMR has also emerged as a promising tool for the diagnosis, and investigations and follow-up of a number of brain disorders.

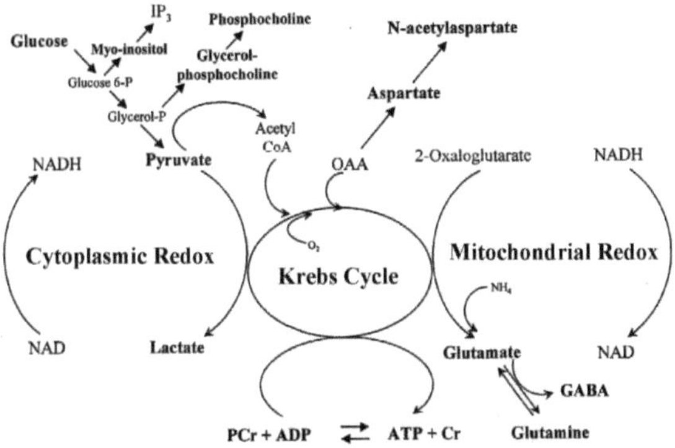

Figure 13.5: Metabolic pathways in brain. The pathways which are unique to brain have been shown in bold.

3.1 ^1H NMR Spectroscopy

The major metabolites which are present in millimolar concentration in brain and are easily detected by ^{1}H NMR are Cr, Glu, Gln, m-Ins, Cho, PCr, Asp, γ- amino butyric acid (GABA) and N-acetyl aspartate (NAA). Several of the metabolites containing methyl group(s) can be detected with better sensitivity due to larger number of protons in such molecules. Lac can be observed under certain pathological conditions, where lack of oxygen prevents oxidation of pyruvate thus leading to increased lactate formation. Methyl and methylene groups from fat and lipids appear as broad resonances. These signals can be suppressed by using suitable editing techniques.

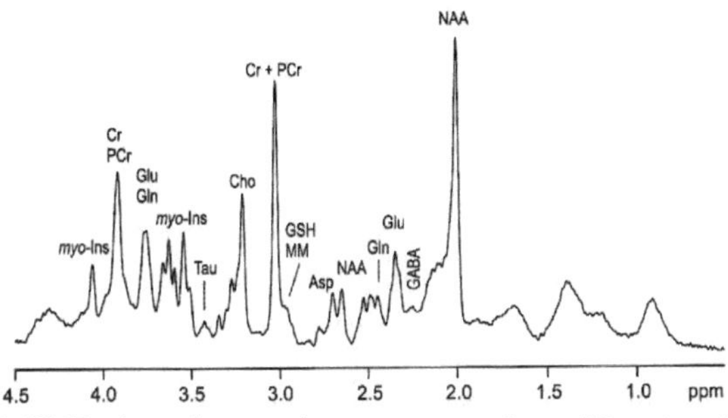

Figure 13.6: ^1H NMR from human brain measured at 7T using STEAM sequence. Experimental Parameter: TE = 6 ms, TM = 32 ms, TR = 5 s, VOI = 8 ml, NT = 160, resolution enhancement by shifted Gaussian function (gf = 0.15 and gfs = 0.08). T. Ivan, P. Andersen, G. Adriany, H. Merkle, K. Ugurbil, R. Gruetter, In vivo ^1H NMR spectroscopy of the human brain at 7T, Magnetic Resonance in Medicine 46, 451-456 (2001). (Reprinted with permission from John Wiley & Sons).

Figure 13.6 shows a typical water suppressed ^{1}H NMR spectra from brain. The bottom traces in panels A and B show spectra acquired from a 27-ml volume encompassing the visual cortex of a normal adult, while the top spectra are from an

8-year-old patient with *end-stage liver disease*. The spectra are taken on the same day, at 1.5 and 4 T, respectively. The difference spectra generated by subtracting the patient spectrum from that of normal adult spectrum are shown in panel C of Figure 13.4. The well-defined baseline of the difference spectra illustrates the excellent, reproducible quality and line-shapes achieved using FASTMAP at 4 T. It may be noted that most of the resonances of interest usually lie in the range of 2-4 ppm.

The molecule NAA is distributed throughout the cerebral cortex. It is among the most abundant compounds in human brain. It is present in the range of 7-15 mM concentrations. NAA exhibits the strongest ^{1}H signal in brain, which is located at 2.1 ppm (Figure 13.6). A small amount of N-acetyl aspartyl-glutamate (NAAG), a neuro-modulator, also contributes to this signal. NAA is believed to be present only in neurons, therefore considered to represent health as well as population of neurons. Any decrease in the concentration of NAA is indicative of neuronal loss or death. It has been proposed that NAA is the source of acetyl groups for lipid synthesis, regulator of protein synthesis, storage form of acetyl CoA or a breakdown product of NAAG. While the precise role of NAA has not been established, it has been suggested that NAA is involved in lipid synthesis during myelination.

The pathological situations, which give rise to a decrease in NAA signal intensity are infarctions, tumours, hypoxia, epilepsy, Alzheimer's, ischemia, diabetes mellitus, AIDS, multiple sclerosis (MS), and several other neurodegenerative diseases. On the other hand, the levels of NAA are high during postnatal development and axonal recovery.

One of the other strong signals arises from the -$N(CH_3)_3$ group present in glycerolphosphocholine, phosphocholine, and phosphotidylcholine. This resonance signal is collectively labelled as the Cho signal. These methyl protons resonate at 3.2 ppm. The group of Cho compounds are involved in membrane synthesis and degradation. The normal level of Cho is around 1-2 mM. The higher intensity of the signal is due to the fact that nine protons contribute to this resonance. Cho levels are higher in glial cells, tumours, demyelinating diseases such as MS, trauma, stroke and post liver transplant. Its level is lower in patients having asymptomatic liver disease and non specific dementia.

The –$N(CH_3)$ group of Cr and PCr is responsible for a strong resonance at 3.03 ppm. Minor contribution to this peak may also arise from GABA, Lys and glutathione. The total concentration of Cr/PCr in normal brain is 8-12 mM. This signal is used as a concentration reference for brain metabolites. The level decreases in tumours, hypoxia, stroke, lymphoma, and which increases with age. As we have seen earlier, Cr/PCr equilibrium is crucial in the maintenance of energy dependent systems in brain and muscles.

Brain cells function under aerobic conditions. Therefore Lac is usually present in low concentration (<1 mM). Appearance of high Lac signal is indicative of deranged oxidative metabolism. This happens under conditions where the energy is derived using only glycolysis. Such situations exist in ischaemia, hypoxia, tumours and mitochondrial diseases. Presence of Lac results in the appearance of a sizable peak at 1.33 ppm due to the CH_3 resonance of Lac. This signal overlaps with the broad lipid methylene resonances at 1.3 ppm. The signals from lipids and Lac can be easily distinguished by double quantum coherences of Lac, which forms an AX_3 spin-coupled system with a J-coupling of 7 Hz.

The lipid signals manifest only when mobile lipids accumulate. This can happen due to pathological conditions such as lymphoma, actively growing tumour, inflammatory diseases, damaged membranes of the degenerative cells, and blood barrier damage.

A peak due to myo-inositol (mI) appears at 3.56 ppm. mI is believed to be localized in the astroglia and therefore reflect the health of these cells. It also controls the osmolarity and is a possible degradation product of myelin. In normal brain the concentration of mI is around 5 mM. The levels are higher in patients suffering from Alzheimer's, diabetes mellitus and hepatic encephalopathy.

The concentration of Glc in normal brain is around 1 mM. Attempts have been made to resolve its resonances by the use of fast STEAM methods. Glc is continuously cycled in the brain, as it does not have its own pathway to generate glucose. Another set of molecules present in large quantities in the brain is amino acids; some of the typical concentrations are: Glu (~ 10 mM), Gln (~5 mM), Asp (~3 mM) and GABA (~1 mM). These molecules are involved in neurotransmission pathway in the brain. Because of their importance in neural transmission, this subject is discussed in a separate section.

3.2 ^{13}C NMR: The Glu, Gln, GABA Cycle

The transfer of information in brain is mainly carried across synapses by certain small molecules, which act as neurotransmitters. The rate of release and recognition of neurotransmitters and its relationship to brain metabolism can be considered as a link between neurochemistry and cognitive sciences. Several neurotransmitters have been identified. These include dopamine, serotonin, acetylcholine, norepinephrine.

Two of the neurotransmitters, glutamate and GABA are the major excitatory and inhibitory neurotransmitters respectively in the central nervous system. The two molecules are present in millimolar concentrations in brain. These compounds have opposite biological activities. While Glu excites neurons, GABA inhibits neural activity. Glutamine (Gln) is not a neurotransmitter, but it serves as an intermediate in the Glu neurotransmitter cycle (Figure 13.7). The neurotransmitter cycle (Glu/Gln and GABA/Gln) involves close interaction between neuronal and astroglial cells (Figure 13.7).

In the mammalian cortex glutamatergic and GABAminergic neurons contribute to more than 90% neurons. They share a common metabolic partner Gln. The high concentration of Glu and Gln makes it feasible to carry out ^{13}C NMR studies on rat and human brains, using ^{13}C labelled substrates as the source for metabolism.

Quantitative assessment of the Glu-Gln cycle and neuronal Glc oxidation has important implications in the understanding of brain function and neuro-imaging. Contrary to earlier belief, it has been established that Glu release and recycling is a major pathway of Gln synthesis in astroglia. Glu released from presynaptic neurons passes to astrocyte for Gln synthesis. ^{13}C studies on rat brains reveal that there is a 1:1 stoichiometry between the rate of neuronal glucose oxidation and the cycling of glutamate neurotransmitter above isolectricity. The results suggest a tight coupling between the release of glutamate neurotransmitter and cortical Glc consumption.

Almost 60-80% of the energy from Glc oxidation in non-simulated cerebral cortex is used to support events associated with Glu neurotransmission. The increase in neuronal Glc oxidation measured during functional activation is strongly related to the Glu/Gln cycle. MRS results have significant implications for the psychological interpretation of data from functional imaging techniques.

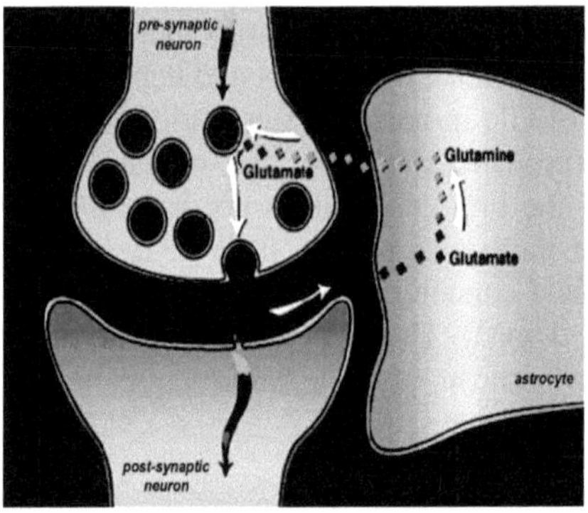

Figure 13.7: Glu released by neurons is transported out of the synaptic cleft by glial cell membranes and converted to Gln. R.G. Shulman, F. Hyder and D.L. Rothman. Q. Rev. Biopyhys. 35, 287-325 (2002). (Reproduced by permission of Cambridge University Press).

3.3. Age and Disease Related Changes in MRS of Brain

The profile and levels of different metabolites depend on a number of factors, but the age and disease are two most important parameters. Both single voxel methods (PRESS and STEAM) and chemical shift imaging (CSI) have been used to detect such changes. MRS and MRSI have been used in a number of brain diseases, such as multiple sclerosis, several white matter disorders, reversible posterior leukoencephalopathy syndrome, axonal loss, demyelination and acute disseminated encephalomyelitis.

MRS has emerged as a safe tool to obtain biochemical status of the brain tissues. This information is complementary to the information obtained on anatomical and structural features from MRI. For example, Figure 13.8 shows an example of the ^{1}H MRSI of a patient suffering with acute disseminated encephalomyelitis (ADEM). In particular the reduction in the NAA signal may be noted.

4. MRS OF OTHER ORGANS

Multinuclear NMR spectroscopy has been used in studies, of a number of other human and animal organs. The power of the technique has been illustrated by giving a few illustrative examples.

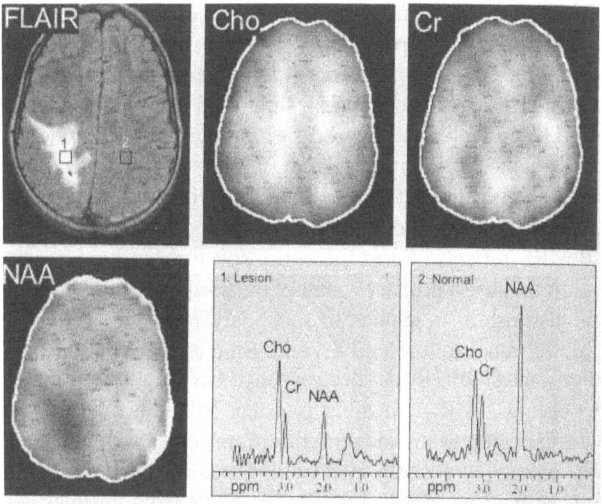

Figure 13.8: Proton MRSI in acute disseminated encephalomyelitis (ADEM). The lesion is characterised by low levels of NAA and a smaller reduction in the Cr leval. (Thanks to P.B. Barker for providing this figure).

4.1 Muscle

In vivo ^{31}P spectroscopy provides important information about the energy status of muscle cells (Figure 13.9). The ratios of the concentrations of PCr and PCr/P_i at rest, during exercise and during recovery, help in understanding the mitochondrial function. Disorders of the metabolic pathways such as glycolysis and glycogenolysis can be distinguished using ^{31}P and ^{1}H NMR.

A large amount of literature is available on the investigations of bio-energetics of muscles under a variety of conditions. These conditions range from nutrition, fatigue,

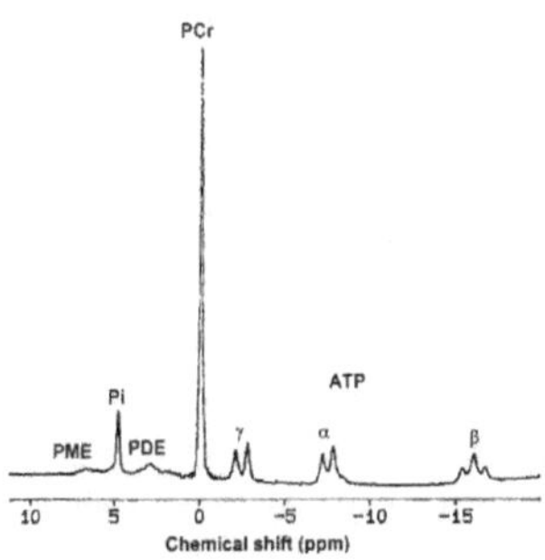

Figure 13.9: ^{31}P in vivo MR spectrum of calf muscle of a normal volunteer using a surface coil with a repetition time of 3 sec. (Thanks to N.R. Jagannathan for providing this figure).

training of athletes to persons suffering from metabolic and mitochondrial diseases. Results from such studies have provided valuable insights into pathological changes and muscle contraction, in relation to energy production and consumption. A number of muscle systems such as, forearm, wrist, thigh and calf muscles have been investigated using NMR. The technique however requires several hours of studies since a typical rest-exercise recovery protocol requires almost an hour. The non-invasive nature of NMR technique allows repeated measurements to be made on humans.

^{31}P NMR spectroscopy is the technique of choice for muscle studies. In view of the fact that ^{31}P MRS reflects almost the entire spectrum of bioenergetic pathways, it has the potential to delineate primary metabolic disorders, which affect muscle energy production. It is also useful in studies of secondary problems of muscle metabolism, such as heart failure, renal failure and thyroid. Studies of CNS and its disorders have also been studied using ^{31}P NMR spectroscopy. CNS and neuromuscular disorders may coexist. Coupling NMR with other physical techniques helps in further understanding of abnormal changes in bioenergetics of muscular systems.

At rest, the energy to muscle is provided mainly by mitochondrial oxidative phosphorylation, with a low metabolic rate. ^{31}P NMR from muscles usually contains three signals from ATP, and one each from PCr and P_i. In normal muscles, one also observes additional peaks from PME and PDE The ratio of PCr/P_i is 6-12 in normal human muscle. The levels of unbound ADP are low. The concentration can however be estimated using CK based equilibrium equation. The intracellular ADP thus estimated is 10-25 μM. This is below the Michaelis constant of ~30 μM which indicates a regulatory role in this system.

Pathological changes in muscles, using ^{31}P MRS have been established for a number of human muscle diseases. In particular, disorders associated with impairment of energy metabolism have been investigated fruitfully. The diseases are often categorized in terms of changes in the ATP levels, PCr depletions, intra-cellular acidosis and rate of PCr recovery, all of which can be quantified. For example, neuromuscular diseases are important cause of neurological disorders. Abnormality in skeletal muscle metabolism is common in such diseases and may provide biochemical markers for diagnosis. Other medical situations include mitochondrial myopathy, PFK deficiency, renal failure, chronic respiratory, congestive heart failure and thyroid diseases. In muscles of patients with mitochondrial diseases, a high level of resting P_i is observed. This is also the case with a number of other disorders such as muscular dystrophy. The ^{31}P NMR technique has been used to monitor course of mitochondrial diseases, during treatment and therapeutic trials. This is particularly true for disorders with proven primary and secondary metabolic defects.

During exercises, ATP used is continuously refurbished by glycolysis and oxidative phosphorylation through the CK reaction. This reaction regenerates ATP from PCr. Intense exercises produce intracellular acidosis. As a consequence, there is a change in the chemical shift of the P_i signal (pH tends to be < 6). During recovery from exercise, glycolysis ceases but oxidative phosphorylation continues to replenish PCr, and the pH returns to the rest level. The recovery rates of PCr, ADP and P_i and their inter-relationship with pH changes provide quantitative information on mitochondrial function.

The ability of ^{31}P to monitor metabolism makes it suitable for investigating metabolic aspects of muscle fatigue both in healthy subjects and in patients. Muscles with impaired glycolysis (due to enzymatic blocks at or distal to the PFK stage) lead to accumulation of glycolytic intermediate. This is reflected in increase level of PME during work. Patients with glycogenolysis disorders (such as, McArdle's disease) show an increase in pH due to consumption of protons associated with PCr hydrolysis. There is higher than normal end-exercise P_i. In both cases, muscle lacks the ability to produce lactic acid during exercise.

In recent years, ^{1}H MRS has been used to study several aspects of glycolysis and lipid metabolism in muscles. Techniques for the detection of Lac signal in exercising muscles have been developed. The spectra from the muscle tissues show separate lipid signals from intra- and extra-myocellular compartments of lipids. The lipid metabolism may be affected by certain muscle oriented diseases such as polio. There is a reduction or complete absence of Cr, carnitine and Cho metabolites in severely paralysed patients. Unfortunately, the information obtained from *in vivo* spectra is limited due to broad resonances. Analysis of perchloric extracts of muscle tissues, reveal a large number of metabolites such as amino acids, sugars and membrane precursors. The concentrations of several of these metabolites change during neuromuscular diseases.

4.2 Insulin Regulation of Glycogen Metabolism

A good example of the power of ^{13}C spectroscopy is the work of Shulman and his group on the role of insulin in glycogen metabolism. Insulin is excreted from pancrease following a meal, which activates metabolism in muscle and liver. In both organs, glycogen synthesis is stimulated by insulin and high levels of Glc. Three enzymes are involved in the metabolic pathway between plasma glucose and muscle glycogen; Glc transporter, hexokinase and glycogen synthase. These reactions are regulated by insulin. For patients suffering from non-insulin-dependent diabetes mellitus (NIDDM), the ability to maintain a balance between Glc absorption, metabolism and the secretion of insulin is lost. This results in elevated level of Glc in blood and consequent cellular damage.

Previously, it was believed that the enzyme glycogen synthase is responsible for NIDDM. The use of ^{13}C signals from glycogen and ^{13}C labelled Glc have helped to study the metabolic pathway in considerable detail. These experiments, in conjunction with ^{31}P NMR, have shown that the rate of glycogen synthesis in the muscle is controlled by the Glc transporter enzyme. A defect in the activation of this enzyme is responsible for impairment of this pathway. This defect is present prior to the development of diabetes and may be the primary cause of NIDDM. It has also been shown that the defect may be corrected by simply by exercise since exercise and insulin stimulated glycogen synthesis occur by different mechanisms.

4.3 Breast

The strong ^{1}H NMR signals from human breast normally consist of water and lipids (Figure 13.10). Their relative amount of these two molecules is different in healthy and cancerous tissues. The breast tumours are characterized by high water to fat ratio (W-F) as compared to the values obtained for control and the unaffected contra-lateral breast tissue of patients. The relative intensities of the two signals have been used to identify clinical cases, localisation of affected volumes and for following therapeutical changes. For example, the W:F ratio shows a significant reduction in patients undergoing chemotherapy along with a reduction in the size of the tumour.

In vivo MR spectroscopy has the ability to provide biochemical information from a well-defined region of interest or voxel in breast cancer. It is well documented that total Cho is specific to malignancy and can be used to differentiate cancerous from benign tissues. The utility of *in vivo* MRS in distinguishing benign lesions from malignant ones in young women has been demonstrated. Methods for the suppression of both the water and lipid signals simultaneously, to improve the detection of choline resonance have been developed. The specificity of the Cho signal for diagnostic purposes is limited by the fact, that an increase of this signal may also take place in other situations. For example, an increased Cho signal has also been detected in mothers who were breast feeding at the time of examination. Recently, the potential of proton MR spectroscopic imaging (MRSI) in breast cancer has been reported. It has been indicated that the addition of quantitative ^1H MR spectroscopy to the breast MR imaging examination may help to improve the radiologist's ability to distinguish benign from malignant breast lesions and in following response to chemotherapy.

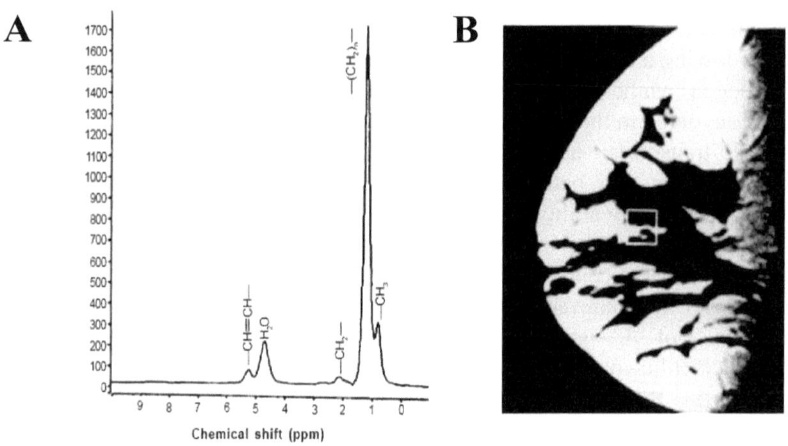

Figure 13.10: (A) Proton NMR spectrum from a 2×2×2 cm^3 voxel positioned in the normal breast tissue of a volunteer using STEAM pulse sequence. The spectrum was recorded at an echo time of 135 ms with a repetition time of 3 s. (B) T_1-weighted (TE=15 ms and TR= 650 ms) sagittal MR image of the same patient showing the voxel location from which the ^1H spectrum was obtained. (Thanks to N.R. Jagannathan for providing this figure).

^{31}P NMR spectroscopy has shown that in human breast tumours, large signals from PME and PDE are observed. A decrease in PME signal is associated with a stable or responding condition. However, an increase of PME is linked with progression of disease. The PME signal is associated with PE and PC, with the ratio of the two molecules ranging from 1.3 to 12. The PC content of high-grade tumours has been found to be higher. Significant changes in the metabolic levels have been observed in patients undergoing hormone, chemotherapy or radiotherapy.

4.4 Prostate

Prostate is one of the leading cancers in males. Prostate is a sex gland. It secretes prostate-specific acid phosphatase, a factor responsible for liquefying semen. The development of prostate cancer does not present any specific symptoms, which makes it difficult to detect prostate cancer. The currently used methods such as digital rectal examination, transrectal ultrasound, prostate specific antigen (PSA) and sextant biopsy for detection of prostate cancer, have limited efficacy.

A combination of MR techniques such as imaging, contrast enhanced MRI (Figure 13.11 A), diffusion weighted imaging, and MRS with endorectal coil have shown tremendous potential in diagnosis of prostate cancer. MRI provides multi-planar images with excellent depiction of prostatic zonal anatomy. With MRI, the current status has low specificity for diagnosis of prostate cancer. It is generally useful for advance stage for detection of extra capsular extension, seminal vesicle invasion, nodal and bone marrow metastasis. Contrast enhanced dynamic MRI technology has been reported to improve tumour assessment. Diffusion weighted imaging is yet another technique which has shown potential in differentiation of malignant and normal tissues.

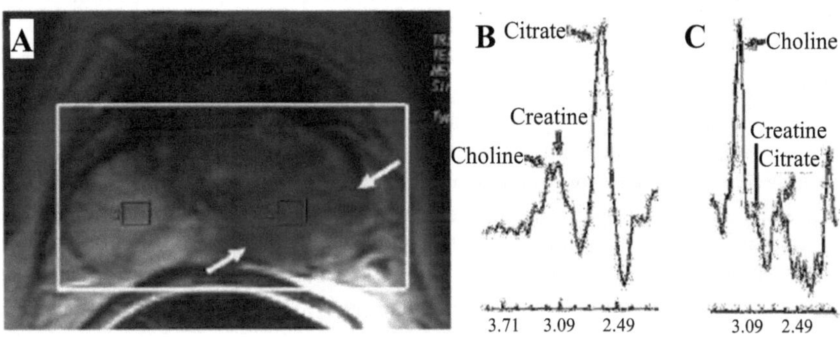

Figure 13.11: (A) Axial T2-weighted magnetic resonance imaging (MRI) and (B and C) MRSI of the normal gland and prostatic cancer. MRSI was acquired from the whole gland (white rectangular box in panel A) showing a large low signal intensity lesion (highlighted with two arrows) in the left peripheral gland. The right voxel from the peripheral gland shows the high signal intensity with the normal metabolic profile of the prostatic gland as seen in panel B (see text). The spectrum from the left voxel shown in panel C, shows increased choline and reduced citrate, which is indicative of a metabolic profile of prostatic tumor. J.C. Vilanova and J. Barcelo. 32(2) 253–261 (2007). (Reprinted with permission from Springer Science+Business Media).

Prostate cancer can be differentiated from BPH and normal tissue using MRS (Figure 13.11 B and C). MRS of the prostate provides useful information about the various metabolites present in the gland. The ^{1}H MRS displays concentrations of citrate, creatine and choline at 2.6, 3.0, and 3.2 ppm, respectively. The citrate has been most valuable in distinguishing benign and malignant tissues. Normal prostate tissue contains high levels of citrate, while the malignant tissues are characterized by reduced or undetectable citrate level with simultaneous increase in choline level. The creatine peak which is close to the choline peak remains same both in healthy and cancerous tissue. Thus, in clinical practice, the ratio of (choline+creatine)/(citrate) (abbreviated as CC/Ci ratio) is used for spectral analysis.

3D ^{1}H magnetic resonance spectroscopic imaging (MRSI) provides spatial distribution (voxel size 0.24 cm^3) of metabolites from the entire prostate resulting in improved sensitivity and specificity in detecting prostate cancer. MRSI can be used for early detection, residual or recurrent cancer or response to therapy. It is possible to perform MR-guided biopsies. Recently, the use of MRSI has been reported to target needle biopsies under transrectal ultra sound (TRUS) guidance, for detection of prostate cancer at and early stage.

Further, MR investigations at higher field strengths (typically 3T) may allow more specific localization of prostate cancer with better signal to noise ratio. The wide spectrum of MR methodologies seems to be the most promising tools for diagnosis of prostate cancer.

4.5 Cardiovascular Disorders

MRS is a useful technique not only for studying cardiac morphology, function and perfusion but also for the assessment of cardiovascular disorders. In particular, ^{31}P NMR shows significant alterations in high energy phosphate metabolism (PCr:β-ATP ratios) in a number of cardiac disorders. These include ischemic heart disease, heart failure and dilated and hypertrophic cardiomyopathies.

For example, in heart failure, the failing myocardium is characterized by the reduced levels of both phosphorylated and free creatine. The myocardial PCr to ATP ratio is reduced due to impairment of the energy metabolism. Ischemic injury results in the alteration of HEP metabolism with decrease in both the ATP and PCr levels. The use of decrease in the PCr:ATP ratio has been suggested as a means for guidance of the timing for valve replacement surgery.

5. MR IMAGING AND SPECTOSCOPY USING OTHER NUCLEAR SPINS

We have extensively discussed the application of ^{1}H and ^{31}P spins in MRS. The use of ^{13}C and ^{15}N has been relatively less frequent due to their low natural abundances. In particular, ^{13}C NMR can provide valuable metabolic information in brain and other tissues, similar to what has been discussed with respect to applications in body fluids and cells. It has already found applications in studies related to Alzheimer disease and epilepsy. ^{15}N NMR has potentialities for study of ammonia transport and glutamine metabolism. Extensive use of both ^{14}N and ^{15}N spins have been made for studies of primary nitrogen metabolism in plants, though the applications in animal studies have been somewhat limited.

A number of other isotopes are suited for applications in *in vivo* studies. Their use has been reported for several specialized applications. NMR experiments using spins other than ^1H and ^{31}P have the usual problems of lower sensitivity, broader resonances and often the low natural abundance of such nuclei. Secondly unlike the natural isotopes ^1H, ^{13}C, ^{15}N and ^{31}P, the concentrations of molecules containing such spins are often lower than 10 mM in living system.

5.1 Difficulties in vivo Spectroscopic Studies using other Spins

Most of the applications using other nuclear spins *in vitro* MRI and MRS studies are in development stage. These nuclei are often more useful in *in-vivo* spectroscopy rather than in imaging, and are not yet used widely in hospital environment. In particular, ^2H, ^7Li, ^{19}F, ^{23}Na and ^{129}Xe have found useful applications. Applications of certain other nuclei such as, ^{17}O, ^{10}B, ^{119}Pt have been reported. Both endogenous and exogenous molecules containing such spins have been used.

Multinuclear capabilities are now available on advanced level NMR machines developed for research on both animal and human systems. This has opened up the area of multi-nuclear NMR both for imaging and spectroscopy. While the general considerations in the use of MRI and MRS for such nuclei are the same, certain hardware adjustments have to be made to account for different resonance frequencies and gyro-magnetic ratios of such spins. Additional transmitter and receiver coils, tuned to the frequency of the particular nucleus may be required. Therefore, nucleus-specific RF coils have to be built.

The image resolution (pixel size; δx), in a single dimension is given by $1/(\gamma Gt)$, where t is the sampling time. Since γ is smaller for hetero-nuclei, to achieve the same resolution, either the gradient G or the sampling time t, need to be increased. However, the sampling time is limited by the T_2 of the nucleus. The software has to be modified to take into account this effect.

Clinical applications of hetero-nuclei, are slowly getting alleviated with the availability of high field machines with multi-nuclear capabilities. Developments of coils with higher sensitivity, and novel processing methods allow more rapid measurements and better spatial resolution.

5.2 ^{19}F

^{19}F (I=1/2) has 100% abundance. Its sensitivity is very close to ^1H (83%) and it has short relaxation times. However, fluorine is not an endogenous element. This can be used as an advantage for *in vivo* studies of drugs and other endogenous molecules. Further, the chemical shift range for ^{19}F is large, allowing good resolution of metabolic products following administration of a fluorine compound. The spin has a moderate spin-lattice relaxation time in biological systems. New pharmaceutically active fluoro compounds are being introduced on a regular basis. Further, fluorinated drug analogues may also be used. Applications of ^{19}F studies therefore cover a wide range such as anaesthetics, fluorinated blood substitutes, antibodies and markers of ionic concentration and body pH.

The ^{19}F based MRI and MRS based experiments have therefore found wide applications in studying body uptake and distribution, metabolism of the fluorine containing drugs. Such studies have served as complementary tools to studies based on ^{1}H and ^{31}P. It may be noted that tissue heterogeneity, magnetic field inhomogeneity and restricted molecular motions considerably broaden ^{19}F resonances under *in vivo* situations and all the metabolites may not be observed. Further, because of the binding of the drug molecules with macromolecules and membranes, several of the metabolic products may not be observed.

More than 150 fluorinated compounds have been used clinically as drugs. Several fluorine compounds are used as psychoactive compounds. These include the anti-depressant drug fluoxetine. It is a serotonin reuptake inhibitor. Compounds such as trifluoperazine and fluophenazine are used in the treatment of schizophrenia. The concentrations of these drugs in brain (10-200 μM) are significantly higher than the serum level. The level of the drug continues to increase long after clinical effects are manifested. ^{19}F spectroscopy has been used to monitor the level of these drugs in humans and animal models. One of the major uses of ^{19}F resonance is in spectroscopic studies and *in vivo* monitoring of these compounds and for studies of drug metabolism. For example, even after 3-5 weeks of treatment by fluoxetine, a significant level of this compound has been detected in the brain. Localized ^{19}F spectra have been particularly useful for such studies (see Figure 13.12 for an illustrative example).

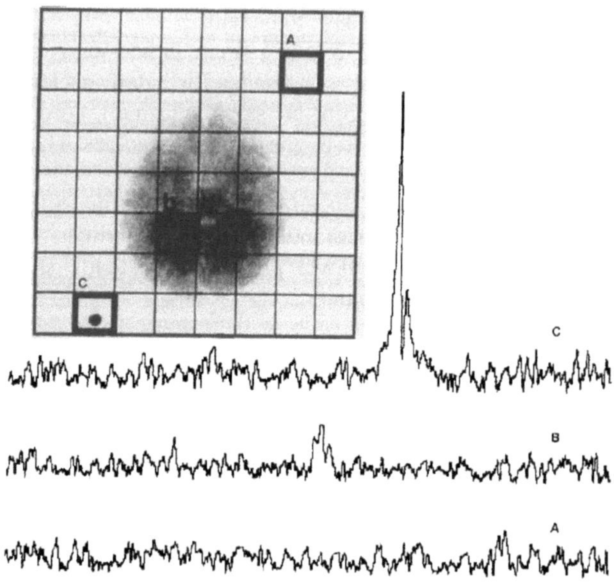

*Figure 13.12: Localized ^{19}F MRS from a patient who has been administered the drug fluoxamine. (A) Region containing no signal (B) A brain region showing the distribution of the drug. (C) Spectrum from a region containing a vial of trifluoroethanol used as an standard. N.R. Bolo, Y. Hode, J.F. Nedelec, E. Laine, G. Wagner, J.P. Macher, Neuropsychopharmacology, **23**, 428–438, (2000). (Reproduced by permission of the American College of Neuropsychopharmacology).*

Perfluorocarbons have been used in the assessment of tumour vascularity and in studies of kinetics of tissue perfusion. The ^{19}F relaxation times of these compounds are sensitive to the concentration of oxygen in the tissue. This property has been used to measure local oxygen tension in living tissues. One application involves use of gases such as C_2F_6 and SF_6 to image lungs and measure volumes and wash-out kinetics.

Several inhaled anaesthetics contain fluorine. ^{19}F MRS has been used in pharmacokinetic studies on such molecules. Other applications of ^{19}F NMR include measurement of lung volume, oxygen partial pressure, tumour oximetry, catheter visualization, imaging of gastrointestinal tract and metabolism of anti-inflammatory drugs.

A large number of bio-active molecules containing ^{19}F are known. New fluoro compounds are under development. The fact, that typically only one fluorine atom is present in the parent molecules, makes it a convenient spin for studies in drug metabolism. As an example of the type of information it can provide, one may take the case of 5-fluoro uracil (5-FU), which is widely used as an anti-cancer drug. The *in-vivo* levels and its metabolism in target organs have been monitored by ^{19}F NMR. Improvement in localization techniques has allowed the *in-vivo* detection of 5-FU with a time resolution of 4 minutes. Such studies are valuable not only in understanding the action of the drug but also in the development of new pharmacologically active molecules. However, most drugs are present in small quantities in the organs. However, with the higher sensitivity, such molecules may be detected. The observations may be coupled to MRS studies using ^{1}H and ^{31}P, to obtain drug related changes in the metabolism.

5.3 ^{2}H

As discussed earlier, even though ^{2}H has spin 1, it gives fairly good NMR signals. A promising methodology for *in vivo* applications is to inject deuterated compounds and then study blood flow. Applications in both cerebral blood flow, lipid metabolism and other metabolic processes in kidney have been demonstrated. However, the applications are limited due to the low natural abundance of ^{2}H and the problems associated with isotope effects on metabolism.

5.4 ^{7}Li

Lithium has two NMR active isotopes, ^{6}Li and ^{7}Li. Both isotopes have quadrupole moments. The more abundant isotope, ^{7}Li has spin 3/2, is 92.6% abundant and has a relatively good NMR sensitivity. It normally gives sharp resonances in most biological systems. The range of its chemical shift is narrow, and spectra usually show a single resonance (Figure 13.13). High quality spectra can be obtained with in a few minutes. Also, imaging has been done to characterize the distribution of Li in rat brain.

A disadvantage is the long spin-lattice relaxation time, particularly for quantification. Also higher concentrations can be toxic. Li concentrations in the brain may be a better measure of the efficacy and neurotoxicity than the serum levels. Hence there is a need for *in vivo* studies, particularly since there is an absence of a correlation between serum and brain Li levels.

Li is administered for manic-depressive illness in the form Li_2CO_3. There is considerable interest in monitoring its clinical effects and pharmo-kinetics and to obtain brain to serum ratio, using MRI. The Li distribution in human brain may shed light on the mechanism of Li action in bipolar disorder. Attempts to obtain good spatial resolution in human brain are being actively pursued. At this stage, ^{7}Li NMR remains primarily a research tool and is tied to treatment of a single disorder. Li studies are best carried out in animal models such as rats, where substantially higher concentration of Li can be used.

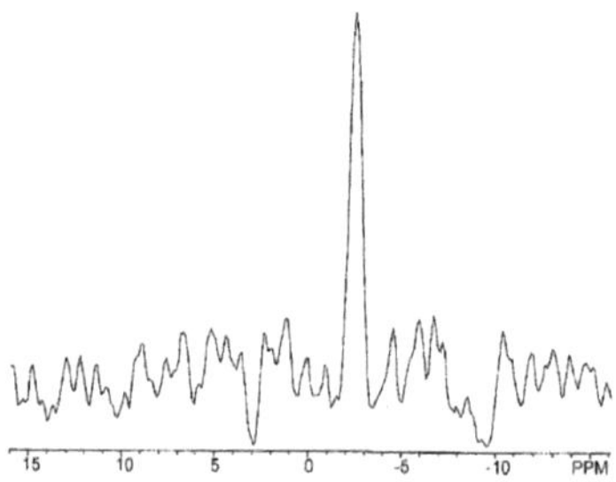

Figure 13.13: Localized ^{7}Li MRS spectrum from a 0.76-ml volume in rat brain acquired at 4.7 T using point resolved spectroscopy (PRESS) sequence, with pulse repetition delay of 5 s, echo time of 6.8 ms, and 3600 acquisitions. The brain concentration was 0.22 mM. (Thanks to R.A. Komoroski for providing this figure).

5.5 ^{23}Na

With spin 3/2 and 100% abundance, ^{23}Na MRI is a powerful technique for obtaining distribution of Na^{+}, between intra- and extra-cellular environments. Significant changes in Na ion concentration take place in certain situations such as oedema development, cerebral redistribution following ischemia, myocardial infarction and cartilage degeneration. Such changes have been monitored by ^{23}Na based studies. The low sensitivity and the larger line-widths limit the applications of this spin in biological and biomedical studies.

5.6 ^{3}He and ^{129}Xe

Lungs are filled with air and have very little water. The water MR images of lungs are therefore poor, and MRS studies are limited. Techniques have been developed to hyperpolarize the nuclear spins of ^{3}He and ^{129}Xe using optical pumping. The hyperpolarized gases are then pumped into lungs. A gain of about 10^5 is thus achieved in spin magnetization, which provides strong NMR signals. Both nuclei have spin ½, and their relaxation times are long. Since spin polarization in this case is not a result of the magnetic field, the available signal intensity decreases with each RF pulse. Therefore, by using small flip angles, images can be obtained over a long time period. Several images can be obtained with a single polarization by using fast imaging techniques.

The main uses of polarized spins have been to study pulmonary ventilation and to visualise lung defects. For most applications, He is preferred over Xe due to its better NMR properties and virtually no side effects. On the other hand, Xe is soluble in lipids and water and shows large chemical shift, when it is dissolved or absorbed. These properties have been exploited for studying lung function, perfusion, blood oxygenation and detection of tumours.

5.7 Other Nuclear Spins

A number of other nuclei have been used in MRI and MRS studies on humans. Applications include ^{27}Al for measuring gastro-intestinal transit times, ^{10}B for monitoring neutron capture agents, ^{17}O for cerebral metabolic rates of oxygen and ^{195}Pt for monitoring the metabolism of anti-cancer drugs such as carbo-platin.

6. APPLICATIONS OF NMR IN DRUG DEVELOPMENT

An important challenge before the pharmaceutical industries is to bring out in the market new molecules for human health, which have lower side effects, are more potent and less expensive. However, research and development (R&D) in pharmaceutical sciences is an expensive proposition. A new compound has to undergo several levels of clinical trials. Only a small number of molecules reach the market. The chances of success for a new molecule are less than 5%. The cost of discovery and introduction of a new molecule is around US $ 800 million. The industries have to spend large amount of money in R&D developments and clinical trials. Some of the molecules, which had been used for several years were found to have disastrous side effects. These drugs had to be withdrawn with heavy penalties on pharmaceutical industries. Only the more powerful companies can afford such financial constraints.

For these reasons, the pharma industry concentrates on drug discovery for diseases from which they can earn profits. The new products are marketed at fairly expensive prices, as the industry tries to make maximum profit during the years that the molecules are protected by patent rights. Attempts have therefore been made to improve the process of drug-discovery. NMR has proved to be an effective tool in drug discovery.

6.1 Stages in Drug Development

The process of drug-design usually proceeds through the following steps.
(i) Choice of the target (receptor molecule) based on the particular disease that has to be tackled (called the target validation).

(ii) Identification of fragments or ligands with low molecular weight that can bind selectively to the so called *hot spot* on the selected receptor (lead identification and screening).
(iii) Use such a selected fragment as scaffold for design. The fragment is modified with the appropriate functional groups. A core molecule is thus derived, which may be able to bind selectively to the hot spot on the receptor with good affinity, and thereby modify its functional property (the process is called lead optimization).
(iv)A second molecule is explored, with somewhat weaker binding properties to serve as a second binding site near the hot spot (lead development). The two molecule fragments are joined with a linker such that the designed molecule binds simultaneously to the hot spot and a second binding site, with good affinity.
(v)The molecule thus selected is subjected to preclinical studies for toxicity, metabolism, and activity.
(vi)This is followed by clinical studies on animals.
(vii) The final stage is the human trials, which typically take place in three phases.

NMR is the only technique, which can be used for all the above steps involved in drug discovery. In fact, NMR is the most versatile tool for such studies, since it can be used to explore all levels of biological systems. It has been employed in a variety of diverse ways. We will discuss some of the more important frontiers in NMR based drug discovery.

6.2 Target Validation and Receptor Identification

One recalls that in biological systems, control of activity is possible at various stages of the information transfer chain. Small exogenous molecules (often referred as ligands, but mean potential drugs in the current context) can bind to macromolecules and supramolecular structures in biological systems (called targets or receptors), and change their structure and function. In this way, ligands can control, activate or inhibit a specific biological step. The target molecules may be DNA (for control at the level of expression and replication), RNA (to control of translation), specific proteins (for control specific functions and growth of foreign organisms), supramolecular structures such as membranes (resulting in the modulation of functional action of membrane proteins through changes in the dynamic behaviour). Specific cell lines (to control their growth) as well as whole organs and systems can also be targeted. As discussed earlier, each of these levels of biological systems is amenable for NMR investigations. Therefore, there are a number of ways in which ligand binding studies can help in discovery of new biologically active ligands.

The targets are large biomolecules, which are keys to the control of a particular activity in the body. The targets can be identified and characterised using knowledge of genomics, proteomics and sometimes by therapeutic identification. A crucial aspect is the determination of 3D structure of the target molecule. This may not be possible in all cases.

The most common targets used in NMR based screening are large proteins. A major impetus to such studies has been the knowledge of 3D structures of functional proteins. For very large proteins, screening can also be carried out using molecular fragments of the functional regions. Inhibition of proteins may lead to a low or zero activity of a critical enzyme. The choice of the target protein is thus based on the disease for which the search is being made. For example, in an attempt to design molecules for treatment of type II diabetes and obesity, a critical enzyme in insulin-signalling pathway, namely protein tyrosine phosphatase 1B, has been used as a target. Calcium binding proteins can be used as targets for certain bacterial diseases. Compounds that inhibit protein synthesis are potential anti-microbes.

Recently, functionally important RNA structures have been used as targets. A variety of biochemical processes such as replication, translation, viral transcription and control of gene expression are influenced by binding of ligands to such targets. As a result of such intermolecular interactions a non-functional conformation of RNA may be stabilised. Alternatively, the interaction of RNA with other macromolecules involved in the function may be blocked. Inhibition of an RNA molecule can prevent production of several proteins.

DNA is another possible target. Drugs can bind to DNA either in the grooves through electrostatic interaction with the DNA bases or through stacking interactions. In some cases, such interactions are sequence specific. DNA is the usual targets for anti-cancer molecules. Membrane bound proteins are also possible targets. The changes in membrane fluidity are closely related to the function of lipid bound proteins. Changes brought about by certain drugs in the membrane may be closely related to their ability to inhibit or activate a particular pathway.

6.3 Structural Approach to Drug-design

The number of potential molecules, which may be useful for a given disease, is enormous. Knowledge based libraries for the purpose of drug-design, are being continuously developed. Further, it is possible to synthesize a large number of novel and potential compounds, for example by using the techniques of combinatorial chemistry. Biotechnological tools and natural products provide another source.

One can screen such compounds by looking at their binding to a target molecule responsible for the disease. High through-put *in silico* based screenings (HTS) are available for such investigations. However, even if the search is restricted to let us say a million compounds, the cost and labour involved in such an effort is enormous. Clearly, such hit and trial approaches are no longer an attractive proposition for development of novel biological active molecules.

Rationale approaches for drug-discovery are therefore being developed and explored continuously. One of these is a semi-empirically approach based on known biological active molecules and their structural properties, which serve as lead compounds. One looks at changes in the chemical structures, which may make the compounds more useful while retaining the basic elements in the structural motifs responsible for the biological action. Computer based drug-design, such as Quantitative Structure Activity Relationships (QSAR) fall in this category. Similar approaches are available to check on the toxicity. A knowledge based library may be constructed using computational screening of data-bases of known compounds.

A more rationale approach uses the potentialities of new molecule to target the 3D structures of appropriate biopolymers through computational methods such as docking. Such studies help to optimise molecular structure in lead discovery. High throughput methods have been developed to screen binding of drugs to biomolecular motifs. The major problem in such approaches is that the 3D structures of most of the potential targets have not yet been solved. The initial phases of the development have to be followed by looking at the metabolism of the new compounds and clinical trials.

6.4 Drug-Receptor Binding

Recall that two molecular entities are involved in drug-receptor interaction. First is the target molecule, which can be a protein, a segment of RNA or a particular sequence in DNA. By inhibiting or activating sites of the target molecules, a small molecule (ligand) may alter the biological events. Interaction between a biological macromolecule (target receptor) and a lead compound, involves intricate recognition events. These are influenced both by the structure and dynamics of the target. Since the ligand and receptor must adapt to each other to form a strong complex, a detailed knowledge of ligand receptor binding helps in drug-design. Obviously, knowledge of 3D structures of relevant macromolecular targets from genomic data has the potentialities to accelerate drug-discovery.

It has been shown that the affinity of a ligand to its receptor increases with the number of non-hydrogen atoms, but plateaus around 15 heavy atoms. This implies that there are small regions on the target surfaces, which provide the binding surface to ligands. Such regions are called the *hot spots*. The hot spots can be explored by looking at the binding of compounds to be screened either by following the NMR spectrum of the target or the ligand molecule. This approach is complementary to the traditional high-throughput analysis, but offers considerable advantages. NMR based screening can detect binding directly and is more efficient. The information can then be used to as a basis for knowledge based design strategies around the binding moiety.

In the so called linked-fragment approach, the drugs are usually developed based on two-site interactions with the target. Two peripheral sites are used for the interaction of the drug on the receptor. One is the actual site involved in the activity of the receptor, i.e. the hot spot. The second binding site is usually close to the hot spot and is used to provide greater stability and specificity to the to the drug-receptor interaction.

6.5 Lead Identification: Ligands for Hot Spot

The process of drug-discovery begins with the screening of thousands of potential compounds. This process is aided by synthetic and combinatorial chemistry, biotechnology and pharmacology. Compounds that show activity are identified and a profile of their physical, chemical and biological properties is obtained.

High throughput NMR structures have been used to identify weakly binding small molecules fragments from chemical libraries. This information in turn, helps to guide the development of such fragments into larger core compounds. Automation of NMR spectral accumulation provides means for fast screening of potential ligands. These approaches can be coupled to QSAR, computational chemistry and X-ray crystallography. The identification of flexible regions of biopolymers by NMR helps to serve as targets for such ligands. Ligand-receptor binding studies can be done either by looking at the receptor or the ligand spectrum. Each approach has its own advantages and limitations. Such methodologies have been the subject of intensive research.

The changes in the chemical shifts of the target molecules is a reliable indication of binding of the ligand at the first site, and can be used as a method to screen potential candidates from a large library. Specific ligand binding can be detected by following either the 1D ^{1}H NMR spectrum of ligand or the 2D ^{15}N-^{1}H and ^{13}C-^{1}H correlation spectrum of the receptor. Such a screening is typically done with mixtures of several potential compounds. The NMR of the mixture that produces large perturbations in the chemical shifts can be deconvoluted to identify the lead compounds.

The observation of chemical shift perturbations of the target is a strong evidence for a specific and well defined binding. The map of the chemical shifts provides ligand binding sites. With the use of ^{15}N and ^{13}C labelling of the target, such screenings can be carried out in a fast and cost effective manner for proteins and RNA of molecular weight of is 40 kDa. Note that however, most therapeutically important targets have much larger sizes. Deuteration and use of TROSY experiments can push the molecular weight limits to a higher level. However, the high cost in labelling is a disadvantage.

An alternative approach is to use screening based on ligands. One can exploit changes in relaxation rates, diffusion rates, NOE, saturation transfer etc. to monitor the binding of the ligand to the macromolecular target. The methods rely on fast exchange between bound and free ligand. High affinity ligands are likely to be missed in this approach. It is impossible to distinguish between specific and non-specific binding. These problems can be avoided to a certain extent by using competitive screening methods. Here a probe molecule with moderate affinity is first bound to the target. The test compound is then checked for its ability to displace the probe molecule.

The methods discussed above, not only help in the screening of potential molecules out of a large library, but are also useful to obtain binding affinity of the fragment and for obtaining information on the locations in the hot spot involved in binding. If the 3D structure of the target is known, then such methods can prove even more valuable.

Once the lead compound for the binding scaffold has been obtained, it may be necessary to make structural modification to optimise it by functional groups desirable from the view of pharmaceutical action, affinity, solubility and size. With these modifications one may have a core molecule with desirable characteristics.

6.6 Second Binding Site: The Linked-Fragment Strategy

The simultaneous binding of a second fragment, at a site close to the hot spot, can be exploited to achieve further selectivity and specificity. This is the basis of the linked–fragment approach of drug-design. The second fragment can also be explored by NMR, by repeating binding experiments in presence of the core molecule bound to the hot spot. The fragment for this second peripheral site should have a much lower affinity. The methods used in search are similar to those employed in the search of the first binding, except for the inclusion of the first-ligand. Fragments having competitive binding with the first ligand should be avoided.

Once a good second fragment has been obtained, the final phase involves designing a linker to connect the two fragments, in a manner such that the two fragments in the final molecule can simultaneously bind to the two sites.

6.7 Lead Optimization

Even with the most careful design of potent molecules, the designed molecule may not show *in vivo* biological activity. Metabolism, and *in vivo* drug binding studies using NMR, can be used in cell lines during early screening of potential compounds. Testing compounds for toxicity and cell metabolism are important aspects at this stage. The ability of a potential molecule to cross cellular membranes, metabolism, relative binding to target receptor as opposed to other molecules, toxicity, or conformational differences of the target and drug interaction under *in-vivo* as opposed to *in-vitro* conditions, are aspects which are crucial in drug discovery. The compartmental distribution (i.e. intra- and extra-cellular) is important as the receptors probably reside in a specific compartment. All these issues are amenable to NMR studies.

6.8 Pre-clinical studies: Drug Metabolism using Body Fluids

The next stage in drug development is to study the effect of the potential molecule through animal studies and preclinical studies on animals. The chemical process is also scaled up at this stage. Metabolism of drugs can be studied by using biological fluids such as sweat, urine, aqueous humour, amniotic fluid, seminal plasma, CSF, synovial fluid and blood plasma. This provides information on metabolic disorder, biochemical basis of drug action, xenobiotic metabolism, and serves as a pathological tool for studies of toxicological processes. Exogenous molecules may interact at a series of organizational level in biological system, ranging from changes in genetic expression, biochemical regulation, or by inducing synthesis of drug metabolizing enzymes. In such studies, genomic and proteomic approaches may be useful. In many cases, toxic effects of a drug may be indirect, but drug induced perturbations will result in changes in concentrations of some key molecules. The analysis of body fluids also helps in detection of such changes in metabolism.

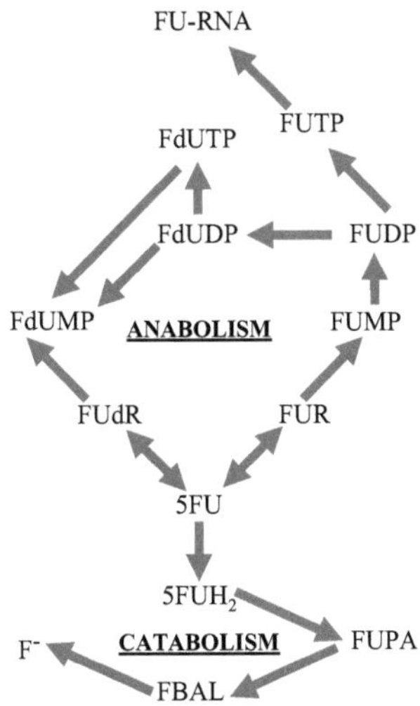

Figure 13.14: Metabolism of 5FU.

Even though the parent molecule may be useful to attack a particular disease, its metabolic product may be injurious to the body. Therefore a detailed metabolic analysis of the drug molecule in question is an important part of drug-development process. The metabolism in tissues and cells may be an involved process resulting in several metabolic products as is the case with 5-fluoro uracil (Figure 13.14).

The strength of the pharmacological effect of a molecule depends on the concentration at the receptor site. Drug levels in blood plasma may not reflect the concentration at the active site. MRS can be used to measure drug levels in tissues as well as to probe drug metabolism. Pharmaceutical applications of NMR also include study of drug metabolism and the metabolite distribution in body fluids. The most convenient fluids for analysis are urine and blood plasma, as these can be obtained in sufficient quantities. The only problem is the large background of endogenous urinary components. Plasma samples require a step of protein precipitation. Bile on the other hand contains micellar components and high bile salt concentration.

6.9 Human Trials

This is the most crucial aspect of drug discovery since even with the best possible analysis on the basic molecule, the metabolites may be toxic and may have side-effects. The final stages of drug-development involves trials first on animals and then on a selected groups of human volunteers. This is the most expensive and difficult phase of drug-discovery. MRI and MRS based techniques have been helpful in such efforts.

The process of human trials goes through three levels of testing. In the first phase, the molecule is tried on healthy volunteers. In phase II the trials are conducted on 500-600 persons with the particular indication of disease. In the last phase, the trials are scaled to cover about 5000 volunteers.

6.10 Drug purity: Interfacing NMR with LC and MS

As discussed, the major sources of novel compounds for drug discovery are obtained through combinatorial chemistry or natural products. The power of NMR in analysis of such products as well as the drug metabolites in body fluids such as urine, bile and plasma is greatly enhanced by coupling it with liquid chromatography (LC) for the purpose of separation, and mass spectrometry (MS) to obtain molecular weights. Additionally, interface with solid phase extraction, enhances the power even further. The availability of commercial LC-NMR-MS system in the 300-800 MHz, has provided ideal platform in the quest of novel biologically active compounds. The ability to rapidly identify known as well as undesirable compounds is crucial for pharmaceutical industries.

Interfacing LC, NMR and MS had problems in the past, because of the difficulties in setting LC and MS too close to the NMR magnet. The availability of shielded magnets has solved this problem. Purified sample can be split such that 95% can be used for NMR and the rest for MS. In these modes, about 100 ng of material can be detected. Use of micro-coil or cryo-probe can help to improve the sensitivity. 2D NMR experiments such as TOCSY and ^{1}H-^{13}C, HSQC further enhances the power of the approach and such experiments can be performed in reasonable time in such an interfacial system.

We had earlier discussed the metabolic fate of the drugs, which is crucial in drug-discovery. The LC-NMR-MS systems make it convenient to study metabolism and pharmacokinetics of potential drugs, by analysing body fluids. The advantages of the three techniques coupled together and applied in an automated fashion, makes it a powerful technique in rational drug design strategies.

6.11 Use of Solid State NMR for Drug Powders

NMR is routinely used to characterize and analyse the composition and purity of drugs. The potency of drug molecules is often influenced by the way capsules and powders are prepared and formulated. Often, because of solubility, drug delivery to target systems and other reasons, the drugs may be bound to matrices such as cyclo-dextrin for delivery to human organs. High resolution, solid state NMR provides information on aspects such as packing, the polymorphism and conformation of drugs, their hydration pattern, in solids, powders and capsules. Different solid forms of a drug may display significantly different properties such as stability, dissolution and bioavailability. Physical characterization of the active ingredients, excipients, physical mixtures and dosage therefore forms an integral part of drug development.

A number of spectroscopic and thermodynamic methods, including IR, Raman, and X-ray crystallography have been used for this purpose. An advantage of solid state NMR is that the analysis can be achieved with only minor influence on the particle size. It can differentiate between different solid-state forms and provides structural details of the material. NMR has an advantage in studies of materials, which cannot be crystallised for X-ray studies. Comparison between the structures in solid-state and solution is also important. The study of polymorphism of pharmaceutical compounds is the most important aspect of such studies. In most cases, ^{13}C CP/MAS NMR has been used for such studies. However, use has been made of other nuclear spins. Use of 2D NMR spectroscopy and micro-imaging are new developments in such studies. Advances in technology are bringing solid-state NMR as a quality tool for pharmaceutical industries.

7. REFERENCES

7.1 Further Reading

D.I. Hoult, S.J. Busby, D.G. Gadian, G.K. Radda, R.E. Richards, and P.J. Seeley, *Observation of tissue metabolites using ^{31}P nuclear magnetic resonance*, Nature **252**, 285–287 (1974).

P.A. Bottomley, W.A. Edelstein W.A. Edelstein, J.H. Foster et al., *In vivo solvent suppressed localized proton NMR spectroscopy: A window on metabolism*, Proc. Nat. Aca. Sci. USA **821**, 2148–2153 (1982).

P.A. Bottomley, T.H. Foster and R.D. Darrow, *Depth resolved surface coil spectroscopy (DRESS) for in-vivo ^{1}H, ^{31}P nad ^{13}C NMR*, J. Magn. Res. **59**, 338–342 (1984).

P.A. Bottomley, *Spatial Localization in NMR Spectroscopy in vivo*, Ann. N.Y. Acad. Sci. **508**, 376–385 (1987).

M. Ala-Korpela, A. Korhonen, J. Keisela, S. Horkko, P. Korpi, L.P. Ingman, J. Jokisaari, M.J. Savoloinen and Y.A. Kesalniemi, *^{1}H NMR-based absolute quantitaion of human lipoproteins and their lipid constents directly from plasma*, J. Lipid Res. **35**, 2292–2304 (1994).

M.O. Leach et al., *Measurement of human breast cancer using magnetic resonance spectroscopy: A review of clinical measurements and a report of localized ^{31}P measurements of response to treatment*, NMR Biomed. **11**, 314–340 (1998).

N.R. Jagannathan, M. Singh, V. Govindaraju, P. Raghunathan, O. Coshic, P.K. Julka and G.K. Rath, *Volume localized in vivo proton MR spectroscopy of breast carcinoma: Variation of water-fat ratio in patients receiving chemotherapy*, NMR Biomed **11**, 414–422 (1998).

Z. Argov, M. Lofberg and D.L. Arnold, *Insights into muscle diseases gained by phosphorus magnetic resonance spectroscopy*, Muscle Nerve **23**, 1316–1334 (2000).

N.R. Bolo, Y. Hode, J.F. Nedelec, E. Laine, G. Wagner, J.P. Macher, *Brain pharmacokenitics and tissue distribution in vivo of fluvoxamine and fluoxetine by fluorine magnetic resonance spectroscopy*, Neuropsychopharmacology **23**, 428–438 (2000).

D.E. Bugay, *Characterization of the solid-state: spectroscopic techniques*, Adv. Drug Deliv. Rev. **48**, 43–65 (2001).

T. Ivan, P. Andersen, G. Adriany, H. Merkle, K. Ugurbil, R. Gruetter, *In vivo ^{1}H NMR spectroscopy of the human brain at 7 T*, Magn. Reson. Med. **46**, 451–456 (2001).

Z. Serber, R. Lewidge, S.M. Miller, and V. Dotsch, *Evaluation of parameters critical to observing proteins inside living Escherichia coli by in-cell NMR Spectroscopy*, J. Am. Chem. Soc. **123**, 8895–8901 (2001).

R.G. Shulman and D.L Rothman, *^{13}C NMR of intermediary metabolism: Implications for systemic physiology*, Annu. Rev. Physiol. **63**, 15–48 (2001).

R.G. Shulman, F. Hyder and D.L Rothman, *Biophysical basis of brain activity: Implications for neuroimaging*, Q. Rev. Biopyhys. **35**, 287–325 (2002).

O. Corcoran, and M. Spraul, *LC-NMR-MS in drug discovery*, Drug Discov. Today **8**, 624–631 (2003).

U. Sharma and N.R. Jagannathan, *Potential of in-vivo MR spectroscopy in medicine*, Proc. Indian Nat. Sci. Aca. **70A**, 555–577 (2004).

T.W. Scheenen, G. Gambarota, E. Weiland, D.W. Klomp, J.J. Futterer, J.O. Barentsz and A. Heerschap *Optimal timing for in vivo ^{1}H-MR spectroscopic imaging of the human prostate at 3T*, Magn. Reson. Med. **53**, 1268–1274 (2005).

J.C. Vilanova and J. Barcelo, *Prostate cancer detection: magnetic resonance (MR) spectroscopic imaging*, Abdom. Imaging, Spinger Science+Business Media, LLC **32**, 253–261 (2007) (online).

7.2 Books and Reviews

M.P. Deutscher (ed.), *Guide to protein purification*, Methods Enzymol. **182**, 1–894 (1990).
J.M. Berg, J.L. Tymoczko and L. Stryer, *Biochemistry*, 5th International edition, W.H. Freeman and Co., (1995).
U. Holzgrabe, *NMR Spectroscopy in Drug Development and Analysis*, Wiley (1998).
M.A. Brown and R.C. Semelka, *MRI: Basic Principles and Applications*, 2nd Edition, Wiley Liss (1999).
R. de Graaf, *In Vivo NMR Spectroscopy, Principles and Techniques*, John Wiley and Sons (1999).
A.L. Lehninger, *Principles of Biochemistry*, W.H. Freeman and Co. (2000).
N.R. Jagannathan (ed.), *Proceedings of the International Workshop: Biomedical Magnetic Resonance*, JayPee (2005).
N.R. Jagannathan (ed.), *Recent Advances in MR Imaging and Spectroscopy*, JayPee (2005).
U. Sharma, R. Sharma and N.R. Jagannathan, *Characterization of breast lesions by magnetic resonance imaging (MRI) and spectroscopy (MRS),* Current Med. Imaging Rev. **2**, 329–340 (2006).

CHAPTER 14

STRUCTURE AND METABOLISM OF PLANTS

1. INTRODUCTION

Following recent developments in molecular genetics, there have been several path breaking discoveries in plant biochemistry, as has been the case for animals. Plant molecular biology has provided new opportunities for investigating metabolic pathways. New tools for quantitative analysis and modelling of biochemical pathways have provided an integrated view of plant metabolism.

From practical view-point, there are several reasons, why plant metabolism has received renewed interest in recent years. The most important of these is related to the issue of rising world population and ever increasing demands on plant products. Progress in plant sciences therefore has a close link with man's constant need for food, fodder, wood, fibre and medically active compounds. With limited agricultural land, there are massive efforts for developing high yielding crops, balanced supply of plant nutrients and agronomic practices including irrigation. There are also needs for new developments in horticulture and floriculture. Often such efforts have to be balanced with environmental issues arising from soil degradation and water management and contamination.

NMR has contributed immensely to the recent developments in plant sciences. Several of these advances have not received as much attention among the general NMR community as the studies on biological molecules, animals or humans. On the other hand, most of plant biologists are unaware of the contributions and potentialities of NMR in plant sciences.

Several of the metabolic pathways in plants are different from those in animal systems. In this Chapter, some aspects of plant NMR, which are distinct from those of animals have been discussed. Further, since biological macromolecules have been discussed in some detail earlier, the focus in this Chapter will be first on cells, tissues, sub-cellular components of plants and then on whole plants.

1.1 Cell Structure of Plants

The structures of plant and animal cells are significantly different (Figure 14.1). The most important difference is that the plant cells have a wall, which is absent in the animal cells. This cell wall is located outside the cell membrane and is composed of cellulose along with other polysaccharides and glyco-proteins. Most mature plant cells have large vacuoles (empty appearing areas found in the cytoplasm), that are relatively small in animal cells. The vacuoles in a matured plant may occupy as much as 90% of the cell-volume and serve to store food that a cell may need to survive.

Another unique feature of plant cells is the presence of plastids. Plastids contain chlorophyll and other photosynthetic pigments that are responsible for the important reaction of photosynthesis. Plant metabolism is regulated to a greater extent by intracellular compartments.

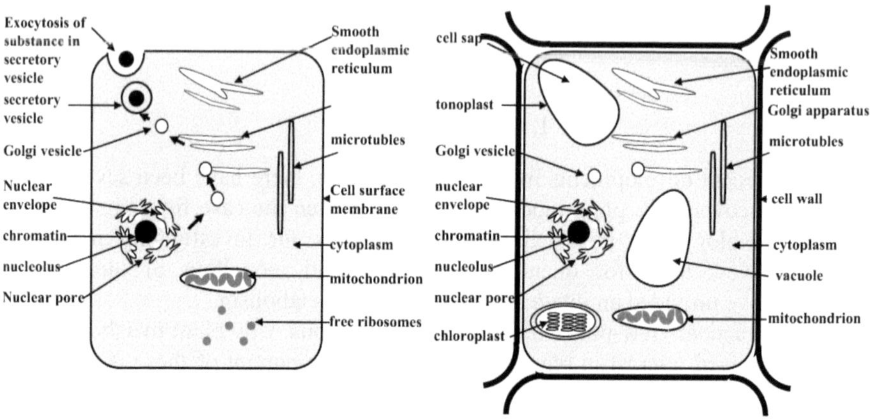

Figure 14.1: An illustrative view of animal (left) and plant (right) cells.

1.2 Biochemistry of Plants

The common features of animals and plant cells are the manner in which the genetic information is stored, replicated, transcribed and translated. Both types of cells have nucleus, ribosomes, mitochondria, endoplasmic reticulum and Golgi complex.

The basic biochemistry of the animal and plant cells has certain common features. However, plant cells have several unique metabolic pathways, which have been discussed in the next section. Another unique feature of the plant cells is that they are *totipotent,* a term which means that a single vegetative plant cell has the capacity to give rise to an entire plant. The components of plant system such as roots, stem, leaf flowers and fruit arise from a common genome but each is distinguished by its specific role, which arise from different gene expression. Each part has a unique function to play in the development and function of the plant. The actual functioning of individual cells is regulated for the coordinated function of the entire plant.

1.3 Importance of NMR in Plant Biochemistry

Research in plant biology has been revolutionised by the development of techniques to generate transgenic plants. These transgenic plants have specific perturbations in their the metabolic pathways. NMR can easily provide a snap-shot of the metabolic changes induced by the introduction of a foreign gene in the plant genome. This has provided an opportunity for the exploration of involved metabolic networks. Analysis of this type can be valuable in several diverse applications of plant metabolic network and design of new varieties of plant.

2. UNIQUE METABOLIC PATHWAYS IN PLANTS

The metabolism in plants differs from that of animals in several ways. Several of the metabolic pathways in plants have been studied using NMR techniques.

2.1 Photosynthesis

Plants depend on the available light, CO_2, water, nutrients and favourable temperature to meet their energy requirement and for their growth, maintenance and reproduction. A major feature of plant metabolism is the fact that the plants are able to trap and convert light energy into chemical energy through a process called photosynthesis. Photosynthetic organisms have two principle pigments which absorb light: the lipid soluble chlorophylls and carotenoids, and the water-soluble phycobiliproteins. Chlorophylls have diverse chemical structures, but the common feature is a porphyrin ring bound to Mg^{+2}. Besides harvesting light energy, the photo-pigments mediate a number of other photochemical reactions.

Photosynthesis is unique to plant-kingdom, photosynthetic bacteria and certain algae and requires light energy. Photosynthesis results in fixation of carbon leading to the synthesis of sugars. This reaction provides food for both the plants and animals. The light dependent reactions allow incorporation of carbon, hydrogen, oxygen, nitrogen and sulphur in the form of organic molecules, namely the *phytomass*. The phytomass in turn, is used for synthesis of various macromolecules, and for the construction, maintenance and turnover of different plant organs. The rate of solar energy conversion determines the yield of phytomass.

2.2 Nitrogen Metabolism

The second important aspect of plants is nitrogen metabolism. Nitrogen is the key-nutrient for plant growth. It is taken up by the plants in the form of nitrate, organic amines, dinitrogen (N_2), ammonium and oxides of nitrogen. Nitrogen metabolism results in the synthesis of amino acids and other nitrogen compounds. The carbon and nitrogen metabolism are closely inter-linked. Thus the growth of plants depends, both on supply of energy in the form of sunlight and the assimilation of nitrogen from compounds present in the soil.

The metabolic patterns of plants show that light plays an important role in the regulation of enzymes and metabolic behaviour. For example, plants assimilate nitrate at a much faster rate under light as compared to dark conditions. Light has an indirect but important effect on the transcription and modulation of enzymes such as nitrate reductase and nitric reductase, two of the key enzymes involved in nitrate assimilation.

2.3 Photorespiration

A more recent discovery is a process called photo-respiration in certain plants. In animals and bacteria, one encounters only the so called *dark respiration*, which does not depend on light. This process is equivalent to the so called respiration in animals. In several green plants, there is an additional mechanism involving light-driven uptake of oxygen and evolution of CO_2. This series of reactions depend on the photochemical activity of chloroplasts and biochemical activity of peroxisomes and mitochondria. Not all plants possess the activity of photo-respiration. There are significant differences in the cells of plants that possess the activity of photorespiration.

2.4 Sulphur Metabolism

Sulphur metabolism is another important feature of plant biochemistry. Nitrogen and sulphur are assimilated primarily into proteins in a molar ratio of 20:1. There exists a close interaction between the uptake, reduction and assimilation of these two elements. Nitrogen and sulphur are mostly assimilated in the plant roots in the form of nitrate and sulphate, respectively. Such co-regulation between nitrogen and sulphur metabolism is important for protein synthesis.

Modern biology has played a major role in the current understanding of plant metabolism and has led to new developments in plant sciences. These include discoveries of novel pathways in photosynthetic CO_2 assimilation and photorespiration in some green plants. Likewise, metabolic pathways during nitrogen assimilation and conversion have been investigated extensively. The findings have a strong bearing on plant productivity and photosynthetic efficiency.

3. NMR OF PLANTS: GENERAL FEATURES

NMR investigation of plant metabolism is not a straight forward extension of the methods discussed for animals and organisms. Plants exist in a wider variety of shapes and sizes. The environmental factors have a greater influence on plant physiology. The lower water content of certain plant tissues, intercellular air-spaces, presence of substantial quantities of structural materials and lower populations of metabolites can cause serious difficulties in NMR investigations. Thus, to study whole plants, plant cells and tissues under physiological conditions require special considerations and proper design of the probes, as is the case with animal cells, tissues and organs.

3.1 Handling Whole Plants

It is important in such experiments that the physiological requirements of the plants are met during metabolic studies using NMR. This is not trivial, because of the various shapes and sizes of the plants, their physiological needs and environmental factors, all of which can have a profound effect on plant metabolism. A special factor which has to be taken care is that of gravitropism. Plants are sensitive to gravity for growth. Therefore horizontal bore magnet used in animal and human studies are not satisfactory for plants. Proper illumination is another factor in differential metabolism during dark and light conditions.

Special probes and equipment have been designed to study different aspects of whole plant NMR. The conventional high field magnets used for spectroscopy, have serious limitations in studies of whole plants, because of the lower sensitivity of the technique and problems of the compatibility due to shape and sizes of plant and the magnet bore size. However, studies have been conducted using conventional spectrometers on small plants and seedlings. Most NMR spectrometers used for plant studies in green house environment use low magnetic fields, and specially designed probes. For example, the Weningen centre has designed a system for studies on whole plants using a 3T magnet, with a 60 cm bore (Figure 14.2A).

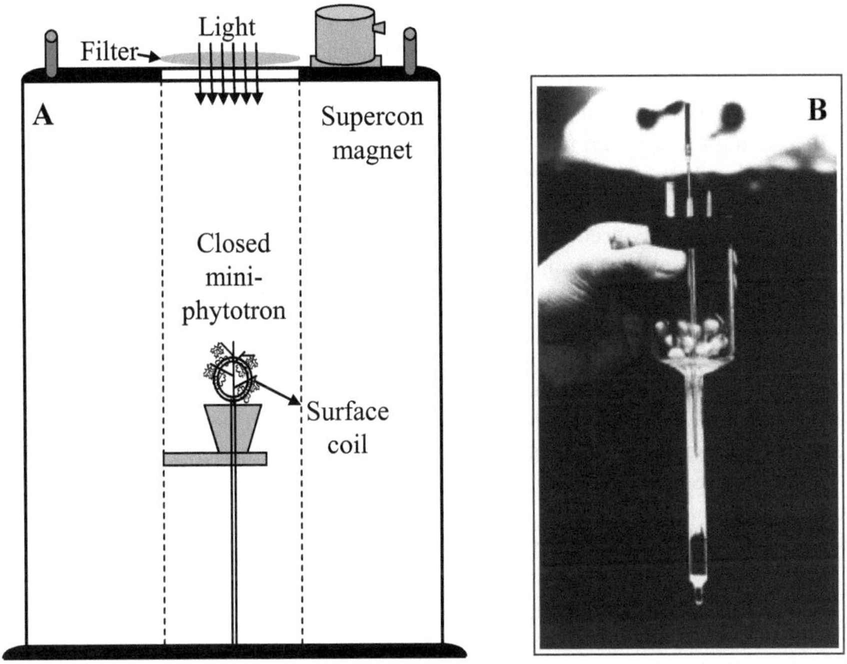

Figure 14.2: An experimental arrangement for recording in-vivo NMR spectra: (A) A typical whole plant system; the bore size of super-conducting magnet is choosen (~60cm) to accommodate the plant under study; the phytotron regulates humidity, gas, composition and temperature. The root system is temperature and nutrient regulated (B) From the roots of intact maize seedlings J.K.M. Roberts and J.H. Xia. Methods Cell Biol. 49, 245-258 (1995). (Reprinted with permission from Elsevier). The modified NMR sample tube fits in the probe head of an 11.7 T magnet. The suspending medium is pumped through the tube from the bottom to the top. The probehead is used without a dewar, and a hole at the base of the RF coils allows the tube to be positioned such that it is possible to record spectra from different parts of the roots.

Often a balance between the low sensitivity of NMR and the requirements of the experiment can be achieved using a large number of small plants, packed in a conventional probe and a high field magnet. In such experiments, a compromise has to be reached in terms of the requirement of the physiology of plants. While spectra of abundant spins can be obtained in a few minutes, those from rare spins may take several hours.

3.2 Tissues and Small plants

Whole plant NMR studies pose serious difficulties in handling. Therefore, such studies have often been limited to small plants and seedlings. In most studies, use has been made of leaf pieces and root tips, with suitable perfusion medium.

Several probe designs have been suggested to maintain plant roots, leafs or other parts in a physiologically viable and controllable state (Figure 14.2 b). Provision is made to remove waste product and to maintain supply of oxygen and various nutrients. Effect of perturbations such as changes in circulating media (effect of sucrose, phosphates, amino acids and removal of oxygen) can thus be studied conveniently. Aerobic and anaerobic conditions can be controlled. The composition of the circulating medium may be modified to study effects of starvation or lack of oxygen. However, plant samples give relatively broad signals, with limited information. To control the circulating media and environmental factors, low field magnets are generally used for studies on whole plants.

As in the case of animals, studies on whole plants are often coupled or replaced with studies on plant cells, tissues and extracts. This considerably simplifies interpretation of the spectra by reducing the number of tissue types in an experiment. There can be considerable differences between NMR spectra from different plant tissues. Plant tissues can be maintained in the magnet for much longer periods of time as compared to animal tissues. To match physiologically relevant conditions, the cells and tissues are supplied with appropriate nutrients, oxygen and carbohydrate. Special probes have been designed to achieve such conditions. The concentration of paramagnetic ions such as manganese in plant growth media should be restricted, to prevent broadening of NMR signals.

3.3 General Features of Plant NMR

As compared to micro-organisms, animals and humans, studies on plants present unique challenges. These have been outlined below:
(i) Metabolism in plants is more complex and has several unique features, which are absent in animals.
(ii) Plants have several active sub-cellular compartments. The presence of large vacuoles causes serious difficulties in NMR investigations arising from the loss of sensitivity and effects due to internal changes in bulk susceptibility.
(iii) Plastid is an organelle unique to plants. This organelle complicates interpretation of metabolic data from NMR.
(iv) Physiological conditions should be met during NMR studies. These include regulation of temperature and nutrients, control of temperature, humidity, light and other factors. These factors have to be controlled within the confines of the magnet.
(v) The metabolism in light and dark can be significantly different. Studies of light driven processes require special arrangements.

(vi) Cellular heterogeneity and consequent differences in sensitivity. This can be avoided by careful tissue selection or using cell suspensions.
(vii) Plants exist in a wider variety of shapes and sizes, and can adapt to environmental changes. This complicates interpretation of the data.
(viii) Intermolecular interactions can broaden resonances, beyond levels of their detections. This effect is particularly severe for ^{14}N resonances.
(ix) Presence of paramagnetic ions can severely hamper the quality of the spectra.

In spite of these difficulties, NMR based spectroscopic and imaging techniques have been used extensively. Such studies have thrown valuable light on plant metabolism, composition and physical state of cell-walls, identification of active pathways and quantification of metabolic fluxes. Studies have also been conducted to test hypotheses about the effects of metabolically engineered traits on plant functions. Special NMR based microscopic techniques have been developed to study thin structures such as leaves and roots. Solid state NMR techniques have found wide applications in studies on hard tissues, seeds and plant reactions which occur in plant cell organelles.

3.4 Scope of Plant NMR

Successful NMR studies on plant anatomy, structure and function have been carried out at all levels: whole plants, parts of plants such as roots and stems, excised plant tissues and cells, cell extracts, cell organelles, extracted plant products and isolated macromolecules. The whole arsenal of NMR techniques, such as imaging, whole body spectroscopy, and multi-dimensional high-resolution spectroscopy on isolated plant products have been used.

There is extensive literature on measurements of water, ions and metabolites and components of cell walls. The results have given new insights on plant physiology and metabolism. Special probes and perfusion systems have been designed for each type of material and for different types of applications. A wide spectrum of metabolic-activities has been studied using NMR. These include areas such as photosynthesis, nitrogen, carbon and sulphur metabolism, and phosphorylation.

3.5 Nuclear Spins used in Plant NMR

As usual, ^{1}H NMR provides the most valuable information. A typical spectrum is shown in Figure 14.3. As is the case with animals, the spectra are highly complex and suffer from serious spectral overlap. Besides H_2O, lipids and fat, strong signals can also arise from sugars in plants. Very few signals can be directly assigned from 1D ^{1}H NMR. The use of multi-dimensional ^{1}H NMR and more sophisticated techniques, therefore becomes imperative.

Among other nuclear spins, ^{31}P NMR, with its good sensitivity and presence in lesser number of compounds, provides important information on metabolically important molecules. As in the case of animals, ^{31}P NMR is a good indicator of intra- and inter-cellular pH, and the respiratory state of the tissue. It has also been used as a probe for the detection of Mn^{+2}. Saturation transfer experiments allow studies of turnover rates of metabolites such as ATP or Glu-6-P. 2D NMR spectroscopy based on exchange reactions during unidirectional enzyme reactions have proved extremely informative for studies involving phosphorus metabolism. In this way, several unidirectional enzymatic reactions have been followed.

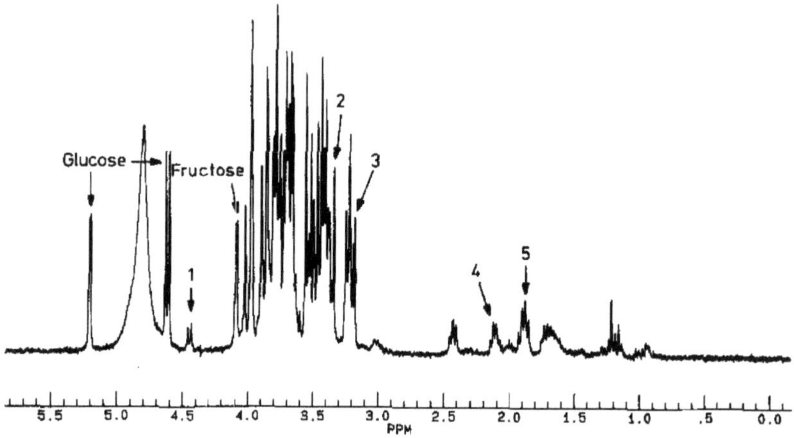

Figure 14.3: 300 MHz ^{1}H-NMR spectrum from a cell-extract of Tabemaemontana divaricata plant. This spectrum was obtained from cell material 11 days after subculture. Indicated are the signals used for the quantification of compounds. Signal 1 is from an unknown compound. Signals 2, 3, 4 and 5 are from betalne, choline, glutamine and arginine, respectively. J. Schripsema, C. Erkelens, and R. Verpoorte. Plant Cell Reports 9, 527-530, (1991) (Reprinted with permission from Springer Verlag).

^{13}C spectra have been very useful in metabolic studies. The spin has also been used extensively in tracer experiments using NMR, by using labelled substrate as starting point in a metabolic reaction.

One of the important spins used in plants is the more abundant isotope of nitrogen, ^{14}N. In spite of its large quadrupole moment, several metabolites can be detected using ^{14}N. In fact, the range of nuclear spins used in plant NMR is quite large. Beside ^{1}H, ^{13}C, ^{15}N and ^{31}P, use has been made of ^{2}H, ^{14}N, ^{19}F, ^{23}Na, ^{27}Al, ^{35}Cl, ^{37}Cl and ^{39}K. Some of these spins provide important information on the ionic balance in the intracellular media. Such an approach based on the use of several spins, provides greater potentialities in interpretation and analysis of data on plant metabolism and alterations arising from environmental stresses on plant biochemistry.

The biological information is obtained using spins in natural abundance as well as by making use of labelled substrates with NMR spins such as ^{13}C, ^{15}N and ^{2}H. The labels thus introduced move to the subsequent metabolites during biochemical reactions and provide valuable clues on the metabolic pathways. For example ^{13}C can be introduced in the form of acetate, amino acids and carbohydrates and ^{15}N in the form of nitrate or ammonium ion. Treating plant tissues with $^{2}H_2O$ provides clues to the role of hydrogen during metabolic pathways. Further, it helps in monitoring various transport processes within a given plant.

3.6 Interpretation of Spectra:

Plants pose the usual problems in the assignment of the NMR signals. The liquid phase spectra particularly from ^1H and ^{13}C are fairly complex. As in the case of animal cells, extensive use of 2D spectroscopy, pH dependent shifts and spiking methods help in resolving ambiguities. In particular, extensive use has been made of both homo- and hetero-nuclear 2D experiments to study plants. Indirect detection of rare spins is preferred for reasons of sensitivity. Thus, the techniques of HSQC, HMQC and HMBC have been applied extensively to analyse metabolite concentrations in plant extracts.

It may be pointed out that several factors may make a NMR signal undetectable, even when the metabolite is present in sufficient quantities. Paramagnetic ions such as Mn^{2+} are one of the major sources for line broadening in plants. The immobilization of metabolites is another reason. The metabolite signals may be obscured by strong signals from water and other molecules.

3.7 Compartmentation in Plant Cells

In favourable cases, it is possible to obtain interpretable signals on spatial location of NMR detected metabolites. It is often possible to obtain distinct chemically shifted signals from cytoplasm and vacuole and from inside and outside of the cells. Such differences arise because of the differences in the media in different regions. In particular, pH differences are very well reflected in terms of signals due to P_i. A typical illustrated example is shown in Figure 14.4, for a transgenic plant containing gene for pyrophosphate.

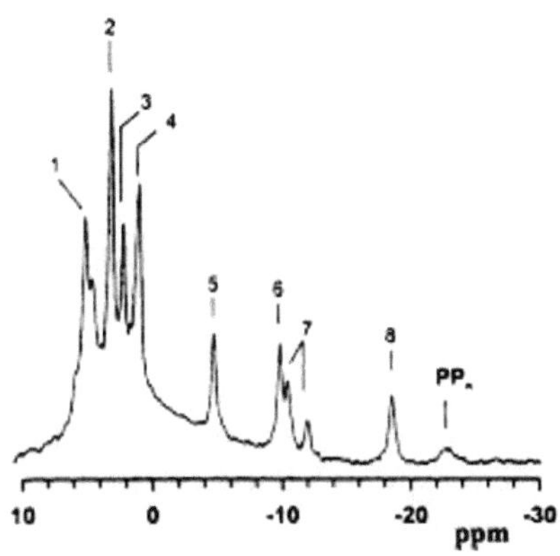

Figure 14.4: ^{31}P NMR spectrum of excised leaf material from potato plants, which are transformed with a bacterial polyphosphate kinase gene. The spectral resonance assignments are as follows: 1. phosphomonoesters; 2. cytoplasmic P_i; 3. P_i of the medium; 4. Vacuolar P_i; 5,6 and 8 are the γ–, α– and β–phosphates, respectively, of nucleoside triphosphate; 7. nucleoside-diphospho-hexoses and NAD(P)(H). PP_n is arising from the non-terminal phosphates of polyphosphate. T. van Voorthuysen, B. Regerer, F. Springer, C. Dijkema, D. Vreugdenlid, and J. Kossmann, J. Biotechnol, 77, 65 (2000). (Reprinted with the permission from Elsevier).

Favourable pK values for other acids and ammonium ion present in the cells can also be used for studies of pH in different compartments. In addition, separate signals for the Cl^- from cytoplasmic and vacuolar pools have been detected. Similarly, in Caesium incubated cells, ^{133}Cs signals from the two pools have been detected separately. In fact, ^{133}Cs signals have also provided means for probing cation transport.

4. PLANT METABOLISM USING NMR

Studies of metabolism in plants have been stimulated by several factors. Application of techniques for creating mutant and transgenic plants require detailed understanding of metabolic pathways. This information in turn, is directly related to the development of strategies for metabolic engineering. Techniques for metabolomic analysis and relating gene-function to metabolic phenotypes has stimulated interest in developing suitable methodologies for such studies. NMR provides a suite of analytical approaches for probing metabolic pathways and changes in metabolism as a result of genetic and environmental factors. Such investigations have thrown valuable light on several aspects of molecular biology of plants. These include, plant anatomy and root architecture using imaging, synthetic pathways in metabolism, identification and quantification of plant products, and macromolecular structures of molecules of plant origin. The applications of such developments in agriculture and horticulture are obvious.

Two major techniques have been used for metabolic analysis of plants. As with cellular studies in animals, both mass spectroscopy and NMR have provided major opportunities for detailed exploration of metabolic pathways. Though NMR is considerably less sensitive than mass spectroscopy, it has the advantage that spectra can be obtained directly from functional-tissues. The metabolic response of tissues to physiological perturbation can be observed directly.

High-resolution NMR studies have been reported over a wide range of metabolic activity. These include photosynthesis, nitrogen metabolism, carbon metabolism and phosphorylation. Often it is advantageous to use ^{13}C, ^{15}N or ^{2}H labelled substrates in such studies. The flow of such isotopes, during metabolic pathways provides a good insight on the mechanisms of biochemical reactions.

In this section, the results obtained from high-resolution NMR of cells, tissues and extracted materials have been discussed. Applications which can be carried out using traditional high-resolution NMR spectrometer have been discussed in the next two sections. Imaging, microscopy, solid state NMR and spectroscopy of whole plants has been discussed in later sections.

4.1 Tissues, Cells and Cell Extracts

For reasons explained in the next section, most *in-vivo* studies have been carried out on small leaf pieces and root tips. Special probes have been designed for this purpose. These probes have provision for the supply of oxygen, nutrients, light and removal of waste products such as ethanol. The composition of the circulating medium may be changed to provide various stress conditions such as sucrose or oxygen starvation or introduction of phosphate. The tissue types are carefully selected. The packing of the sample is adjusted to obtain good S/N ratio and to simultaneously meet the physiological needs of the sample.

The most widely used tissues in NMR studies are the roots. Studies have also been made on stems, leaves, seeds and fruits. One of the problems in such studies is the presence of intra- and inter-cellular air spaces, which has considerably detoriate the quality of signals. Suitable precautions are taken to make the conditions to reflect the physiological situation, as close as possible. The excised tissues are kept in a medium with suitable nutrition and oxygen. The information from NMR experiments can be increased further, by changing the conditions under which the experiments are performed. NMR methods have been very helpful in probing the flow of spins during metabolic processes.

While studying cell suspensions, oxygen and nutrients have to be supplied continuously. Paramagnetic ions, particularly Mn^{+2} which exist in plants can produce serious line-width problems and have to be avoided. Mn^{+2} is a micronutrient for plants at low concentrations but it is toxic when present in large amounts. It is preferentially localised in the large vacuoles. NMR studies have also been made on cell organelles such as chloroplasts, vacuoles and mitochondria.

4.2 Nitrogen Metabolism

A unique feature of plants is the nitrogen metabolism. Both ^{14}N and ^{15}N NMR has been extensively used in studies on plant metabolism. As discussed, most nitrogen containing metabolites give broad signals for the more abundant spin ^{14}N. However, in some cases such as in NH_4^+, the ^{14}N signals are fairly sharp. This is not the case for most of the other products of metabolism. In such cases, attention has to be turned to the less abundant isotope ^{15}N, in spite of the weaker signals. For studies on metabolic pathways, it is advantageous to use ^{15}N labelled substrates to avoid such difficulties, and to follow its movement to various metabolic products.

Nitrogen metabolism is always coupled to metabolic pathways which include hydrogen and carbon, and often phosphorus. Therefore, NMR active spins such as ^{1}H, ^{13}C, ^{31}P and ^{2}H have been used extensively, to complement the information obtained from nitrogen NMR. As in the case of animal systems, multi-dimensional NMR has provided valuable information on the detection and assignments of metabolic products. Inverse detection has been used for *in-vivo* studies on ^{13}C in natural abundance. Such experiments are not possible for nitrogen as it is often bonded to an exchangeable proton.

The central themes in nitrogen metabolism are: uptake, accumulation and assimilation of nitrate and ammonium and the amino acid metabolism (Figure 14.5). Of the three forms of nitrogen available to plants (ammonium, nitrate and dinitrogen) nitrate is usually of major importance. Plants such as rice and legumes can also assimilate nitrate. The relative degree of assimilation depends on a number of factors including soil conditions. The process of nitrogen utilization involves its uptake, assimilation in the roots, storage in roots and translocation of excess nitrogen to shoots. In the roots, the most active zone in the process is the tip.

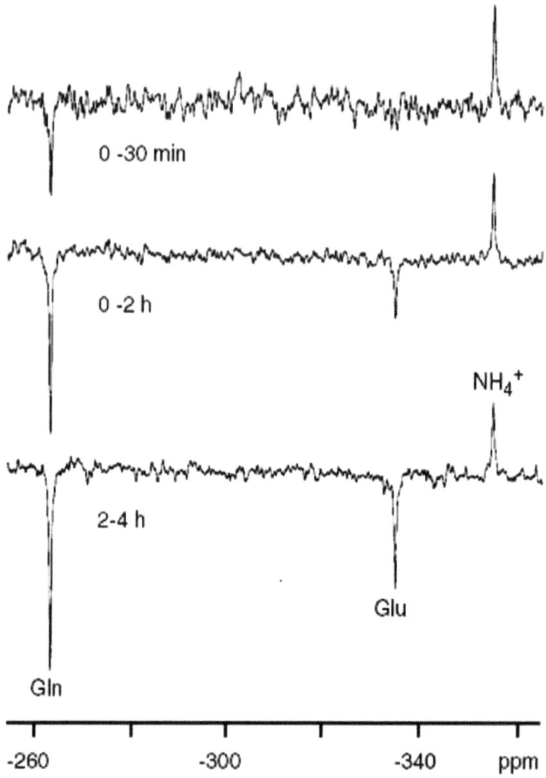

Figure 14.5: In vivo ^{1}H-decoupled ^{15}N NMR spectra recorded from an oxygenated carrot (Daucus carota) cell suspension immediately after the addition of 20 mM [^{15}N] ammonium chloride. The spectra were recorded in either 30 min or 2 h during the intervals indicated, and the detection of the glutamine amide signal before the appearance of the glutamate amino signal is consistent with the assimilation of ammonium via the GS/GOGAT pathway. Adapted from A.D. Carrol, G.G. Fox, S. Laurie, R. Phillips, R.G. Ratcliff and G.R. Stewart. Plant Physiolo. 106, 513-520 (1994). (Reprinted with permission from the American Society of Plant Biologists).

The arsenal of NMR techniques offers unique ways to explore nature of metabolic pathways involving nitrogen. Metabolic changes due to physiological and chemical perturbations can be followed directly by monitoring the fate of isotope labelled substrates. The metabolic changes in genetically modified plants can also be studied in this way. Both these themes can be conveniently followed by using NMR of the two naturally occurring nitrogen isotopes. Both nitrate and ammonium give sharp ^{14}N signals which can be readily detected under *in-vivo* conditions. Studies have been carried out on different plant components such as root, leaf and cell suspensions, under a variety of stress conditions (for example, nitrogen starvation and ammonium accumulation). There is a good correspondence between the results obtained from NMR and other analytical techniques. Further, distinction between the various cellular pools of nitrogen (e.g. cytoplasm and the vacuole) has been possible.

The use of ^{14}N NMR to probe assimilation of nitrate and ammonium is difficult since the end products such as amino acids give very broad signals. The use of ^{15}N labelled nitrate and ammonium have been successful in carrying out a variety of metabolic studies on plant tissues (Figure 14.5). Such investigations provide the redistribution profile of ^{15}N in real time spread over a time course of several hours. Amino acids Glu and Gln can be detected in real time. In this way, the relative contribution of metabolic pathways involving glutamine synthetase, glutamate synthase and glutamate dehydrogenase during the assimilation of ammonium have been established.

4.3 Amino Acid Metabolism

Both ^{13}C and ^{15}N NMR have been used to study amino acid metabolism in plants. Several aspects of glutamate metabolism have been investigated. In particular, the relative role of conversion of amino acids to GABA, deamination to α-ketoglutarate and conversion to proline under salt stress during glutamate metabolism have been followed, using NMR. The conversion to GABA is observed during ^{15}N ammonium assimilation. However, the flux is altered by cytoplasmic pH. An increase in pH leads to lower GABA production. Deamination of glutamate has been followed using both ^{13}C and ^{15}N NMR. The oxidative deamination of glutamate by glutamate dehydrogenase has been established in potato tubers. The synthesis of proline from glutamate is another pathway, which has been followed using NMR.

The metabolism of glycine has also been followed extensively. The metabolic fate of amino nitrogen is followed using ^{15}N NMR, while the metabolism of the carbon skeleton can be followed using ^{13}C NMR.

4.4 Photosynthesis

Photosynthesis is the most important process for sustaining both animal and plant life. The total result of the complex sets of chemical reactions involved in this process can be represented by the following equation:

$$6CO_2 + 12\,H_2O + Light \rightarrow C_6H_{12}O_6 + 6H_2O + 6O_2 \qquad 14.1$$

The whole process is completed in two main steps, which are carried out in the presence and absence of light. Each step involves a series of complex chemical reactions.

The first stage results in the production of reduced coenzyme NADPH, H^+ and ATP, through two separate cooperative reactions. The plant photosynthesis reactions are activated by two light driven electron pumps, commonly referred as photosystem I (PS I) and photosystem II (PS II). The primary electron donor of PS II P680 is a strong oxidant in its radical cation form. It is able to oxidise water. The primary electron donor of PS I (P700) allows reduction of CO_2 to biological matter. In the second stage, the energy rich compounds NADPH and ATP are used for synthesis of carbohydrates through the formation of intermediates such as phosphoglyceric acid and oxalacetic acid.

Both these sets of reactions are fairly complex. At this stage, the mechanistic steps involved are yet to be fully understood. The cell organelles involved are fairly large and are generally not amenable for analysis by structural methods. NMR studies of photosynthesis are therefore often limited to studies on model systems.

For studying photosynthesis it is necessary to illuminate the plant suspensions. The work on cell suspensions normally uses high concentrations of tissue, resulting in a high optical density. This is a major constraint in NMR studies of photosynthesis. A compromise has to be made between the quantity of material used and the intensity of light used, along with maintenance of oxygen balance. Investigations using tissue extracts are much more profitable. One of the important applications has been to analyze site specific isotope fractionation in the end products of photosynthetic metabolism. It has been possible to analyze differences in the metabolic behavior of plants following different carbon pathways during photosynthesis such as in the C3 and C4 plants. The differences between the reactions in the dark and light reactions have been followed in a similar way.

4.5 Carbon Metabolism

Carbon has a special role in all metabolic reactions. Therefore, ^{13}C NMR has been widely used in such studies. It has been widely used for following carbon metabolism in nitrogen assimilation as well as in photosynthetic and non-photosynthetic reactions.

The use of ^{13}C labelled substrates further help in obtaining insights into the metabolic pathways. For example, ^{13}C labelled bicarbonate has been used to study synthesis of malate and citrate. However, the most commonly used substrate is ^{13}C labelled Glc. Some of the important investigations made in this way are:

(i) Reversible interconversion and cycling of hexose and triose phosphate in cytosol.
(ii) Glycolysis, sucrose turnover and polysaccharide synthesis.
(iii) Pathways of intermediate metabolism.
(iv) Respiration and pentose phosphate pathways.
(v) Effect of oxygen deprivation in energy production and pH changes.
(vi) Studies on secondary biosynthetic pathways such as the alkaloid metabolism.

The ability to study such diverse reactions by NMR holds good promise for further advances in studies on plant NMR and plant biochemistry.

4.6 Sulphur Metabolism

Another important feature of plants is their role in metabolic transformation of sulphur in biosphere. The available source of sulphur is normally sulphate. Most animals lack the enzymes for conversion of sulphate into cysteine and methionine.

However, direct studies of sulphur metabolism by NMR are limited due to lack of a suitable NMR probe. The information on sulphur metabolism is therefore obtained indirectly, using other NMR spins involved in such reactions.

4.7 Phosphorus Metabolism

The relative ease with which ^{31}P signals can be monitored, both during the *in-vitro* and *in-vivo* experiments, and its importance in plant metabolism, have provided incentives for using ^{31}P as an additional probe. The signals can be easily obtained at the level of leaves, cell suspensions and subcellular systems. Such studies have been used to obtain information on phosphorus nutrition. Signals of inorganic phosphate from metabolic active pool in cytoplasm and the storage pool in vacuole under different physiological conditions have provided information on the dynamics of pools of inorganic phosphate and their interaction.

Valuable information on metabolic pathways can be obtained using 1- and 2D ^{31}P NMR. Levels of phosphorylated metabolites and intermediates such as ATP, NAD^+, $NADP^+$, NADPH have been monitored directly. Accumulation and compartmentalisation of the phosphorylated products have been investigated. The studies have provided information on fluxes in a number of metabolic pathways in plants. These include ATP synthesis and hydrolysis, UDP-glucose and glucose-1-phosphate hexose phosphate isomerase, galactose, choline metabolism and phosphorylation of glycerol and homoserine. The steady state fluxes between bio-energetically important phosphorylated metabolites have been used to gain information on metabolic pathways and their control. The host-fungal interactions have also been studied.

4.8 Other ions and Substrates

There have been several studies using other resonances as well. Typically, both the presence of potassium chloride and borate can be detected using NMR of the respective ions. An interesting ion is paramagnetic Mn^{+2}, which is an important nutrient for plants at lower concentration. Its presence has been measured indirectly, using the changes in relaxation rates of water, phosphate and other metabolites. The pH dependence of the P_i signal provides information on differentiation among various cellular compartments, particularly cytoplasm and vacuole. The pH difference between the two compartments can also be used to identify ^{15}N signals from NH_4^+, ^{35}Cl signals and ^{39}K signals from the two pools. These resonances therefore provide an effective tool in several studies, including plant nutrition.

4.9 Paramagnetic Reagents

A number of applications in plant NMR have used paramagnetic reagents such as Gd^{+3} to obtain additional information. In particular, contrast reagents are helpful in studies of ion transport and distribution. Non-permeant agents help to distinguish intra- and extra-cellular distribution signals in roots and stems. Such reagents are also helpful in transport studies.

4.10 Effect of Physiological Conditions

The effect of perturbations in environmental conditions during plant growth is an important aspect of plant physiology. The results are influenced by variations in light, diurnal changes, weather, soil and geographical variations. These may consist of variation in supply of nutrients (phosphate, nitrate, ammonium or potassium), growth factors, oxygen deprivation, salt conditions, water stress, temperature and pH. NMR studies in controlled condition provide valuable insights on processes that may occur in agricultural fields, where such conditions are difficult to meet and there are difficulties in the study of their effects. Studies have also been conducted to study the effects of metabolic inhibitors for metabolic pathways. However, the most fruitful area is the study of metabolic changes in mutants and transgenic plants. NMR has been used to assess the magnitude and specificity of gene expression in genetically modified plants.

4.11 Secondary Metabolic Pathways

NMR methods have also provided valuable information on the secondary nitrogen metabolism. The complex chemical structures of alkaloids, which have multiple asymmetric carbon centres, provide challenges. The use of 2D NMR provides opportunities for a definitive structural and pathway analysis and for studies of the stereochemistry of the plant products.

4.12 Plant-Fungal Relations

One of the important factors in plant metabolism is the symbiosis with fungi. Separate signals from the host and the fungus have enabled studies on *mycorrhizal* metabolism. Information on the carbon metabolism during the fungal life cycle has been obtained using ^{13}C NMR. This enables studies of principle pathways of fungal metabolism and storage compounds.

5. METABOLIC FLUXES AND PLANT METABOLOMICS

Products of metabolic activities of plants have major implications for the utilization of plant products for human health and welfare. The final products of metabolism depend on a number of environmental factors, some of which may be difficult to control. Small changes in light conditions, diurnal changes in photosynthesis, geographical and seasonal variation, the stage of the growth of the plant and harvesting procedure can change the final outcome. The control of environmental conditions is more difficult in the case of plants. As in the case of human studies, NMR in conjunction with techniques such as GC, MS and FT-IR has emerged as potential tools for such analysis. The metabolic profile provides direct link between metabolic network and gene sequences in plants.

5.1 Analysis of Metabolic Fluxes

As described already, NMR has been extensively used to investigate a large number of metabolic pathways in plants. An important part of such studies is to obtain quantitative measurements of metabolic fluxes. Such fluxes can be measured from the fractional enrichment during labelling experiments. Fluxes between different systems or under different physiological states can be compared. This leads to important insights into the regulation and integration of metabolic fluxes. The steady state distribution of a particular spin and its fractional enrichment provides valuable information, which has been used to analyse metabolic network.

5.2 Secondary Products of Plant Metabolism

The secondary plant products include terpenes, phenols and alkaloids. Several thousand of secondary metabolites have been found in plants. Some of these products, such as flavonoids and phenolic compounds may be toxic. However, there are a large number of compounds which have proved useful in our life.

The wide variety of secondary products present in plants is in sharp contrast to the situation in animals. In the later case, such molecules are often excreted as waste products. In plants these are retained and deposited in plant-structures such as vacuoles, resin canals and cell walls. Some of these compounds protect the plants from mammals, insects and fungi. Compounds such as rubber and certain alkaloids are produced by only a few plants. Other metabolites are found to be more abundant. The commercial importance of such products in human life has activated NMR studies.

5.3 Plant Metabolomics

Metabolonomic studies have also been carried out in plant systems. Even in green house environment, control of environmental variants for studies on plants, are difficult. Small changes in light conditions, diurnal changes in photosynthesis, geographical and seasonal variations, and harvesting procedures can affect the biochemistry of plants, and the production of small molecular weight metabolites. Nevertheless, NMR is a useful technique to analyse variations in the metabolic behaviour of plants.

6. IMAGING AND MICROSCOPY IN PLANTS

Both MRI and MRS have found wide applications in studies on whole plants. MRI techniques used in plants and animals are similar. However, the use of MRI in plant sciences has yet to develop to the same popularity as that in animals.

6.1 NMR Microscopy

Optical microscopy has been an important and traditional source of information for the histology of plants. The technique involves a long and destructive process consisting of cutting the specimen into small pieces, fixing it, treatment with a polymer matrix, sectioning and staining. 3D images by optical methods require several days of efforts. Confocal microscopy has led to major advances in optical microscopy of plants.

Hardware has been developed for NMR based microscopy on thin sections of plants. The principles of NMR microscopy are similar to that of imaging of large structures of plants. The images thus obtained can be compared with those obtained from optical microscopy. An advantage of NMR microscopy is that dynamic and density information can be obtained using diverse types of weighted images. Most images use ^1H, though examples of the use of other spins exist in literature. Several factors such as diffusion, susceptibility differences, vacuoles, magnetic transfer due to chemical exchange and the presence of paramagnetic substances influence quality of images. While NMR microscopic images can be obtained much more conveniently, they suffer from a lower resolution as compared to confocal microscopy. Typically in-plane resolution of μm^2 can be achieved using high field NMR spectrometers.

The techniques of optical and NMR microscopy are complementary. However, applications of NMR microscopy in plants are not as wide spread as those of optical methods. Part of the reason is that the NMR based techniques of microscopy and imaging are less well understood by plant biologists, as compared to the zoologists and medical scientists. It is likely that the technique will become more widely accepted in future by professional botanists.

6.2 Studies of Plant Water

Most NMR studies concentrate on studies of water in the plants, using low magnetic fields. Both the water content and its flow are important for studies of plant physiology. Such experiments can be conducted on small plants. Portable NMR magnets allow studies on plant-water relationship in greenhouse environment. Water signals can be obtained with much higher sensitivity than those of metabolites. Therefore, imaging has been widely used on seedlings, plant products such as seeds and fruits and whole plants. The water content and flow can be measured in small plants. Systems have been designed to measure water signals in plants grown in green houses, to obtain information on plant-water relations. NMR can easily establish a correlation between changes in water in plants, its flow, tissue anatomy, effects resulting from water stress and heat injury.

As in the case of animals, different imaging protocols have been used on plants as well. Diffusion and compartmentalisation is reflected in the relaxation properties of water in such studies. More advanced NMR techniques such as diffusion and flow weighted images provide information on water fluxes during growing seedling, internal recirculation of water and to the relevance of transpiration to plant growth.

6.3 Studies of Roots

Water images using NMR has been used extensively for investigation of architecture of roots and changes during growth. The opacity of the soil makes observation of roots with clay impossible to study. MRI studies allow observations involving soil columns. The studies can be divided in the following main groups.
(i) Spatial distribution of water, to study the hydration level of the tissue and physiological heterogeneity.
(ii) Water diffusion and flow in plant vascular system.
(iii) Spatial distribution of water relaxation rates over the volume of the sample.
(iv) Root architecture, growth and water changes occurring within the roots.
(v) Microbial development and directed transport of carbohydrates to tips of the roots.

Some of the difficulties in studies of roots are due to heterogeneous nature of the soil and in some cases, the presence of high concentration of paramagnetic compounds.

6.4 Stems, Leaves and Flowers

The linear nature of stems provides opportunities to image 2D transverse images with thick slices. Changes in the anatomy induced by plant diseases have been studied in this way. Development of plants, which can stand environmental stress, is of great importance. The use of stem imaging has been made to study the effect of temperature, susceptibility to cold damage and freezing. Similarly, the effect of environmental pollution has been reported.

6.5 Fruits and Seeds

Recall that the first NMR images were reported on fruits (Figure 14.6). MRI studies on damage due to water, temperature and pests on fruits and seeds have been popular. A detail understanding of changes in the tissue structure during its development and ripening is important to understand effects of fungal infection, insect metamorphosis, transport, and storage. Such considerations are of great importance for insect control and fruit preservation.

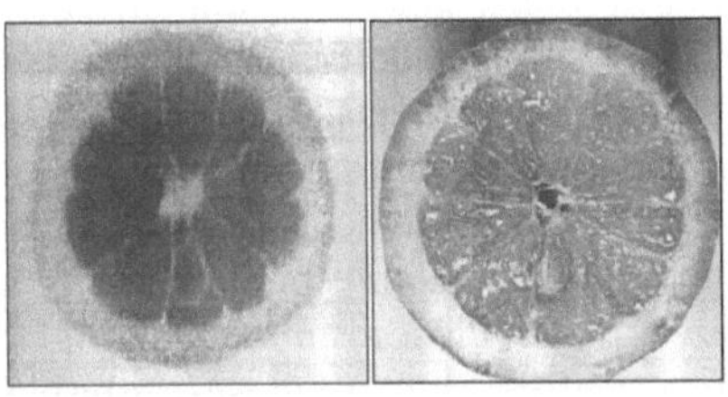

Figure 14.6: Thin transverse section proton NMR image of an intact lemon (left) and a photograph (right) of the actual section cut subsequently. E.R. Andrew. Philosophical Transactions of the Royal Society of London. Series B, Biological Sciences. The Royal Society. Pp. 471-481 (1980) (Reproduced with the permission of the Royal Society of London.).

6.6 Chemical Shift Imaging

In addition to water, a number of molecules are present in plant tissues in large quantities. Application of chemical shifts imaging allows one obtain images of some of the more abundant molecules, such as sucrose and glucose, amino acids such as glutamine/glutamate, Lys and Arg, and sodium in tissues from plants grown in high salt environment. Imaging has been used as a potential tool for investigating metabolism and distribution of the more abundant molecules.

A number of imaging protocols have been used for this purpose. The most commonly used sequence is CHESS. In particular, metabolite mapping has proved very useful for nitrogen metabolism.

One of the more promising areas is the metabolic images of ^{13}C and ^{31}P. The former is achieved by indirect detection of ^{1}H of the ^{13}C labelled metabolites.

7. SOLID-STATE NMR STUDIES

When working with tissues, roots, fruits and seeds one encounters relatively broad NMR signals. Recently, high-resolution solid-state techniques based on MAS, cross polarisation and dipolar decoupling, have been used on semi-solid plant systems. Using such techniques, decent spectra for both ^{13}C and ^{15}N have been obtained. The results have found wide applications, particularly in detection of amides, amino acids and amines in plant tissues. Another example is application of NMR to the seeds, which are solids and give broad lines. In such cases, MAS has been used to obtain decent NMR signals.

7.1 Studies on Plant Organelles

Several tissues with thick cell walls are also amenable for solid state NMR studies. In such cases use has been made of high-resolution solid-state NMR techniques to obtain spectra from proteins, carbohydrates, lipids and plant metabolites. Due to high speed used in rotors, it is difficult to manipulate physiological conditions. However, certain plant tissues can stand the high rotor speeds and can be conveniently studied.

The relaxation rates T_1, T_2 and $T_{1\rho}$ have been used to obtain information on ordering of polymers in cell walls, and to obtain information on hydration. As discussed, even though nitrogen metabolism in plants have been followed by recording spectra from tissues and extracts, solid state ^{13}C and ^{15}N spectra from seeds and thick tissues provide valuable information.

Recent solid-state ^{13}C NMR measurements of intact soybean leaves labelled by $^{13}CO_2$ have led to the conclusion that photorespiration is 17% of photosynthesis for a well-watered and fertilized plant. Photorespiration refers to the oxygenation of ribulose-1,5-bisphosphate (RuBP) by ribulose-1,5-bisphosphate carboxylate/ oxygenase and the events that are stimulated by the immediate production and subsequent rescue of a two-carbon molecule that is not compatible with the Calvin cycle. This NMR report of photorespiration is the first direct assessment of the level of photorespiration in a functioning plant. The $^{13}CO_2$ labeling of a soybean leaf has been carried out using a compact-disc case as a labeling chamber. The labeling gas contained 21% O_2, either 200 or 300-ppm $^{13}CO_2$, and the remaining N_2. This labeling gas entered at the bottom left through a copper pipe closed at the end and with multiple exit holes along the sides (Figure 14.7 A). At the end of the labeling period, the leaf was cut from its stem, immersed in liquid nitrogen, and subsequently lyophilized. Such lyophilized leaf were chopped into 1mm fragments by hand with a razor blade, packed into a magic-angle spinning rotor, and examined by SSNMR (Figure 14.7 B).

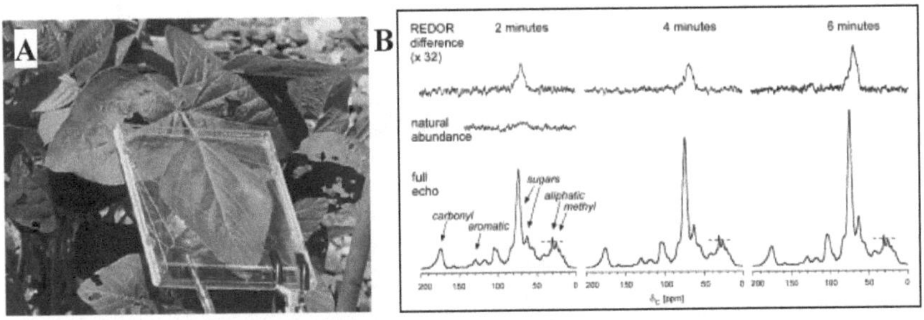

Figure 14.7: (A) The $^{13}CO_2$ labeling of a soybean leaf using a compact-disc case as a labeling chamber. (B) $^{13}C\{^{31}P\}$ eight rotor-cycle rotational-echo double-resonance (REDOR) spectra of soybean leaves labeled by 300-ppm (by volume) $^{13}CO_2$ for 2–6 min. The REDOR difference spectra (S_o- S) are at the top of the figure and the full-echo spectra (S_0) at the bottom. The full-echo spectra have been normalized by their natural-abundance methyl-carbon peaks (dotted lines). The REDOR difference arises from Calvin-cycle intermediates source and acknowledgment with ^{13}C labeled phosphorylated carbons. L. Cegelski and J. Schaefer. J. Magn. Res., 178, 1-10, (2006). (Reprinted with permission from Elsevier).

The $^{13}C\{^{31}P\}$ eight rotor-cycle rotational-echo double-resonance (REDOR) spectra of such leaves labeled by 300-ppm (by volume) of $^{13}CO_2$ for 2–6 min are shown in Figure 14.8 B. The REDOR difference spectra (S_o- S) are at the top of the panel and the full-echo spectra (S_0) are at the bottom. The REDOR difference arises from Calvin-cycle intermediates with ^{13}C labeled phosphorylated carbons. These spectra indicate that by 2 min after the start of labeling, gas exchange has replaced unlabelled CO_2 within the leaf and the Calvin cycle intermediates are already at least partially labeled. With such methods it should be possible to establish whether photorespiration varies from plant species to another. The answer to this question could help in deciding whether the transformation of crop plants with foreign Rubisco to enhance net CO_2 assimilation would indeed result in increased productivity.

7.2 Solid-State NMR on PS I and PS II

As mentioned earlier, the plant photosysnthetic assembly is fairly complex. In a novel technique for exploring the electronic structures of P680, the photochemically induced dynamic nuclear polarisation (photo-CIDNP) in photosysnthetic reaction centers (RC) has been coupled to MAS NMR. In the photo-CIDNP, nuclear polarizations of several orders of magnitudes higher than the Boltzmann distribution, are observed. The electron polarisation decays in sub-microsecond time-scale. However, the nuclear polarisation is maintained for several seconds. This allows studies on the ground state after changes induced by the photocycle. The selective detection of several NMR signals in these complex systems (Figure 14.8) provides information on the electronic structure of the radical pair.

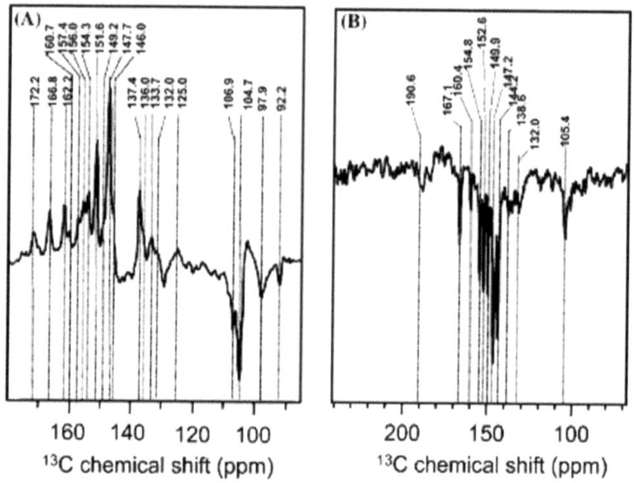

Figure 14.8: ^{13}C spectra of (A) PS II and (B) PS I, reaction centers, obtained with continuous illumination of white light at 223 K, at a magnetic field of 9.4 T and MAS of 9 kHz. A. Diller, E. Roy, P. Gast, H.J. van Gorkom, J. Zaanene, H.M.J. Groot, C. Glaubnitz and Matysik. 84, 303-308 (2005). (Reprinted with permission from Springer Science+Business Media).

The absorptive (positive) of the aromatic rings in PS II, appear in the range 120 to 170 ppm and can be assigned to resonances arising from the radical pair state $P680^{*+}$ in Chl a. The signals from PS I arise mainly P2 cofactor of P700, and can be assigned to a single set of Chl a resonances. From these spectra, it has been concluded that both P680 and P700 are monomeric species. The different signs of the two spectra has been ascribed to a predominantly radical pair mechanism in PS I, with strong matrix interaction. The radical cation of PS I appears to be Chl a cofactor, which is essentially undisturbed.

7.3 Chlorosomal Bacteriochlophylls

Solid-state NMR techniques have also been used to study the self-organization of bacteriochlorophylls in the extra-membraneous light harvesting antenna complexes called chlorosomes. Chromosomes are composed of bacteriochlorophyll (BChl) c molecules (or BChl d and e in some bacteria), surrounded by lipids and some proteins. In the model systems used, the natural metal ion Mg^{+2} is replaced by Cd^{+2}. Extensive studies using ^{13}C and ^{15}N enrichment in the ring moieties and a consolidated approach based on several nuclear spins has been used to study the nature of aggregation and molecular control of such chlorophylls. For example, Figure 14.9 shows the ^{113}Cd CP-MAS spectra of aggregated chlorines.

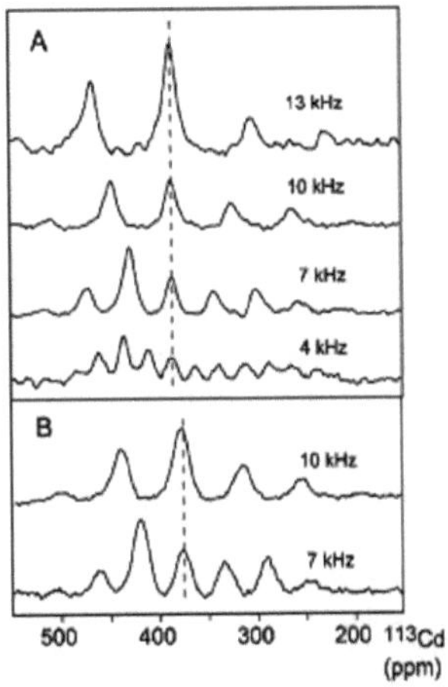

*Figure 14.9: ^{113}C spectra of aggregated (A) chlorin **1** and (B) chlorin **2** at 17.6 T with different spinning frequencies. I. De Boer, J. Matysik, K. Erkelens, S. Sasaki, T. Miyatake, S. Yagai, H. Tamiaki, A.R. Holzwarth and H.J.M. Groot. J. Phys. Chem. 108, 16556-16566 (2004). (Reprinted with the permission from The American Chemical Society).*

The ^{113}Cd CSA tensor is sensitive to the coordination state of the Cd^{+2} ion. In case of above complexes, the high value of σ_{33} of 626 ppm is indicative of low electronic shielding. The other two elements of the shielding tensor, reflect axial symmetry of the electronic system. The NMR results indicate a five coordination to the metal ion. The Cd coordination is very similar to the Mg system in natural systems.

The individual molecules stack in the form of planar layers leading to anomalously large ^{1}H ring current shifts of up to 10 ppm. Molecular modelling and ring-current shift calculations show the highly stacked nature of BChl c moieties in chlorosomes. The 3D arrangements of the layers are also confirmed by ^{13}C-^{13}C correlations experiments. The sheets of chlorin molecules are oriented in a head to tail fashion in the aggregates. The results show that the chlorosome structure is compact, robust and stable.

8. REFERENCES

8.1 Further Reading

E.R. Andrew, NMR, *Imaging of intact biological systems*, Philos. Trans. R. Soc. London, Ser. B, **289**, 471–481 (1980).

J. Schripsema, C. Erkelens and R. Verpoorte, *Intra- and extracellular carbohydrates in plant cell cultures investigated by lH-NMR*, Plant Cell Reports **9**, 527–530, (1991).

A.D. Carrol, G.G. Fox, S. Laurie, R. Phillips, R.G. Ratcliff and G.R. Stewart, *Ammonium assimilation and the role of g-aminobutyric acid in pH homeostasis in carrot cell suspensions*, Plant Physiol. **106**, 513–520 (1994).

J.K.M. Roberts and J.H. Xia. *High-resolution NMR methods for study of higher plants*, Methods Cell Biol. **49**, 245–258 (1995).

R.G. Ratcliffe and Y. Sachar-Hill, *Probing plant metabolism with NMR*, Ann. Rev. Plant Mol. Biol. **52**, 499–526 (2001).

Y. Shachar-Hill, *Nuclear magnetic resonance and plant metabolic engineering*, Metab. Eng. **4**, 90–97 (2002).

M.R. Viant, E.S. Rosenblum and R.S. Tjeerdema, *NMR-based metabonomics: A powerful approach for characterizing the effects of environmental stressors on organism health*, Environ. Sci. Technol. **37**, 4982–4989 (2003).

I. De Boer, J. Matysik, K. Erkelens, S. Sasaki, T. Miyatake, S. Yagai, H. Tamiaki, A.R. Holzwarth and H.J.M. Groot, *MAS NMR structures of aggregated cadmium chlorins reveal molecular control of self-assembly of chlorosomal bacteriochlorophylls*, J. Phys. Chem. **108**, 16556–16566 (2004).

A. Diller, E. Roy, P. Gast, H.J. van Gorkom, J. Zaanene, H.M.J. Groot, C. Glaubnitz and J. Matysik, *Photo-CIDNP solid-state NMR on Photosystems I and II: What makes P680 special?* Photosynth. Res. **84**, 303–308 (2005).

W.B. Dunn, N.J.C. Bailey and H.E. Johnson, *Measuring the metabolome: Current analytical techniques*, Analyst **130**, 606–625 (2005).

L. Cegelski and J. Schaefer, *NMR determination of photorespiration in intact leaves using in vivo* $^{13}CO_2$ *labeling*, J. Magn. Res. **178**, 1–10 (2006).

8.2 Books and Reviews

P.T. Callaghan, *Principles of Nuclear Magnetic Resonance Microscopy*, Clarendon Press, Oxford (1991).

H.D. Kumar and H.N. Singh, *Plant Metabolism*, 2nd Edition, East-West Press Pvt. Ltd. (1993).

Y. Shachar-Hill and P.E. Pfeffer, *Nuclear Magnetic Resonance in Plant Biology*, American Society of Plant Physiology (1996).

J.A. Chudek and G. Hunter, *Magnetic resonance imaging of plants*, Prog. NMR Spect. **31**, 43–62 (1997).

R. Bligny and R. Douce, *NMR and plant metabolism*, Curr. Opin. Plant Biol. **4**, 191–196 (2001).

R.G. Ratcliffe, A. Rosher and Y. Shachar-Hill, *Plant NMR Spectroscopy*, Prog. NMR Spect. **39**, 267–300 (2001).

F. Mesnard, and R.G. Ratcliffe, *NMR Analysis of Plant Nitrogen Metabolism*, Photosynth. Res. **83**, 163–180 (2005).

INDEX

FOCUS ON STRUCTURAL BIOLOGY

1. S.R. Kiihne and H.J.M. de Groot (eds.): *Perspectives on Solid State NMR in Biology*. 2001 ISBN 0-7923-7102-X
2. J.N. Housby (ed.): *Mass Spectrometry and Genomic Analysis*. 2001 ISBN 0-7923-7102-X
3. C. Dennison: *A Guide to Protein Isolation*. 2003 ISBN 1-4020-1224-1
4. O.M. Becker and M. Karplus (eds.): *Guide to Biomolecular Simulations*. 2005 ISBN 1-4020-3586-1
5. G.S. Rule and T.K. Hitchens: *Fundamentals of Protein NMR Spectroscopy*. 2006 ISBN 1-4020-3499-7
6. G. Govil and K.V.R. Chary: *NMR in Biological Systems*. From Molecules to Humans. 2008 ISBN 978-1-4020-6679-5

MIX
Papier aus verantwortungsvollen Quellen
Paper from responsible sources
FSC® C105338

If you have any concerns about our products,
you can contact us on
ProductSafety@springernature.com

In case Publisher is established outside the EU,
the EU authorized representative is:
Springer Nature Customer Service Center GmbH
Europaplatz 3, 69115 Heidelberg, Germany

Printed by Libri Plureos GmbH
in Hamburg, Germany